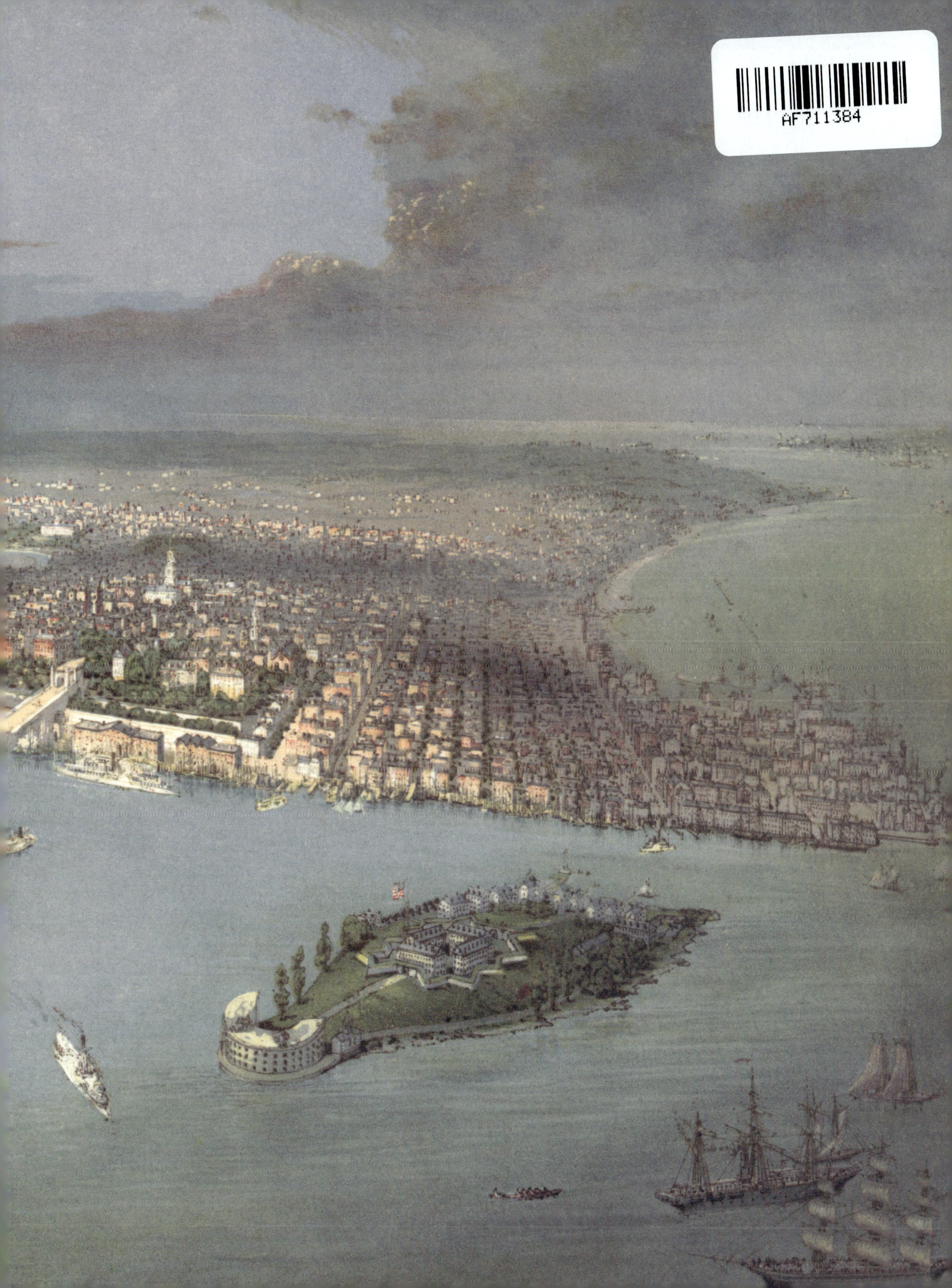

Building the Brooklyn Bridge
1869–1883

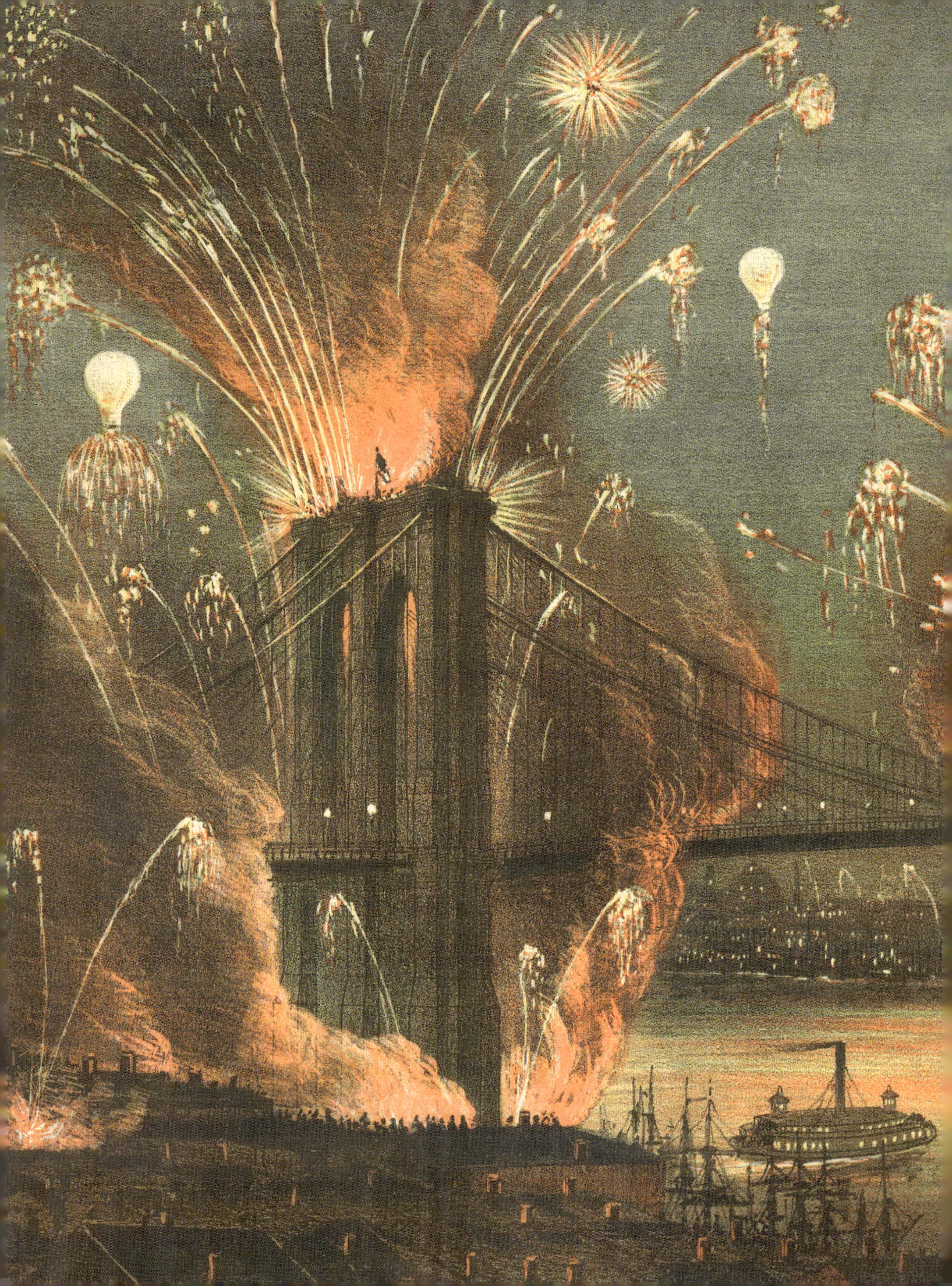

BUILDING THE Brooklyn Bridge 1869–1883

An Illustrated History with Images in 3D

Jeffrey I. Richman

Essays by Richard Haw and Erica Wagner

BAUER AND DEAN PUBLISHERS, NEW YORK

IN ASSOCIATION WITH The Green-Wood Historic Fund

To Cole, Jude, Eloise, and Arley, who will grow up and be inspired by many things—including, perhaps, the Brooklyn Bridge

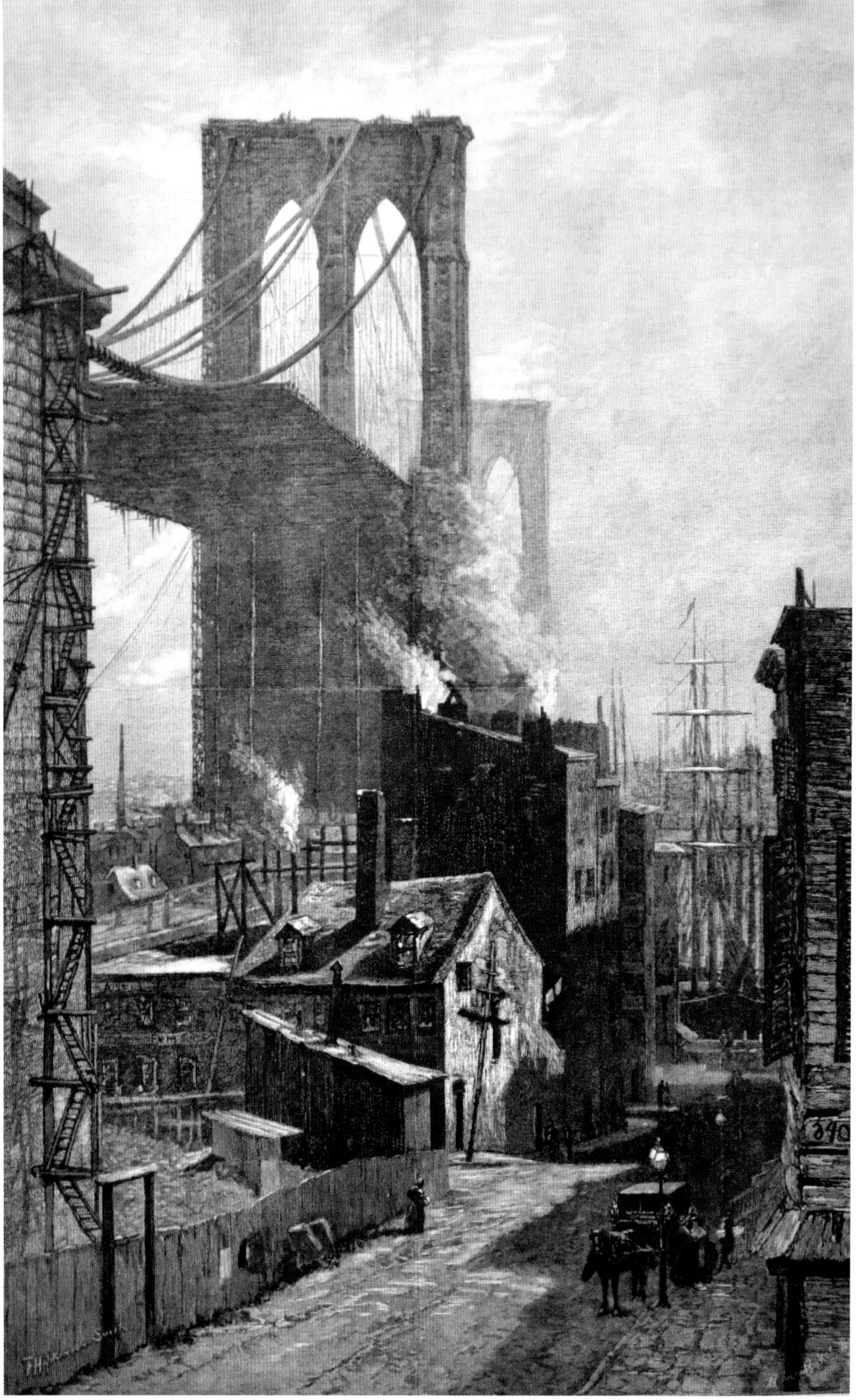
340

Contents

> “It so happens that the work which is likely to be our most durable monument, and to convey some knowledge of us to the most remote posterity, is a work of bare utility; not a shrine, not a fortress, not a palace, but a bridge.”

— MONTGOMERY SCHUYLER
HARPER'S WEEKLY, MAY 24, 1883
OPENING DAY OF THE BROOKLYN BRIDGE

Preface

As a long-standing and devoted fan of the Brooklyn Bridge, I have endeavored to tell the story of its construction in the pages that follow, relying largely on more than two hundred images. These include magnificent drawings by the bridge's engineering staff, superb photographs taken by those long forgotten, and crisp woodcut prints that appeared in newspapers throughout the fourteen years—1869 to 1883—that it took to build the bridge. Many of these images have never been published on the printed page before. They are gathered here, for the first time together, to tell the story of one of the wonders of the world.

Over the years, several names would be given to the most famous bridge connecting Brooklyn and Manhattan: Empire Bridge, New York Bridge, East River Bridge, Roebling Bridge, Great Bridge, and Brooklyn Bridge. Chief Engineer Washington A. Roebling, the man primarily responsible for its construction, was himself inconsistent in how he referred to it. On a single page of specifications for the bridge's cable wire, he referred to it as the "East River Suspension Bridge," the "New York & Brooklyn Suspension Bridge," and the "East River Bridge." In this book it will be the Brooklyn Bridge, the name by which we know it—and honor it—today.

As a native New Yorker, I remember, from an early age, the bridge capturing my imagination. In 1983, as a young man, I attended the celebration of the hundredth anniversary of the bridge's opening. Standing atop the nearby Eagle Warehouse, I watched in awe as fireworks exploded above the Brooklyn Bridge—just as they had on May 24, 1883—transporting me back to a different world and time. Years later, as the historian at Brooklyn's Green-Wood Cemetery, I discovered that many of the men who led the effort to build the Brooklyn Bridge are interred there. Over the years, I have researched those men: three of the six assistant engineers, virtually all of the bridge's Brooklyn trustees, the Brooklyn mayors (until 1898, Brooklyn was a separate city) who served during its construction, and even a shipbuilder whose graveside memorial singles out his work on the Brooklyn Bridge as the proudest accomplishment of his life.

I have been a collector my entire life. I started with baseball cards, then moved on to antique tools, duck decoys, cookie jars, photographic paperweights, and stereographic views (hereafter referred to as "stereoviews"). A stereoview consists of two photographs mounted on cardboard, each taken from a slightly different angle. When viewed through a stereoscope, or stereo viewer, a three-dimensional effect is achieved. These stereoviews, primarily produced in the latter half of the nineteenth century, made it possible for people to travel the world without leaving their nineteenth-century parlors.

Since the 1970s, I have also collected woodcut prints of the Brooklyn Bridge (as well as other bridge ephemera) that were published as it rose above what were, at the time, the separate cities of New York and Brooklyn. I still remember the antiquarian book fair at which I bought an original program for the opening ceremonies of the bridge, as well as invitations to those same ceremonies and to walk onto the bridge before the public opening; all of these are reproduced on the following pages.

About twenty years ago, when eBay was in its infancy, I came across a listing for glass lantern slide photographs detailing the construction of the Brooklyn Bridge. As I remember it, the seller was a television repair shop in Tennessee. I had no idea how the proprietor had come by these photographs, and the few images posted gave scant indication of what, exactly, the slides showed. Nevertheless, I decided to leave a substantial bid and won the lot. I was not disappointed: they were astonishing! I later excitedly showed them to Robert Vogel, who, at the time, was the curator of the Smithsonian's Division of Mechanical and Civil Engineering at the National Museum of American History in Washington, D.C. We first met when he organized the museum's 1983 exhibition marking the Brooklyn Bridge's centennial, to which I lent several stereoviews. Vogel thought that these glass slides had likely been used for instruction in a college engineering course and were probably sold off when they were no longer relevant to current practices.

As one might imagine, given the popularity of the Brooklyn Bridge, the competition for exceptional photographs of its construction has long been intense among collectors. My own collection of stereoviews of the bridge being built has grown to dozens over forty years, though I (like every collector) still remember the ones that got away. In 2018, by happenstance, I met Michael E. Ach, a collector who had outbid me on a spectacular lot of Brooklyn Bridge construction stereoviews, and he invited me to his home to see them in person, and in 3D. They were fantastic! Inspired by Michael's collection—and by my own—I realized that I was well positioned to write a book about the bridge's construction that would feature these terrific images, as well as others. Once Michael generously gave me permission to use his stereoviews—many of which have never been published on the printed page before—this book was on its way.

"Bird's-eye View of the Southern End of New York and Brooklyn, Showing the Projected Suspension Bridge Over the East River, From the Western Terminus in Printing House Square, New York—Drawn by Theo. R. Davis," *Harper's Weekly*, November 19, 1870. This view looking southeast toward Brooklyn, an independent city at the time, envisions what the Brooklyn Bridge will look like when completed.

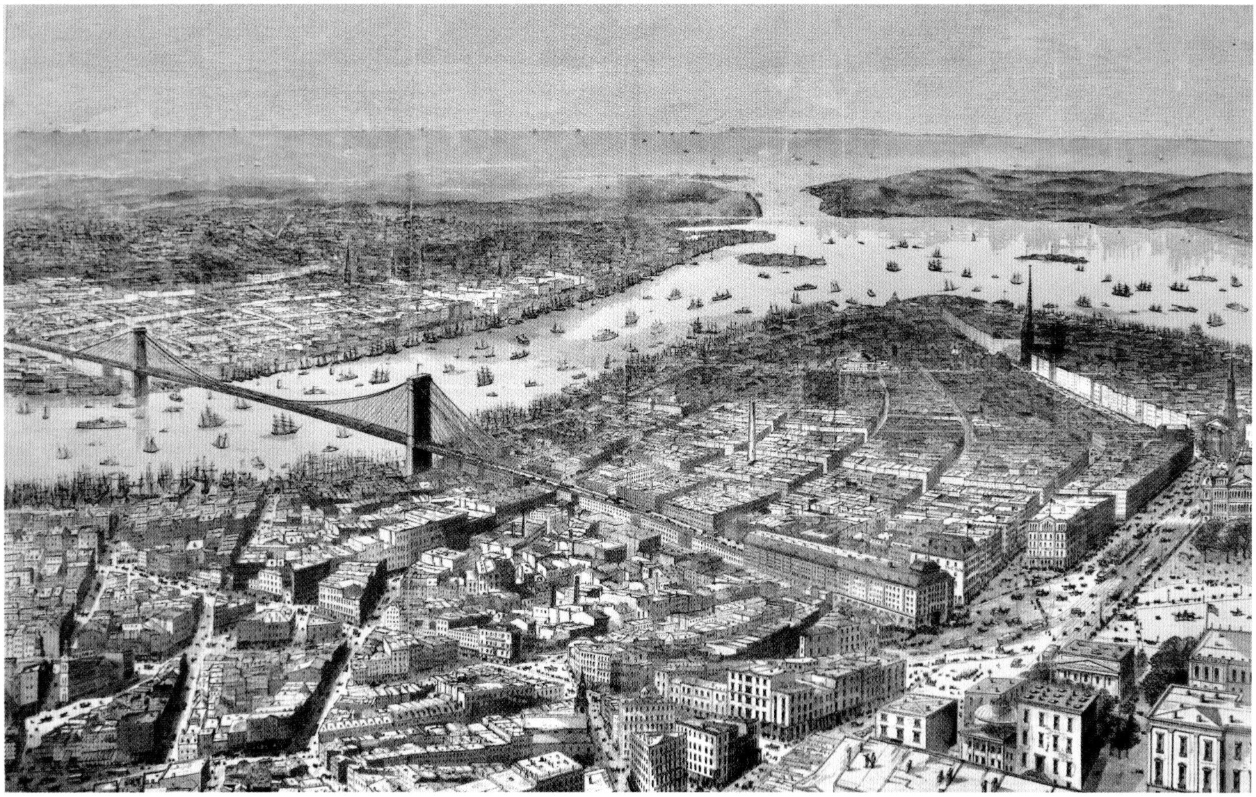

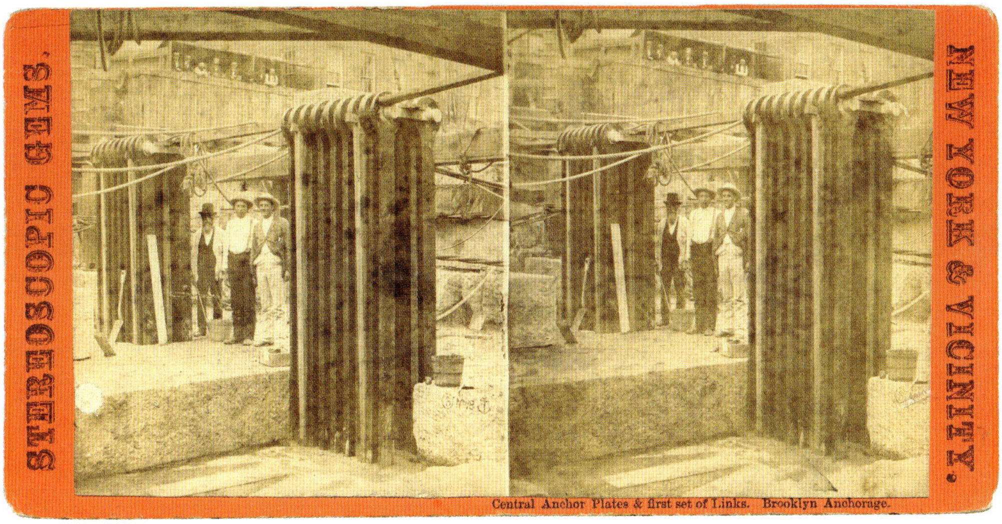

This is a stereographic view—two photographs mounted on a card side by side. These are commonly known among collectors as "stereoviews." Each photograph is taken from a slightly different perspective so that when both are looked at together through a viewer, the eye is "tricked" into seeing the image in 3D. This same scene appears again as an "anaglyph" (a computer-generated 3D image, derived from the stereoview) on page 130. Courtesy of The Green-Wood Historic Fund, Jeffrey I. Richman Stereoview Collection

In considering other sources for images, the first name that came to mind was that of collector extraordinaire and dealer Jeff Kraus, whom I have known for about forty years. Jeff has what is likely the best collection of stereoviews of New York City (some of which appear on his website, antiquephotographics.com). I contacted him, and as I suspected, he owned many wonderful views of the bridge. He kindly agreed to let me use those images for this book, and many of them, like Michael's, have never appeared in print before.

I also scoured the leading public collections of images of the Brooklyn Bridge being built: New York City's Municipal Archives, the Museum of the City of New York, and the Roebling Collection at Rensselaer Polytechnic Institute (RPI), where Chief Engineer Washington Roebling had studied engineering and to which he donated his personal and priceless collection of drawings, specifications, and photographs, many of which relate to the construction of the Brooklyn Bridge.

This book also greatly benefited from conversations with, and suggestions from, Erica Wagner, author of *Chief Engineer: Washington Roebling, the Man Who Built the Brooklyn Bridge*, the thorough and engaging biography of Washington Roebling. In addition, Erica generously reviewed the manuscript, sharing her expertise. Erica, who seems to know everyone who knows anything about the Brooklyn Bridge, introduced me to engineer Frank Griggs, who specializes in restoring historic bridges and who shared his expertise on their construction. Thanks are also due to Richard Haw, author of three stellar books on the Brooklyn Bridge, who kindly shared his vast knowledge of the bridge and the art that it has inspired, including his collection of woodcuts from newspapers of the time. I also am most grateful to Richard and Erica for sharing their insights in this book.

Like so many, I owe a great debt to David McCullough, whose classic book *The Great Bridge* has been an inspiration since it was published in 1972. I have included several wonderfully descriptive excerpts from it to help the reader better understand the building process. His chronology of that construction, which I expanded upon, formed the basis for the one in this book.

I also wish to acknowledge two wonderful books that were published in 1983 for the centennial of the Brooklyn Bridge's opening: *The Great East River Bridge, 1883–1983,* published in conjunction with a magnificent exhibition at the Brooklyn Museum, and Mary J. Shapiro's pioneering *A Picture History of the Brooklyn Bridge.* Both were inspirations and helped me home in on the best images of the bridge's construction.

Appreciation goes to Bob Zeller, longtime president of the Center for Civil War Photography, who pioneered the modern publication of nineteenth-century 3D images. It was Bob who urged me to use anaglyphs—images that miraculously transform into three-dimensional scenes when viewed through red/cyan glasses—rather than reproducing the stereoviews themselves. I also want to thank John Richter, director of imaging for the Center for Civil War Photography, who informed me about a free online program, StereoPhoto Maker, which allowed me—or, rather, dedicated Green-Wood volunteer Jim Lambert—to transform stereoviews into the anaglyphs that are reproduced in this book. John also most helpfully guided us through the use of that program and critiqued the results.

Kudos to Beth Daugherty of Bauer and Dean Publishers, whose persistence, patience, and professionalism made this a much better book. And thanks to Richard J. Moylan, Green-Wood Cemetery president, who championed the publication of four of my books. Lisa Alpert, Green-Wood's vice president of development and programming, kindly supported this project. Stacy Locke, manager of collections for The Green-Wood Historic Fund, was, as always, very helpful and quick to produce the scan or photograph that I needed from those collections. Jenifer Monger at RPI is a credit to that institution. All material obtained from RPI is credited as "MC 4, Roebling Collection, Institute Archives and Special Collections, Rensselaer Polytechnic Institute, Troy, NY," abbreviated in the captions as "RPI." Ken Cobb at New York City's Municipal Archives was a pleasure to work with, even when he could not get into his office because of the Covid-19 pandemic. And thanks go to Lindsay Turley at the Museum of the City of New York, who was a great help. All images courtesy of the Museum of the City of New York are identified in the captions as "MCNY." Much credit is due Kathy Hart for the design of this book. I am grateful for her dedication and unstinting efforts to make this book as close to perfect as possible. Thanks to Julian and Sylvia Wolff for their encouragement. Great appreciation to Joan Teshima for her love, wisdom, and, of course, her eagle eye.

I want to acknowledge the readers who provided such valuable feedback on the book: Paula Bennett, Jon Brown, Carolyn Coulter, Janet Daugherty, Cliff Fernandes, José Franjul, Debbie Freundlich, Marilyn Friedman, Betsy Gillen, Jennifer and Peter Gyr, Peggy Haine, Joan Harrison, Carol Hart, Michael Kathrens, Joanie Mackie, Rich Moreland, Barbara Page, Rob Richard, Ethel Rompilla, Viva Stowell, and Joan Teshima.

On the pages that follow, I share with the reader the joy of viewing amazing scenes in 3D of the Brooklyn Bridge as it was being built. Using the software described above, we converted over forty stereoviews of the bridge's construction into anaglyphs. Each is noted by the symbol "3D" before its caption. And a pair of red and cyan glasses—which will allow readers to see these images in 3D—is found in the pocket at the back of the book. These glasses should be used to view the anaglyphs and to fully experience the extraordinary scenes. Just don't forget to take them off to read the book—and to enjoy all the two-dimensional images that are equally spectacular!

— JIR

Building the Brooklyn Bridge, One Image at a Time

BY RICHARD HAW

Few nineteenth-century engineering projects were as well documented as the Brooklyn Bridge. Engineers sat patiently at drafting tables, working on detailed drawings devoted to every aspect of the work. Photographers roamed up, over, and around the construction site. Stereoscopic cards of the half-built bridge were sold across the United States. Woodcut prints regularly appeared in illustrated newspapers during the years the bridge was being built, documenting the endeavor and helping sell the idea of the bridge to a still somewhat skeptical public. Once the bridge edged close to completion, printmakers worked up triumphant images to capture the majesty—and immensity—of the project. Together these images form one of the richest visual histories of any construction project.

As a technological feat, the Brooklyn Bridge was a gargantuan, almost unimaginable, leap forward. It would be the longest suspension bridge in the world. Its towers would be the tallest man-made structures in North America. It was the first bridge to use steel wire and the first to be lit up at night by electric light. The timing of its construction was also fortuitous, as it coincided with the rise of photography, illustrated newspapers, and lithography, all of which helped make the bridge famous before it had even opened. Closely followed by professionals and the general public alike, the bridge's ongoing construction, started in 1869 and completed in 1883, was an object of fascination around the world.

The construction drawings reproduced in *Building the Brooklyn Bridge* are both functional and charming. They are handsome works of applied science, handicraft, and imagination. From the initial tower designs to the color-coded cable cross sections and stone placements, every aspect of the bridge was meticulously and artistically drawn. The complex inner workings of the immense caissons, hoisting machines, anchor plates and chains, the graceful arches of the bridge's approaches, and electric lampposts—all were laid out in hundreds of renderings. No computer program assisted in the creation of these designs, just steady and meticulous hands infused with the creative spirit of civil engineering.

In the years between the first design sketch of the bridge and its opening to the public lies the drama of its construction. Hundreds of photographs were taken as the bridge was being built. Some tell stories of ambition and vision—of size and scale—while many others tell the no-less-important stories of the day-to-day work at the bridge site. They help us understand the minutiae of what was done, the invention and technique and how it all fit together. They tell a story of creation and the equipment that made it possible, of problems faced and problems solved. They help us follow the construction, day by day, year by year, just as those who were alive as the bridge was being built followed its progress. Above all, these photographs remind us that the bridge was made by the hands of many men: the blocks of granite used in the approaches and towers were shaped by hand, the cables were hand wrapped, and the roadway was hung by an army of laborers. Many of these workers look at the camera—frozen for a moment in time.

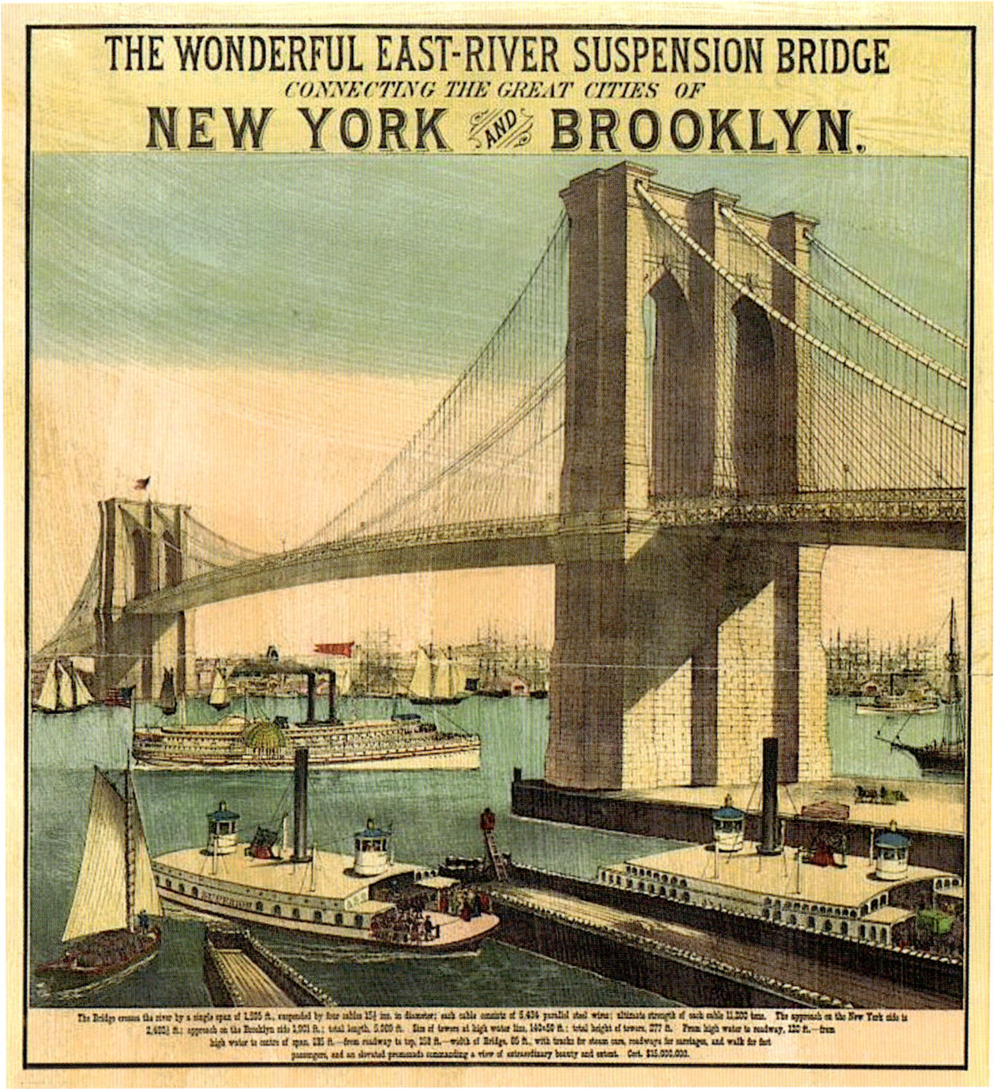

A colored print, ca. 1885, advertises the bridge, with its specifications noted in the text at bottom.

If these photographs of the construction are snapshots of time, the woodcut prints that populated the illustrated newspapers during construction brought the essential drama of building the bridge to life, allowing the public to follow the progress in close to real time. Dozens of such images were published while the bridge was being built. The sketch artists who captured these scenes worked where photographers could not—down in the caissons, where hundreds of shirtless men toiled in mud to dig the tower foundations, so dimly lit that photographers could not capture an image. These illustrations also re-created key events, both the triumphs and the tragedies. In 1876, at the height of the country's centennial summer, seven years into the construction, Master Mechanic E. F. Farrington's jubilant ride over the East River on a boatswain's chair was publicized nationally with prints in magazines and newspapers. Two years later, the *New-York Illustrated News* published a print of the sad fate of Harry Supple, a twenty-five-year-old riggers' foreman and a six-year veteran of the bridge work. That illustration captures the moment when a wire rope under great tension snapped and struck Supple square in the chest, sending him careening off the New York anchorage to his death below on a cobblestoned street.

Each of the illustrated weeklies focused on particular aspects of the construction. Rather unsurprisingly, *Scientific American* concentrated on the technical aspects of the work. *Harper's Weekly* featured grand panoramas and full-page illustrations capturing the construction against the backdrop of urban New York. *Frank Leslie's Illustrated Newspaper* took a less ambitious approach, often focusing on discrete tasks, such as taking apart the bridge's footbridge—a temporary hanging walkway that allowed workers to reach the cable wires and rearrange them as needed. *The Daily Graphic* offered both—the grand and the mundane. With some compositional license, the illustrated press did an excellent job of depicting the massive clearance work needed to build the bridge's approaches, capturing a semblance of what Brooklyn and Manhattan looked like before hundreds of their buildings were torn down to make way for the bridge. But also, figments of artists' imaginations appeared in the illustrated press—in *Scientific American, Harper's Weekly,* and *The Illustrated Christian Weekly*—widely publicizing the Brooklyn Bridge, though sometimes getting the details wrong.

As the construction proceeded, well-established printmakers such as Currier & Ives took up the mantle of imaging the bridge. Most of these renderings were pure civic boosterism, helping sell the bridge as a peerless icon of American technological vision and achievement. Many of these artworks were fantasy, showing the East River chock-full of maritime commerce, the cities in the background both affluent and orderly. The bridge's towers loom over New York like a modern-day Colossus of Rhodes. The bridge was huge, of course, but not quite as large as it appeared in many of these images. But accuracy was hardly the point. These artists and the commercial establishments they worked for had a story to tell, one of municipal greatness and American progress. In their images, the bridge was a harbinger of the future, of the city to be. Ultimately, the completed bridge fulfilled the promise of its promotion, ushering in an era of achievement, expansion, and prosperity, for Manhattan, Brooklyn, the entire country, and the world.

Images—whether drawings, photographs, or prints—are evidence. They can sharpen our understanding of what happened. They can even allow us to grasp what was planned but never realized, what was envisioned versus what was, in fact, done. They can show how places are altered and even remade. Between 1869 and 1883, hundreds of images recorded the design and construction of what was initially known as "The Great Bridge." They tell the story of how this American icon came to be, one stone block and one steel wire at a time. The images gathered in this book—as they have never been gathered before—allow us to witness what happened long before we were born. They present us with the extraordinary story of the building of the Brooklyn Bridge.

— Richard Haw is Professor of Interdisciplinary Studies at John Jay College of Criminal Justice (CUNY) in New York City and author of *The Brooklyn Bridge: A Cultural History* (Rutgers University Press, 2005), *Art of the Brooklyn Bridge: A Visual History* (Routledge, 2009), and *Engineering America: The Life and Times of John A. Roebling* (Oxford University Press, 2020).

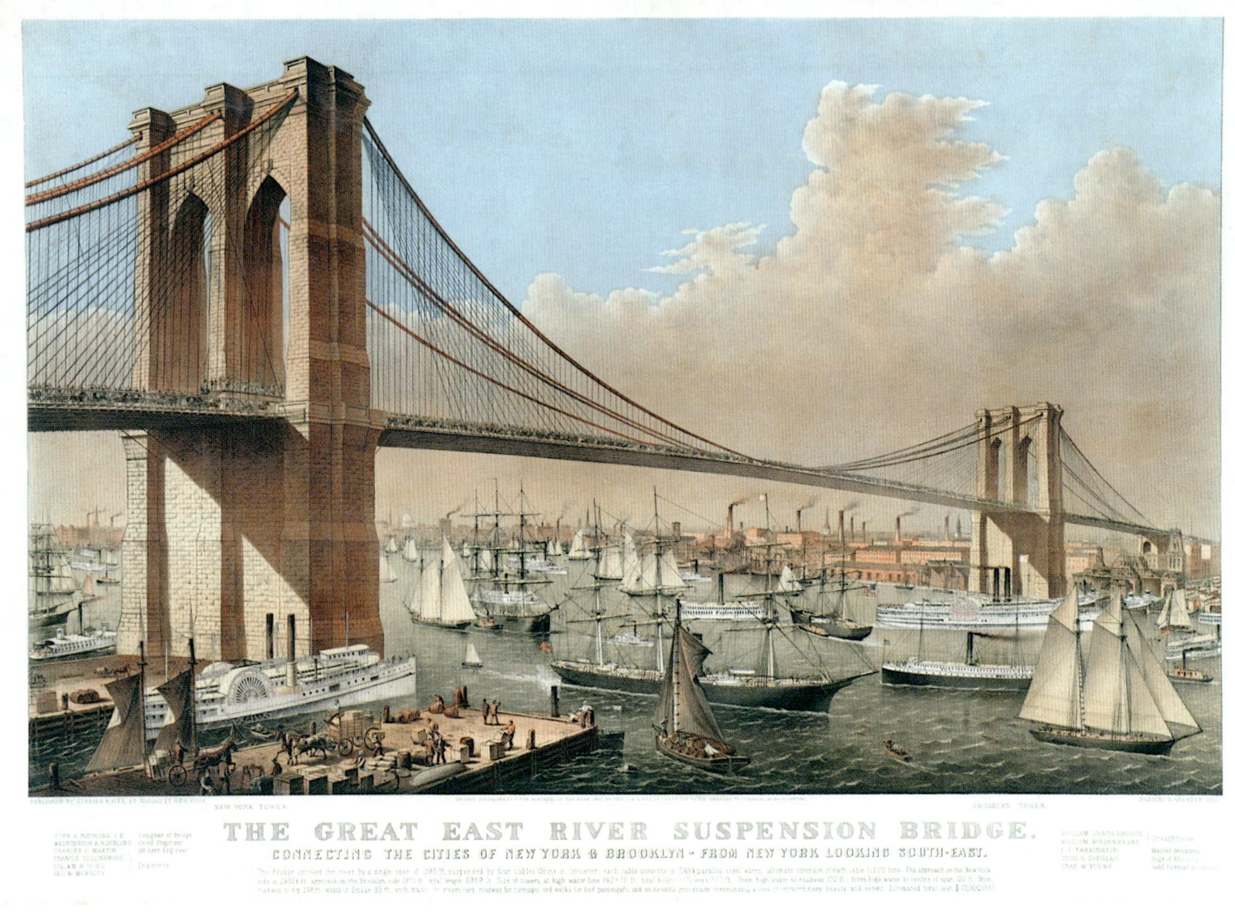

"The Great East River Suspension Bridge. Connecting the Cities of New York & Brooklyn—From New York Looking South-East," published by Currier & Ives, 1877. Note that the Fulton Ferry Terminal in Brooklyn is visible at the far right and the placement of the white steamers at far left are characteristic of the Manhattan side. Therefore, this view is, in fact, looking northeast, not southeast, contrary to the print's title.

The Brooklyn Bridge: A Love Story

BY ERICA WAGNER

The Brooklyn Bridge is a radical, visionary structure. Now that it has been an essential element of New York's infrastructure for nearly a century and a half, this fact can—perhaps—be hard to recall. Seen from a distance, the bridge can look, in the twenty-first century, a little antiquated. From the water side of Brooklyn Heights its stone towers are dwarfed by the shimmering glass structures that crowd lower Manhattan. Closer up, the Gothic Revival arches of its granite and limestone towers seem to speak to the past rather than the future. Yet step onto the bridge itself—onto the wooden promenade that rises over the river, over the traffic, between those towers, walk into the cradle created by the great steel cables and the fanned spread of wire suspenders—and the genius and courage of its builders become immediately apparent. It is vast and delicate, solid as stone and light as air: the contrast between those cathedral arches and the web of wire creates not only a bridge between Brooklyn and Manhattan but a bridge between the past and the future, between the old world and the new.

The first time I set foot on the Brooklyn Bridge was in the year of its centennial, though I was unaware of that anniversary at the time. I was sixteen in 1983. It's difficult to describe what happened to me that bright December morning: I suppose you could call it an epiphany or, perhaps more plainly, a sense of recognition and understanding. It wasn't just that I was struck with wonder at the beauty and power of a structure I'd never considered before. I felt—and this is the mystery of art—that I belonged there, on the bridge. That this was my place. That I understood the bridge, and the bridge understood me. Strength, harmony, flexibility, utility: they were qualities I wanted in my life, and I found them on and in the bridge.

As I stood that morning on the Brooklyn Bridge, my hand wrapped around one of the slender suspender cables, the movement of the roadway, the towers, and the wires felt in the vibration of my palm, my fingers, my arm, my own ignorance astonished me. How had this structure come to be? When was it built? And perhaps most important: who built it? Even as a teenager I like to think I was a little ashamed that I'd never given any of these questions a moment's thought. A wonder of the world was in my backyard, and I'd never even noticed.

And so I began to read. The first book I read was David McCullough's *The Great Bridge*, his landmark account of its construction, first published in 1972. And it was in that book that I first met Washington Roebling, chief engineer of the bridge. McCullough remarked—with some surprise—that no one had ever written a full biography of Washington; despite my youth I determined to do so, and so I went on to delve in archives, poring over his official reports, letters he'd written from the battlefields of the Civil War, private musings he'd jotted in pencil in a neat, sloping hand. I came to feel I knew the chief engineer as well as anyone ever had. It's worth remarking that Washington's calling was hardly his choice: the bridge was designed by his father, John Roebling. The family fortune was founded on John's development of wire rope technology, used not only in bridges but, as the nineteenth and early twentieth century progressed, in practically every aspect of the modern industrial and urban world: telegraphs, telephones, elevators, automobiles, airplanes—Roebling wire could be found in both the Wright brothers' *Flyer* and Lindbergh's *Spirit of St. Louis*.

But John died of tetanus just as work was commencing on the bridge, in the summer of 1869. It was left to his eldest son to actually construct what his father had envisioned, a bridge far larger than any in existence, using experimental technology (the sinking of underwater caissons for the bridge's foundations) and brand-new materials (the Brooklyn Bridge was the first bridge built with steel rather than iron). Over the course of fourteen years of construction, Washington, too, became gravely ill; the work nearly took his life as it had taken his father's, and he was fortunate to be more than ably assisted by his wife, Emily Warren Roebling. I determined to tell the story of Washington's remarkable life: my book *Chief Engineer* does him, I hope, a little justice.

The subtitle of my book is *Washington Roebling, the Man Who Built the Brooklyn Bridge*, but that's something of a canard or, at least, a synecdoche. Washington's own labor was extraordinary and came at great personal cost, but he did not labor alone. One of the speakers at the grand celebration of the bridge's opening on May 24, 1883, was the prominent clergyman Dr. Richard S. Storrs, who knew that the chief engineer would wish to express his debt to those who had served under him. "He would be the first to acknowledge," Storrs said, "the intelligent, faithful, indefatigable service rendered in execution of his plans by those who have been associated with him, as assistant engineers, as master mechanics, or as trained, trusted, and experienced workmen. On their knowledge and vigilance, their practiced skill and patient fidelity, the work has of necessity largely depended for its completed grace and strength. They have wrought the zealous labor of years into all parts of it; and it will bear to them hereafter, as it does to-day, most honorable witness."

One of the most striking features of Jeff Richman's *Building the Brooklyn Bridge* is the way in which it brings us into the presence of those who actually did construct it: workmen from all over the world, the vast majority of their names forever lost to history. And yet these

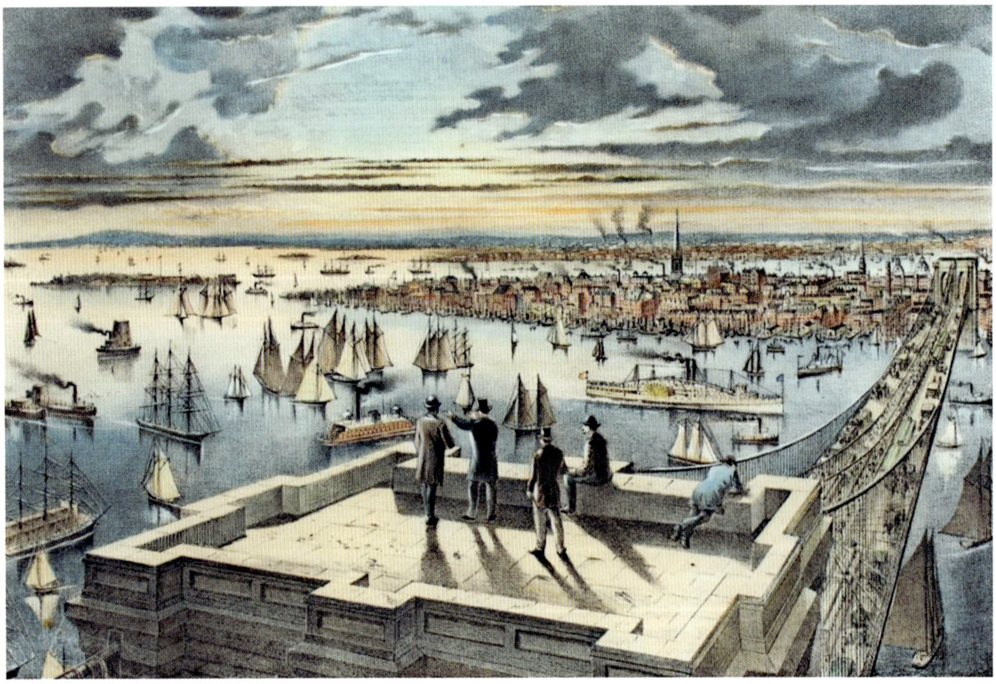

"The Harbor of New York, From the Brooklyn Bridge Tower—Looking South West." Printed by Currier & Ives, this lithograph is a dramatic view with men atop the tower looking out over the harbor, the finished bridge below bustling with activity.

men are not quite lost, for they appear in even the earliest drawings for the bridge. John Roebling's designs would not be complete without his neatly sketched workforce: a man toiling away with a sledgehammer on the deep riverbed, even his over-the-knee work boots perfectly detailed; above him another man hauls himself up a steep ladder. Their living counterparts appear in these photographs: here is a group of three men, in hats, rumpled clothes, suspenders, standing by the massive iron eyebars that hold the bridge's four main cables fast in an anchorage to have their picture taken. Those eyebars have been hidden from view since the nineteenth century, and those men are long in their graves, but now they stand before us again. A scattered crowd of them poses atop one of the half-built towers, completed as far as the roadway: look closely and you'll see that one of them seems to be waving at us, a greeting across the ages. In other images the men ignore the camera, refusing to stand still for the long exposure times required in those days, and so they slip like ghosts across the frame, their shadowy figures an eerie metaphor for the passage of time itself. These images allow us to feel what it might have been like to be there, over those fourteen years of construction. To sense the vastness of the towers as they rose into the New York sky. To see the river finally spanned, wire by wire. To watch as the roadway rose over the water, forging a connection where none had ever been before.

And that is what bridges do: they speak to our human desire for connection. We can walk on water if we set our minds to it: we can choose to build a bridge. *Building the Brooklyn Bridge* shows us how it can be done, step by fascinating step. "I know that nothing can be done perfectly at the first trial," Washington Roebling said. "I also know that each day brings its little quota of experiences, which, with honest intentions, will lead to perfection after a while."

—Erica Wagner is consulting literary editor of *Harper's Bazaar* and a contributing writer for the *New Statesman*. She was previously a literary editor of the *London Times* for seventeen years, and her most recent book is *Chief Engineer: Washington Roebling, the Man Who Built the Brooklyn Bridge* (Bloomsbury, 2017). Born in New York, she lives in London and is a lecturer in creative writing at Goldsmiths College. She will cross any bridge she finds.

A Note on Using 3D Glasses

The 3D symbol below appears throughout the book to indicate an anaglyph that can be viewed in three dimensions using the 3D glasses found in the pocket at the back of the book. The engaging 3D effect may be improved by standing directly over an image, with the book open on a flat surface. Variations in lighting may alter results. To adjust the fit, the 3D glasses have two fold marks on the arms. If you wear glasses to read, keep them on and simply place the 3D glasses over them, with the red lens over the left eye and the cyan lens over the right.

3D

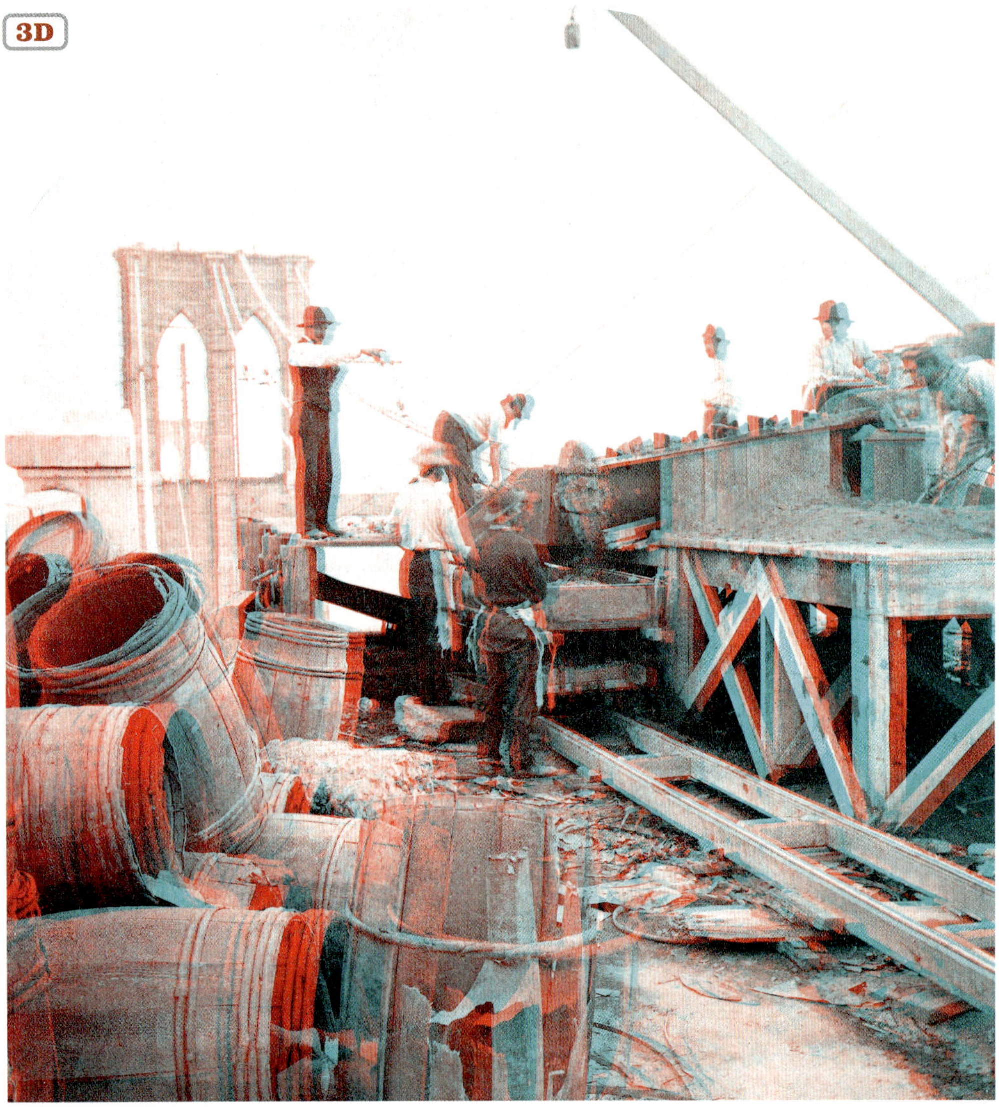

The same spectacular view of ongoing Brooklyn Bridge construction is reproduced here in three formats: an anaglyph (above), a stereoview (opposite, top), and a print from a lantern slide photograph (opposite, bottom). The anaglyph was generated by software from a scan of the stereoview, specifically for inclusion in this book.

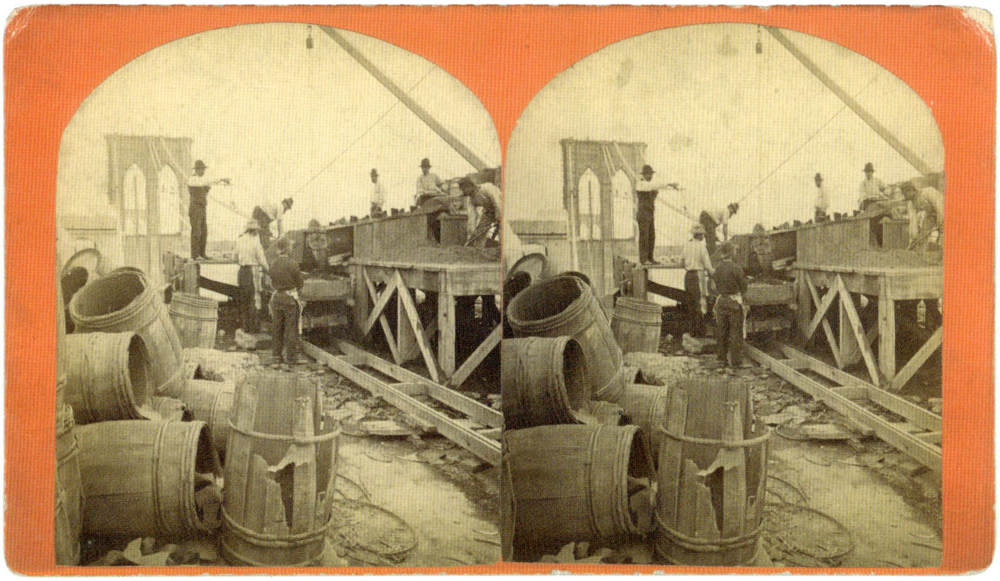

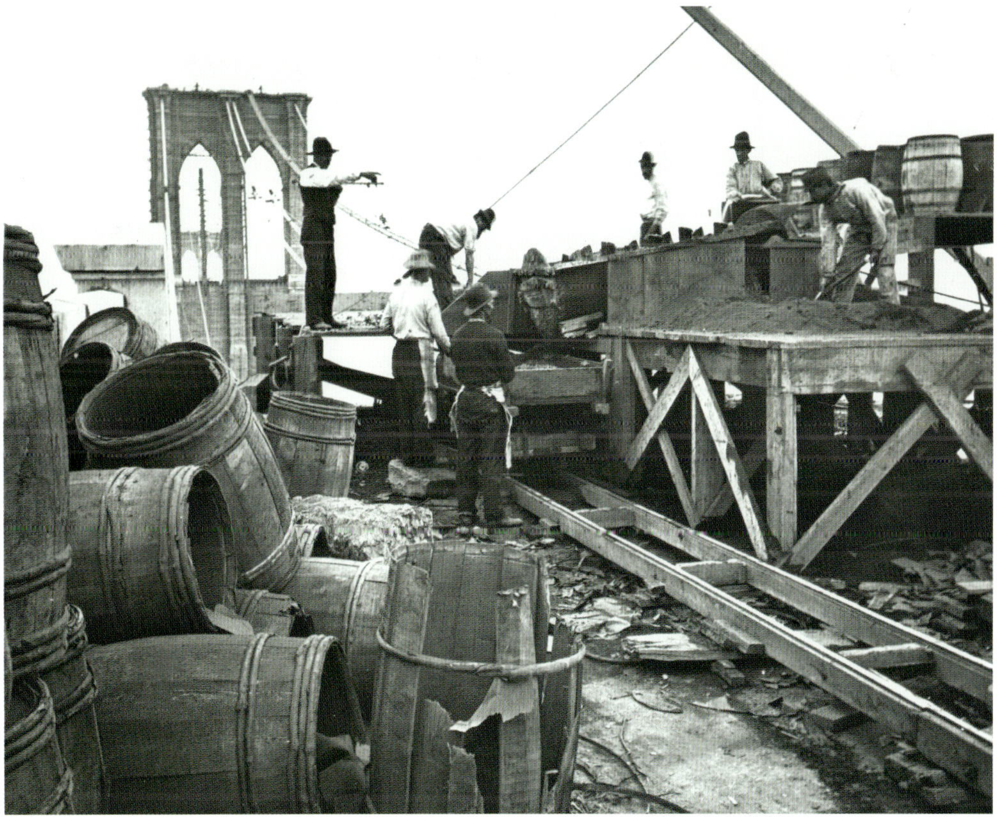

These images, ca. 1876, are very well composed to maximize the 3D effect, with elements in the foreground, middle ground, and distance. Empty wooden barrels at lower left frame the view, hatted workers stand in the middle, and the bridge towers are in the background. Note the small-scale wooden rails at lower right—this is a temporary wooden railway used to transport building materials. Two men stand next to the supply cart that sits on the rails. Bottom photograph from author's collection. Stereoview at top courtesy of the Michael E. Ach Collection.

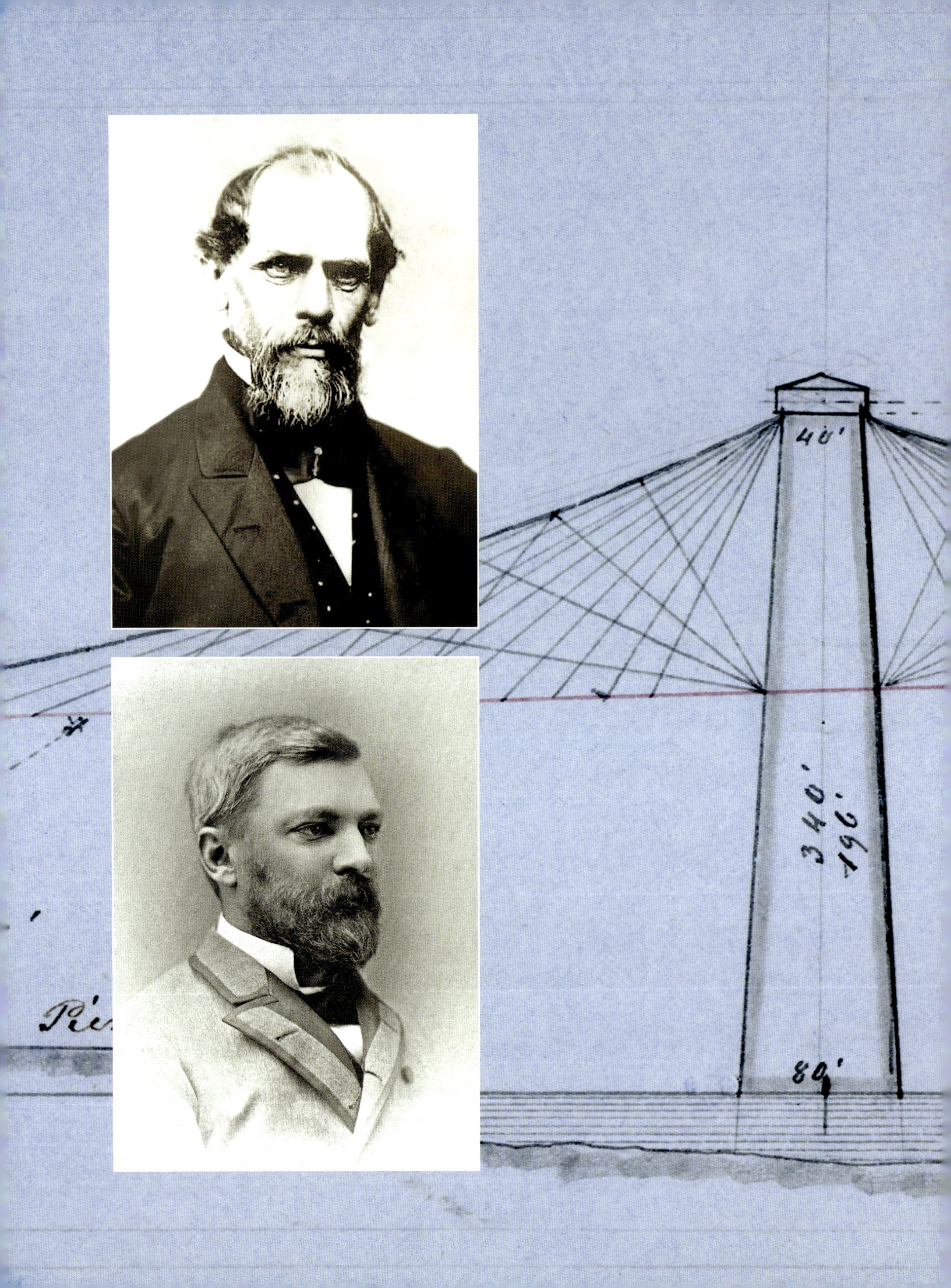
40'
340'
196'
80'

CHAPTER 1

Engineers

1869–1883

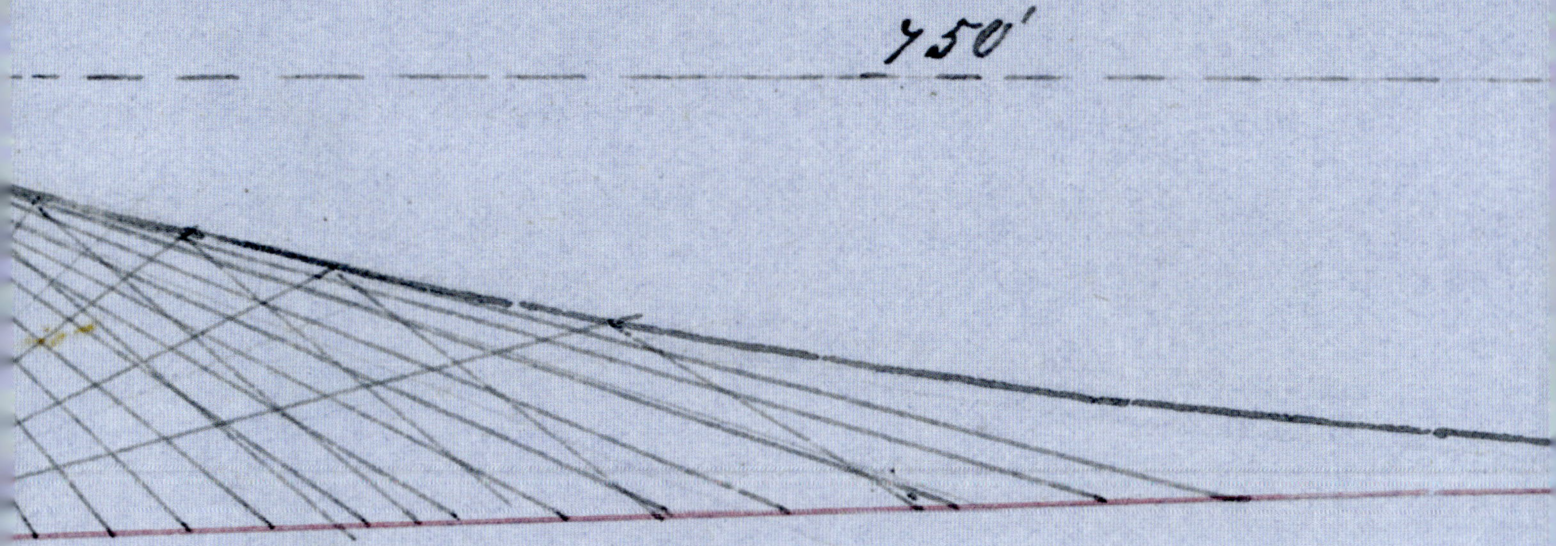

"...it may be stated that the East River Bridge will connect the two cities of New York and Brooklyn, which are separated from each other by an arm of the sea, called the East River."

—CHIEF ENGINEER WASHINGTON A. ROEBLING

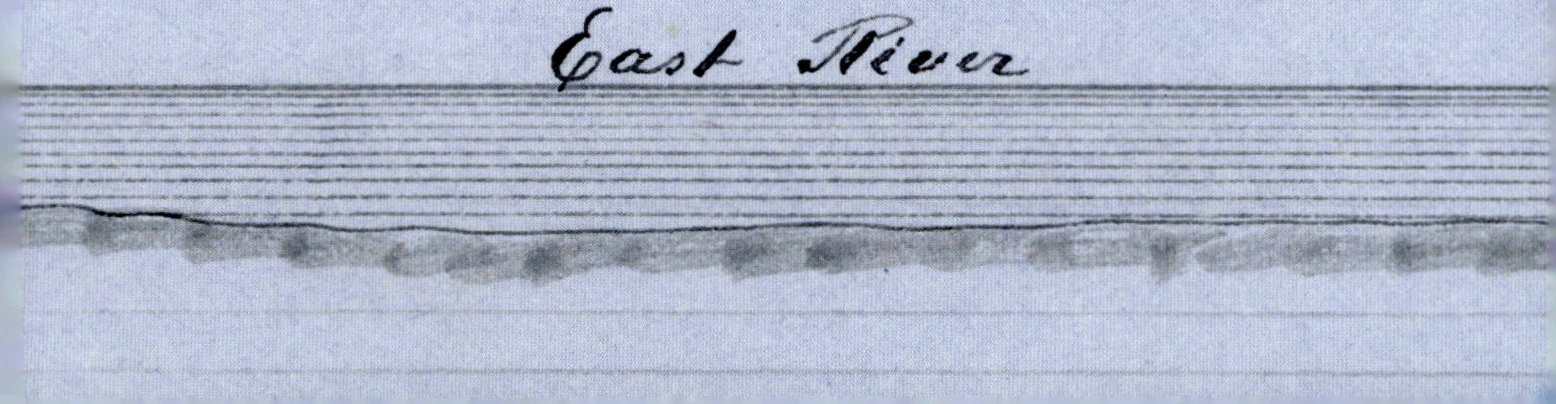

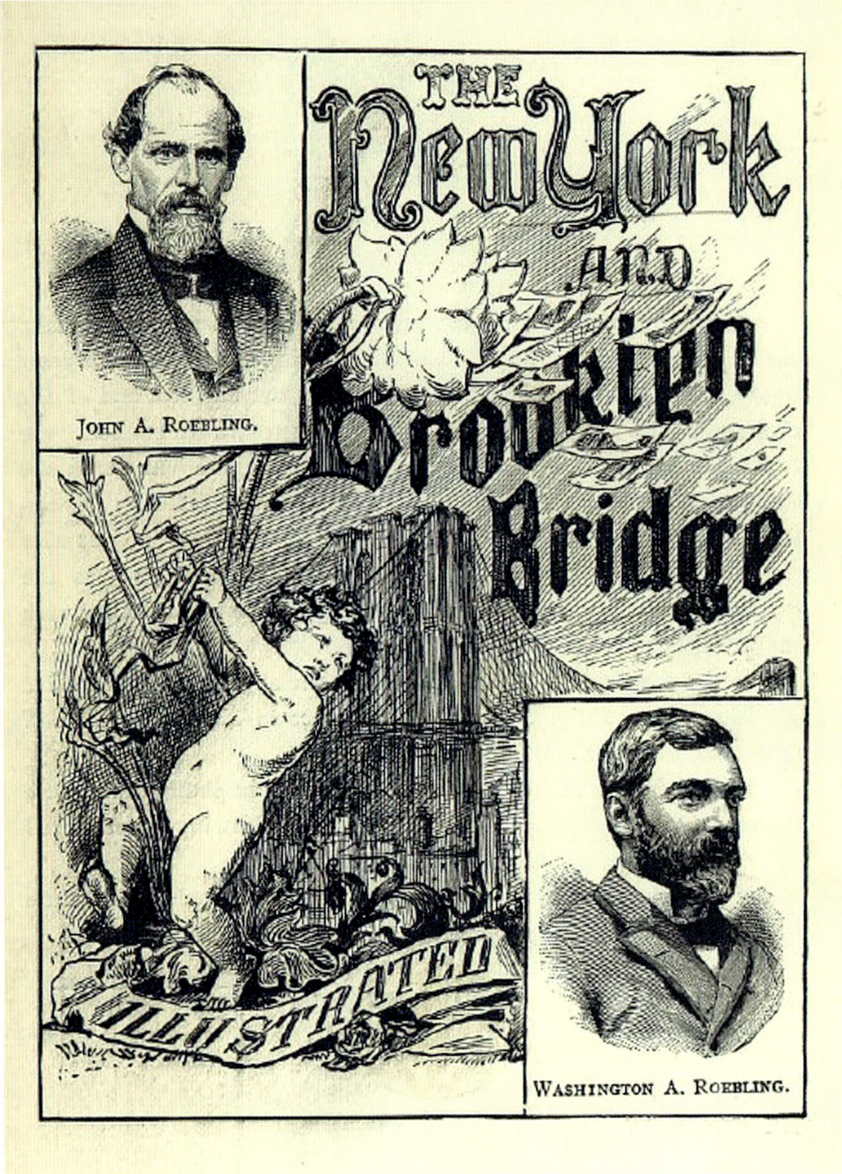

The Brooklyn Bridge had two chief engineers. The first was John Augustus Roebling, the leading suspension bridge builder in the world at the time. When John died in 1869, before construction even began, his son Washington Augustus Roebling, whom John had trained, took over and led the massive bridge construction to completion.

PREVIOUS SPREAD: Detail of John A. Roebling's 1857 plans for a bridge across the East River at Blackwell's (now Roosevelt) Island. RPI. Portraits of John A. Roebling (top) and Washington A. Roebling (bottom). Author's collection.

The first serious proposal for building a bridge across the East River—uniting what were then the cities of Brooklyn and New York—was made as early as 1800. And many more such plans were to follow. Half a century later, in 1849, the editor/publisher of the *New-York Tribune*, Horace Greeley, long an influential voice for local and national causes, advocated for the bridge, writing in his newspaper, "New York and Brooklyn must be united."

To some living in New York City, Brooklyn was a backwater across the East River, and they may have questioned Greeley's assertion. But the fact was, Brooklyn was a boomtown. It had the larger port and was growing faster than New York City (essentially the island of Manhattan), which by mid-century was already the most populated city in the country. In 1800, Brooklyn had just five thousand inhabitants (a thousand of whom were enslaved), but by 1869, the year construction of the Brooklyn Bridge began, its population had shot up to four hundred thousand, making Brooklyn the third most populated city in the United States. That same year, ferries were making more than a thousand daily crossings between Brooklyn and New York City, carrying half a million people every weekday. When work on the bridge began, Brooklyn was about seventy square miles—while New York City was just a third that size. Importantly, a good part of Brooklyn was undeveloped, providing substantial space for both commercial and residential expansion.

Further impetus for building a bridge that would connect these two cities was provided by the extremely frigid winter of 1866–1867. It was so cold that the East River repeatedly froze solid, shutting down ferry service, the public transportation system that so many relied upon. At the time, with the ferry service disabled, it was said that a visitor from Albany could make the trip from the upstate capital to New York City (about 145 miles) more quickly than one could get from Brooklyn to Manhattan. A bridge would make travel between the two cities not only safe and convenient, but also reliable, regardless of the weather.

At the end of 1866, while still in the middle of one of the coldest winters on record, Brooklyn's political and business leaders stepped forward to make the bridge a reality. Henry Cruse Murphy, an influential state senator and former mayor of Brooklyn, agreed to sponsor a law in the New York State Legislature authorizing such a bridge, giving the project the influential support it needed. Just a few months later, in April 1867, the legislature approved the charter for the New York Bridge Company (later known as the New York and Brooklyn Bridge Company), a private corporation empowered by the state to purchase real estate and charge tolls.

Within weeks, on May 23, 1867, John Augustus Roebling (1806–1869) was named the chief engineer of the project. Roebling's reputation and experience as a bridge builder were such that no one else was even considered for the job. He was given free rein to design and supervise the construction of the bridge and would be paid an annual salary of $8,000 (a goodly sum equivalent to approximately $150,000 in 2021). Roebling calculated that the bridge would cost almost $6.7 million to build. Total expenditures ultimately came to a little more than $15 million (equivalent to about $385 million in 2021)—more than double his estimate. The preliminary survey of the bridge location was begun in mid-1869, but construction itself did not begin until January 1870, after New York City, under the leadership of the notoriously corrupt William Magear "Boss" Tweed, finally agreed to help fund the work. Roebling was confident that construction would take just five years. In the end, it would take almost three times as long, and, sadly, Roebling would not live long enough to see the work completed—or even to see it truly begun.

OPPOSITE: "The Great Suspension Bridges of the United States." *Scientific American*, May 31, 1879. The Roeblings built all four of these bridges, from top to bottom: the Niagara Suspension Bridge, the East River Suspension Bridge (Brooklyn Bridge), the Covington-Cincinnati Bridge (bottom left), and the Allegheny River Bridge (bottom right). Author's collection.

John Augustus Roebling was born in 1806, in Mühlhausen, at the time part of the Prussian state, which later became Germany. In 1826, he received a degree in civil engineering from Berlin's Royal Polytechnic Institute, where he had studied the construction of suspension bridges, which use cables and suspenders—supported by towers—to hang the bridge spans. Dissatisfied with his opportunities at home, Roebling led a group of immigrants to America in 1831. As he reported, they left because of "an inclination and an ardent desire that our circumstances be bettered." At first, he tried his hand at farming in a utopian community in western Pennsylvania. Bored with that, Roebling turned to his true love—technological innovation—and became one of the leading pioneers and manufacturers of wire rope, a versatile metal rope consisting of multiple wires. He soon made a fortune. Eventually, he returned to engineering and focused on suspension bridge building—a logical step, given the central role of wire ropes in such construction. By the 1850s, only two engineers had built suspension bridges of consequence in America; John Roebling was one of them.

Ultimately John Roebling and his elder son, Washington Augustus Roebling (1837–1926), would successively serve as chief engineer of the Brooklyn Bridge, each leading the effort to build the largest suspension bridge in the world. By the time John was hired in 1867, he was the world's leading suspension bridge builder. He had already completed three with Washington's help: the Niagara Falls Suspension Bridge across the Niagara River, begun in 1851 and completed in 1855; the Allegheny River Bridge in Pittsburgh, built between 1857 and 1859, after Washington had graduated with a degree in civil engineering from the Rensselaer Institute—later known as Rensselaer Polytechnic Institute (RPI)—in Troy, New York; and the Covington-Cincinnati Bridge across the Ohio River, the longest suspension bridge in the world at its completion in 1866. Construction on that bridge took ten years; Washington joined his father on site in 1865 after being discharged from his Civil War service.

While working on the Niagara bridge, the first of his grand suspension bridge projects, John took his teenage son Washington out of school to assist him with the construction. That bridge, torn down in 1897, was 825 feet long, with towers rising 60 feet, four cables (each with a 10-inch diameter), and a truss made of wood. Upon seeing John Roebling's plans for the Niagara bridge, English engineer Robert Stephenson—who in 1846 had built the Britannia Bridge, a series of iron boxes set on stone piers connecting the island of Anglesey and the rest of Wales by railroad—responded, "If your bridge succeeds, then mine have been magnificent blunders." Upon completion, Roebling's Niagara bridge was able to carry heavier loads than Stephenson's, even though the Britannia Bridge was made up of six times more material per linear foot of bridge span. Despite Roebling's achievement, Mark Twain did not appreciate the work at Niagara: "You drive over the Suspension Bridge and divide your misery between the chances of smashing down two hundred feet into the river below, and the chances of having a railway-train overhead smashing down onto you. Either possibility is discomforting taken by itself, but, mixed together, they amount in the aggregate to positive unhappiness."

Father and son next worked together on the Allegheny River Bridge. From end to end, this entire bridge was 1,030 feet long, only about two-thirds of the Brooklyn Bridge's central span (from tower to tower). Due to the nature of suspension bridges, many of the Allegheny bridge's characteristic features echoed those of the Niagara bridge and

SCIENTIFIC AMERICAN

A WEEKLY JOURNAL OF PRACTICAL INFORMATION, ART, SCIENCE, MECHANICS, CHEMISTRY, AND MANUFACTURES.

Vol. XL.—No. 22. [NEW SERIES.] NEW YORK, MAY 31, 1879. [$3.20 per Annum [POSTAGE PREPAID.]

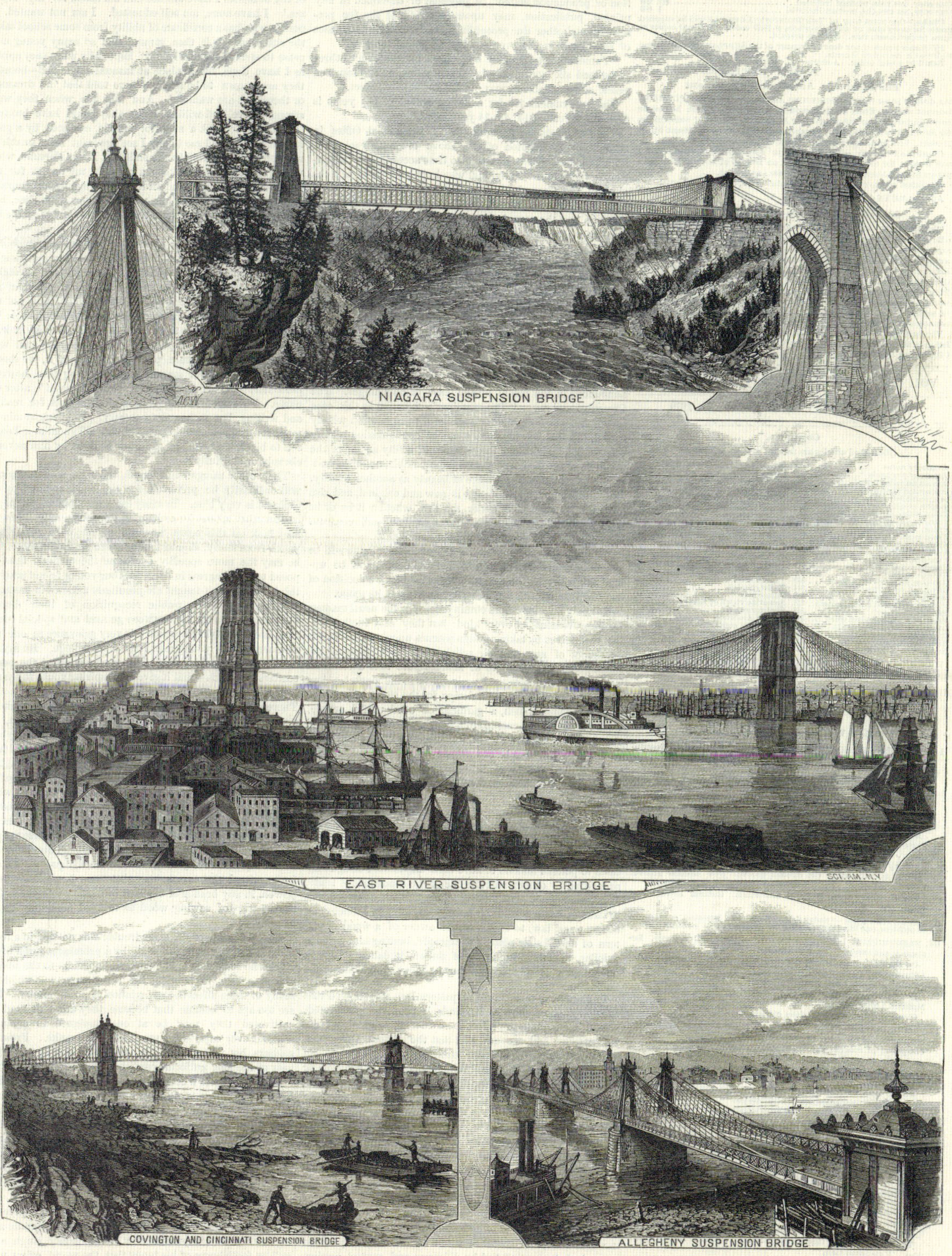

THE GREAT SUSPENSION BRIDGES OF THE UNITED STATES.—[*See page* 337.]

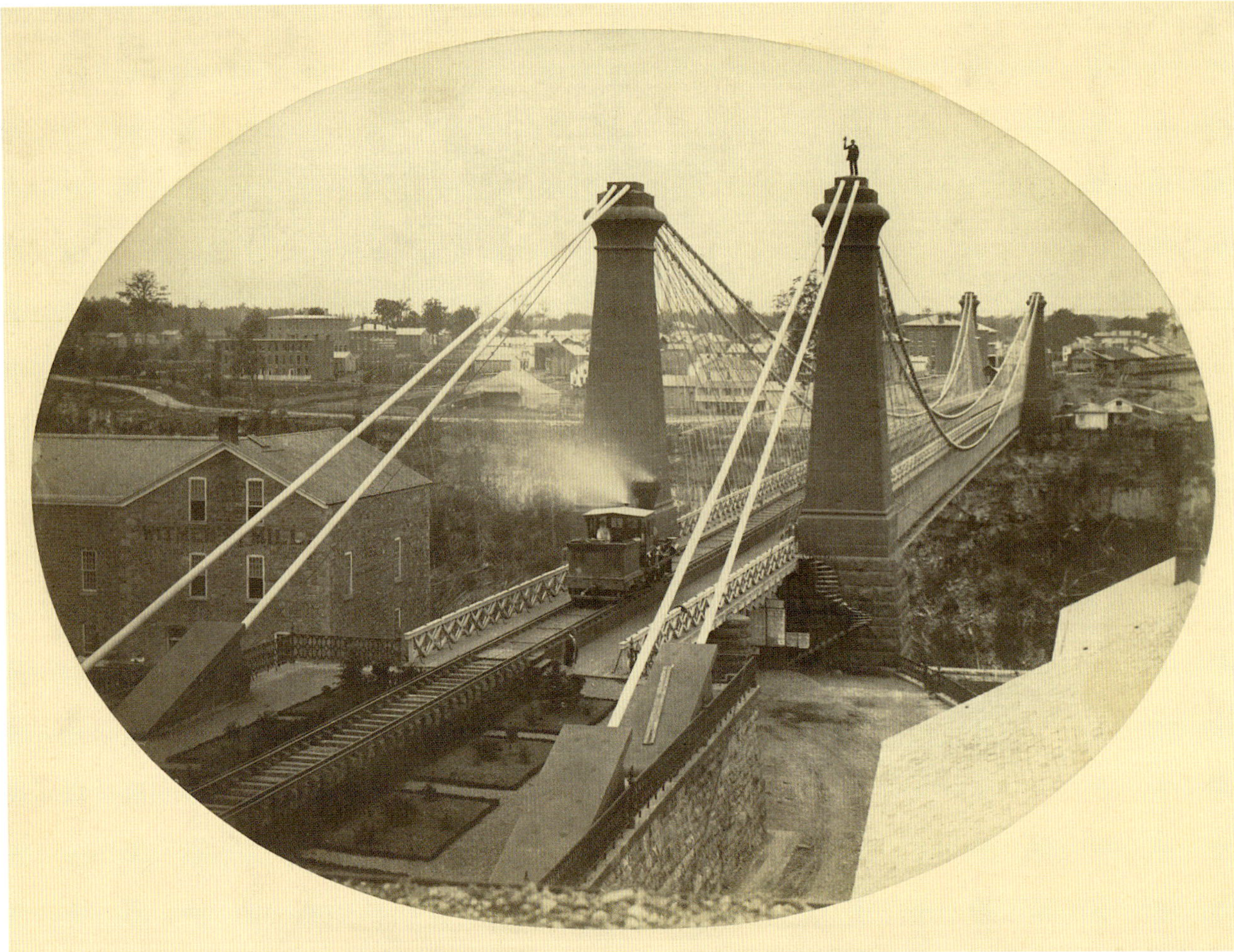

Photograph of the Niagara Falls Suspension Bridge by Silas A. Holmes, ca. 1860. Also called the International Suspension Bridge, it was built by John A. Roebling and completed in 1855. The bridge crossed the Niagara River between Niagara Falls, New York, and Niagara Falls, Ontario, Canada, and had two tiers: carriages and pedestrians traveled the lower level, while railroad trains ran across the upper. The bridge was dismantled in 1897. Note the man standing atop the tower, arm raised, hat in hand.

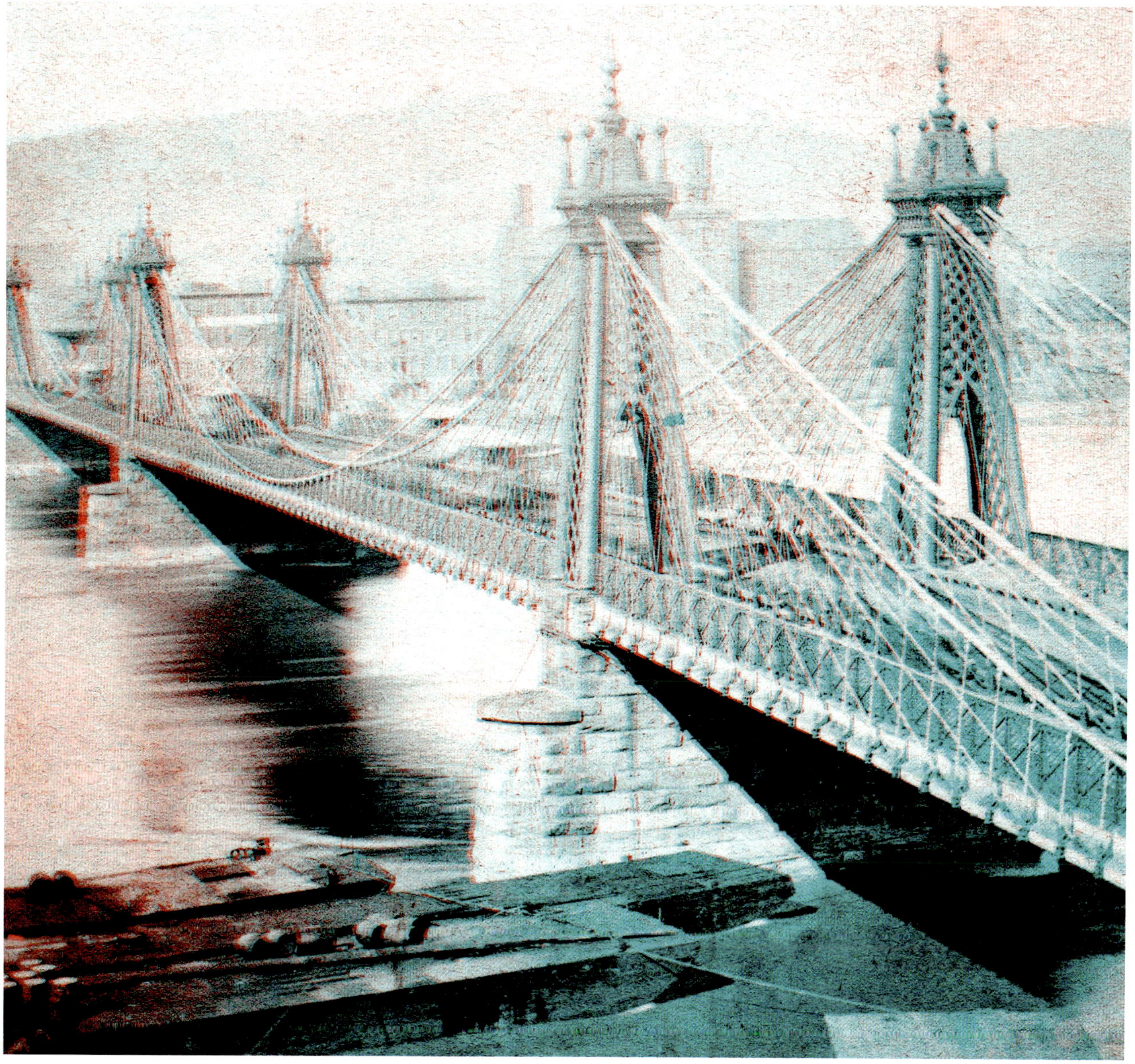

3D Allegheny River Bridge, Pittsburgh, Pennsylvania, ca. 1865. Completed in 1859 by John Roebling, this bridge was demolished in 1892. Courtesy of The Green-Wood Historic Fund, Jeffrey I. Richman Stereoview Collection.

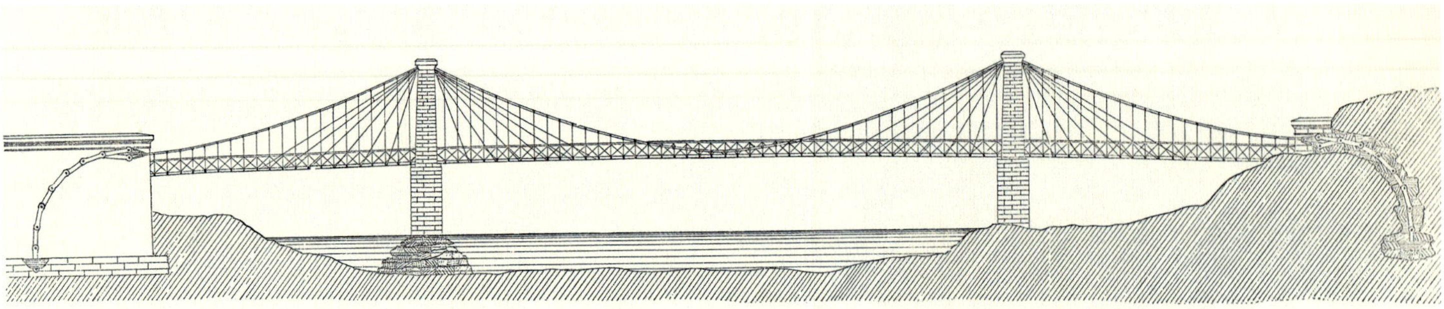

"A Suspension Bridge, showing all the different parts in their relative positions." One of numerous engravings illustrating the 1880 lectures of Brooklyn Bridge Master Mechanic E. F. Farrington. This one shows different means of anchoring the cable ends of a suspension bridge. At the left is a built anchorage, with an anchor plate and anchor bars embedded in it to hold the cable ends. At the right, the alternative to constructing an anchorage is to bury an anchor plate and anchor bars in a natural hill. The Brooklyn Bridge required built anchorages, because no such hills existed along the East River. Also note the key elements of a suspension bridge depicted here: two towers supporting the cables, with suspender wires dropping down from the cables and holding up the suspended superstructure (or deck) of the bridge.

would be seen again on the Brooklyn Bridge: two towers over which the cables ran; a deck hanging from suspender wires that dropped down from four large wire cables that had been "spun," or made, in place; the cables "anchored" by chains of metal bars buried deep in massive anchorages; and a strong superstructure reinforced by inclined stays and wires descending from the towers to enhance the span's rigidity. Washington Roebling, still in his early twenties, was put in charge of the work in Pittsburgh after his father went home to Trenton, New Jersey.

In 1846, almost a decade before completing the Niagara bridge, John Roebling went to Cincinnati to begin work on a bridge over the Ohio River. But construction there did not begin until 1856. Within a year of starting that project, John began work on yet another project—the Allegheny River Bridge in Pittsburgh. No doubt, supervising two large construction projects was a challenge. Soon after the completion of the Allegheny in 1859, the Civil War began. Workers became scarce as men, including his own son, went off to fight.

Early in 1865, Washington was discharged from the United States Army, and his father appointed him assistant engineer of the ongoing Covington-Cincinnati project. But John, knowing it was time to step aside and "leave bridge-building to younger folks," soon placed Washington in full charge of construction. Working for the first time without his father's onsite supervision, the younger Roebling directed the "cable spinning" (making the cables in place), the most difficult aspect of suspension bridge construction at the time. His father's confidence was rewarded in late 1866, when the Ohio River bridge reached its triumphant completion.

The overall length of the Covington-Cincinnati Bridge connecting Covington, Kentucky, and Cincinnati, Ohio, is 2,252 feet, not even half of what the Brooklyn Bridge stretches out to. Its width of 36 feet was less than half of the 80-foot width already being planned for the Brooklyn Bridge. When the Cincinnati bridge opened to the public on December 1, 1866, it was headlined as the "Biggest Bridge in the World." Thomas Kinsella, editor of the *Brooklyn Eagle* and a champion of the Brooklyn Bridge, described the Covington-Cincinnati Bridge as "the stateliest and most splendid evidence of genius, enterprise and skill it has ever been my lot to see." It was the forebearer of the Brooklyn Bridge, though one at a greatly reduced scale compared to what was to come.

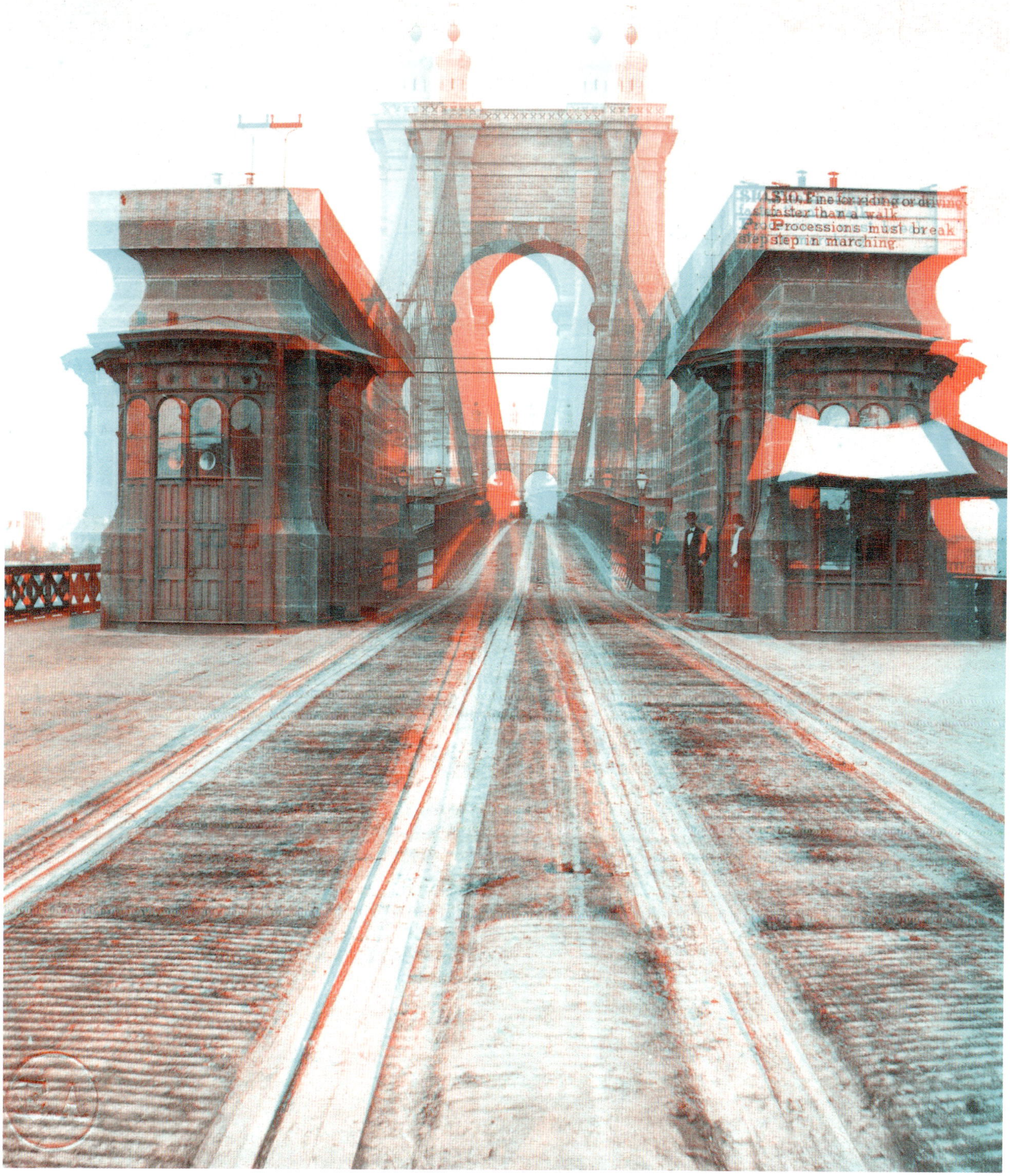

3D The entrance to the Covington-Cincinnati Bridge, ca. 1865. This bridge crossed the Ohio River between Covington, Kentucky, and Cincinnati, Ohio. Note the sign at right: "$10. Fine for riding or driving faster than a walk. Processions must break step in marching." This sign reflected the fear that periodic vibrations might set the bridge to undulating and cause it to collapse. In 1983, the bridge was renamed the John A. Roebling Suspension Bridge. It still stands. Courtesy of The Green-Wood Historic Fund, Jeffrey I. Richman Stereoview Collection.

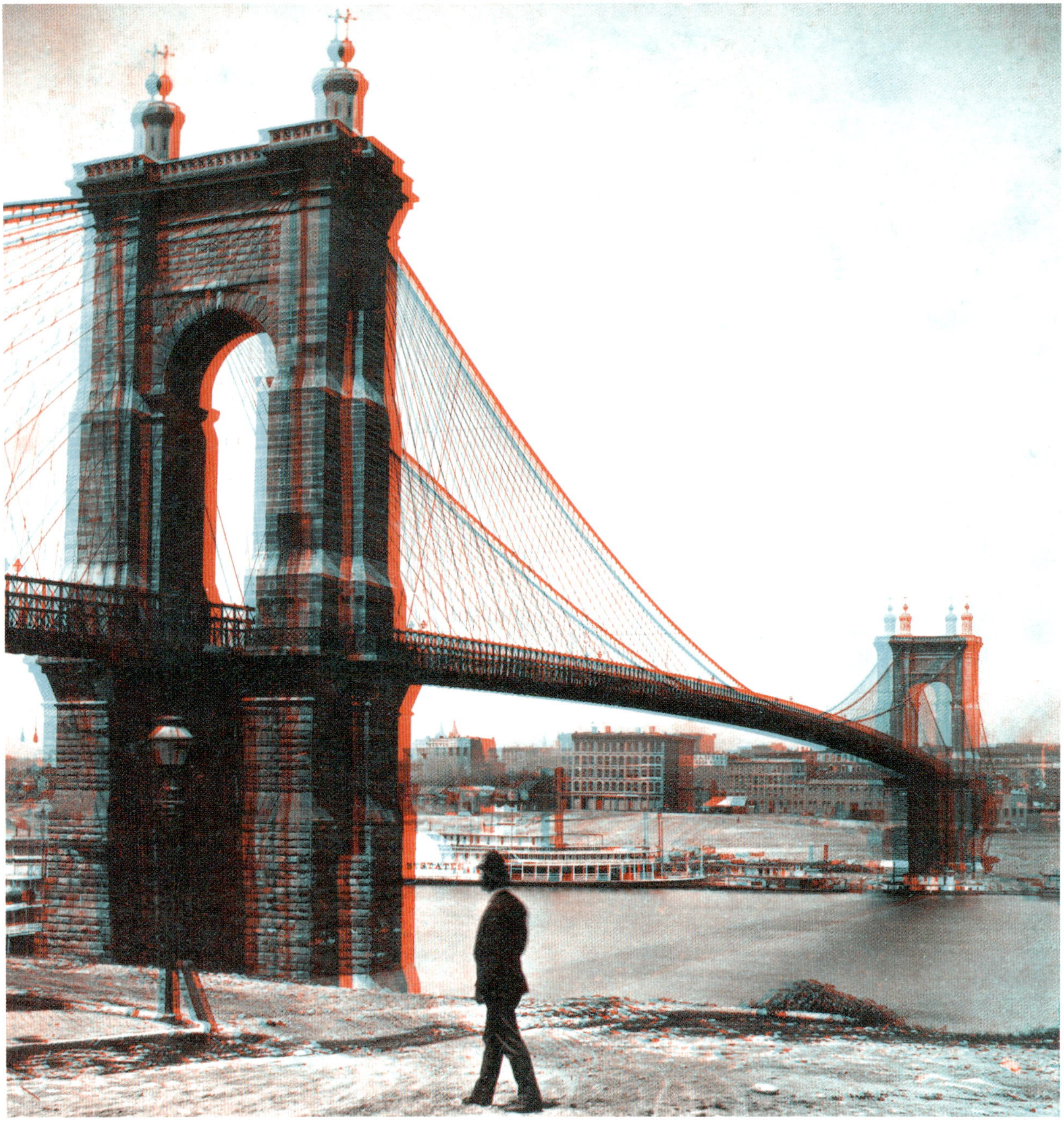

3D

Photograph of the Covington-Cincinnati Bridge, ca. 1870. The largest—and widely considered the finest—suspension bridge in the world at the time of its completion, it was, however, a mere miniature of what John A. Roebling would design and his son Washington would soon build: the Brooklyn Bridge. Courtesy of The Green-Wood Historic Fund, Jeffrey I. Richman Stereoview Collection.

OPPOSITE: Drawing of proposed Egyptian Revival towers for the Brooklyn Bridge, signed by John A. Roebling and dated March 10, 1857. Ultimately, this design lost out to the Gothic Revival arches we see today. RPI.

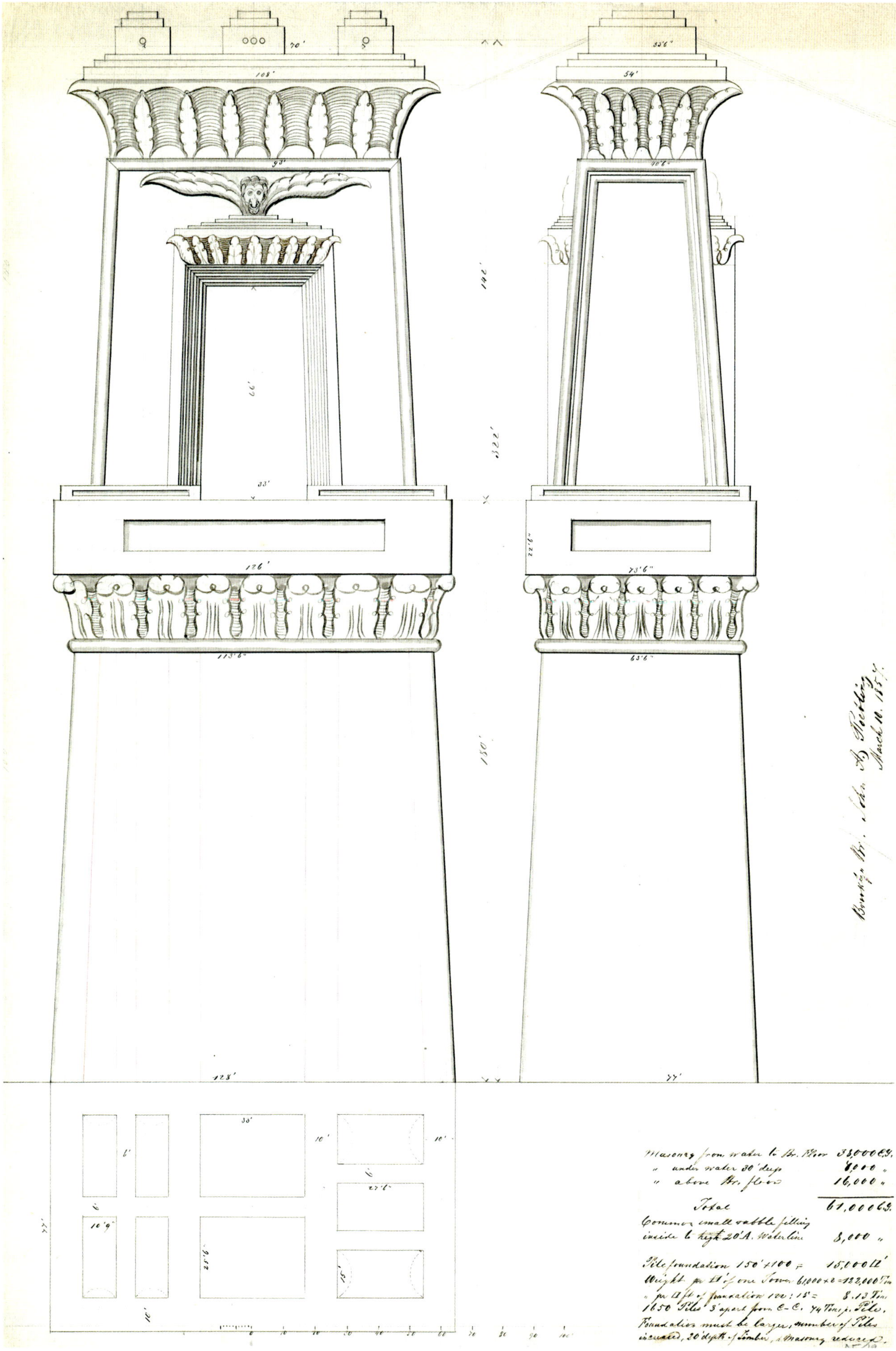
Brooklyn Br. John A. Roebling
March 10. 1857.
Masonry from water to Br. Floor 38,000 C.Y.
" under water 30' deep 8,000 "
" above Br. floor 16,000 "
Total 61,000 C.Y.
Common small rubble filling inside to high 20' A. Waterline 8,000 "
Pile foundation 150' × 100 = 15,000 □'
Weight per □' of one Tower 61,000 × 2 = 122,000 Tons
" per □ ft of foundation 122 : 15' = 8.13 Tons
1650 Piles 3' apart from C-C. 74 Tons p. Pile.
Foundation must be larger, number of Piles increased, 20' depth of Timber, & Masonry reduced.

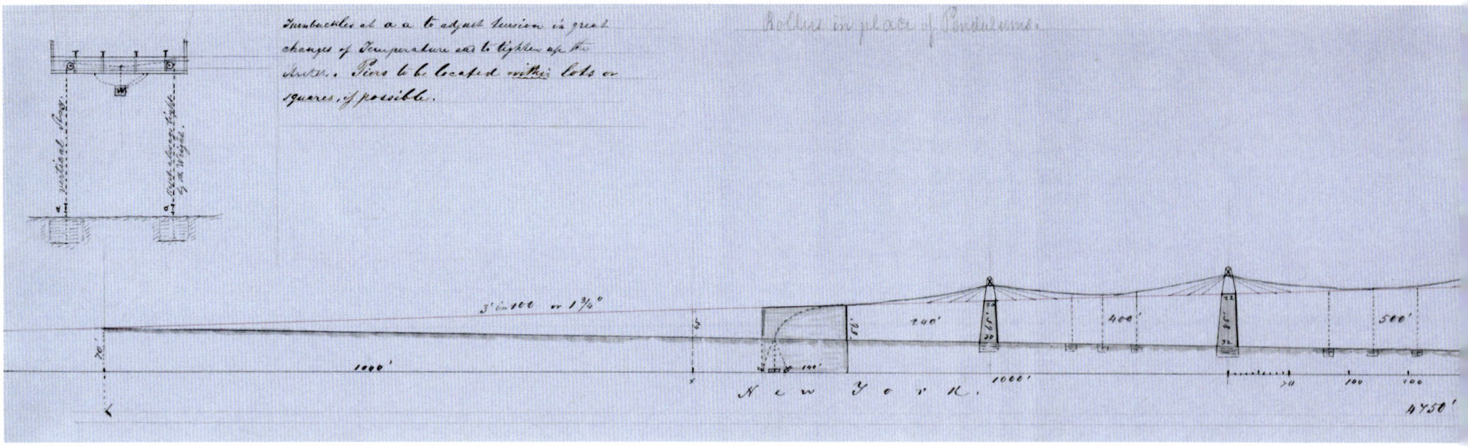

In 1856, the year construction started on the Cincinnati bridge, John Roebling was approached by New York businessmen who wanted his thoughts on the feasibility of building a bridge across the East River. Given the length that the bridge needed to be—a span longer than anything that had yet been built—its construction presented substantial challenges. Furthermore, the East River is tidal, meaning that its water level rises and falls with the tides, and reverses direction, creating other problems. In addition, any bridge spanning a navigable river had to be high enough to allow ships to sail beneath it but with a rise gentle enough to allow horses to pull their load up and over it.

John Roebling initially produced plans for a bridge that crossed the river at Blackwell's Island, now known as Roosevelt Island. As early as March 1857, Roebling, from his home in Trenton, produced four drawings of various designs for the bridge towers: one Egyptian, two Roman, and one Gothic Revival. It was the Gothic that would prevail.

After submitting his designs to the New York businessmen, John Roebling wrote to the prominent newspaperman Horace Greeley, a long-time advocate of an East River bridge. Roebling's letter, published in Greeley's *New-York Tribune* on March 27, 1857, proposed "a wire suspension bridge crossing the East River by one single span at such an elevation as will not impede the navigation." Roebling noted that his bridge would be "without rival" and went on to describe it as a bridge suspended from steel, not iron (the material in all his previous bridges), cables. These would run over two towers, "the great features of the work." Each tower would have two "pointed arches" through which the roadways would run. He was, he wrote Greeley, designing not for ornamentation but for "simplicity, massiveness and strength."

The Roeblings continued their work on the Covington-Cincinnati Bridge, while Brooklyn and New York leaders worked to advance plans for an East River bridge. On January 23, 1867, almost two months after the Cincinnati bridge had opened to the public and while the East River was still frozen solid, John Roebling headed to New York City to secure his next project. John's intention was to get the go-ahead to begin that project, then pass it on to his trusted assistant, his son Washington.

Four months later, John signed a contract making him its chief engineer. That summer, he resumed working on the design of the bridge, revising his initial sketches to create what would be, far and away, the greatest suspension bridge ever built.

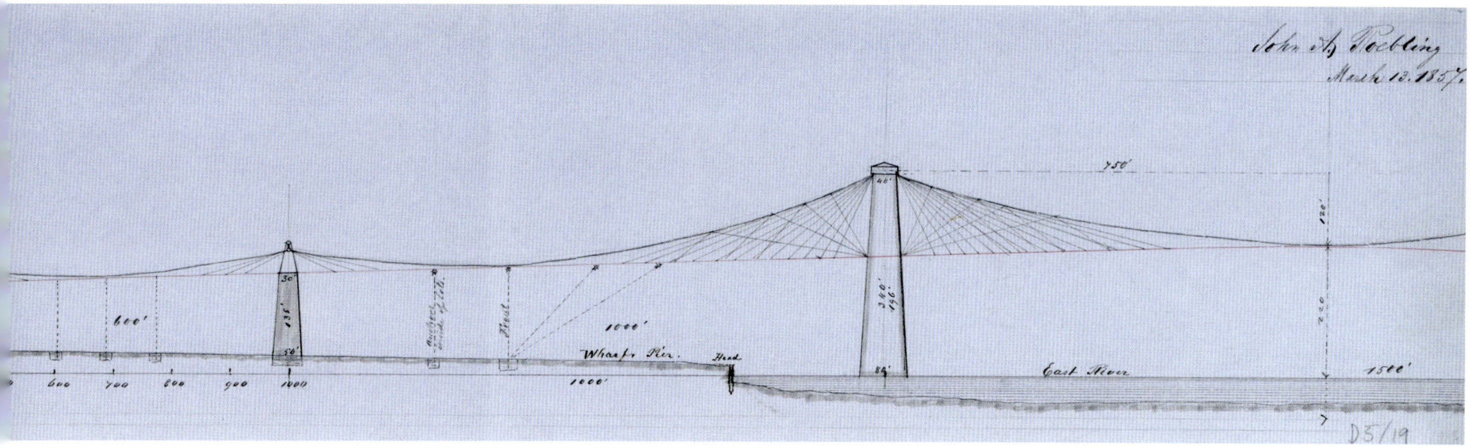

John A. Roebling's initial sketch for his proposed bridge across the East River at Blackwell's (now Roosevelt) Island, dated March 12, 1857. RPI.

In September 1867, John Roebling presented his plans to the bridge trustees, reading from his report:

> The contemplated work, when constructed in accordance with my designs, will not only be the greatest Bridge in existence, but it will be the great engineering work of this continent, and of this age. Its most conspicuous features, the great towers, will serve as landmarks to the adjoining cities, and they will be entitled to be ranked as national monuments. As a great work of art, and as a successful specimen of advanced Bridge engineering, this structure will forever testify to the energy and enterprise and wealth of that community, which shall secure its erection.

As John Roebling envisioned this project, which would become known worldwide as the Brooklyn Bridge, it would stretch for more than a mile (5,862 feet from end to end, to be exact). Instead of building a bridge using a series of iron boxes supported on piles of iron or wood, or one of stone arches, as had been the standards, Roebling decided once again to suspend his bridge spans from cables—a construction technique he had pioneered in America.

Suspension bridges offered both visual appeal—primarily a result of the graceful catenary formed by its cables—and economy, as they required less mass than other types of bridges, and less mass meant less cost. But suspension bridges also had a history of collapsing. John Roebling knew that the stiffer and heavier he made the roadway, the better the prospect of the bridge's survival. He also understood the importance of providing a firm foundation for the limestone and granite towers, essential for elevating the cables to their necessary height. Early on, Roebling planned on using caissons, a new form of foundation, upon which to build the towers. These massive, open-bottomed boxes would be positioned at the sites of the two towers and dug down until each reached solid bedrock. Because the bridge was being built along the shores of a river, compressed air would need to be pumped into the caissons to keep water from flooding their interiors while workers within were digging toward bedrock. Roebling further envisioned the two approaches on either side of the bridge as "small piers of masonry" containing "stores, dwellings or offices and would stretch over existing roads using iron spans." He also proposed railroad cars that would run from one end of the bridge to the other.

John Roebling spent the rest of 1867 and all of 1868 refining his designs. As he worked at his drafting table, he made occasional trips to the site that he had chosen. The bridge would ultimately run from Fulton Street in Brooklyn to City Hall Park in Manhattan, just north of and parallel to the route of the Fulton Ferry—one of the ferry businesses that the bridge was meant to replace. In preparation, test borings gauged the various depths of the rock, sand, mud, and gravel that the workers in the caissons would encounter as they dug their way to bedrock.

By spring 1869, money for the construction of the bridge had been secured and the designs had been reviewed and approved by a committee of engineers. All that was needed was the go-ahead from the United States military (tasked with making sure that navigation along the East River would not be impeded). In the meantime, John Roebling led a tour to Pittsburgh, Cincinnati, and Niagara to show off his earlier projects to the political and business leaders who would soon be championing the Brooklyn Bridge.

As preparations gained momentum, Washington Roebling moved to Brooklyn with his wife, Emily Warren Roebling, and their son. John Roebling stayed at home in Trenton but visited Brooklyn weekly to interview workers and contractors and to meet with the movers and shakers who would help make the bridge a reality.

On June 28, 1869, during one of John Roebling's visits, while he was surveying the site where the Brooklyn tower would be built at the Fulton Ferry dock, tragedy struck. John had stepped back as a ferry came in, but not far enough. His foot was caught between the arriving ferry and a wooden pile at the dock. Four of his toes were instantly severed, and he developed tetanus, or lockjaw, a bacterial infection that soon killed him.

Washington Roebling dryly described his father's death and its aftermath: "It was while engaged in locating the position of the Brooklyn tower that Mr. John A. Roebling met with a lamentable accident—the crushing of his right foot by the shock of a ferry boat against the fender rack of spring piles on which he was standing. Lock-jaw speedily set in, and after sixteen days of extreme suffering, terminated in death."

On August 3, 1869, John Roebling's elder son, Washington, who had played a key role in the construction of the bridges that had made his father famous, was appointed chief engineer of the Brooklyn Bridge. He was just thirty-two years old.

Washington Roebling was well trained for the task at hand. In June 1867, before construction on the Brooklyn Bridge had begun, his father had sent him to Europe to study pneumatic caissons, which would be crucial to the East River project—and its most challenging feature. Pneumatic caissons, defined by their use of pressurized or compressed air, were to be the foundations upon which the bridge towers—and the bridge itself—would rest. Washington, accompanied by his wife, Emily, traveled to England, France, and Germany, where he interviewed men experienced in bridge building, as well as those who had manufactured wire and worked metal. By the time they headed back home in March 1868, Washington knew more about pneumatic caissons than anyone in America.

Upon John Roebling's death, there had been no question about who would take over as the chief engineer of the Brooklyn Bridge. According to William Kingsley, the bridge's general contractor and a key player in its construction, John Roebling had wanted his son

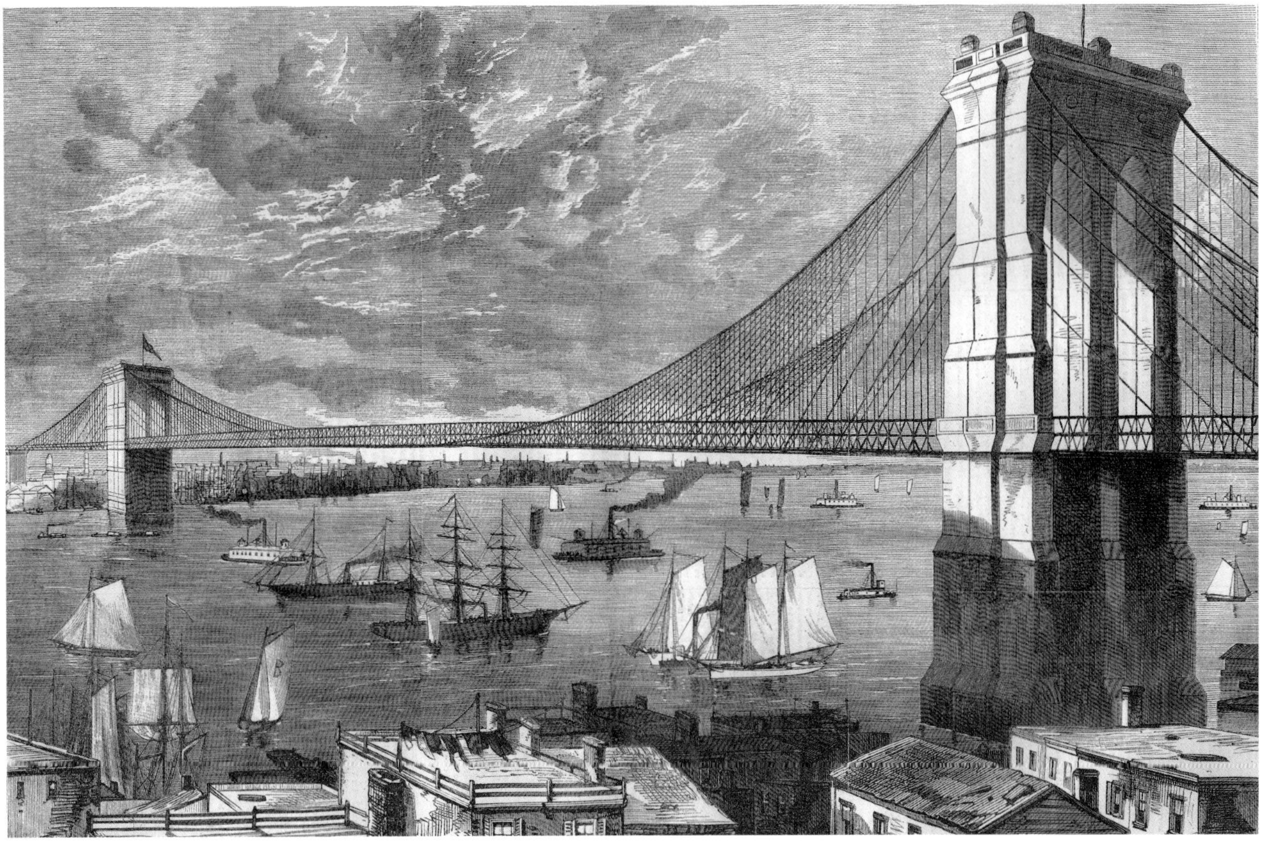

"Proposed Suspension Bridge Across the East River, New York," a woodcut published in *Scientific American*, August 5, 1868. Note the double arches of the towers—in contrast to the single arches of the Covington-Cincinnati Bridge—and the substantial ornamentation at the tops of the towers. Due to cost concerns, these decorative tower caps would not be realized on the Brooklyn Bridge. Courtesy of Richard Haw.

to be in charge from the beginning. But when the bridge trustees insisted that John, with his experience and reputation, be in command, he had agreed, making it clear, however, that his son would take over at some stage. As Thomas Kinsella wrote in the *Brooklyn Eagle*:

> Not long since, before the accident, which led to his death, Mr. Roebling remarked to us that he had enough of money and reputation. And he scarce knew why, at his age, he was undertaking to build another and still greater bridge. His son, he added, ought to build this Brooklyn Bridge—was as competent as himself in all respects to design and supervise it; had thought and worked with him, and in short was as good an engineer as his father.

FOLLOWING SPREAD: "The Great East River Bridge. To Connect the Cities of New York and Brooklyn," published by Currier & Ives, 1872. This view from Brooklyn, looking northwest toward Manhattan, is identifiable by the Fulton Ferry Terminal next to the base of the Brooklyn tower. This print was published three to four years before the towers were completed and does not depict their final details accurately. The tower arches as depicted in the print are not the distinctive Gothic Revival arches of the final bridge, nor did the artist depict the rough-hewn granite or horizontal courses of lighter stone that define the towers today.

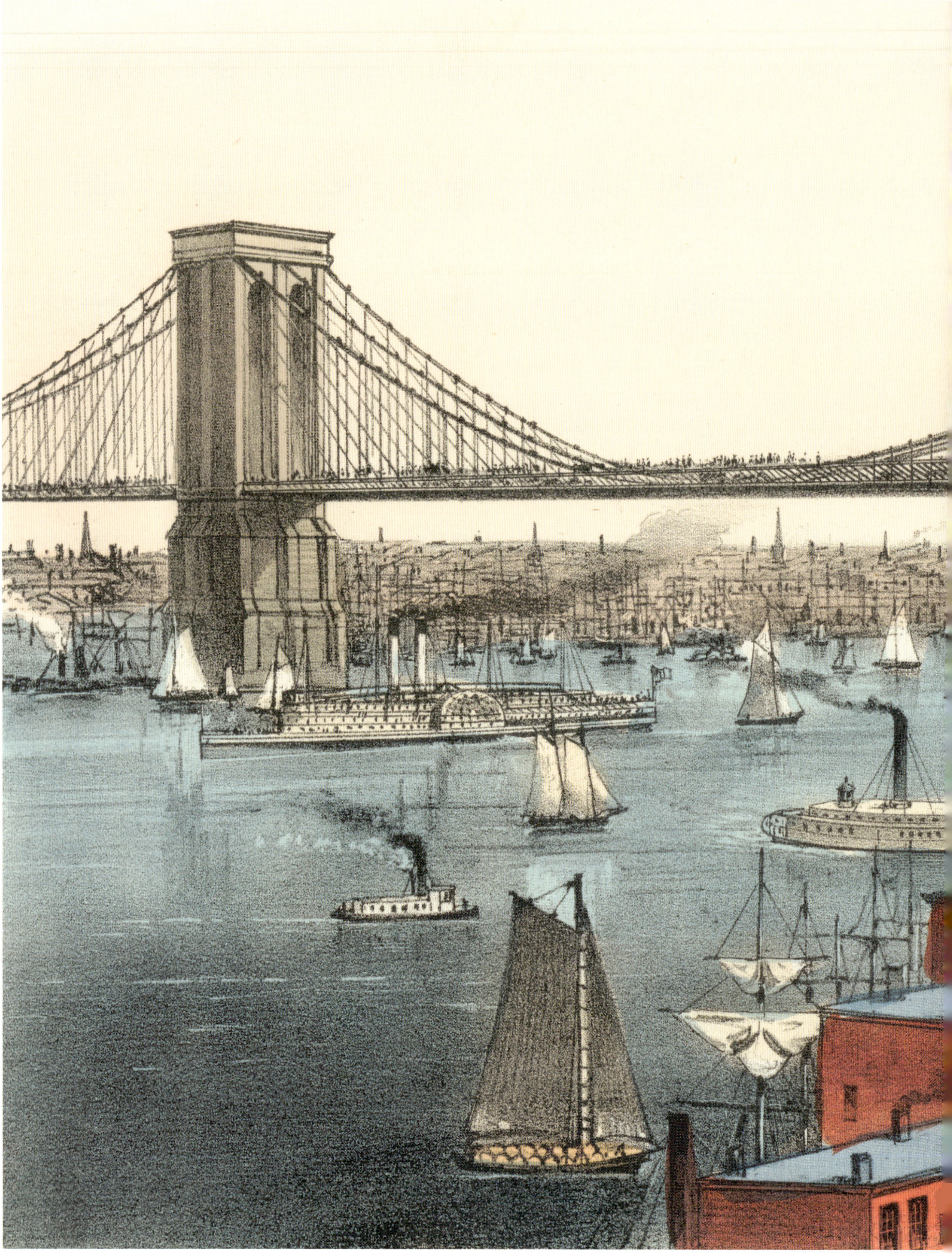

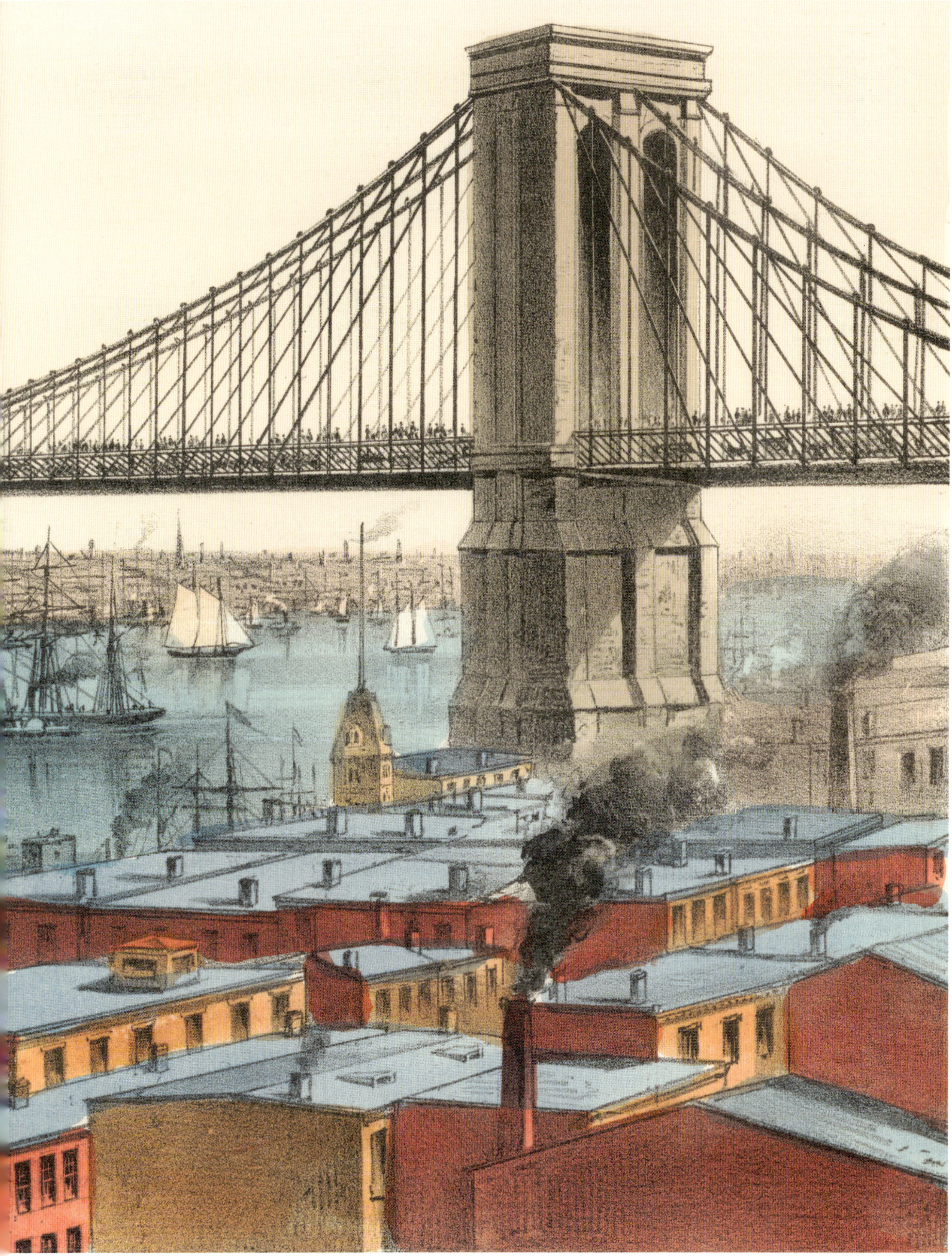

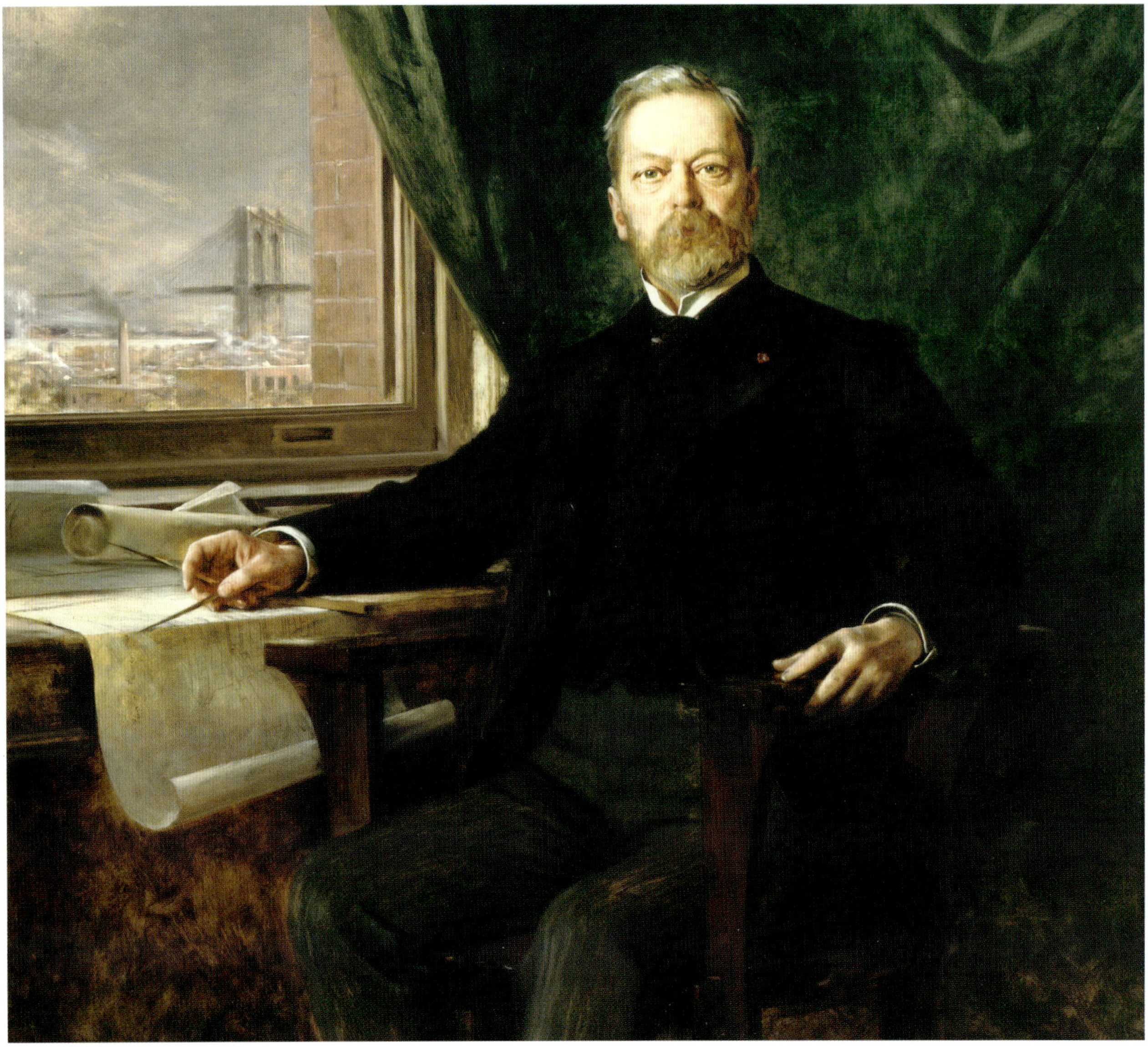

Detail of "Portrait of Washington A. Roebling," 1899, by Theobold Chartran, painted sixteen years after the opening of the Brooklyn Bridge. Brooklyn Museum, gift of Paul S. Roebling.

Years after the bridge's completion, Washington Roebling himself listed three reasons why he had been the man for the job:

> First—I was the only living man who had the practical experience to build those great cables, far exceeding anything previously attempted, and make every wire bear its share.
>
> Second—Two years previous I had spent a year in Europe studying pneumatic foundations and the sinking of caissons under compressed air. When the borings on the N.Y. tower site developed an appalling depth of 106 feet below the water level, all other engineers shrank back . . .
>
> Third—I had assisted my father in the preparation of the first designs—he of course being the mastermind. I was therefore familiar with his ideas and with the whole project—and no one else was.

During his Civil War service, Washington had amply demonstrated the leadership and fearlessness that the chief engineer on such a massive bridge-building project would need. On April 15, 1861—less than a week after the Confederate bombardment of Union-held Fort Sumter launched the Civil War—John Roebling had made it clear to Washington that his time had come to go off and serve his country. The next day, Washington enlisted as a private in the New Jersey National Guard. Over the next four years, he would rise through the ranks to sergeant, second lieutenant, major, and aide-de-camp and ultimately be brevetted (awarded an honorary rank) lieutenant colonel and colonel. According to Colonel Theodore Lyman, with whom Washington served, Washington was fearless: "[H]e goes poking about in the most dangerous places." And Washington was all business, with no time for—or interest in—distractions. As Colonel Lyman noted, Washington was much more interested in doing than in talking. His conversations were typically "curt." In one instance, Major General George Meade, his commanding officer, demanded to know, "What was that redoubt doing there?" Washington's laconic response: "Don't know; didn't put it there."

Washington Roebling was fully aware that the challenges of building the Brooklyn Bridge would be substantial, but by applying his training as an engineer and his experience as a bridge builder, he was confident that he would succeed. On December 15, 1876, in a letter to Board President Henry Cruse Murphy, Washington wrote about wire samples that had been submitted and tested, but his comments could have equally applied to the entire project and, indeed, to his philosophy of life: "I know that nothing can be done perfectly at the first trial; I also know that each day brings its little quota of experiences, which, with honest intentions, will lead to perfection after a while." Washington knew that hard work, and the knowledge acquired in the process, would ultimately result in successfully reaching the goal.

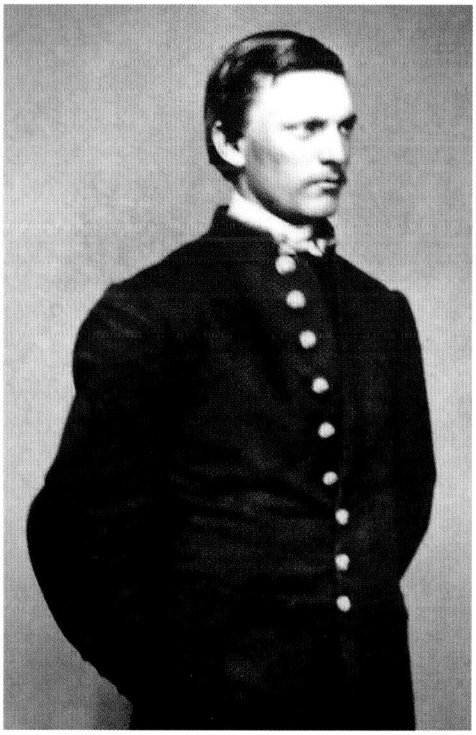

Photograph of Washington Roebling, in uniform during the Civil War, 1864. Courtesy of the Library of Congress.

In late August 1869, the very month that Washington was appointed chief engineer, he presented Francis Collingwood, one of the assistant engineers, with his preliminary drawings and detailed specifications for the bridge construction. Though Washington Roebling performed admirably as chief engineer throughout the almost fourteen years of Brooklyn Bridge construction, he did so despite severe health issues. In the first years of construction, he put so much energy into the work that he was constantly exhausted. Then, late in spring 1872, Washington collapsed from the bends, also called decompression sickness, an affiliation not understood at the time, but now known to be caused by his having left the compressed air inside the New York caisson too quickly, resulting in gas bubbles forming in the blood. He appeared to be near death and was not expected to survive the night at his home on Hicks Street in Brooklyn Heights, the very house where his father had died just three years earlier. But after only a few days he was back on the job.

But Washington's painful attacks of the bends continued into the summer of 1872, and by that September, he was working from his home two to three days a week. Pain and numbness in his legs came and went, and nausea plagued him. He had little energy and less patience. By December, still suffering cramps, dizziness, and vomiting, he was no longer able to go to the construction site.

The doctors told Emily that there was little chance of Washington recovering. He was increasingly convinced that he would not live long enough to see the Brooklyn Bridge completed. Sensing death closing in and fully aware that there was little to guide his assistant engineers once he was gone, he threw himself into documenting all that remained to be done, writing out detailed instructions and sketching indispensable drawings and diagrams for spinning, or making, the cables in place, and assembling the superstructure.

Beginning in May 1873, Washington and Emily spent almost six months in Wiesbaden, Germany, hoping that the warm springs there would revive his health. But he left Germany no better than before. Upon their return to Brooklyn, they bought a house on Columbia Heights, the street in Brooklyn Heights closest to the East River. Its back windows offered Washington a view of his bridge under construction, just a half mile away.

Roebling's condition continued to deteriorate. Intense headaches were frequent, and pain coursed through his stomach and limbs. Emily wrote in her private notes, "There is a popular impression that Colonel Roebling has been for years a helpless paralytic. This is a mistake, as he has never been paralyzed for even one moment and there never has been a time when he has not had the full use of every member of his body." But as Master Mechanic E. F. Farrington, who had known Washington for years, correctly characterized his condition, he was a "confirmed invalid . . . owing to exposure, overwork and anxiety."

David McCullough, in *The Great Bridge*, summed up Roebling's condition:

> The major problem was that his nervous system was shattered. The slightest noise upset him terribly. He was still hounded by visions of his own death before the work could be finished, of disastrous incompetence on the part of some subordinate, of precious days lost at the bridge over some technical problem he could solve in a minute were he there. He felt imprisoned within his own body. He grew extremely short-tempered. When visitors were with him he suffered the whole while. Talk of any kind tired him more than anything else. His eyesight had grown so dim, he could neither read nor write nor sign his own name.

Washington never attributed his physical and psychological difficulties to what was called "caisson sickness" by the bridge workers, what we today refer to as decompression sickness or the bends. Rather, he felt that he had repeatedly pushed himself too hard in the early years of the construction, and his health had suffered for it. But surely his already weak physical and mental state—his exhaustion—only exacerbated the physical manifestations of the bends.

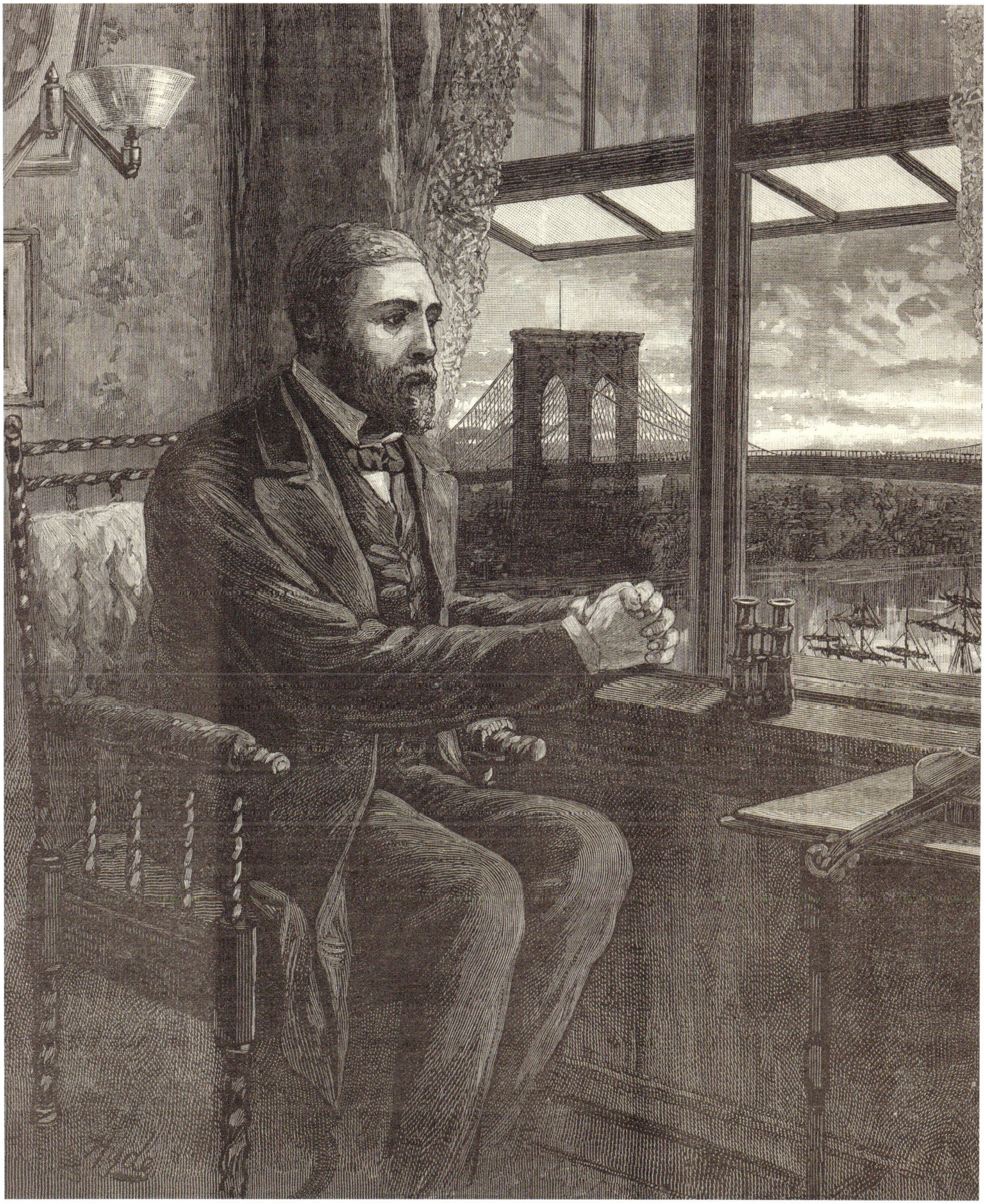

"New York—Washington A. Roebling, the Invalid Engineer-In-Chief of the Brooklyn Bridge, Watching the Progress of the Structure from his Sick-Room on Brooklyn Heights," reads the caption beneath this woodcut published in *Frank Leslie's Illustrated Newspaper*, May 26, 1883. Author's collection.

Sometime later, when Assistant Engineer Francis Collingwood also seemed near collapse from exertions at the bridge, Roebling offered this counsel, based on his own unhappy experience:

> Regarding your health my council [*sic*] would be sit down and keep quiet. . . . Above all don't let a fake ambition lead you on to undertaking tasks that will only break you down all the more. You are no doubt beginning to find out, as I have found out long ago, that nervous diseases are as intractable as they are incurable and only through mental rest of all the faculties and especially the emotions can they even be palliated in the slightest degree.

Work on the bridge was suspended for the winter of 1873–1874, so Washington and Emily returned to Trenton, the Roebling family base, where they would stay for almost three years. During that time, the towers were finished, the anchorages built, and the cable-spinning equipment put in place, all with Washington commanding the work and monitoring its progress from a distance through a constant exchange of letters.

His assistant engineers and Henry Cruse Murphy, the bridge's president, wrote to Washington often. He replied from Trenton, writing in his characteristic longhand specification after specification—for the caisson ironwork, for the footbridge, for timber, for granite, for wire, for whatever was needed. When it came time to make the cables for the Brooklyn Bridge, Washington provided step-by-step instructions to his assistants, across forty-four pages. But, by the middle of December 1875, his letters were being dictated to, and written out by, his wife.

Despite his physical and emotional problems, Washington's mind remained as sharp as ever throughout the construction of the bridge, and his control of the project—and the work of his assistant engineers and all the workers under their command—was complete. As David McCullough wrote in *The Great Bridge*, "He would be the one with the final say. However much staff help he got, every important decision would be his in the end, and at this stage there was seldom any sure way to know which decisions, of all that had to be made, might turn out to be the important ones in the long run."

Roebling often said, "There must be someone at hand to say 'yes' or 'no,' and it often makes a great difference which word they use." He maintained his rule and authority throughout. On one set of specifications for the bridge's ironwork he wrote and initialed, in pencil: "This specification belongs to me and is not to be given away. W.A.R." Another specification made it abundantly clear that he was in charge: "Any question as to the intent of these specifications shall be referred to the Engineer, whose decision shall be final. By the word of the Engineer is meant the Chief Engineer of this work, or his authorized assistant."

In 1876, seven years into construction, Washington wrote that "one mind can in a few hours think out enough work to keep a thousand men employed for years." That same year, *Scientific American*, which had been closely watching the progress on the bridge, published a glowing note about Washington Roebling's work: "The rapidity with which the work had proceeded is evidence that it is conducted by a man who is fully competent to conduct this greatest engineering feat of modern times."

Yet, despite the excellent quality of Washington's work, some newly appointed trustees—who, as Washington well knew, had never even seen him, let alone heard him speak or watched him work—began a campaign in 1876 to appoint a consulting engineer to monitor the work of the chief engineer. Washington opposed the move. Though conceding "that I have been sick for the two years these gentlemen have been members of the Board and even for some time previous," he argued "that up until this time my illness has not in any way interfered with my successfully directing, to the smallest details, the prosecution of the work." Later that year, Washington, frustrated at the continual sniping by some of the board trustees, offered his resignation. It was refused.

With the bridge construction proceeding apace, Washington and Emily moved back to their Brooklyn Heights home. While confined there, he was still able to monitor the bridge's progress out his window—and through his wife, who went out to the construction site as his emissary. By July 1876, settled back in Brooklyn, he was feeling better and discussed with Emily the possibility of his visiting the bridge site. He wrote to Assistant Engineer William H. Paine to share this plan. The cable spinning, which Washington knew better than anyone else from having been in charge of it in Cincinnati, was about to begin. This was the process by which the cables were made, "spinning," or laying out individual wire rope stretching from anchorage to anchorage, which would then be bundled into the cables. But, in the end, it was not to be; in fact, Washington would never set foot on the bridge while it was being built.

Perhaps Washington Roebling's best characteristic was his ability to adapt. As he himself said, "Nothing lasts forever. Only vitality and plenty of it helps you." If rocks and gravel and sand had to be removed from a caisson, he would figure out how to create water shafts and clamshell buckets to do so. If his Brooklyn Bridge needed massive cables of a size that had not been attempted before—each cable was 15.5 inches across (by comparison, the cables on the Covington-Cincinnati Bridge were just 10 inches in diameter)—he would spend hours figuring out the best array of the strands that would form a cable. He drew plans, wrote specifications, detailed, solicited, and set deadlines for bids.

Washington had hired six assistant engineers to work directly under his command. He made sure that they carried out his instructions efficiently until the bridge was completed. They were young men, with an average age of thirty-one when they started: William H. Paine, with whom Washington had served in the Civil War, during which both of them had exchanged their uniforms for civilian clothes to reconnoiter Confederate positions and by so doing risked capture and execution as spies; C. C. Martin, who was the first assistant to the chief engineer and had studied engineering at RPI, like Washington, before becoming Prospect Park's chief engineer; Francis Collingwood, who also had studied at RPI and who had worked as a jeweler; Samuel Probasco, who had tried his hand at gold mining in California before returning east to study engineering; George McNulty, who had studied at the University of Virginia but had never held any job before volunteering to work on the bridge without pay (after he applied for a job and was rejected); and Wilhelm Hildenbrand, a German immigrant with a gift for drawing.

None of the assistant engineers had expected to labor on the Brooklyn Bridge for fourteen years. But they all faithfully stayed on, working with Washington Roebling until the project was done. Even Francis Collingwood, who had only signed up to work thirty days, was still on the job at the end. And George McNulty, who did indeed make it onto the payroll, would name his son Washington—in honor of his boss.

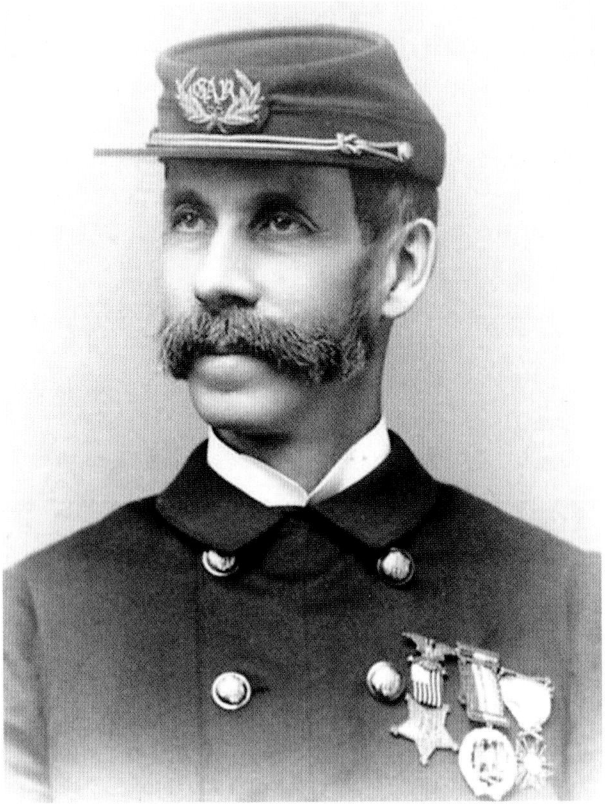

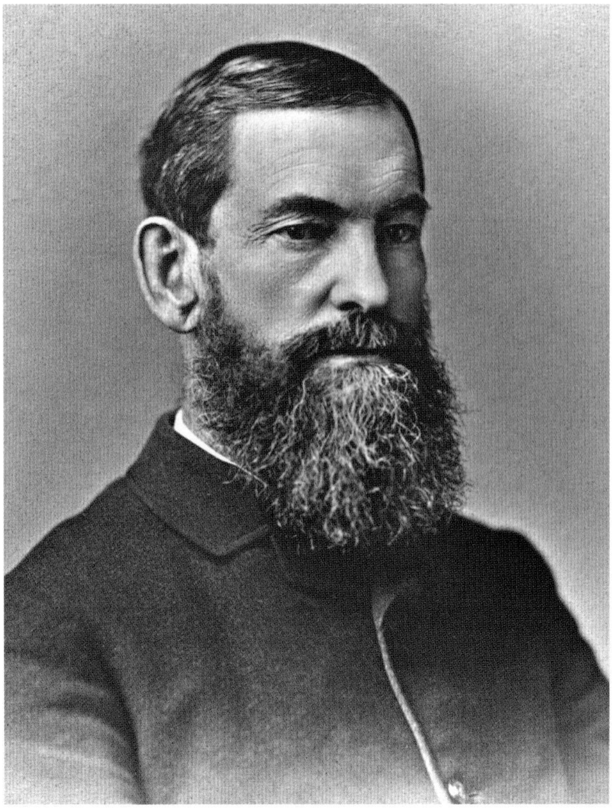

At left, Assistant Engineer William H. Paine, who served with Washington Roebling during the Civil War, in his post-war Grand Army of the Republic uniform, ca. 1880.The brilliant engineer was often referred to as "Colonel Paine" because of his brevet to that rank. Paine was one of the six assistant engineers on the Brooklyn Bridge, along with Charles C. Martin (usually referred to as C. C. Martin), at right. Martin was Washington Roebling's second-in-command on the Brooklyn Bridge construction. Author's collection.

In *The Great Bridge*, David McCullough wrote of the lack of relevant experience that the assistant engineers had, and how the master mechanic, E. F. Farrington, helped fill those gaps:

> Since McNulty, along with each of the other assistant engineers, had had no previous experience building suspension bridges, every step after the stonework was a new one and there might have been costly delays or mistakes had it not been for Roebling's extraordinary written communications and for Master Mechanic E. F. Farrington, the one and only man among them who had ever worked with wire before.

Farrington was in his fifties when construction began on the Brooklyn Bridge and was widely thought of as the best bridge mechanic there was. Born in Massachusetts, he had gone to work at a woolen mill when just nine years old. Afterward, he labored as a farmer, carpenter, seaman, and machinist. He worked with the Roeblings—John and Washington—on the Covington-Cincinnati Bridge and from that experience came to understand wire rope as few men did.

As a result, Washington had tremendous confidence in Farrington: "He is a man of great resource when unforeseen troubles arise, and he has the necessary coolness and perseverance and does not easily get frightened in time of danger." Under Washington Roebling's direction, Farrington built a scale model of the bridge, with five-foot-high towers placed fifteen feet apart. It was used during construction to instruct the assistant engineers and other bridge staff. It was also Farrington who, working with Assistant Engineer Wilhelm Hildenbrand, finalized the plans for the footbridge across the East River that would be needed to monitor—and adjust—the cable spinning. (Hildenbrand also designed the bridge approaches—a huge task, given that each approach involved a distance of almost half a mile.)

Although Farrington played a key role in the bridge construction, he left the job in 1882, a year before the work was completed; hostility between him and C. C. Martin drove him away. According to Washington, Martin had "been trying to get rid of [Farrington] for the past ten years." Farrington eventually wound up back at the Roeblings' Covington-Cincinnati Bridge when it was overhauled, beginning in 1895. He died on that bridge in 1898, run over by a cable car as he worked.

Emily Roebling lauded her husband's assistants and the contributions they made to the bridge:

> It could never have been accomplished but for the unselfish devotion of his assistant engineers. Each man had a certain department in charge and they worked with all their energies to have the work properly done according to Colonel Roebling's plans and wishes and not to carry out any pet theories of their own or for their own self glorification.

C. C. Martin, as the senior assistant engineer, supervised the work at the bridge site. Collingwood was responsible for the Brooklyn tower; Paine, who had created a signal system allowing the workers inside the New York caisson to communicate with those above, was in charge at the New York tower. Once Collingwood completed his Brooklyn tower assignment, he went across the river to take control of the work on the New York anchorage, working with Probasco. Farrington further described their assignments:

> Of the engineers . . . Mr. Martin acts as general superintendent; Colonel Paine has charge of preparing, inspecting and testing the various parts of the superstructure, at the works where they are made. Messrs. Collingwood and Probasco have had charge of the construction of [the] New York anchorage and approach. Mr. McNulty has had charge of the same, on the Brooklyn end of the work.

Early on, Paine and Collingwood, relieved at times by Martin and McNulty, supervised the men inside the caissons. From 1873 through 1875, McNulty was responsible for supervising labor at the Brooklyn anchorage. Later, he was assigned by Washington to run the cable-making apparatus. McNulty subsequently worked on the approaches and later supervised construction of the Brooklyn railway terminal.

In December 1877, after a brick arch on the Brooklyn approach in McNulty's charge collapsed, killing a worker, Washington wrote to Bridge President Henry Cruse Murphy

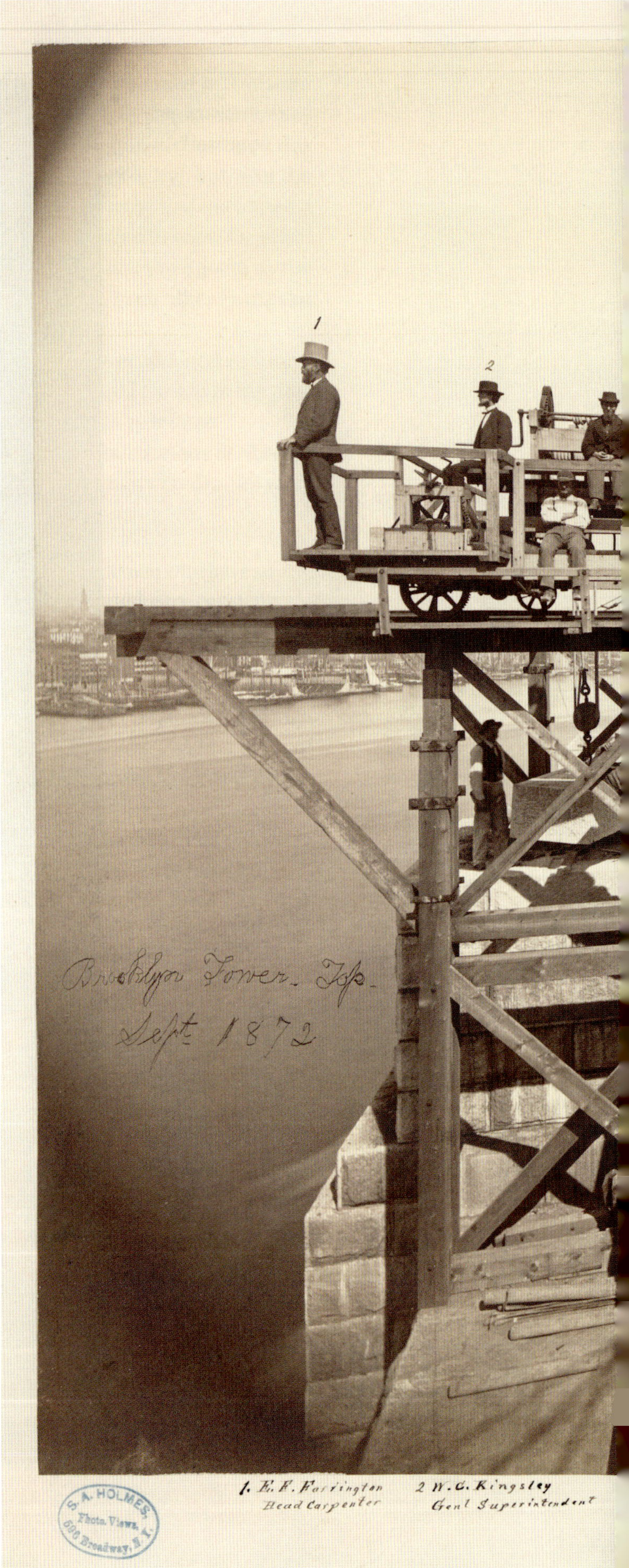

Taken in the fall of 1872, this photograph shows the Brooklyn Bridge trustees and staff atop the bridge's Brooklyn tower, looking north, with the East River and Manhattan at left. Washington Roebling was not present due to his illness. Handwritten on the mat (cropped out): "Top of Brooklyn Pier at 5th Course Above Roadway. Height above Tide 130 ft. Novr. 1872." However, handwritten at the far left is "Brooklyn Tower Top/Sept 1872," creating uncertainty about the month the photograph was taken. Handwritten numbers on the mat, below the image, correspond to those above the mens' heads and identify the individuals from left to right: "1. E. F. Farrington, Head Carpenter [sic]; 2. W. C. Kingsley, Genl. Superintendent; 3. O. P. Quintard, Book Keeper; 4. Alex McKinnon, foreman of masons on this [the Brooklyn] pier; 8. William Paine, Asst. Engr [standing below the others, near the middle of the image]; 5. F. Collingwood, Asst. Engr; 6. G. W. McNulty, Asst. Engr; 7. T. G. Douglass, Head Mason [who died a year after this photograph was taken from injuries sustained while working on the bridge]." The stamp of photographer Silas A. Holmes appears at bottom left. For more on the photographer, see the appendix. MCNY.

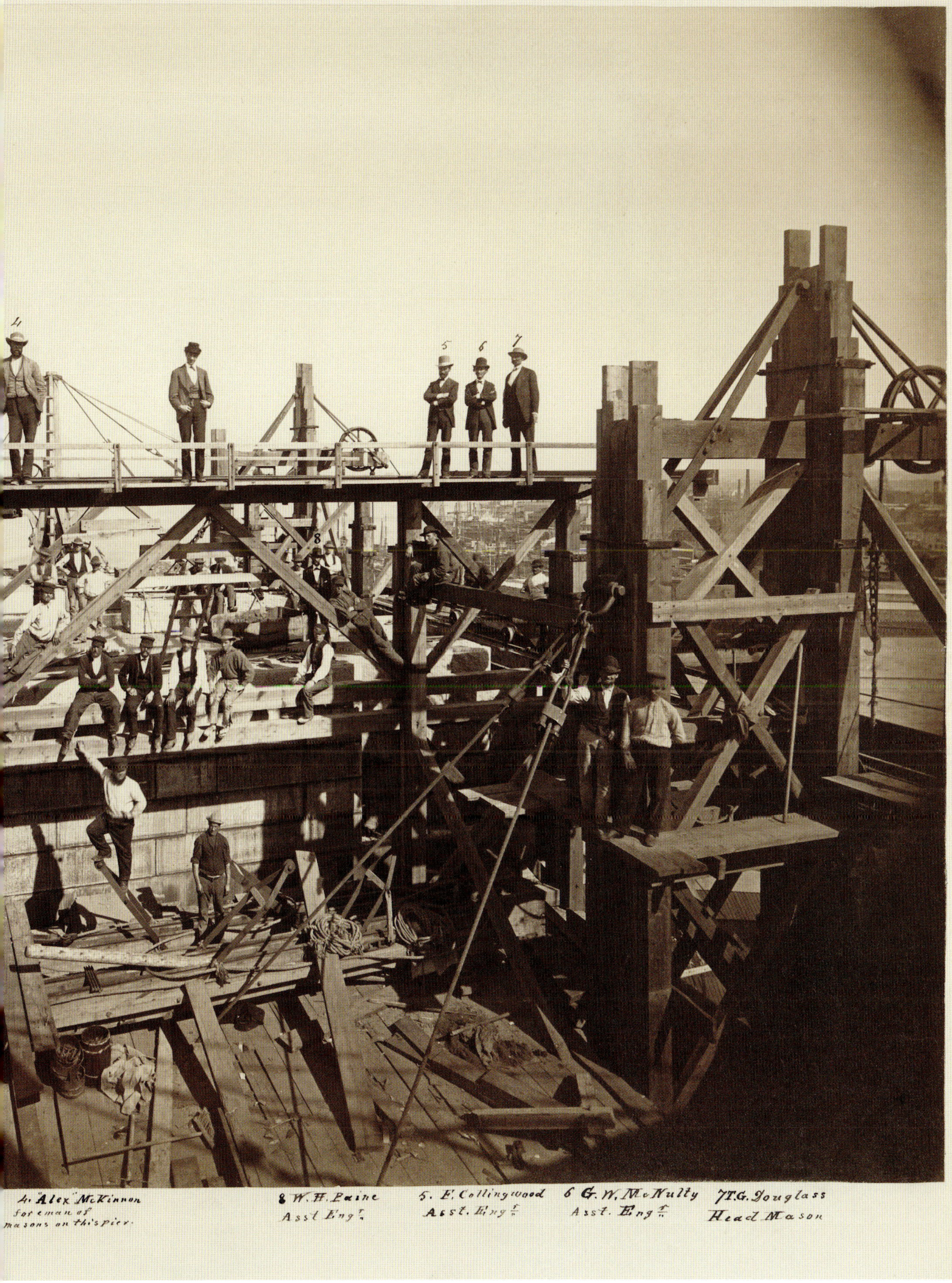

4. "Alex" McKinnon foreman of masons on this pier. 8 W. H. Paine Asst Engr. 5. F. Collingwood Asst. Engr. 6 G. W. McNulty Asst. Engr. 7 T.G. Douglass Head Mason

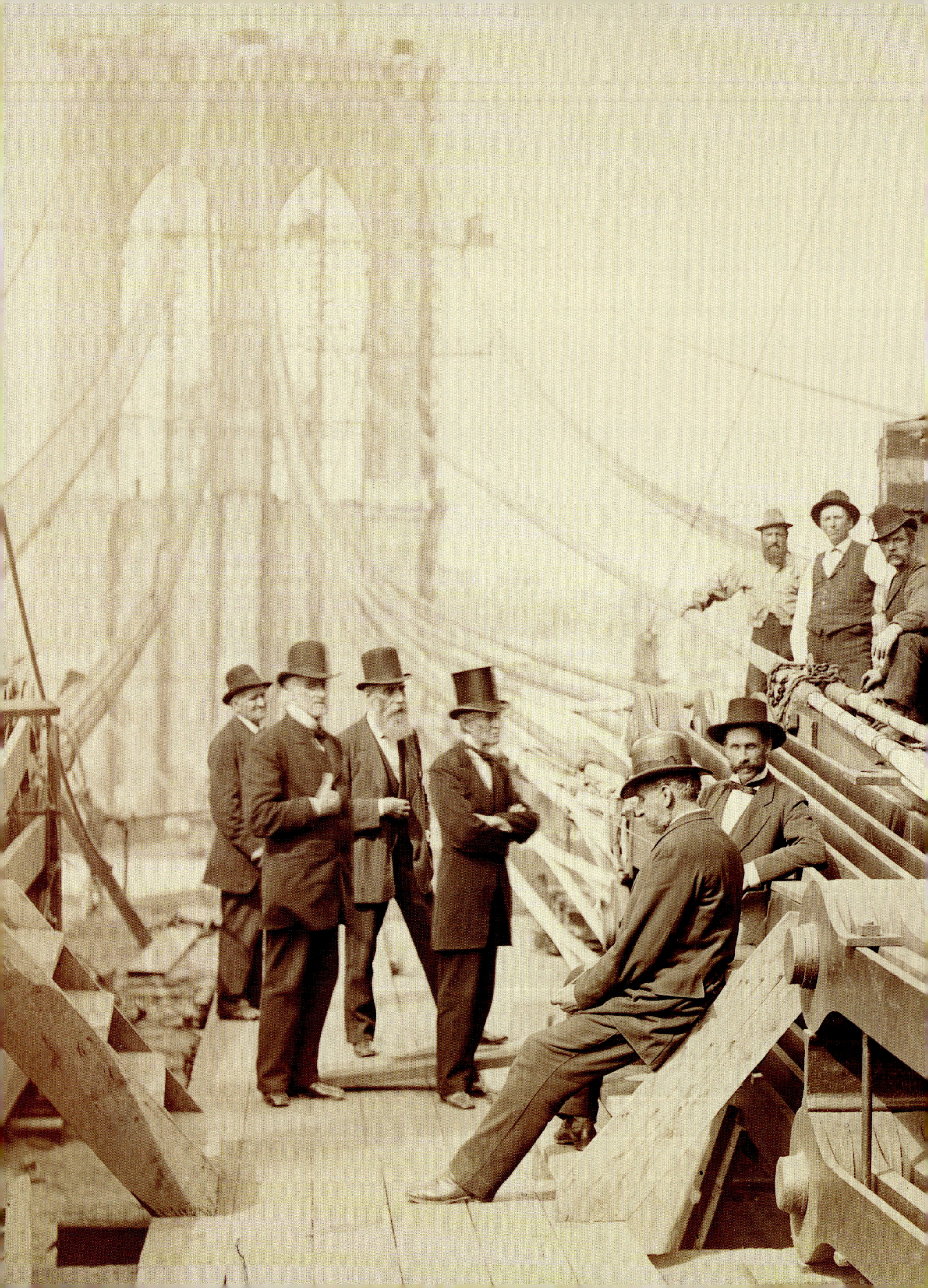

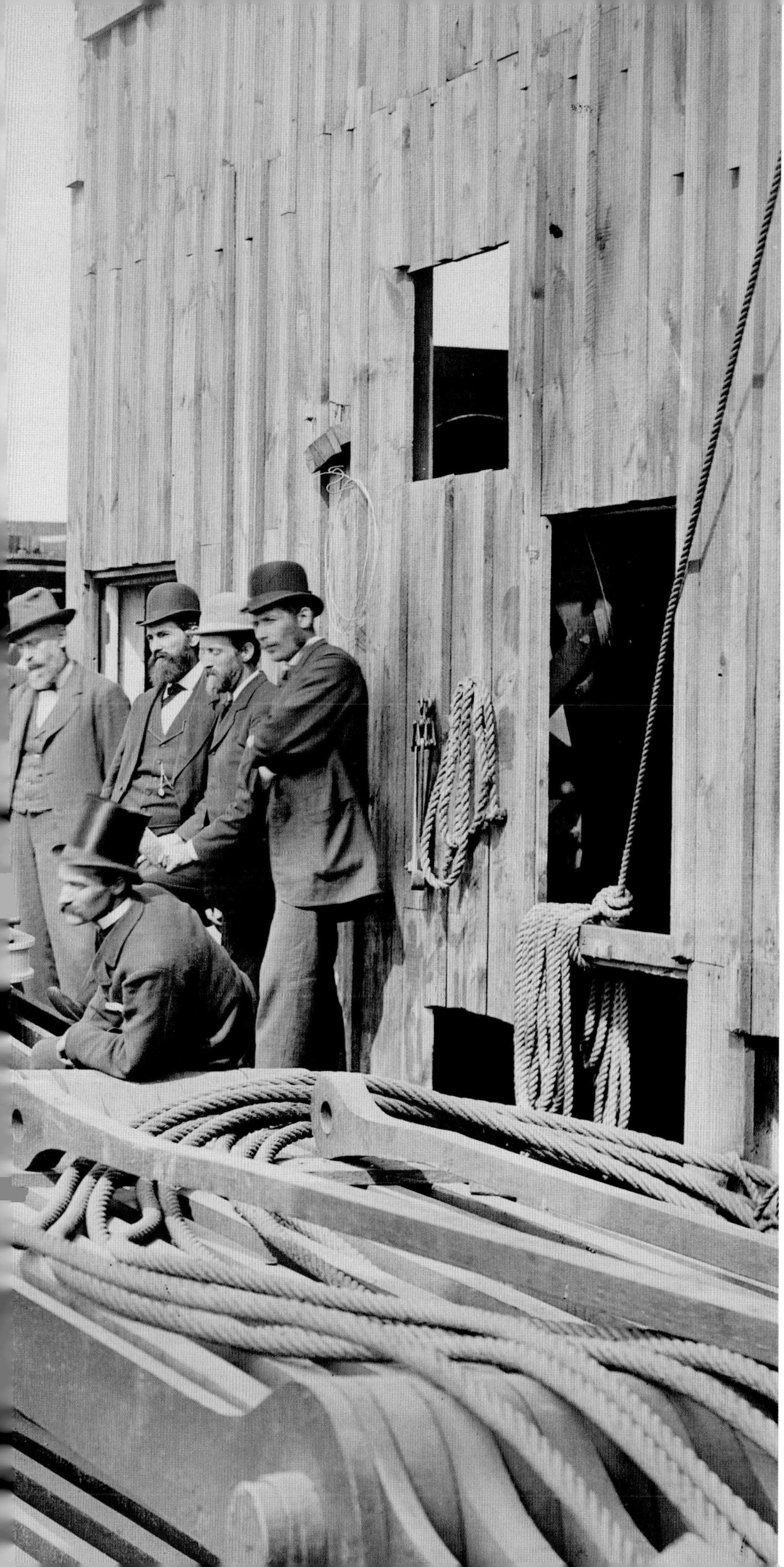

The Brooklyn Bridge officers, engineers, and staff, October 1878. Starting second from left, Treasurer John H. Prentice, Secretary Orestes P. Quintard, President Henry Cruse Murphy, Draftsman William Vanderbosch, Assistant Engineer William H. Paine (looking directly at the camera), and Assistant Engineer George McNulty (seated at right, wearing top hat). At far left in the line of men standing up against the drum shed (the wooden structure at right) is Foreman of the Riggers William Dempsey; sixth from left is Master Mechanic E. F. Farrington; then, proceeding to the right from Farrington are Assistant Engineer C. C. Martin, Assistant Engineer Francis Collingwood, and Chief Draftsman/ Assistant Engineer Wilhelm Hildenbrand. Photograph by G. W. Pach. For more on the photographer, see the appendix. MCNY.

“No trouble was experienced in getting all the labor required, when one man left a dozen were ready to take his place; New York in fact is the best labor market in the country.”

—CHIEF ENGINEER WASHINGTON A. ROEBLING

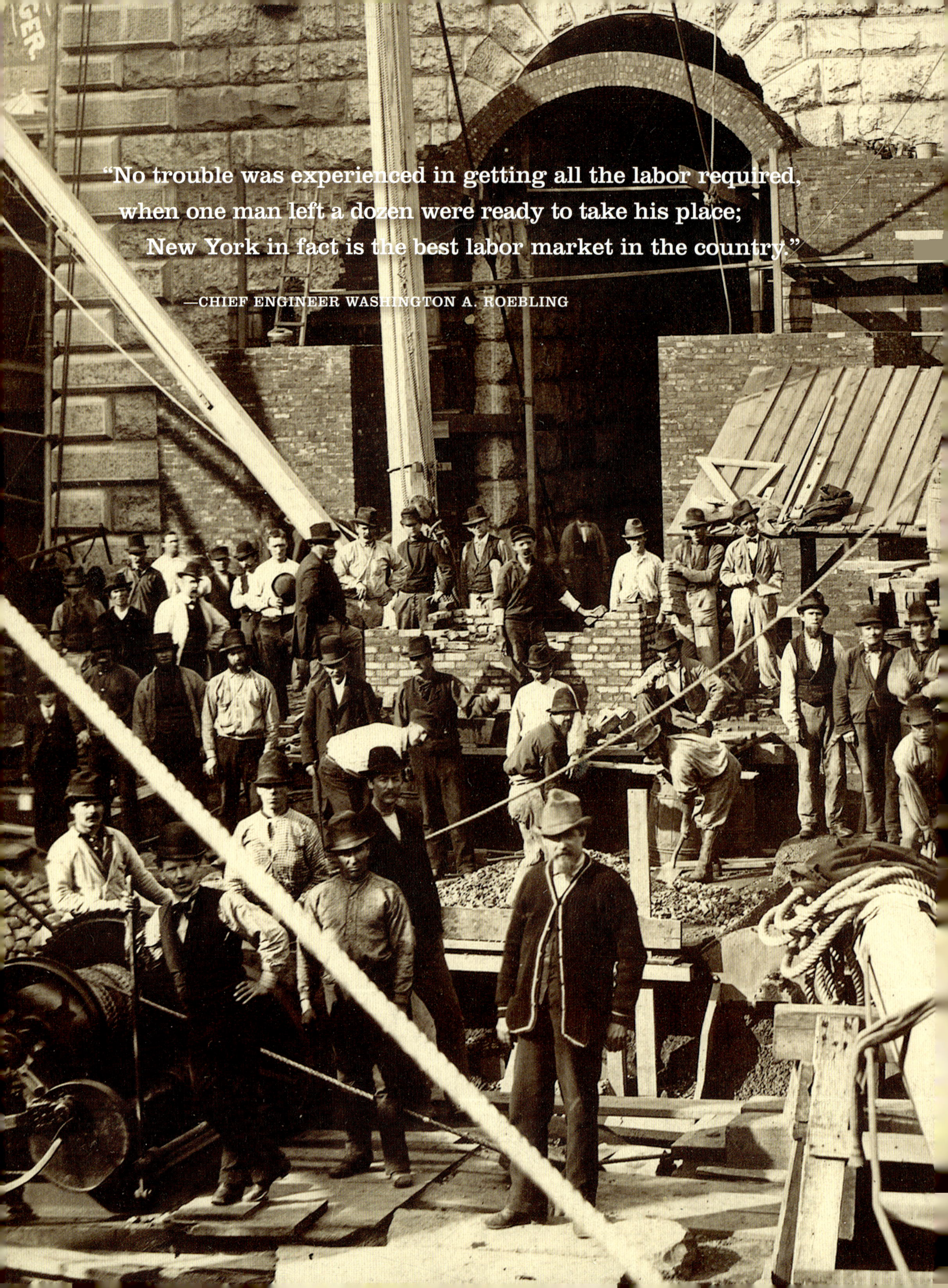

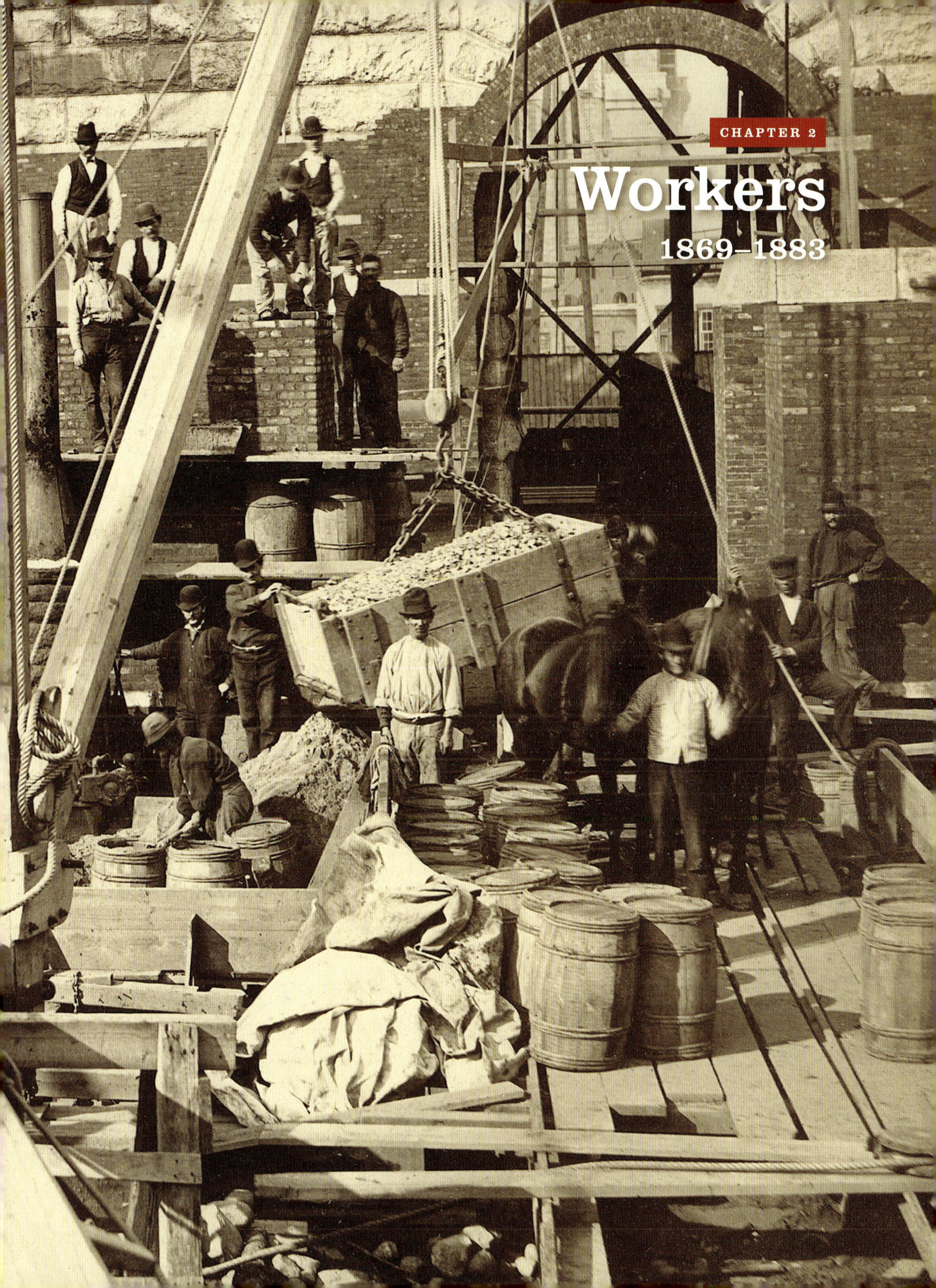

CHAPTER 2

Workers

1869–1883

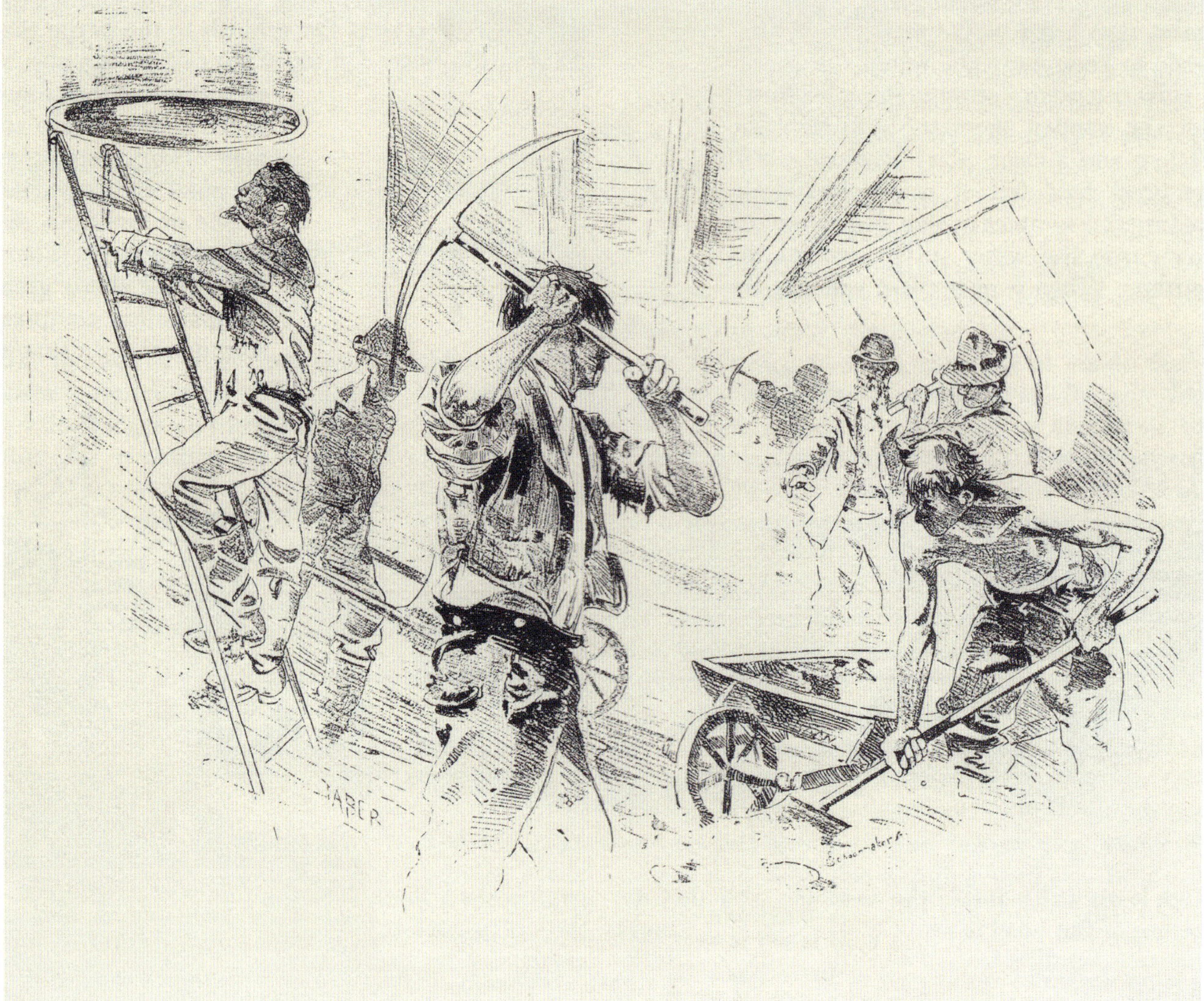

"Inside the Caisson," *St. Nicholas Magazine*, July 1883. Pickaxes, shovels, and wheelbarrows were the key tools for the work inside the caissons.

PREVIOUS SPREAD: A close look at bridge workers standing in front of the Brooklyn anchorage, at the side away from the Brooklyn tower (partially visible through the arch at right). Taken in January 1876, as per a note written on the print's mat, this photograph is a remarkable record of many bridge workers gathered in one place. The image has been carefully composed by the photographer (likely Silas A. Holmes); the men have stopped their work to assemble, most standing still and looking at the camera. Note how the seven workers highest up at right of center have been carefully posed so they frame a derrick's boom. The brick piers of the approach have begun to rise at the base of the anchorage. RPI.

It took thousands of skilled and unskilled laborers, working together, to build the Brooklyn Bridge. Some worked on the project for just a few days; some, for years. Jobs on the bridge were divided into forty-three classifications. Among them, in descending order, according to the number of men doing each task, there were unskilled laborers, masons, carpenters, and foremen. Other specialized workers included divers, blacksmiths, drillers, stone-cutters, wire workers, and draftsmen.

In *The Great Bridge*, David McCullough summarized John Roebling's observations about some of his first hires:

> Charles Kinkel was a German who told Roebling he had had experience with foundations. John Morgan, an English draftsman, wanted employment badly, he said, while William McNamee looked like a "tolerably good man."
>
> And so it went. J. W. Jenkins, a diver, said that if he was hired by the month, he would work for twenty-five dollars a day, do whatever blasting was wanted, and provide his own tools. Two experienced surveyors, Rudolph Rose and Colonel William Paine, each wanted ten dollars a day and both were hired on the spot.

Their work was closely supervised, and Master Mechanic E. F. Farrington, who ran much of the day-to-day operations, had calculated the average time for various tasks down to seconds, with workers expected to meet Farrington's targets. For example, exactly forty-three seconds were allotted to reverse the "traveler." Pulling a wire from one anchorage to the other—across the span and over both towers—was expected to take ten minutes and fifteen seconds.

Some jobs on the bridge were more dangerous than others, and working inside either of the two caissons—the first phase of the construction—was one of the assignments that posed the greatest hazards. Washington Roebling was fully aware of this: "The labor below is always attended with a certain amount of risk to life and health, and those who face it daily are, therefore, deserving of more than ordinary credit."

When work began inside the Brooklyn caisson in 1870, laborers were paid the market rate for New York City and Brooklyn: $2.00 per day (roughly equivalent to $40.00 a shift in 2021). But it soon became apparent that the difficult working conditions—a constant temperature of approximately eighty degrees, impure compressed air, constant seepage of water into the work space, and being confined in a claustrophobic wooden box—discouraged new workers from signing on at the same recompense they could earn at a less taxing job. As a result, the Bridge Company had no choice but to raise the pay.

Two years later, in a report to the bridge trustees on July 1, 1872, Washington wrote that no political influence had affected hiring, and, due to the danger of the work, wages were higher than the standard rate for comparable toil: "It has been supposed that a laboring man would work a caisson for 10 hours a day at $1.75. We are today paying $2.75 for four hours' work, which is at the rate of $7 per day, or an increase of 450 per cent, entailing an increased cost on account of labor alone over $100,000."

In 1873, after work on the caissons had been completed, Washington noted, with respect to the labor force, that "[t]here was one small strike at the beginning, but it amounted to nothing." He also pointed out that, regarding the work in the caissons, conditions had improved somewhat as the work had progressed: "The stages of the work

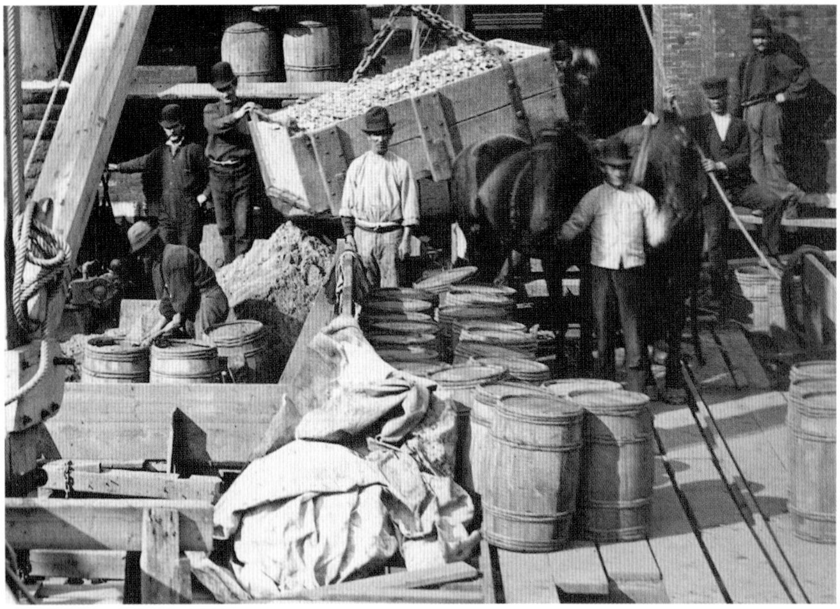

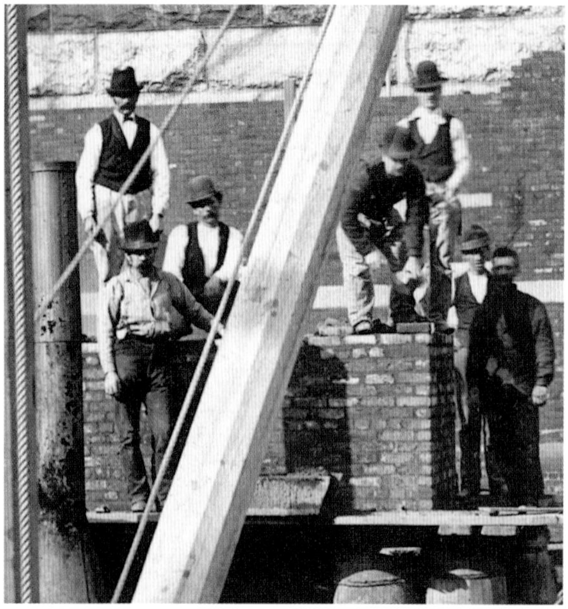

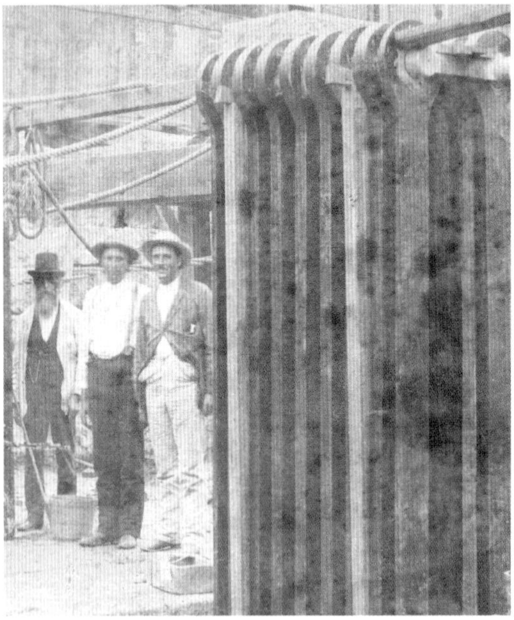

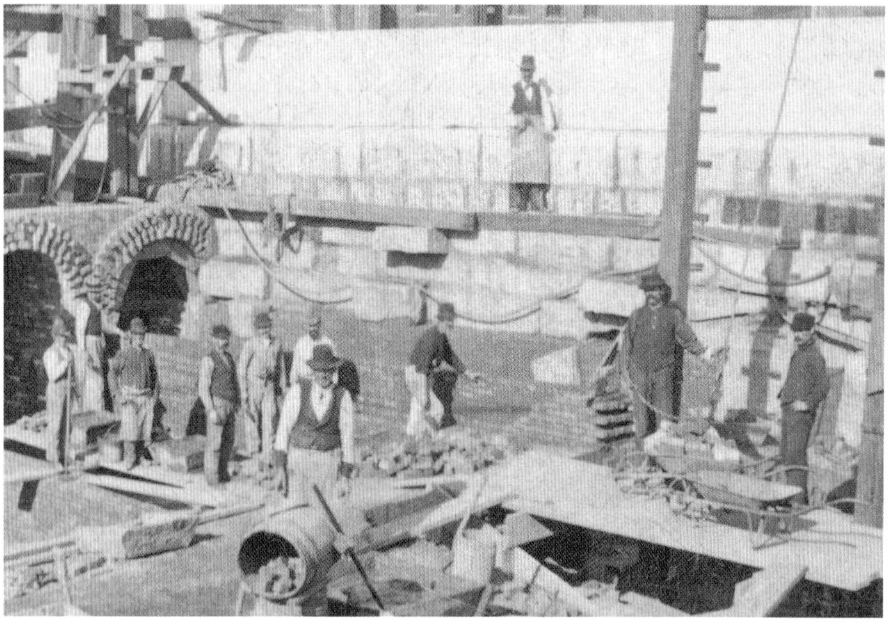

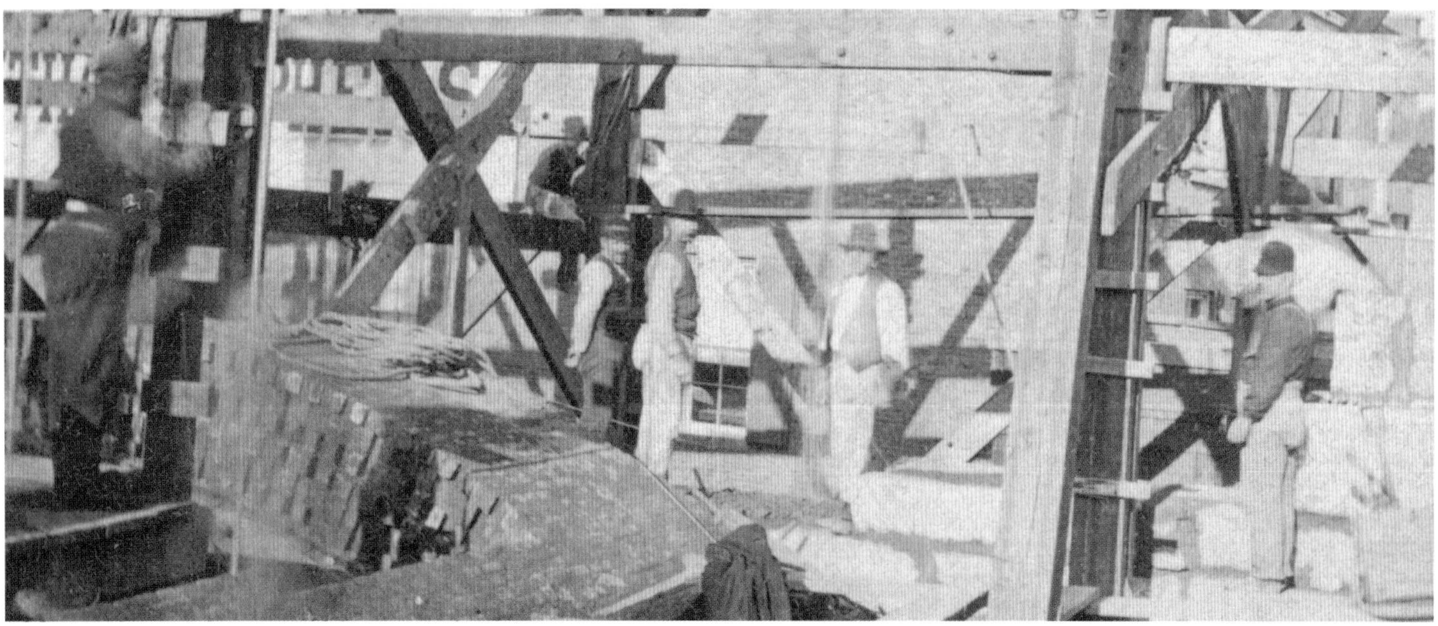

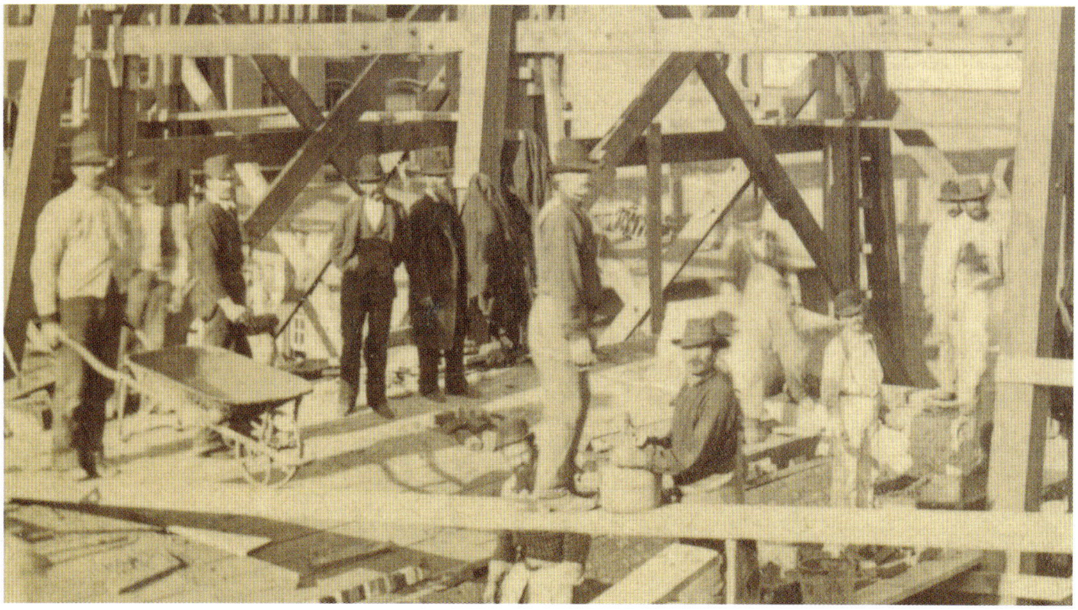

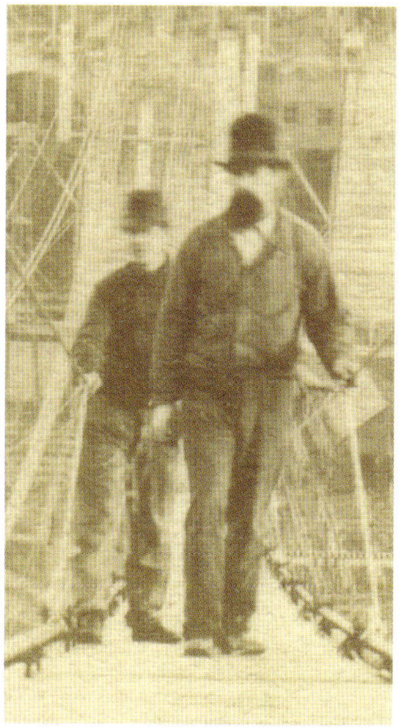

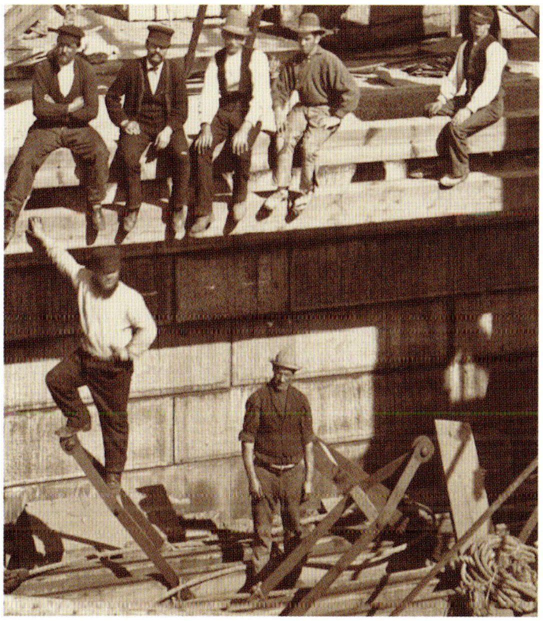

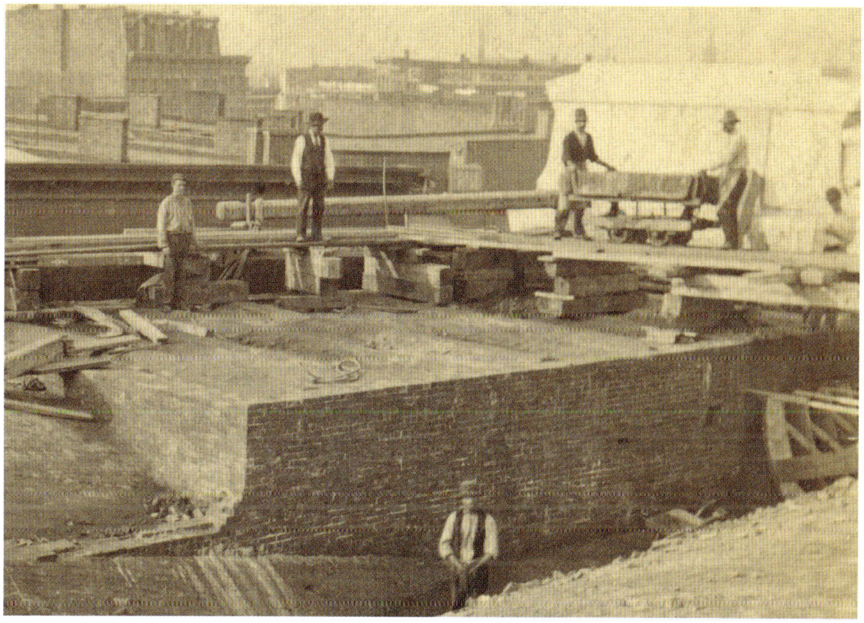

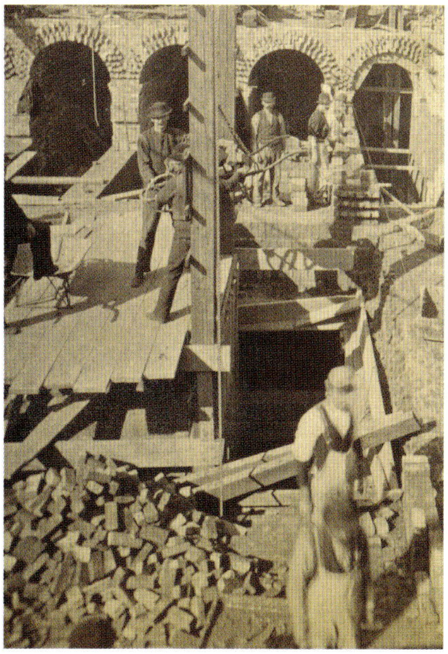

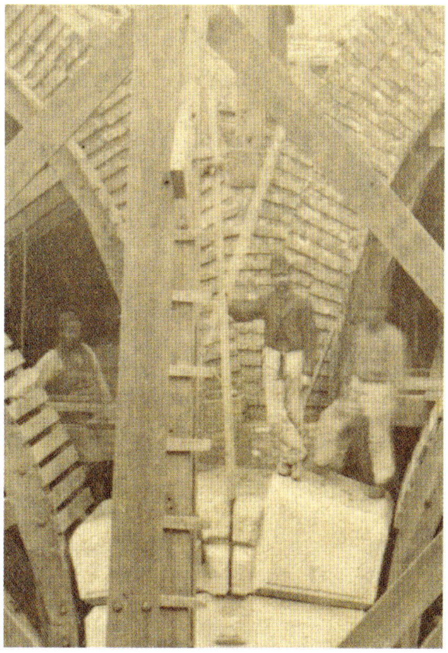

Brooklyn Bridge workers at various stages of the construction process. Thousands of men toiled during the fourteen years it took to build the bridge.

were in reality far more disagreeable than at the end, on account of the constant fog and influx of water, whereas in the latter stages the work was dry."

The Brooklyn caisson, which would serve as the foundation for the Brooklyn tower of the bridge, was built and being dug into the ground before construction of the New York caisson had even begun. Work at the Brooklyn site went on six days a week, around the clock. As the *New York Herald* reported: "The work of the buried bridge-builder is like the onward flow of eternity; it does not cease for the sun at noonday or the silent stars at night. Gangs are relieved and replaced . . . perspiring companies of men follow each other up and down the iron locks, with a dim, quiet purpose."

Day shifts inside the Brooklyn caisson consisted of just over one hundred men; night shifts had about forty workers. In addition, atop the caisson, another hundred workers disposed of debris, operated the two water shafts, as well as the compressors, hoisting machines, and the lighting below. Blacksmiths, carpenters, masons, and mechanics supported those toiling inside the caisson.

Most of the men working below stayed there to eat their meals. Typical fare was bread, cheese or beef, and beer. The workers were furnished with only the most basic of amenities, as reported by Washington: "Two shanties were put up and provided with duly numbered hooks and pegs for the men's clothes, most of which were left above, the temperature below being too warm. Rubber boots were furnished by the company at cost price. In front of the houses are sets of wash-troughs with hot and cold water."

Due to the oppressive heat inside the caissons, most of the men worked shirtless, wearing only pants and boots. The work was exhausting. Smoky conditions produced by the burning lights and blasting caused many workers to develop hacking coughs and turned their spit black; it would remain so months after they stopped toiling in the caissons. The workers lived with the constant fear that they would die in the caisson, either buried alive or drowned by the river water. Even leaving the caisson at the end of the shift was no picnic, particularly when the sweaty men emerged from the eighty-degree heat into the much colder air outside.

When work began at the New York caisson, shifts lasted nine hours, from six in the morning to three in the afternoon, with a one-hour break for breakfast. The next shift ran from three o'clock to eleven at night, with an hour for supper. Then, at eleven o'clock, the "night-gang" reported for work and labored until six the next morning. At night, the work crew concentrated on repairs to the clamshell buckets, cars, engines, and other machinery. The men on the night gang also wrestled with the largest of the boulders unearthed earlier inside the caisson, breaking them up and having them removed.

Washington Roebling noted that at the New York caisson, similar to what he had seen during work on the Brooklyn caisson, "The majority of the men took their meals along and remained down the full eight hours without any injury." As work progressed, the shift increased to 112 laborers and 7 foremen. The night gang had 3 foremen and 40 laborers. The total working in the New York caisson: 14 foremen and 264 laborers.

As was the case at the Brooklyn site, an additional hundred men or so atop the New York caisson supported those below: engineers and firemen running the excavating engines, men operating the engines that pulled the dirt cars, gangs for dumping the dirt cars, engineers on the air compressors and stone-hoisting machines, blacksmiths,

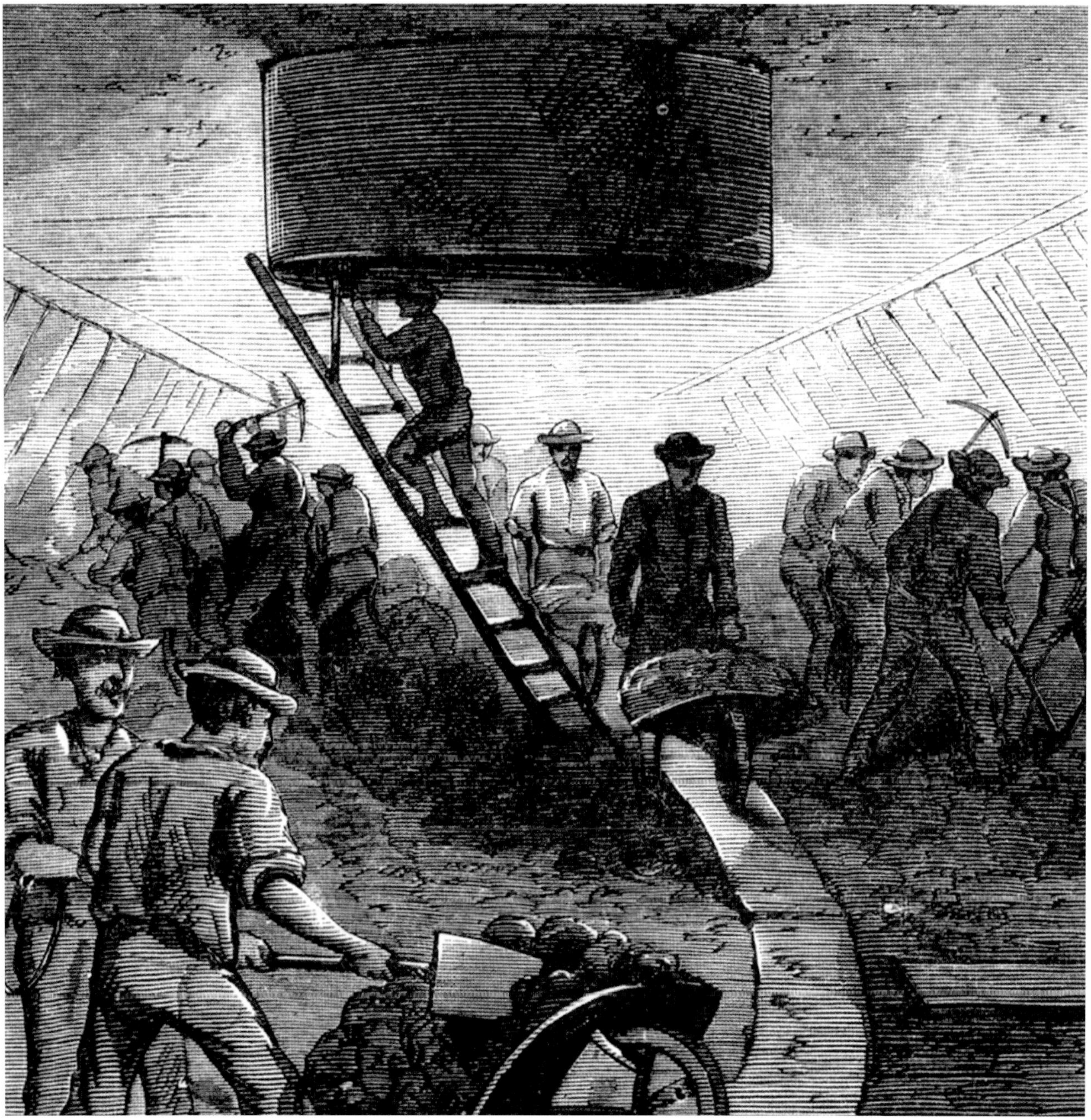

A woodcut of workers using pickaxes, shovels, and wheelbarrows to excavate in the Brooklyn caisson. *Frank Leslie's Illustrated Newspaper*, October 15, 1870. Courtesy of Richard Haw.

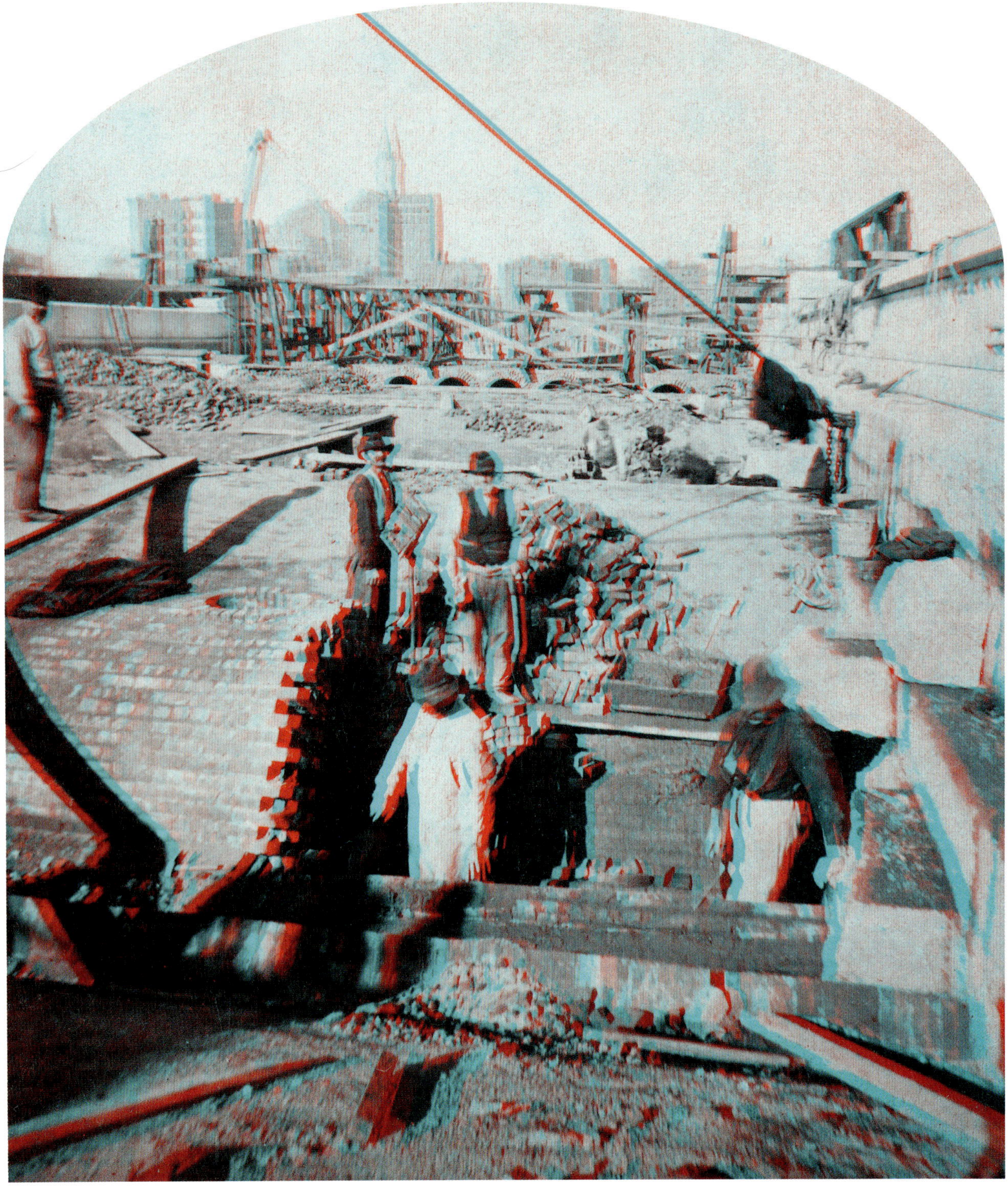

3D Construction photograph, ca. 1880, showing workers laying brick for the New York approach, the part of the bridge that leads to the suspended spans. This approach begins at the land side of the anchorage and extends down to ground level near City Hall Park. The distinctive New York Tribune Building, with its pointed tower, is in the distance, just left of center. Courtesy of the Michael E. Ach Collection.

machinists, gas men, and those assigned to move boulders pulled up by the clamshells. There were also twenty-five carpenters and thirty masons.

Approximately 2,500 men worked at various times on the Brooklyn caisson. David McCullough estimated in *The Great Bridge* that a hundred workers quit every week. But some stayed on. One, Mike Lynch, did not miss a single shift during the almost ten months it took for the Brooklyn caisson to reach bedrock. Working six days a week, he even had the energy to add a day of overtime. General Contractor William Kingsley said of Lynch, "He is strictly temperate and regular in all his habits and is none the worse for his long service in compressed air."

Despite the many workers who quit, plenty of men were eager to fill the bridge jobs. The vast majority of them were immigrants, mostly Irish but also Italian and German. They typically lived in crowded tenements where disease—measles, scarlet fever, the grippe, and diphtheria—flourished. Further, many of them were heavy drinkers. Dr. Andrew H. Smith, a physician hired by Washington to study the physical problems encountered when the workers came out of the compressed air, noted that "the habits of many of the men were doubtless not favorable to health, but everything which admonition could do was done to restrain them from excesses."

The ages of the men working on the Brooklyn Bridge ranged from eighteen to fifty. Some were Civil War veterans. Many had worked as seamen and were experienced in climbing high up in a ship's rigging and masts; they proved particularly adept at working above the East River on the footbridge, the cables, and atop the towers.

There is no question that bridge work was dangerous, and on this bridge, with its unprecedented size, it was particularly treacherous. Precautions were taken throughout construction to avoid injuries. While the engineers had insisted that installing a safety railing atop the towers would not be worth the effort, they did act after three workers fell to their deaths, putting into place a policy that required anyone who felt light-headed to leave the tower and work at ground level. Despite these and other precautions, however, fatal accidents continued to occur.

In 1875, a worker named Reed, who had failed to divulge his history of epilepsy when he applied for the job, fell off a tower during a seizure. Another man, McCann, was at the top of one of the towers when he was knocked off by a box of mortar being swung by a derrick; he fell to his death. A worker by the name of Cope, who had been taught how to keep a rope on the drum of a hoisting machine, reacted poorly when a problem arose: against his training he kicked the rope, got his foot caught, and his leg was mangled as it wound around the drum. He died. A carpenter perished when a large stone fell on him, as did another worker who was struck by a granite block. Yet another was pushing a loaded wheelbarrow high up when it ran off the plank. The man held on to the handles and followed the wheelbarrow down to his death.

There were also narrow escapes. A worker named Frank Harris fell from the top of the Brooklyn tower into its hollow center and plunged what was later determined to be 186 feet. Fortuitously, he landed on a barrel that was floating in three feet of rainwater. As his co-workers started down, no doubt expecting to recover his lifeless body, they heard him calling for them to lower a rope to pull him out. Eight days later, he was back at work atop the tower.

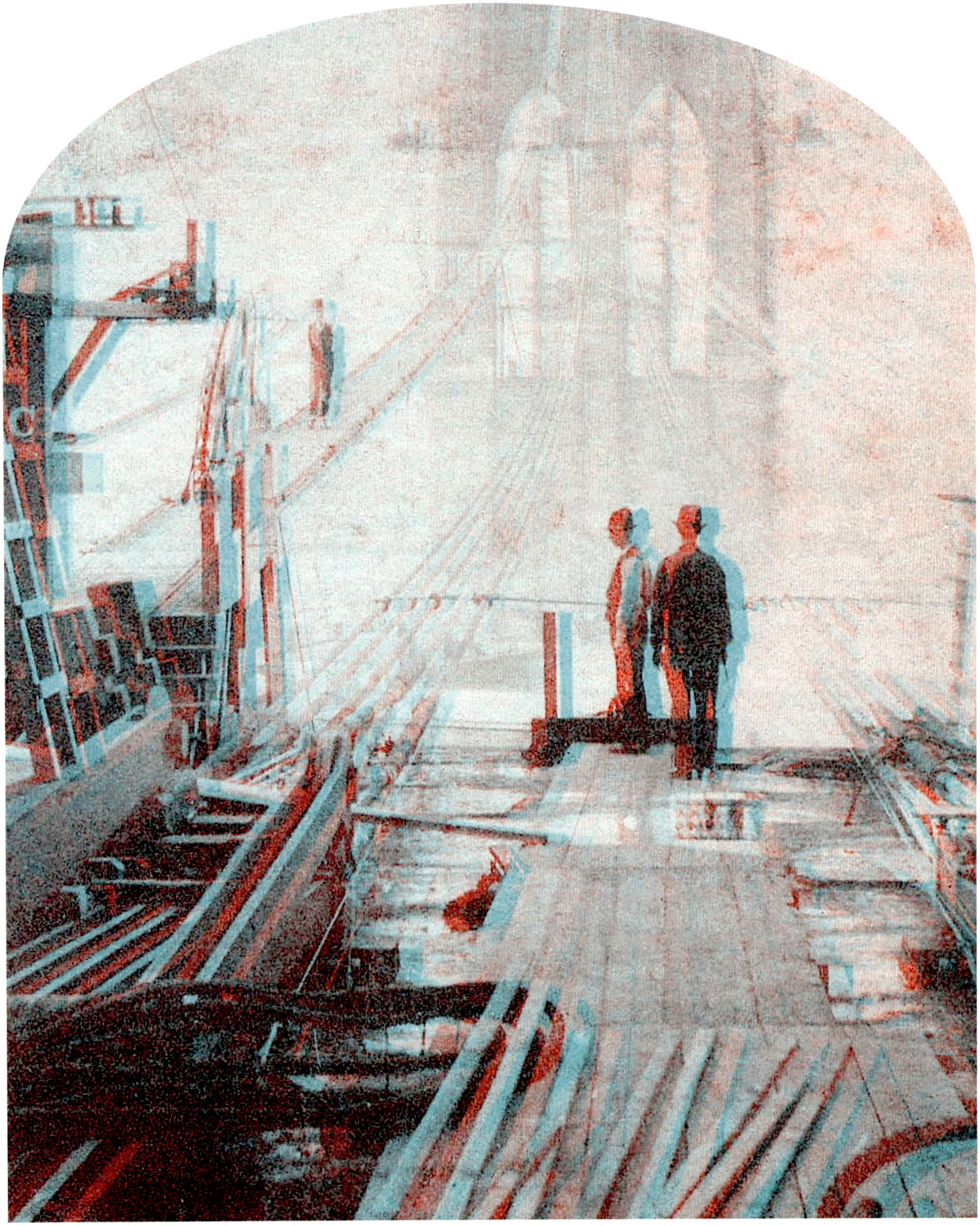

3D Two workers atop one of the anchorages, ca. 1878. At left is a round "carrier" wheel that was used to spin the cables. A bridge tower is in the distance. Note the picnic basket just in front of the two men: lunch! Courtesy of the Jeffrey Kraus Collection.

alike, Washington did not know what caused the bends. He mistakenly believed that, in addition to time exposed to the compressed air, a man's inexperience was the key: "None of the old hands were effected [*sic*] to any extent, even when remaining down eight hours."

Washington continued to downplay the problems resulting from caisson disease: "It is true that scarcely any man escaped without being somewhat affected by intense pain in his limbs or bones and by a temporary paralysis of arms and legs, but they got over it, either by suffering for a few days outside, or by applying the heroic mode of returning into the caisson at once, as soon as the pains manifested themselves, and then coming out very slowly." As he wrote, he believed that embracing a "heroic mode" and returning to work was the answer. Sadly, neither Washington nor anyone else at the time realized that he had hit upon the solution to the bends: "coming out very slowly," or gradual decompression.

Despite downplaying the seriousness of the illness, even Washington could not deny its impact on the construction:

> The fact remains, however, that the effects of compressed air on the human system constitute the principal difficulty attending deep pneumatic foundations [caissons]. Men are somewhat difficult to get, wages are high, and the time of labor becomes so short that the work must necessarily be done under a disadvantage and under an immense amount of supervision, where it is at all difficult or different from ordinary digging in a uniform material.

However, there was indeed a very real impact on the workers. Some died from the bends, others suffered great pain, and still others experienced enormous stress as their co-workers mysteriously fell around them.

Master Mechanic E. F. Farrington was philosophical about injuries and death on a project of such magnitude:

> On a work of this description, running through so many years, many changes and casualties must occur. In the morning you greet your old friend and fellow workman, at night his mangled remains are all that is left to you; and the impressions of familiar faces glide down the vista of time, farther and farther away, like objects seen from the rear of a railway train; and you feel, that the time may not be far away when you too must drop off and be forgotten. Meanwhile, the roar and rumble of the busy world will resound from the bridge. The sun and the stars, will shine upon it; the zephyrs will toy with its stays, and the storms howl through its latticed sides; but through all, and alike indifferent to all, it will stand emotionless and firm, an enduring monument to engineering skill, and daring, patient, laborious effort.

The Bridge Company issued no accounting of the number of men who died while in its employ. This, however, was typical of business concerns of the time, which were subject to little public scrutiny or accountability. Master Mechanic E. F. Farrington estimated that thirty to forty men died during construction. Washington put the number at twenty, as did General Contractor William Kingsley.

Given that the construction of the Brooklyn Bridge was very labor intensive, it fell to thousands of workers to get the job done. That they did, toiling away, often despite great danger to themselves.

"...upon the tower foundations [the caissons] rests the stability of the entire work. The whole enterprise depended on their success."

—CHIEF ENGINEER WASHINGTON A. ROEBLING

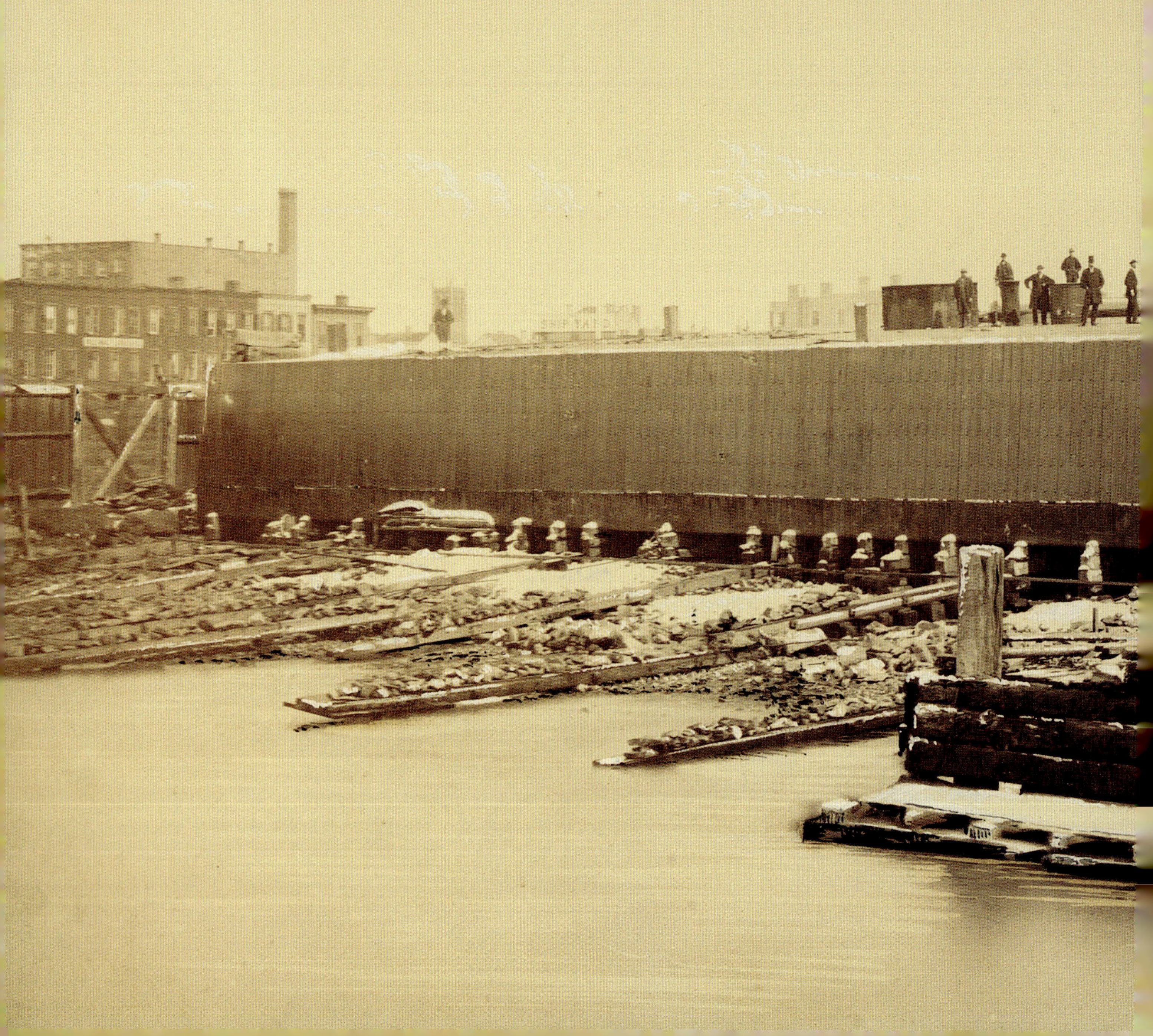

CHAPTER 3

Caissons

1870–1872

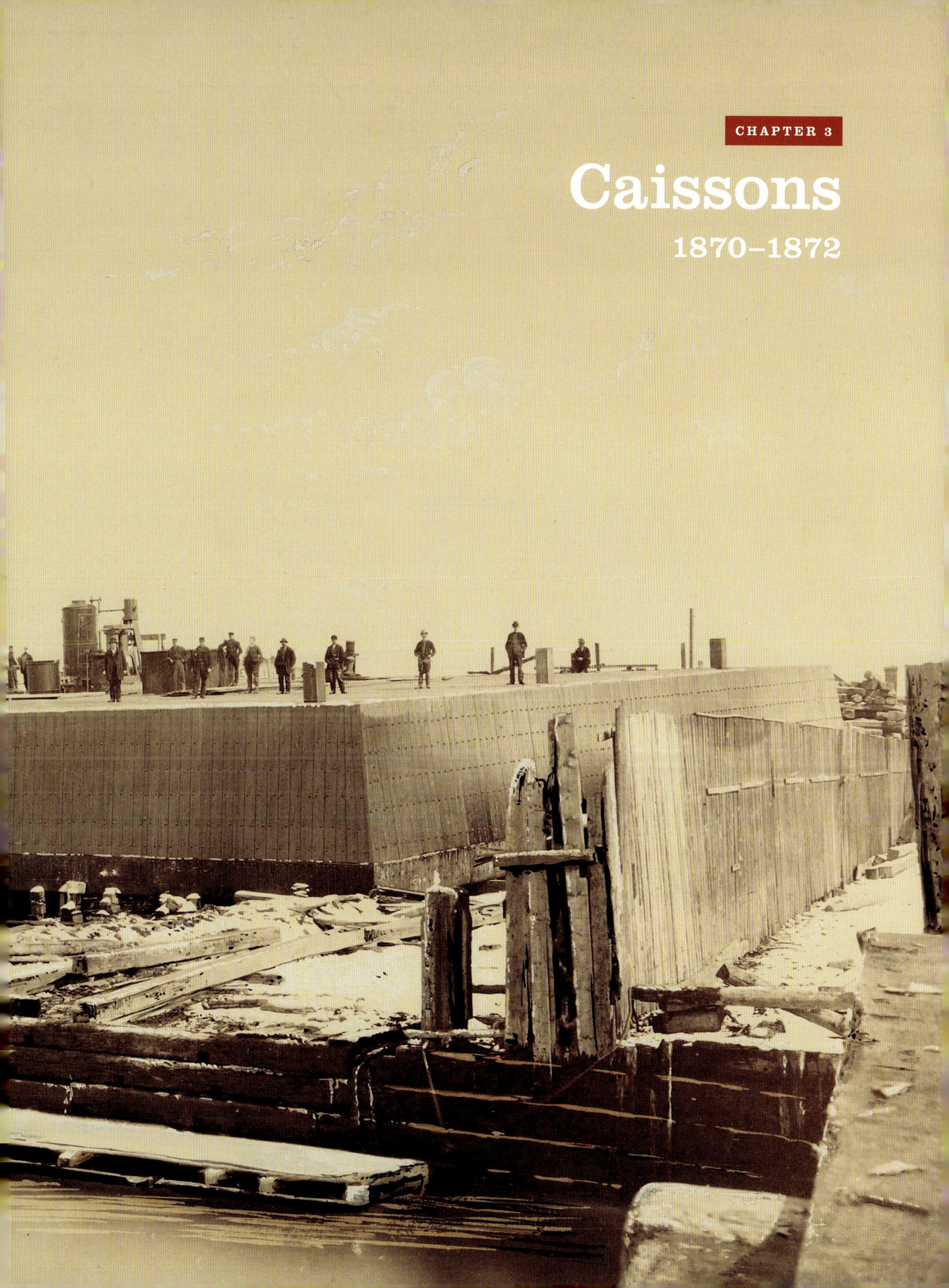

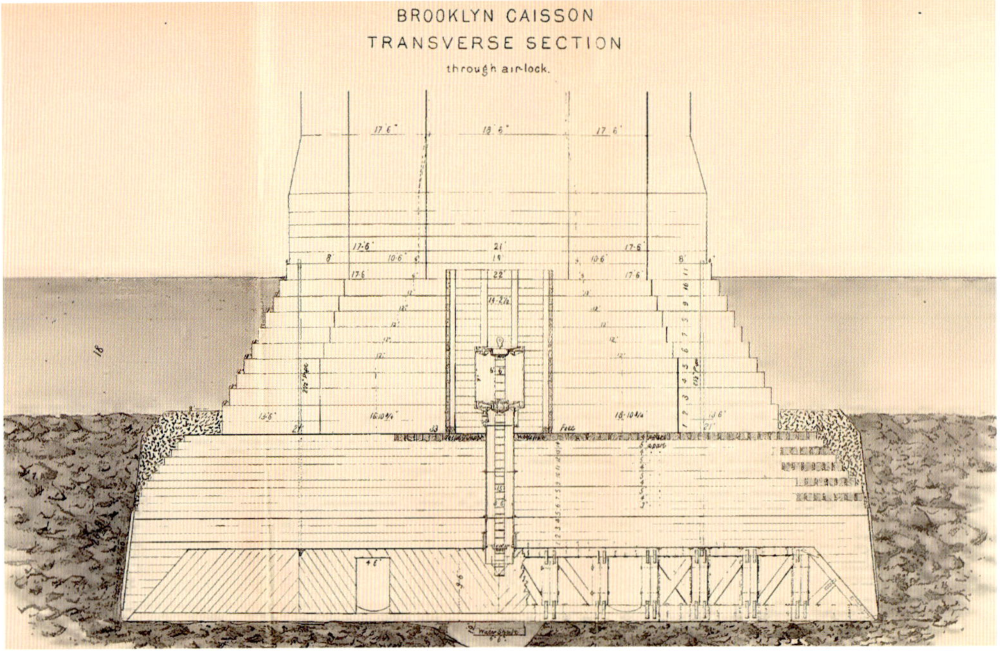

"Brooklyn Caisson/ Transverse Section/ through airlock." An illustration from Washington Roebling's pamphlet *Pneumatic Tower Foundations of the East River Suspension Bridge*, 1873. This print of one of John Roebling's drawings does not show the water, air, and supply shafts that Washington designed and had built into the caissons.

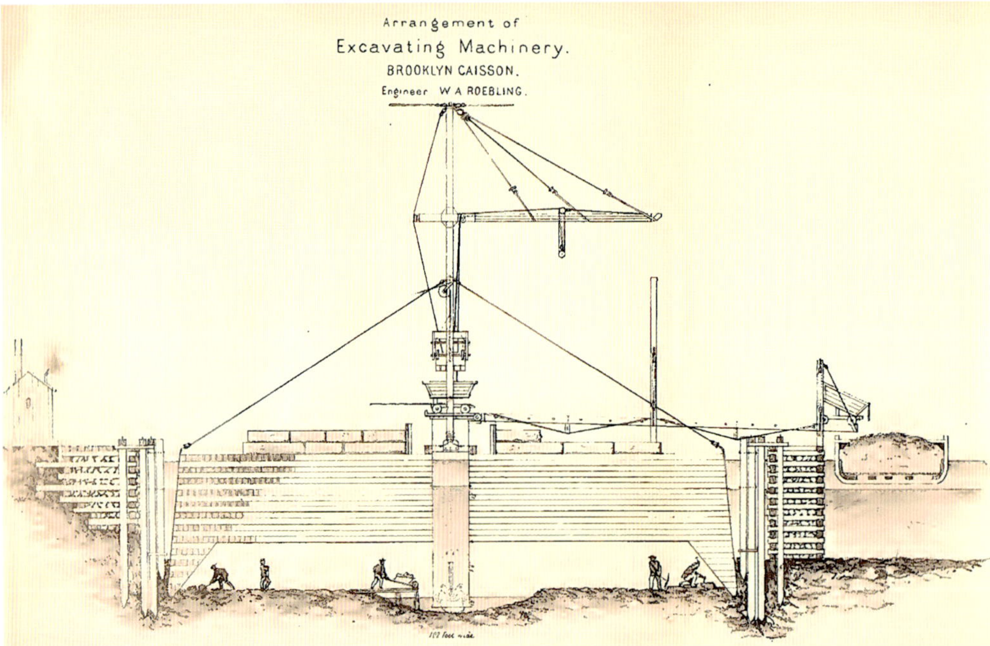

ABOVE: "Arrangement of/Excavating Machinery./Brooklyn Caisson./Engineer W. A. Roebling," an illustration from Washington Roebling's pamphlet *Pneumatic Tower Foundations*, 1873. Inside the caisson's work chambers, laborers use picks, shovels, and wheelbarrows in their digging. On tracks above the caisson, a cart waits to receive debris dredged up from below, and another cart, at right, has rolled down the tracks to dump its contents into a waiting scow. The derrick atop the caisson will maneuver the massive stone blocks for the tower, with parts of two stone courses already in place.

PREVIOUS SPREAD: "Brooklyn Caisson Before Launching." The presence of the people offers a sense of the immensity of the caisson. This photograph was likely taken by Silas A. Holmes, given the similarity of its mat to those of photographs marked with his stamp. MCNY.

The first time a caisson, a large watertight chamber, was used as the foundation for the towers of a bridge was in 1851. So, when construction on the Brooklyn Bridge began in 1869, this use of caissons was still in its infancy. Even though John Roebling was the leading suspension bridge builder in the world, he had no experience with or expertise in their use.

As the Brooklyn Bridge towers were being built, one block of stone at a time, this new form of foundation—the caisson—would provide the stable base needed for this unprecedented undertaking. The Brooklyn Bridge's immense size meant that its two towers would need to be higher and heavier than those of any other bridge built to date. Therefore, the foundations for those towers had to be stronger than any built before. John Roebling recognized this when he wrote: "The foundations for the support of these large masses of masonry must be unyielding." This posed an engineering problem: how best to securely anchor the two massive stone towers in place? The answer was to use caissons. These huge reinforced wooden boxes were filled with compressed air to keep the river water from seeping underneath, while workers dug through mud, gravel, and boulders to drive down the structure to solid bedrock. Piles were erected as a barrier around the caisson while the excavation proceeded inside. Once the digging stopped, the open chambers within the caissons would be filled with concrete to solidify the foundations upon which the towers would rest.

Given that neither John nor Washington Roebling had any firsthand experience with the use of caissons, they no doubt felt much trepidation. In a report to the bridge trustees Washington warned that, with respect to the caissons, "Considerable risk and some degree of uncertainty was necessarily involved in their construction." He later elaborated:

> At the commencement of this enterprise, [the caissons] constituted the principal engineering problem to be overcome, and not until they were accomplished facts could it be said that the project of the bridge was placed upon a firm, immovable footing. The subsequent building of the Towers and of the superstructure is all work which has been done before on a smaller scale. But upon the Tower Foundations rests the stability of the entire work. The whole enterprise depended on their success.

To study how caissons were being used in bridge construction, Washington had gone to Europe in June 1867 with his wife, Emily, in preparation for the most important engineering project of his career. Washington's father had urged him to go, even paying his expenses. In France, at the time of his visit, caissons had already been used for a generation as the foundation of bridge towers. Indeed, the word "caisson" is the French word for "chest." By the time Washington returned from his intensive European study, he had become the leading American expert on bridge caissons.

In his 1873 pamphlet, *Pneumatic Tower Foundations of the East River Suspension Bridge*, Washington described the caissons, or "pneumatic tower foundations," at length, with illustrations. As this phrase indicates, the caissons relied on pneumatics—here, compressed air—to keep the East River out of the chambers where the men worked. The compressed air also helped counter the downward pressure of the river on the caissons, lessening the force on their bottom edges, or "shoes."

In 1883, soon after the opening of the Brooklyn Bridge, writer Charles Barnard published an article in the popular *St. Nicholas Magazine*, a monthly illustrated magazine for "young folks," that explained, in very relatable terms, how pressurized air allowed men to work inside the caissons:

> Get the wooden chopping-bowl from the kitchen and a clear glass of water. Fill the bowl half full of water, and then, holding the tumbler upside down, press it slowly into the water till it touches the bottom. When it rests there you will see there is not water inside the tumbler, and that the bottom of the bowl is nearly dry. The air caught under the tumbler has pushed the water away. If the tumbler were large enough, a man could stand inside and dig out the bottom of the bowl quite comfortably.

Given the size of the Brooklyn Bridge, it is not surprising that the caissons were larger than any ever used before anywhere in the world. Indeed, no structure this large had ever been sunk into the earth. As Washington Roebling noted in his 1873 pamphlet, the Brooklyn Bridge had "the two largest pneumatic foundations extant, as regards area of base" and represented "the first application of timber on a large scale" to caisson construction. The roof of each caisson is more than 17,000 square feet, enough space for four tennis courts and then some. There was 9.5 feet of headroom in the work chambers. Each caisson weighed 3,000 tons and was made from 110,000 cubic feet of timber and 230 tons of iron holding it together. At that weight, the caisson was 50 percent heavier than any ship that had been launched, to date, on the East River. Like most elements of the Brooklyn Bridge, the caissons were designed to withstand stresses and pressures well beyond what the engineers reasonably expected; ultimately they were built four times stronger than required.

In order to construct the caissons, Washington Roebling had to find a company that was skilled in large-scale timber work. He logically turned to local shipbuilders. Fortunately, the East River—along both the Brooklyn and New York shores—had one of the greatest concentrations of shipbuilders in the world. A handwritten note in the Roebling Collection at Rensselaer Polytechnic Institute, dated October 20, 1869, compares two bids for the construction of the Brooklyn caisson, the first of the two caissons to be built. The bids were submitted by shipbuilders T. F. Rowland Company, which had built the famous iron warship *Monitor* during the Civil War, and Webb & Bell, located at the foot of Noble Street in Greenpoint, Brooklyn. Including supply and installation of drift bolts, screw bolts, spikes, and "other forgings," Webb & Bell offered to build the caisson for $90,652 (approximately $1.7 million in 2021). That was far and away the lower bid, and Webb & Bell got the job.

Also in October, a bid came in from Columbian Iron Works, located on Adams Street in downtown Brooklyn, for a good deal of the iron work: air shafts through which compressed air would be fed into the work chambers, water shafts through which the dug debris could be removed without allowing the compressed air to escape, and supply shafts to get equipment and material into and out of the caissons (again while containing the compressed air), as well as the shoe castings, "to be made in the best manner and of the best material," at 12.5 cents per pound. They ended up losing the job to the Burdon Iron Works of Brooklyn, which in an offer dated October 27, 1869, stated that the company would provide water shafts twenty-one feet long, air shafts eleven feet long, supply shafts, shoe castings, and other iron, including galvanized iron bands, at prices ranging from 4 to 15 cents per pound.

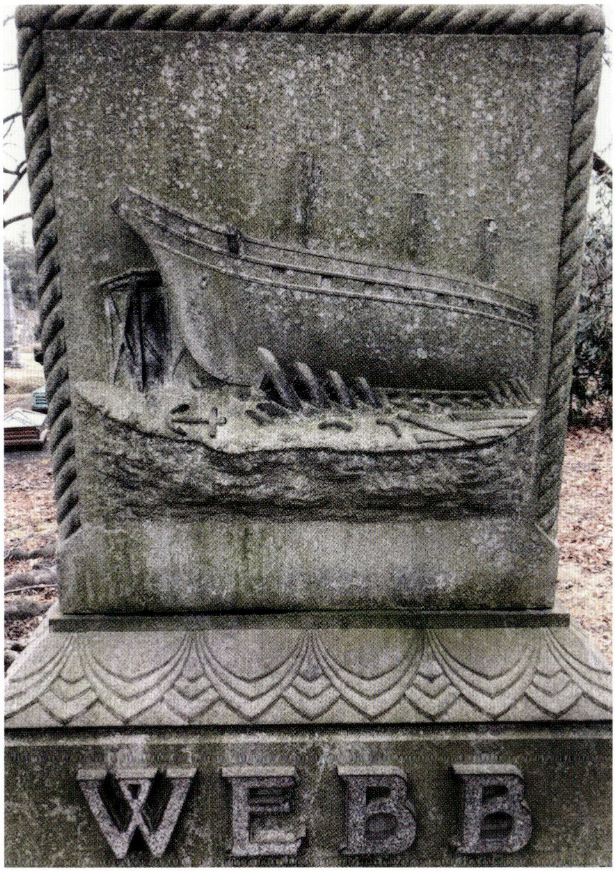

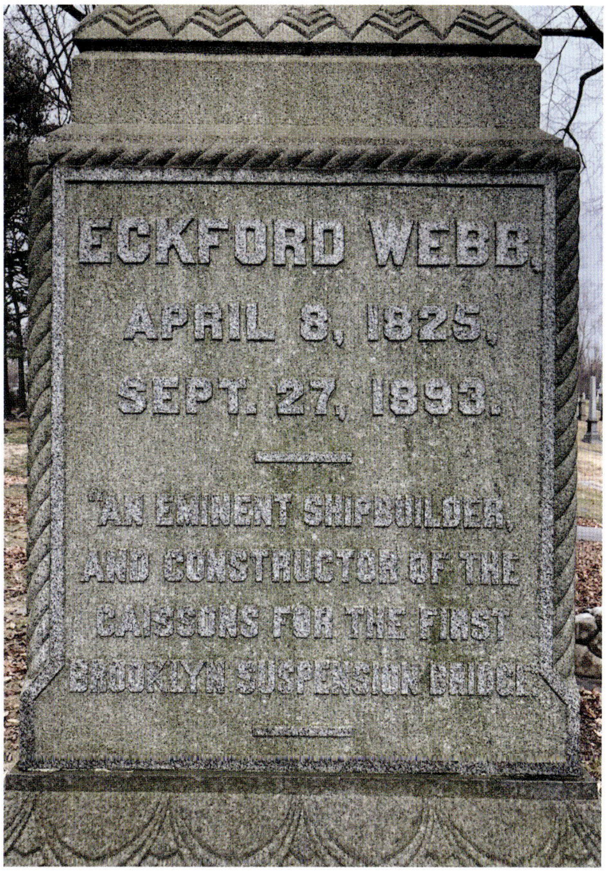

LEFT: The front of the base of shipbuilder Eckford Webb's memorial obelisk at Brooklyn's Green-Wood Cemetery, where he is interred. It shows one of his ships being built. Author's photograph.

RIGHT: On the back of Eckford Webb's memorial at Green-Wood Cemetery, this inscription is carved into the granite: "An Eminent Shipbuilder and Constructor of the Caissons for the First Brooklyn Suspension Bridge." It reflects the pride Webb had in his work on the Brooklyn Bridge. Author's photograph.

Washington specified yellow pine as the primary caisson material, with each timber running several feet in length and a foot square. Headroom inside the work chambers was about 9.5 feet. After being built, each of the two Brooklyn Bridge caissons would be dug into the earth along the shores of the East River one on the east side, in Brooklyn, and the other on the west side, in New York City. John Roebling had envisioned the interior of each caisson as one big open space or, as Master Mechanic E. F. Farrington described it, "a great box turned bottom up." John had drawn the caissons as rather simple constructions, featuring little more than a ladder for the workers to come down through the roof. However, his son had other ideas.

After John's untimely death, Washington, now chief engineer, ordered interior partition walls to be built inside both the Brooklyn and New York caissons, dividing each into six chambers. These walls would allow the tremendous weight of the immense towers to be distributed more widely and evenly, relieving some of the pressure on the caisson's iron shoe. Farrington described these interior walls as "timber partitions two feet thick, strong and well braced." Each of the partitions would have an opening cut into it, tall and wide enough for a worker to pass through it while pushing a wheelbarrow.

The four outer sides of the caissons were V-shaped in cross-section: nine feet thick at the top, where the wall met the wooden roof, then tapering to a thinness of eight inches at its bottom. Iron bolts held the wood frame together and secured the iron shoes along the caisson's bottom lip. The iron shoes protected the wooden sides from boulders encountered as the caisson was being sunk deep into the earth. On the roof, the inside and outside seams

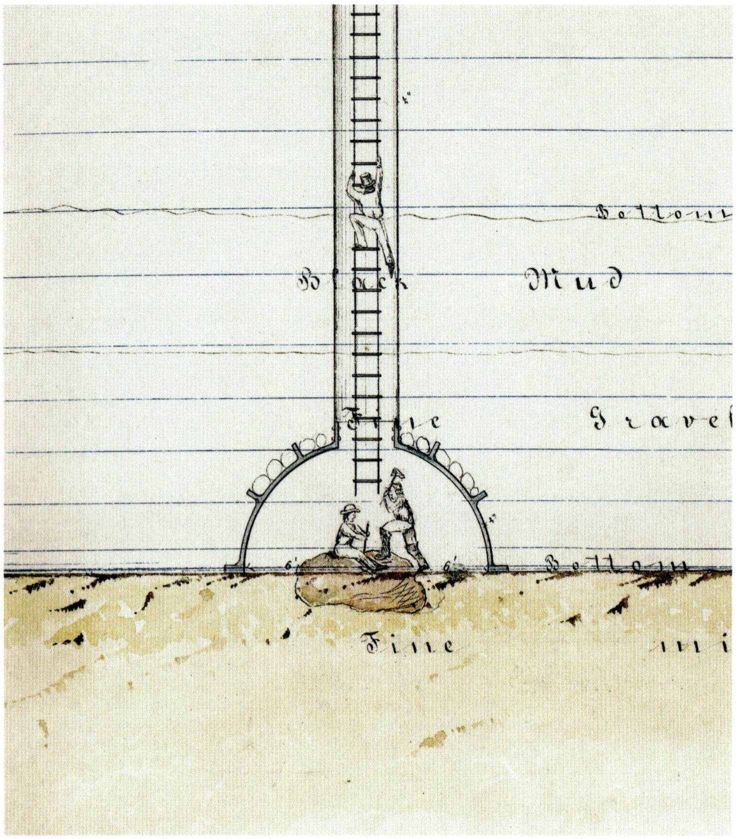

Detail of a drawing showing work being done inside one of the bridge's caissons: a two-man team, using a sledgehammer and iron drill, break apart a boulder, as a third worker, in a top hat, uses a ladder. This was likely drawn by John Roebling, as it does not reflect the modifications Washington Roebling made to the caissons after his father's death. RPI.

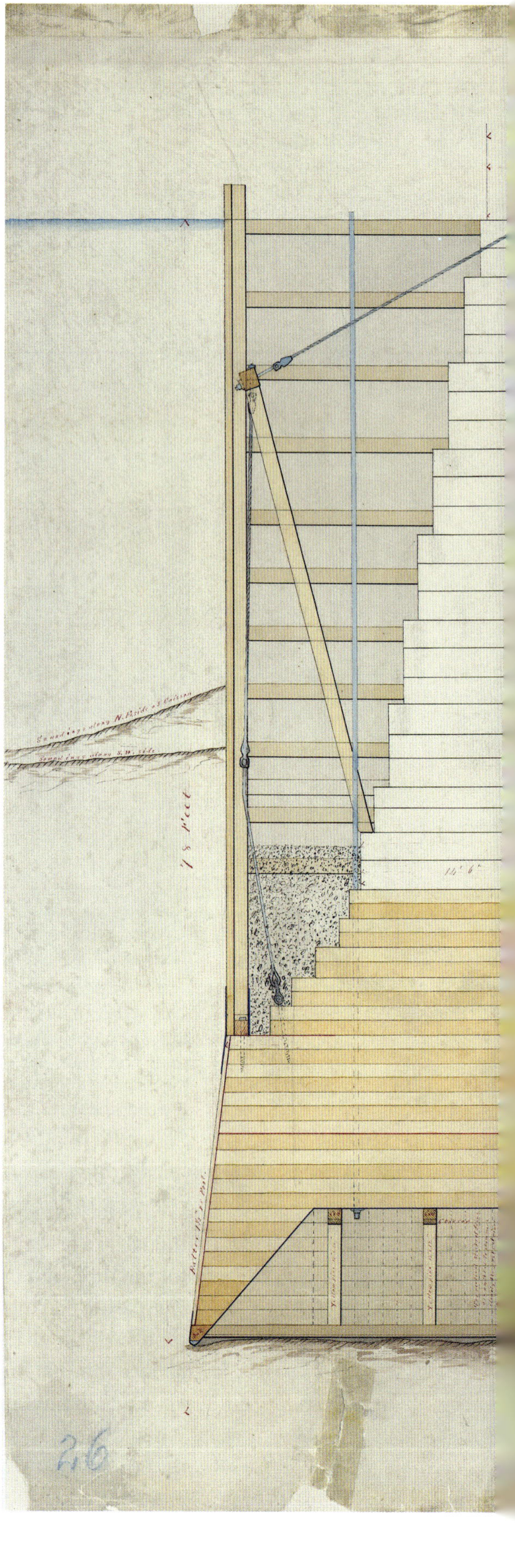

"N.Y. Caisson/Transverse Section/through air-locks," a wonderfully detailed drawing by Assistant Engineer Wilhelm Hildenbrand. The caisson has been dug down, and the tower is partially built atop it. The timber courses are indicated at bottom in light brown, and the stone courses are depicted above them as larger rectangles (the tower tapering as it rises). The spiral staircase at center allowed workers to enter or exit the caisson through the hollow base of the tower. The horizontal blue line at the top of the drawing indicates the level of the river water; midway down the caisson, the riverbed is sketched out. The walls surrounding the caisson and tower were designed to keep the water out. At top center is a mast derrick, with guy wires attached to hold it in place. At right is a detail of a boom derrick. Note the vertical partition walls at the bottom of the drawing, dividing the work space into separate chambers; also the blue pool of water below the water shaft. NYC Municipal Archives.

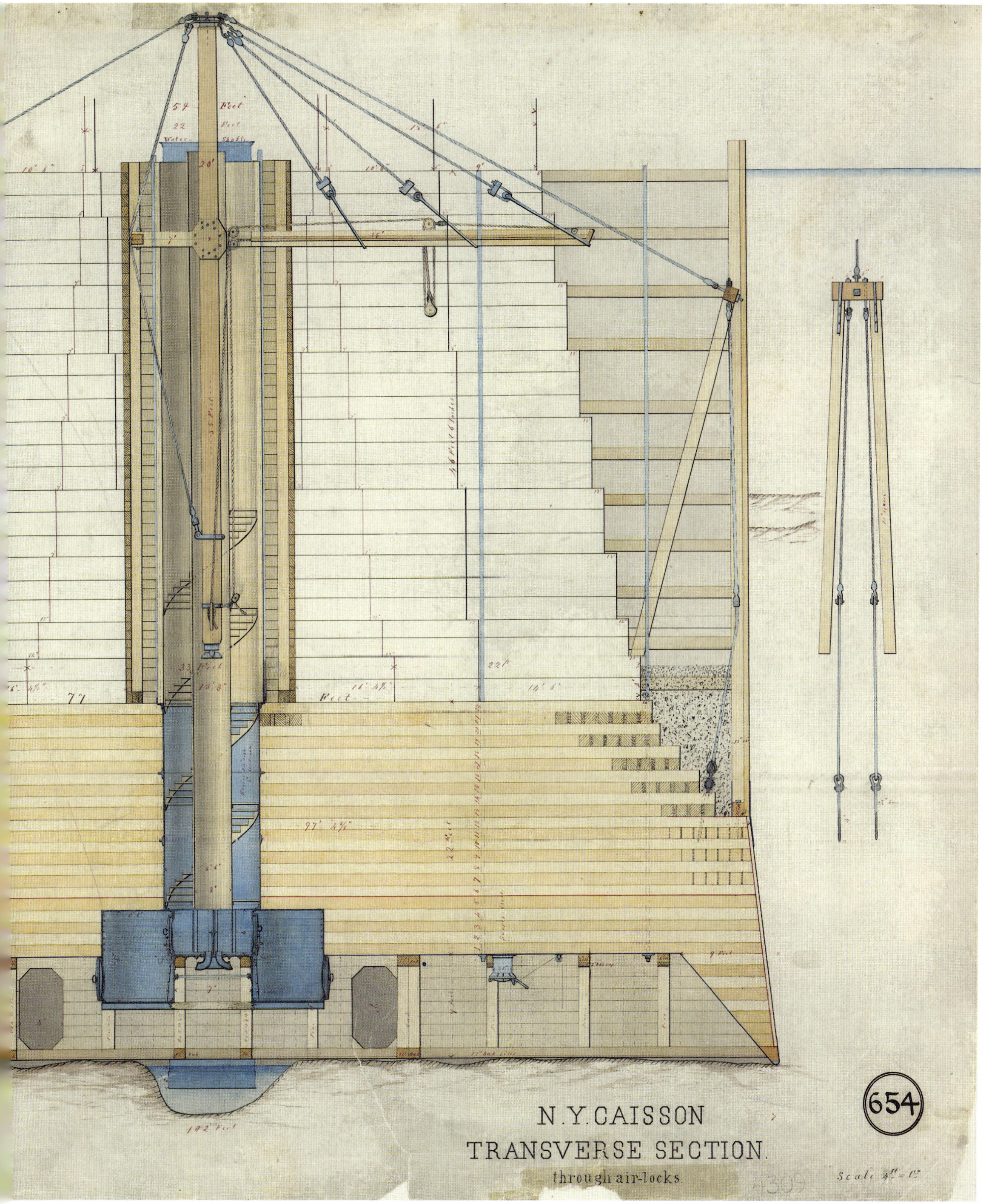
N.Y. CAISSON
TRANSVERSE SECTION.
through air-locks.
654

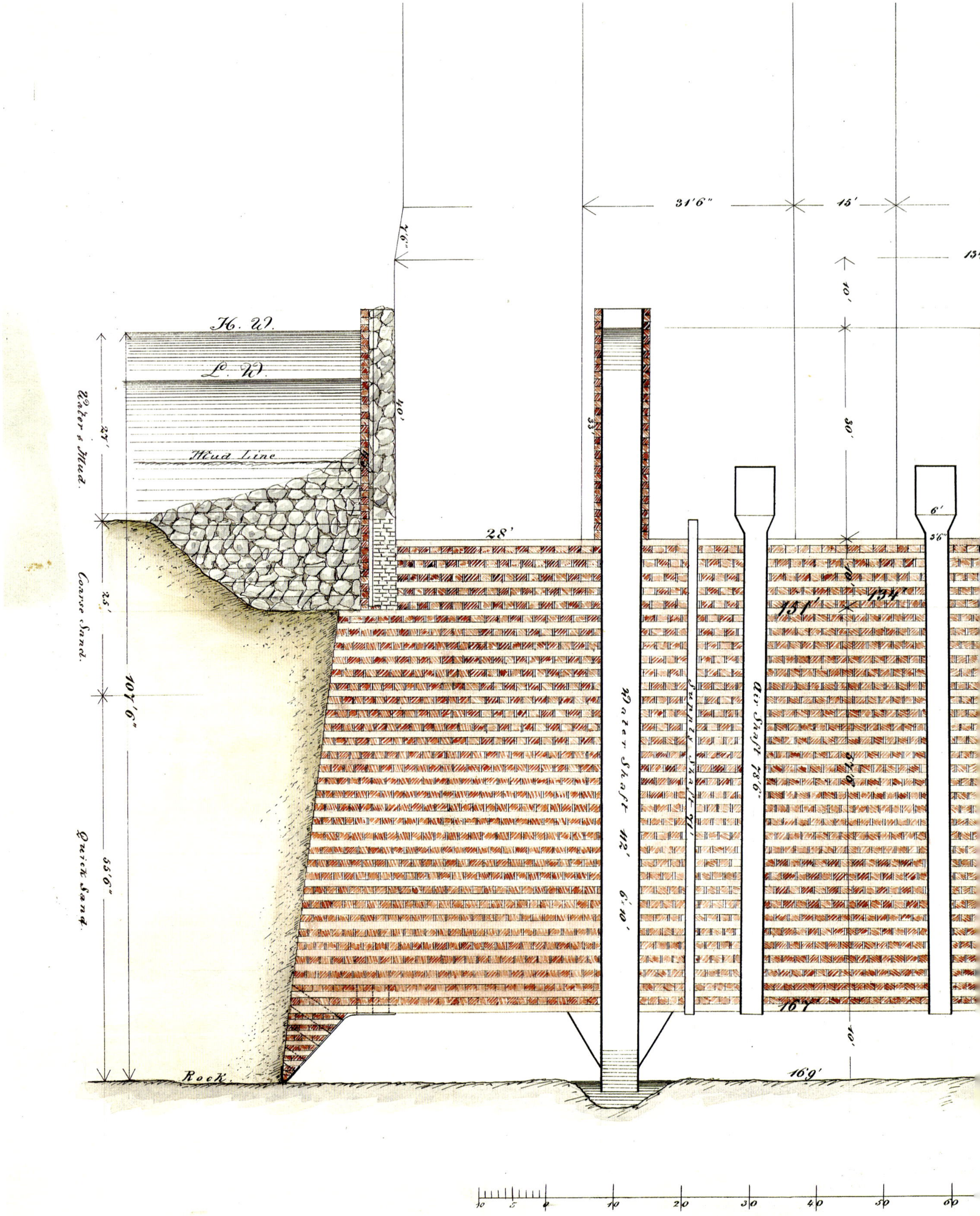
H. W.
L. W.
Mud Line
Water & Mud.
Coarse Sand.
Quick Sand.
Rock.
27'
25'
55'6"
107'6"
7'6"
10'
28'
33'
31'6"
15'
30'
151'
134'
Water Shaft 112' 6'10'
Supply Shaft 71'
Air Shaft 78'6"
6'
5'6"
167'
16'9'

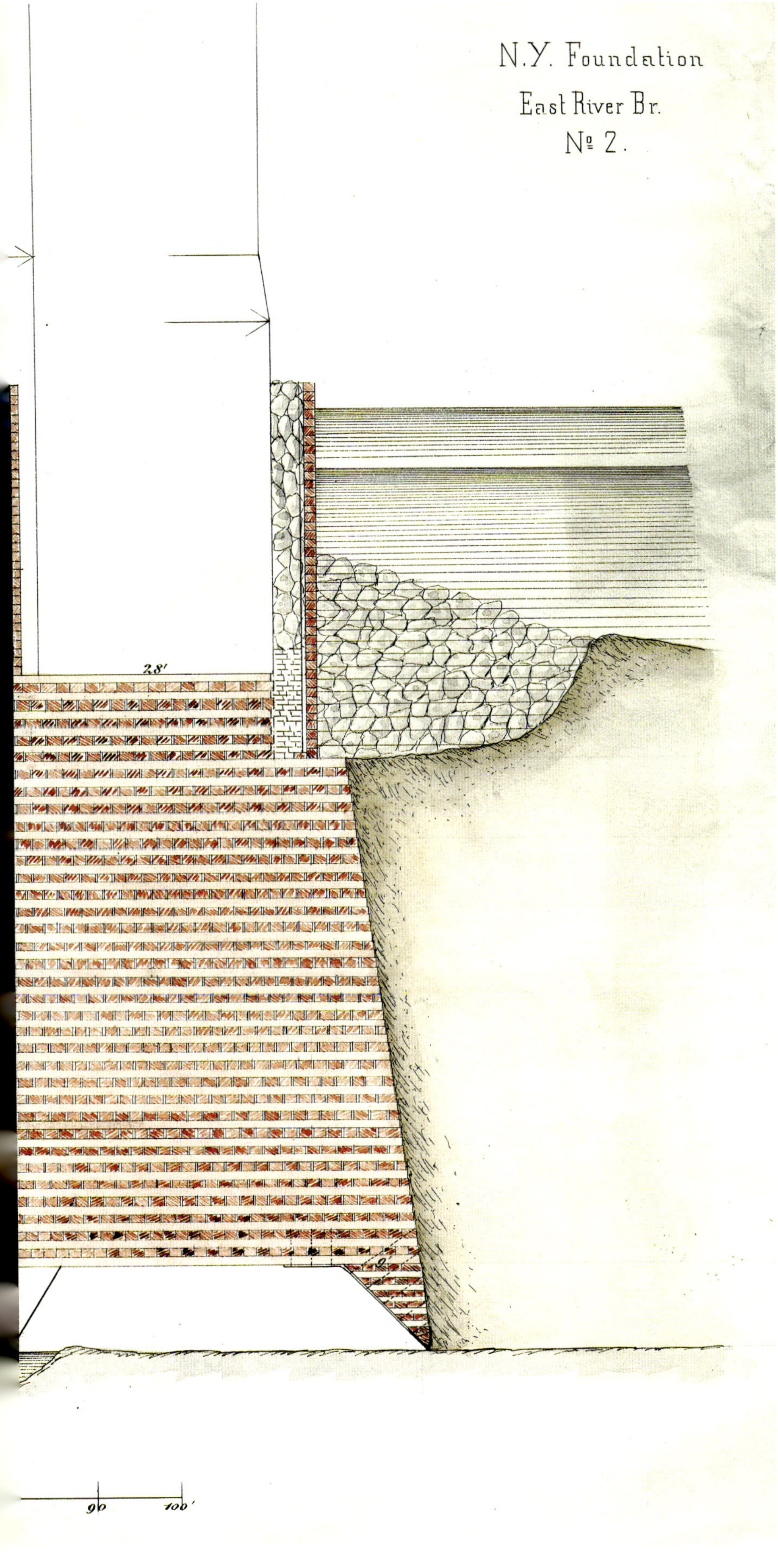

"N.Y. Foundation/East River Br./No. 2." In this early hand-drawn plan, the two water shafts are on the outside; the air shafts, which widen at their tops, are toward the center; and the thin supply shafts are between them. All six run through the timber courses of the caisson's roof; however, the caisson's spiral staircase is not shown. Note the outlines of the rising tower atop the caisson. RPI.

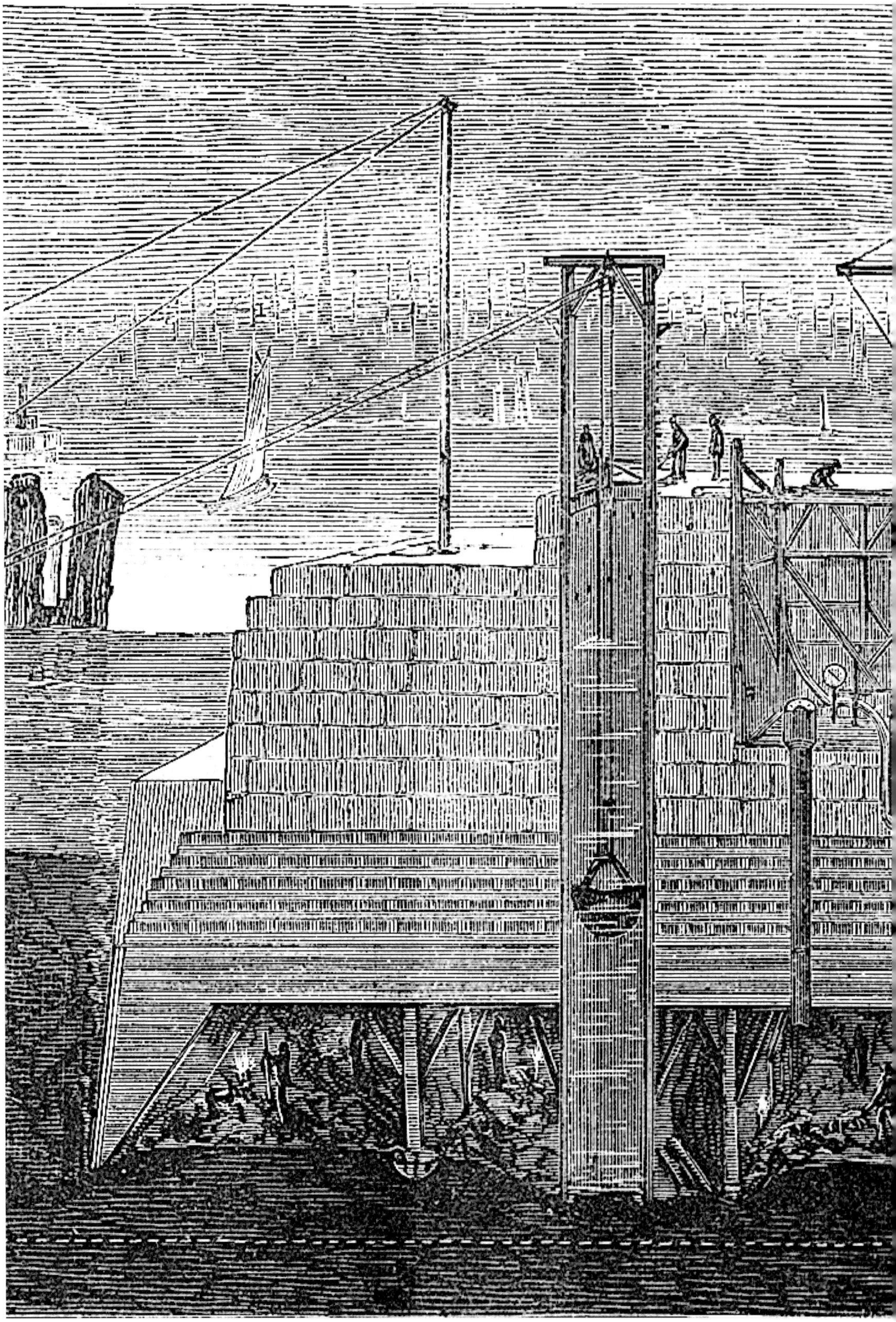

"Sectional View of Foundation, Showing Caisson and Mason Work," *Harper's Weekly*, December 17, 1870. The work at the caisson proceeds, with digging in the partitioned chambers below. Several courses of stone blocks are already positioned atop the caisson as the base of the tower begins to take shape. The clamshell buckets are in operation, and workers are climbing in and out on ladders.

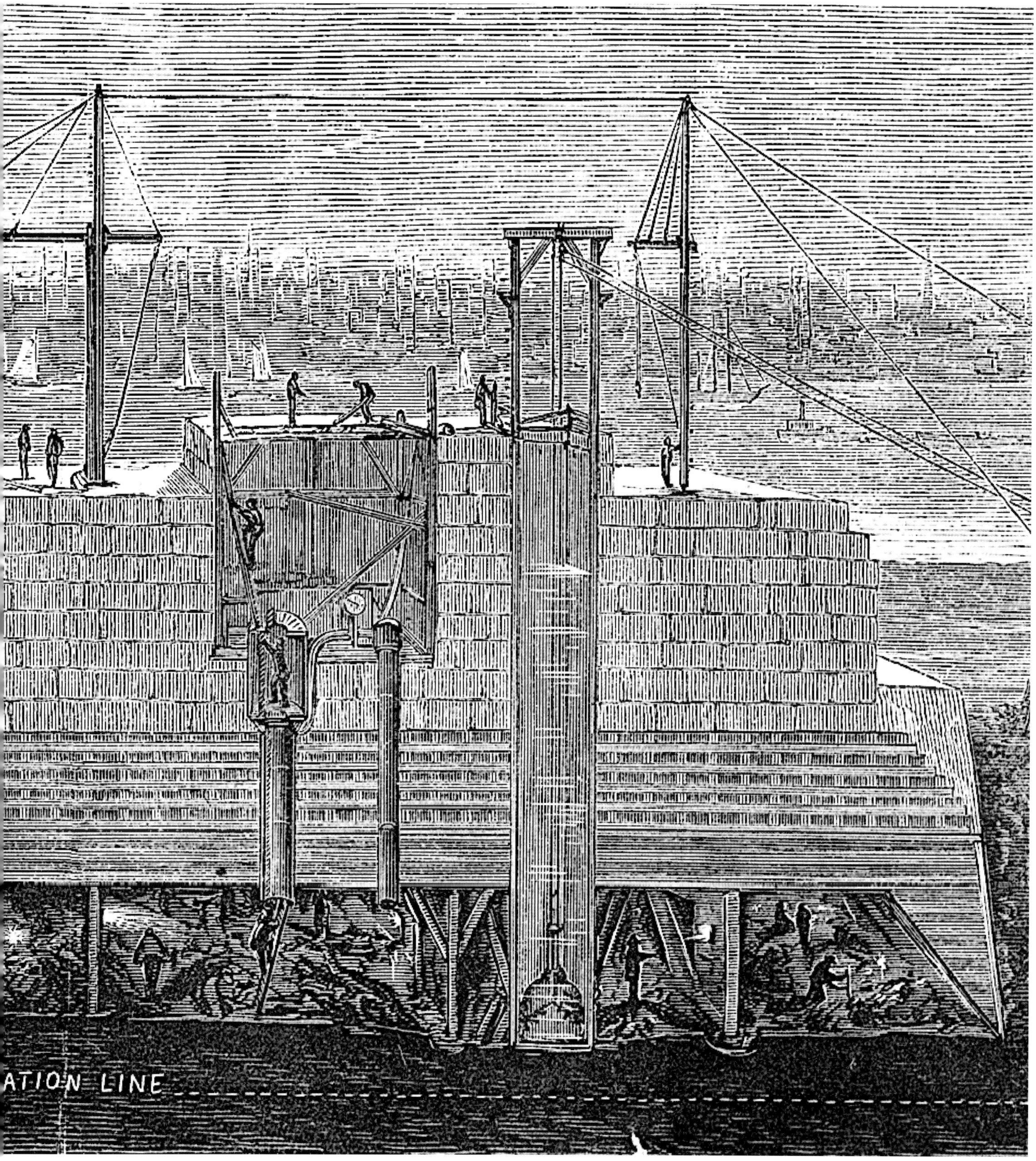
ATION LINE

One of the men who worked on the Brooklyn Bridge recalled the sound in the air lock when the brass valve was opened, allowing the compressed air to rush in from the caisson below: "An unearthly and deafening screech, as from a steam whistle, is the immediate result and we instinctively stop our ears with our fingers to defend them from the terrible sound. As the sound diminished we are sensible of an oppressive fullness around the head, not unaccompanied with pain, somewhat such as might be expected were our heads about to explode."

Each caisson had two water-filled shafts through which the dug debris was hauled up and out. The water was used to prevent the compressed air from escaping. On the Brooklyn caisson, the water shafts were seven feet square, open at the top and bottom, and reached from above the caisson's roof to about a foot and a half below the caisson's bottom edge. As described by Master Mechanic E. F. Farrington, the "lower ends of the water shafts were kept constantly immersed in water to seal them against the escape of air, and a column of water was maintained inside them equal in height to the tide outside, and through this water the dredges hoisted out the excavated material." He wrote that the large "dredge, familiarly known as the 'clam shell,'" worked "much like two human hands united at the wrists—palm to palm—with the fingers somewhat bent but not interlocked." Workers needed to keep the excavated rock and clay loose enough to be scooped up by the clamshell. In order to do so, they used iron bars to mix the debris with water that was continually supplied by a hose.

The water shafts were a clever but not risk-free solution to removing debris from the caissons without losing compressed air. Made of boiler plate, they acted like giant barometers, with the column of liquid inside responding to air pressure. In order to function properly, this column of water had to be just the right weight—heavy enough to prevent the compressed air below from blowing out through the shaft, but light enough so that the pressure exerted by the compressed air would be sufficient to hold up the water column. If the water column were to become heavier than what the compressed air could support, it would collapse and flood into the work chambers. On the other hand, if the weight of the water column were to become lighter than the pressure of the compressed air pushing up against it, there would be a blowout, with the compressed air forcing whatever was in its path upward through the shaft and perhaps hundreds of feet into the sky. There also was another worry: the water column lost weight as its sediment settled out at the bottom. In order to compensate for this loss, workers were assigned to hose more water into the shaft. However, if they miscalculated or neglected to add water when needed, a blowout would also occur. This happened once at the Brooklyn caisson, to dramatic effect, launching a column of mud, rocks, and water five hundred feet into the air.

The supply shafts, as designed by Washington, were iron tubes forty-five feet long and almost two feet in diameter. These were used to supply the workers inside the caissons with construction materials. But their primary purpose would not come into play until after the caisson had reached the depth that the engineers found satisfactory and the digging was complete. Once the men were done working in the caisson, they filled the chambers with concrete in order to create as solid a foundation as possible. This concrete was supposed to be sent down through the shafts. However, by the time the Brooklyn caisson had reached bedrock, the weather was too cold to mix the concrete atop the caisson, so the individual components of the concrete—cement, sand, and fine gravel—were sent down the supply shafts to be mixed inside. When it came time to fill the New

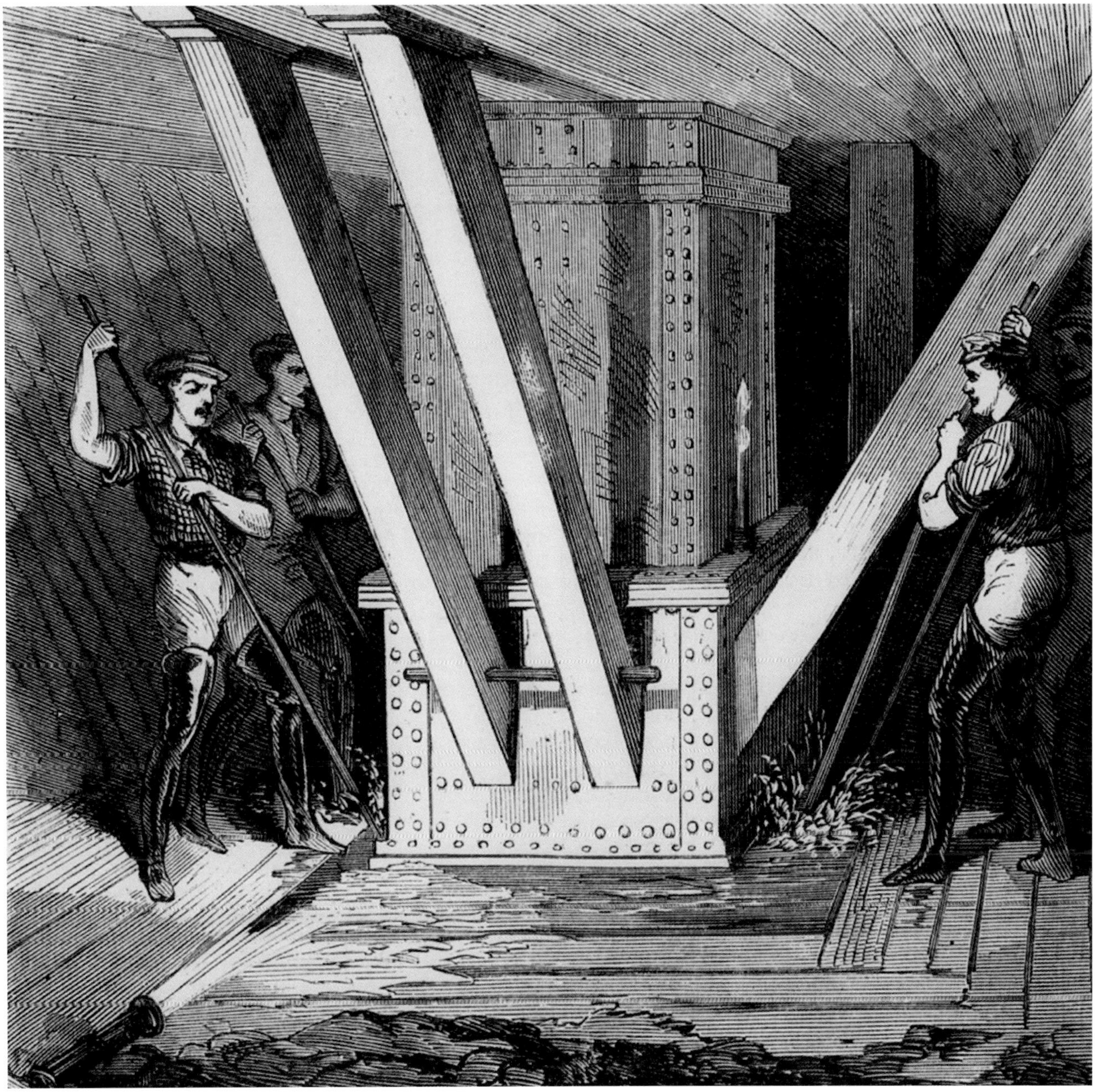

"Feeding the Bucket of the Water-Shaft in the Caisson," a woodcut from *Frank Leslie's Illustrated Newspaper*, October 15, 1870. The lower end of a water shaft extended several feet down into a debris pit. The water pooling there prevented the compressed air in the work space from escaping. Note the hose in the left foreground—it fed water into the pit to keep the debris loose enough to be scooped up by the clamshell bucket and lifted out through the shaft. Courtesy of Richard Haw.

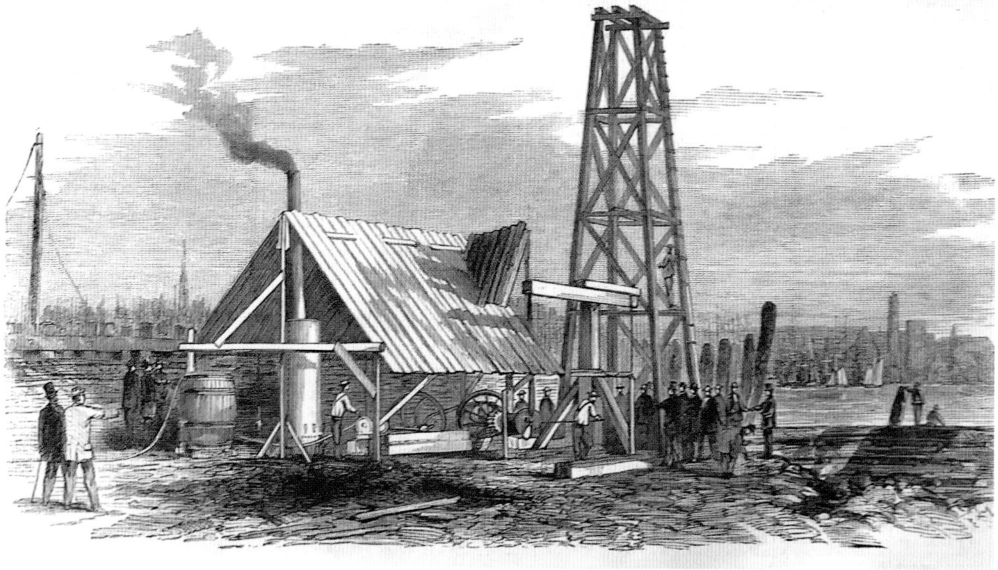

"Boring for Foundations for the New York and Brooklyn Suspension Bridge," *Harper's Weekly,* August 31, 1867. The boring samples informed the engineers about the strata that awaited them as the workers dug the caissons down to bedrock.

York caisson over a year later, the weather was warm, allowing the concrete to be mixed outside and then poured into place through the supply shafts.

Before work on the caissons began, boring samples had been taken at each of the two sites where the caissons would ultimately rest to determine the composition of the strata that would have to be dug through to get to bedrock. The samples established that the bedrock on each side was somewhere between about fifty and one hundred feet down. In that range, both wooden caissons would rest safely below the depth at which sea worms could eat away at the timber. But the test borings on the New York side showed that the bedrock there was significantly farther down than in Brooklyn. This would necessitate greater air pressure inside that caisson to keep the river water at bay and, thus, more compressors to offset the greater water pressure.

For the Brooklyn caisson, compressed air was supplied by six large compressors built by the Burleigh Rock Drill Company of Fitchburg, Massachusetts. They were connected in such a way as to assure that if one failed, the others would continue to function without interruption. Pumps cooled the air once it left the compressors, creating a bit of air conditioning for those working inside the caisson.

On the New York side, there were more than twice as many compressors as had been used at Brooklyn. The largest building used for the bridge construction was the compressor house on the New York side. It held thirteen large Burleigh compressors—four kept the compressed air at the required pressure, six pumped fresh air into the caisson, and three served as backups.

The depth to which the New York caisson would have to be dug to reach bedrock was a worry. Not only for the extra time for digging and the extra weight of the deeper tower needed on that side, but also for the composition of the ground the workers would need to dig through. Washington wrote of the challenges that the Manhattan test borings portended:

> The results showed an extreme difference in the levels of the bed rock of twelve feet, the hole of the least depth touching rock at eighty feet below high water, and the deepest at ninety-two feet. The strata consisted in the main of a black mud deposit of twelve feet, followed by a layer of coarse sand of six feet which overlaid a gravel bed of

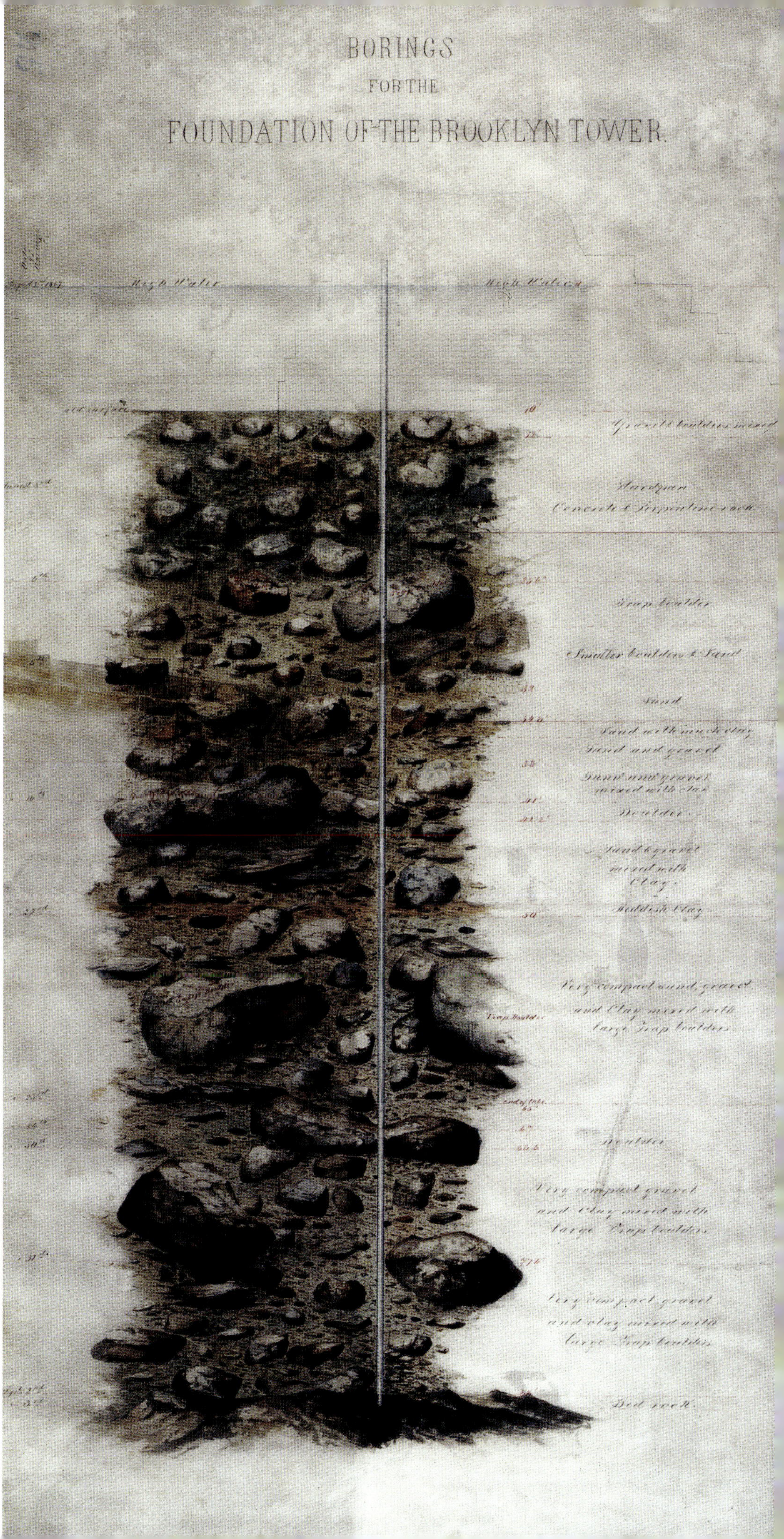

"Borings/for the Foundation/ of the Brooklyn Tower," drawn by William Vanderbosch, 1869. The handwritten notations at right indicate the materials encountered at various depths, marked in feet, including hardpan, "smaller boulders & sand," and "very compact sand, gravel, and clay, mixed with large trap boulders." The dates indicate that boring tests were conducted between August 2 and September 3, 1867. NYC Municipal Archives

air compressor house →

winter

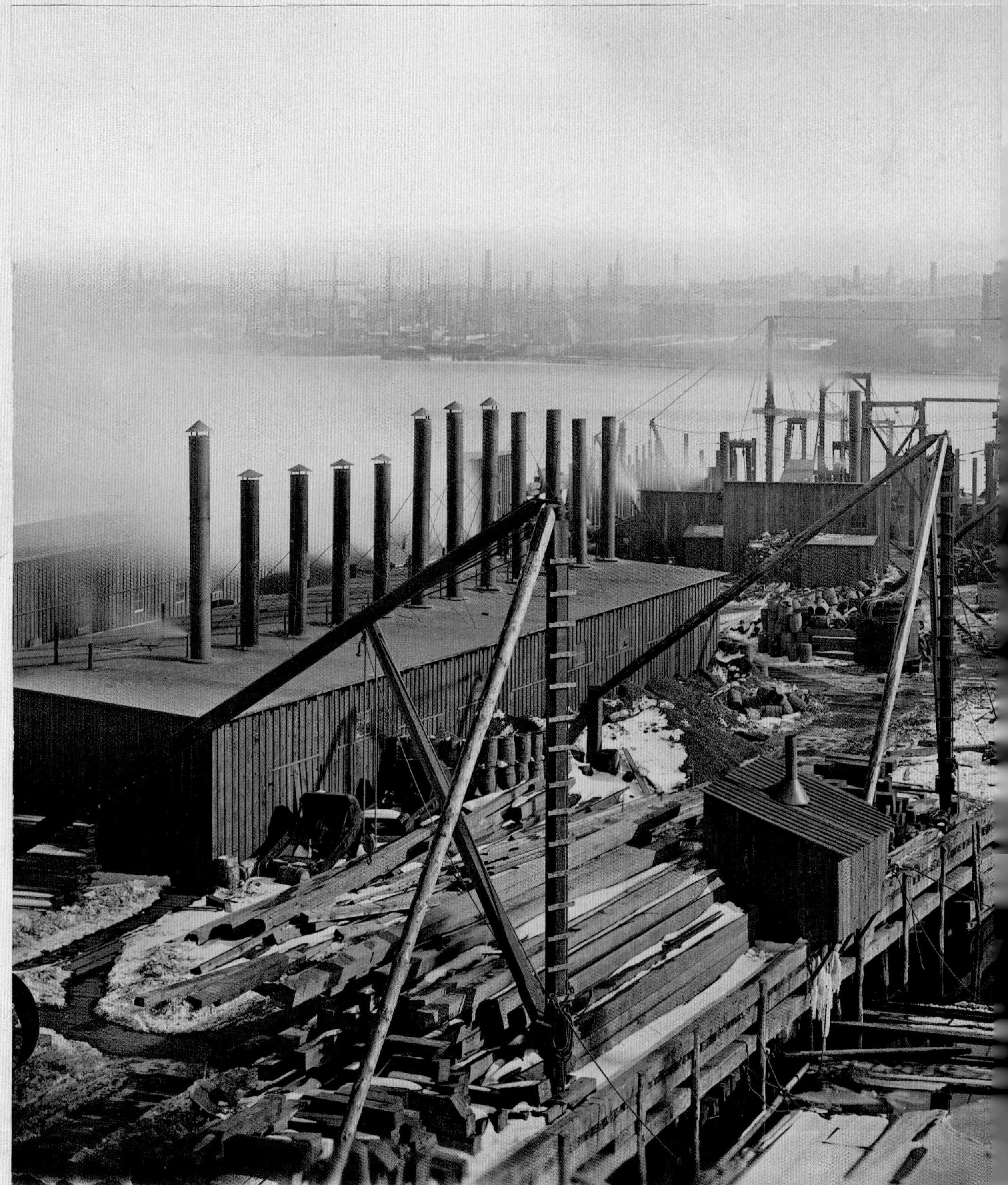

Foundation of
New York Tower, East
March 1872

W. A. Roebling

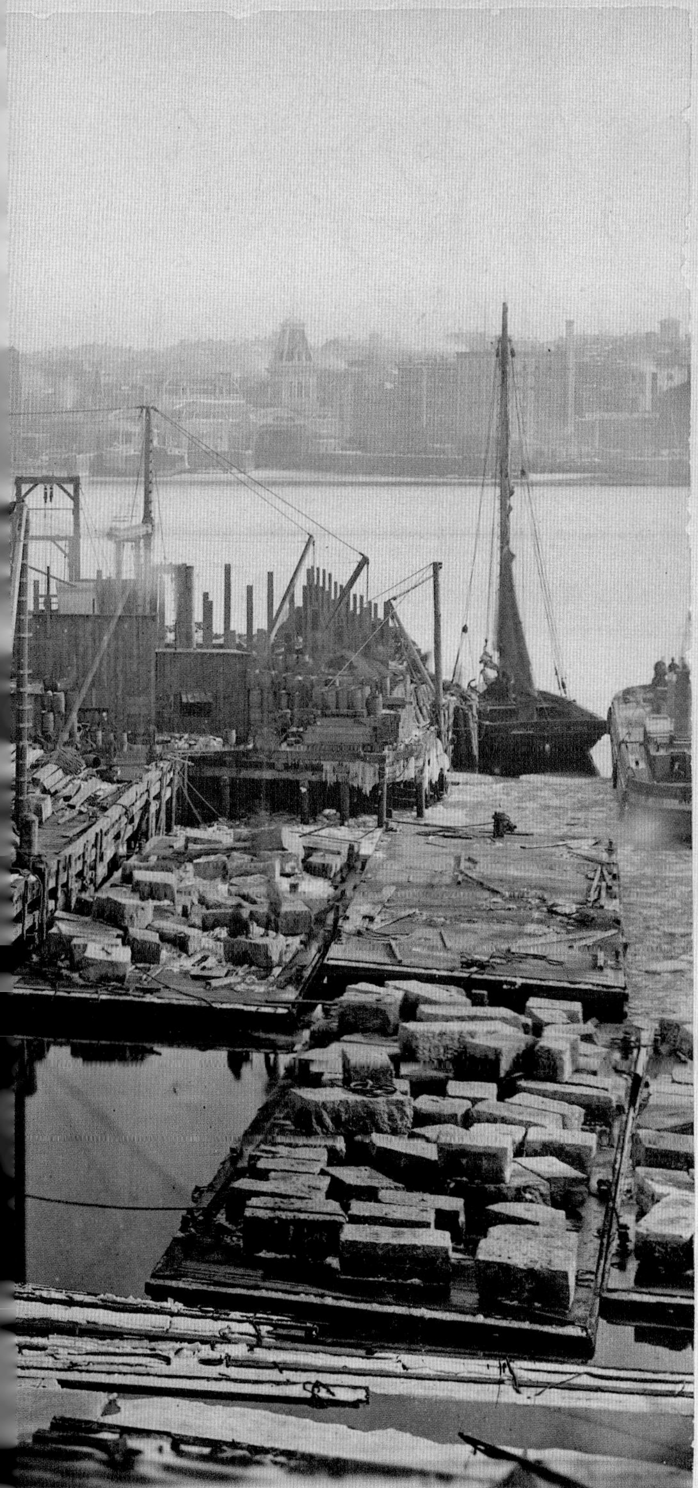

"Foundation of/New York Tower, East River Bridge/March 1872/W. A. Roebling Eng[inee]r," a photograph taken by Silas A. Holmes. Noted in pencil in the left margin of the photograph: "air compressor house." An arrow points toward the low rectangular building with metal stacks rising from it on the left side of the photograph. Another penciled notation below that reads: "winter"—confirmed by the presence of snow. Based on the distinctive handwriting, these notes appear to have been written by none other than Chief Engineer Washington Roebling. Note the two barges at right, loaded with stone blocks that will form the bridge's New York tower. Three mast derricks, connected by guy wires, stand atop the caisson at the end of the pier. They will position the stones that form the base of the tower. Brooklyn is visible in the distance across the East River. RPI.

the same thickness. Beneath the gravel appeared a very heavy deposit of quicksand, varying from fifteen to twenty feet, according to locality, and abounding with boulders in its lower portion, varying from two to five feet in diameter.

Digging down through these strata on the New York side would not be easy. But before that challenge could be tackled, the caissons had to be built. The Brooklyn caisson was the first to be constructed; it took two and a half months of intense labor. Assistant Engineer William H. Paine reported to Washington Roebling on the cumulative labor on the Brooklyn caisson for the week of November 20, 1869: the Webb & Bell shipyard had put 231 days of man labor, 18 days of horse labor, and 18 days of superintendence into the caisson—roughly the equivalent of 39 laborers, 3 horses, and 3 supervisors working a six-day workweek. In that same note to Washington, Paine reported that work was proceeding apace: "The first course of timber had been completed and the second course commenced."

On March 19, 1870, with the work on the Brooklyn caisson done, about three thousand people gathered at the Webb & Bell shipyard to watch its launch. Speeches were given and a band played suitably martial music. Given the daunting size and weight of the caisson, many in the crowd doubted that they were about to witness a successful launch. The caisson had been built to be launched much the way that a new ship, at the time, would have been slid into the water. When the last block holding the caisson in place was split out, the massive and strange-looking wood and iron box began to slowly slide down the slipways into the East River, just as planned. The air trapped below the giant hollow "box" kept it afloat, and a pump on its top pushed air into the empty chambers below, forcing water out and keeping the caisson buoyant. Some air was then allowed to escape so that the caisson floated low in the East River, with just seventeen inches of it visible above the water, exactly as Washington had planned.

The day of the launch, Washington was full of praise for shipbuilders Webb & Bell: "The whole of the launching arrangements, as well as the responsibility of the entire launch, rested with the builders, Messrs. Webb & Bell, who deserve the greatest praise for the successful manner in which they carried out so novel a work. They accomplished the result by simple common arrangements; no money was wasted upon elaborate precautions or fancied contingencies." Coming from the ever-practical Washington Roebling, "simple common arrangements" was a high compliment indeed.

While the Brooklyn caisson was being launched, workers prepared a rectangular basin to accommodate it at its final location on the Brooklyn side of the East River, just north of the Fulton Ferry terminal. As Washington detailed in his *Pneumatic Tower Foundations*, first the ferry slip there had to be removed, as well as wooden sheathing, heavy cribbing filled with stone, and wooden piles (or at least what was left of the piles; sea worms had eaten almost completely through them "between low water and the mud line"). As the basin was dug, mud was removed and a "reef" of boulders was blasted with small charges that totaled thirteen thousand pounds of powder. Gneiss, trap, and sandstone were broken up and hauled away.

Once the basin had been dug down to eighteen feet below high tide, it was ready for the caisson. The six compressors and the steam boilers that powered them were tested, and on May 1, 1870, the caisson, having had two more courses of timber added to its roof after its launch, was floated down from Greenpoint, guided by six tugboats under the direction of a Captain Maginn, to a spot just north of where it would ultimately rest. Finally, on the

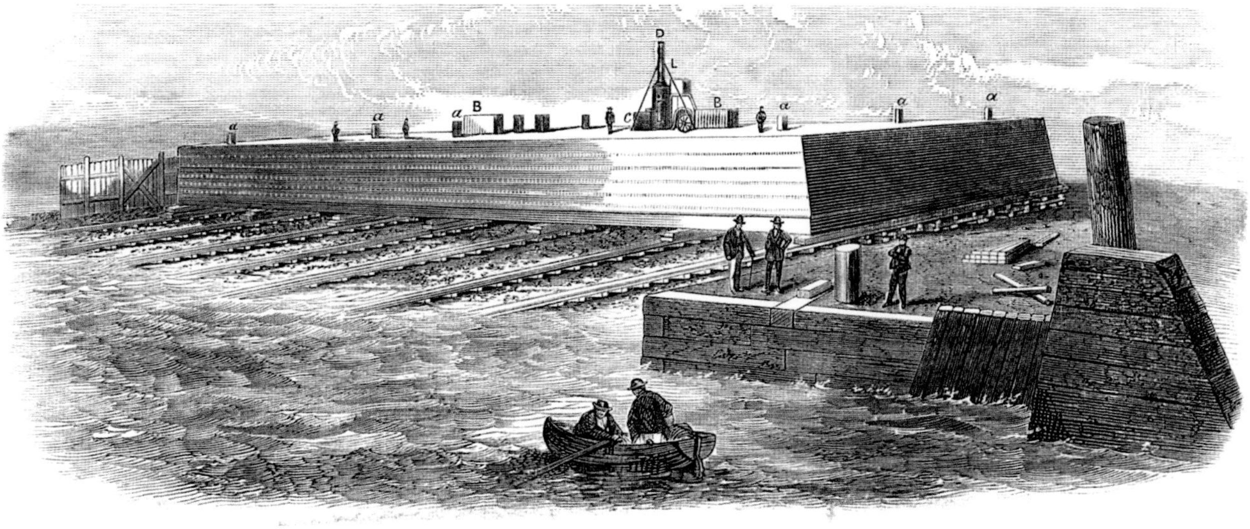

Fig. 2

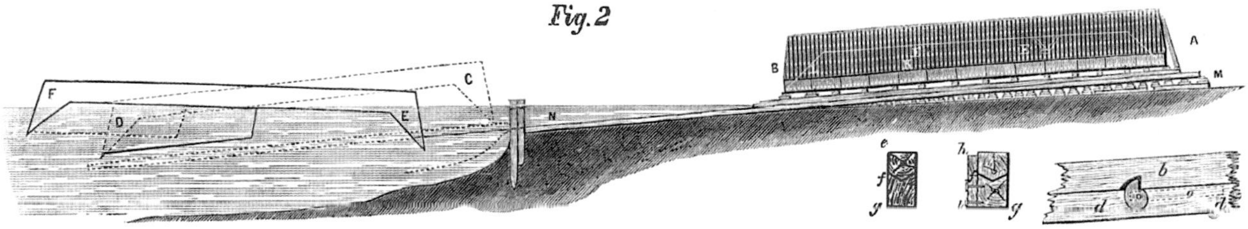

A woodcut published in *Scientific American*, July 9, 1870. The top image captures the launching of the Brooklyn caisson at the Webb & Bell shipyard in Greenpoint, Brooklyn. Below (Fig. 2) is a side view of the slipway—the wooden platform on which the caisson would slide into the water—with the East River at left. Once the blocks holding the caisson in place were removed, gravity sent it down the slipway in the manner of launching a ship. Workers greased the skids to make the largest structure ever launched at the time slide more easily into the river. Author's collection.

morning of May 4, the Brooklyn caisson was positioned in its basin; it took a little more than an hour for workers to get it placed exactly right.

A week later, Washington Roebling and Assistant Engineers William H. Paine and Francis Collingwood went inside the Brooklyn caisson to inspect it. In the ensuing weeks, the temporary air compartment that had helped keep the caisson afloat was removed, and further modifications were made to prepare the chamber for the workers. Ten more timber courses, or tiers, were added to the roof, making it fifteen feet thick, strong enough to support and help distribute the weight of the limestone and granite blocks that would be stacked atop it. Those stone blocks would weigh the caisson down and form the base of the Brooklyn tower. Openings for the two air locks, as well as for the supply, air, and water shafts, were cut through the timber roof, and the shafts were installed. Spaces between the timber courses were filled with concrete, and doorways were cut through the interior partitions that divided the work space into six sections.

Digging the caisson down was a fairly straightforward operation. Washington Roebling had custom designed much of the necessary equipment, including the hoists, derricks, clamshell scoops, and rock drills. Whatever was below the shoe of the caisson—gravel, sand,

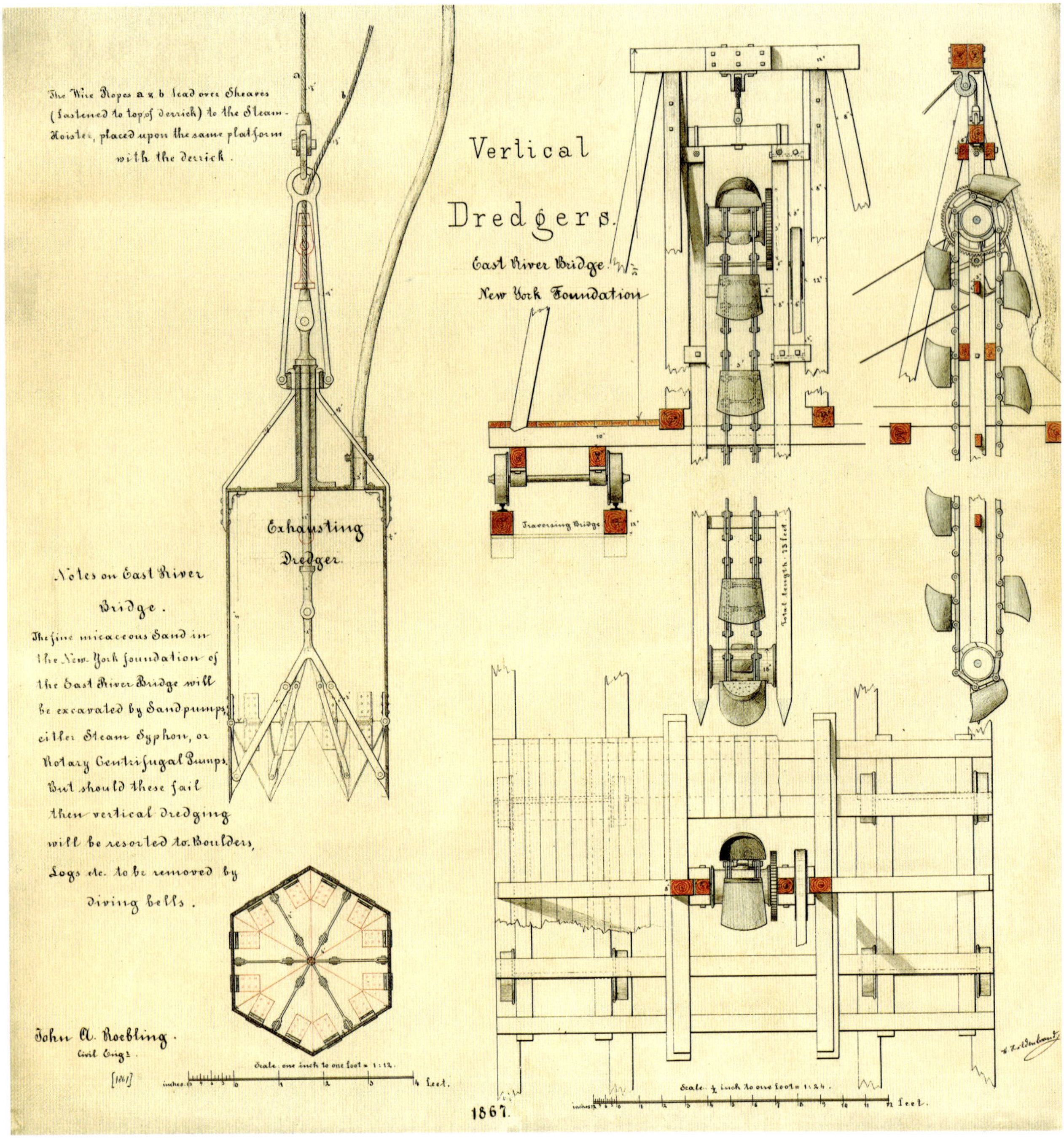

Chief Engineer John Roebling's design for "Vertical Dredgers/East River Bridge/New York Foundation," drawn and signed by Assistant Engineer Wilhelm Hildenbrand and dated 1867. Note that one of the mechanisms is labeled an "exhausting dredger"—terminology that was soon abandoned. Its design varies significantly from the clamshell bucket, or scoop, that Washington Roebling ultimately designed and employed. RPI.

mud, rocks, or boulders—had to be removed; its entire 1,050-linear-foot perimeter had to be clear before the box could slowly and steadily descend to its final depth. Throughout the process, the caisson also had to be kept level without straying from its assigned position, so that the bridge towers ultimately stood exactly where the engineers planned.

The men inside the caisson used shovels, wheelbarrows, sledgehammers, drills, and hydraulic jacks to do their work. Clamshell scoops, after descending through the water shafts, lifted the debris up and out of the caisson. At the same time, courses of limestone blocks were being laid atop the caisson. The weight of these blocks helped steady and push the caisson down toward bedrock.

Assistant Engineer George McNulty was in charge of supervising the placement of the two- to three-thousand-pound limestone blocks atop the Brooklyn caisson, forming the lowest section of the Brooklyn tower (granite was only used higher up in the tower). Three steam-powered derricks were mounted atop the caisson, each with a fifty-foot wooden mast. One of the three masts was always in position to swing a stone block into place. Double hoisting machines were also set up atop the caissons. These hoisting machines performed two tasks. They helped unload stones from one of the four scows—huge flat-bottomed boats Washington had designed to shuttle the tower stones from the Red Hook stone yard. He specified that each scow have a forty-five-foot-long deck, be twenty-four feet wide and six feet high. The hoisting machines, once they had removed the stones from the scows, hauled them up atop the caisson. From there the stones were placed in their assigned position to form the base of the bridge tower. By late June, even before the workers had excavated the first bucketful of debris, three courses of stone had already been positioned atop the caisson. Their weight grounded the caisson and stopped it from bobbing in the water as the river flowed around it. With the compressed air inside the caisson keeping most of the river and underground stream water from leaking underneath the caisson's iron shoe, the digging could begin.

On July 5, 1870, the clamshell buckets—carefully designed by Washington—went into operation for the first time, scooping up rock and mud from the pools below the water shafts, then lifting the debris through the water shaft and depositing it onto carts atop the caisson. Once fully loaded, the carts were moved by steam engine along inclined tracks to the water's edge, where their contents were dumped into scows floating in the river and taken away. If no scow was positioned, the debris was simply dumped into the river, from which dredges would later remove it and load it onto boats for disposal. Some of the debris was used to fill the open area created around the caisson as it descended. Other debris, especially the larger, broken-up boulders, remained on site until the digging was complete, at which point it was used, along with concrete, to fill the caisson.

Five clamshell buckets were on site. However, since there were but two water shafts in the caisson, only two clamshells, at most, were in operation at any given time. The other three were often in the process of being repaired. Each clamshell bucket had impressive, seven-inch-long teeth, but these broke off whenever the buckets got caught under the lower edge of the water shaft—which was often. The stones among the debris also damaged the bucket's teeth. Washington wrote of his frustration, "There was indeed one period when we were almost tempted to throw the buckets overboard." In a report to the bridge trustees, he detailed the challenge: "The buckets were provided with heavy teeth, seven inches long, rivetted [*sic*] fast to steel cutting edges, six by one.

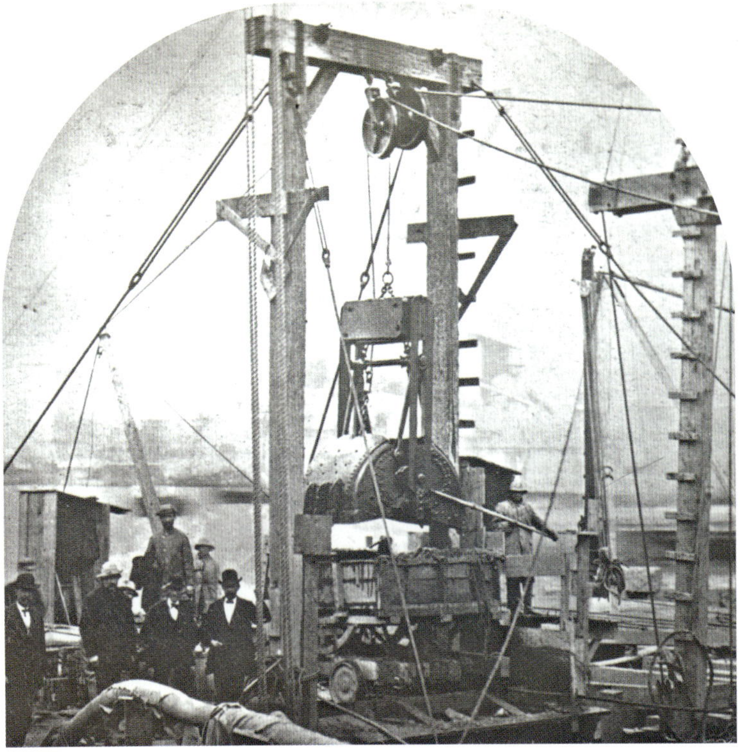

ABOVE: Detail of a lantern slide photograph of a clamshell bucket, used to remove debris as workers in the caissons dug it down to bedrock, as well as the pulleys, ropes, and wood framing that supported the dredge. A cart waits on tracks directly below the open clamshell to transport the debris away from the site. In his "First Annual Report of the Chief Engineer and General Superintendent of the East River Bridge," dated June 12, 1870, Washington Roebling observed that a clamshell scoop was "the only known instrument which possesses the precise action of the human hand in picking up things." Author's collection.

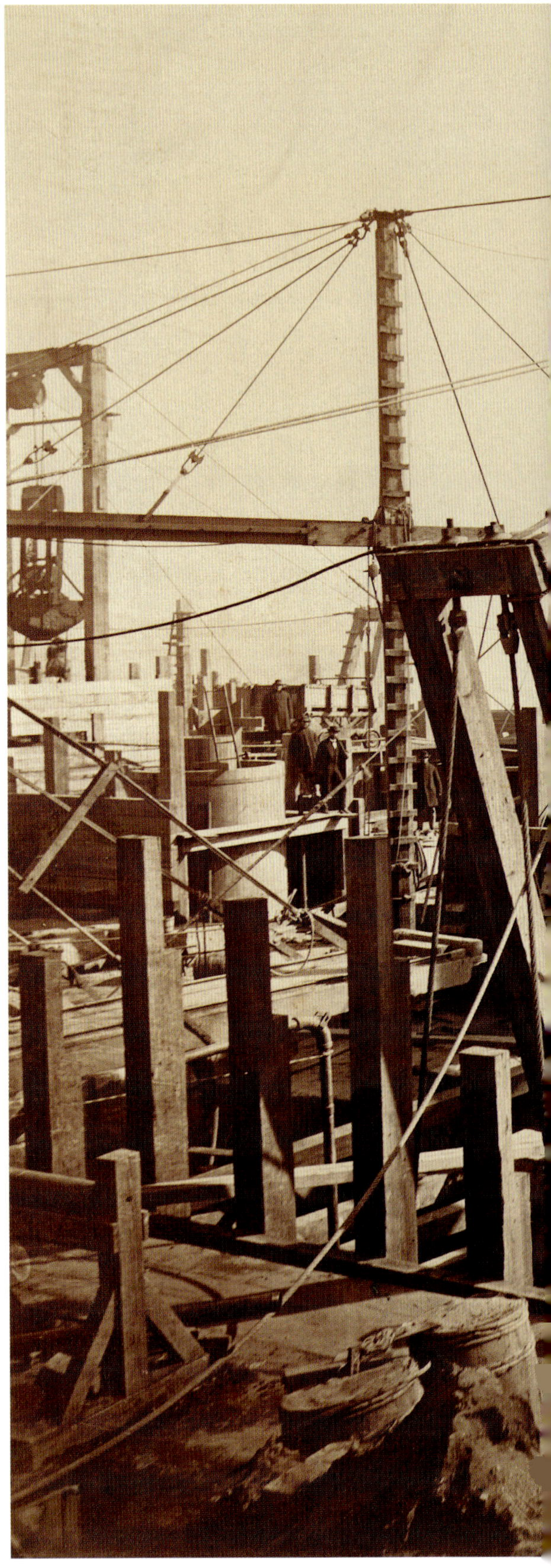

OPPOSITE: "March 1872/N.Y. Caisson Nearly Down," photograph by Silas A. Holmes, showing two of the three large (inverted L-shaped) mast derricks that sat atop the caisson. Their horizontal booms hauled the stone blocks for the tower into place. Also visible are the two clamshell buckets—suspended from wooden frames at far left and middle—and the cylindrical workers' entrance below the derrick at left. Note the worker seated above the clamshell bucket at center, with his feet dangling; it was his job to look down the water shaft and use hand signals to tell the clamshell operator when to lower, raise, and stop the bucket. MCNY.

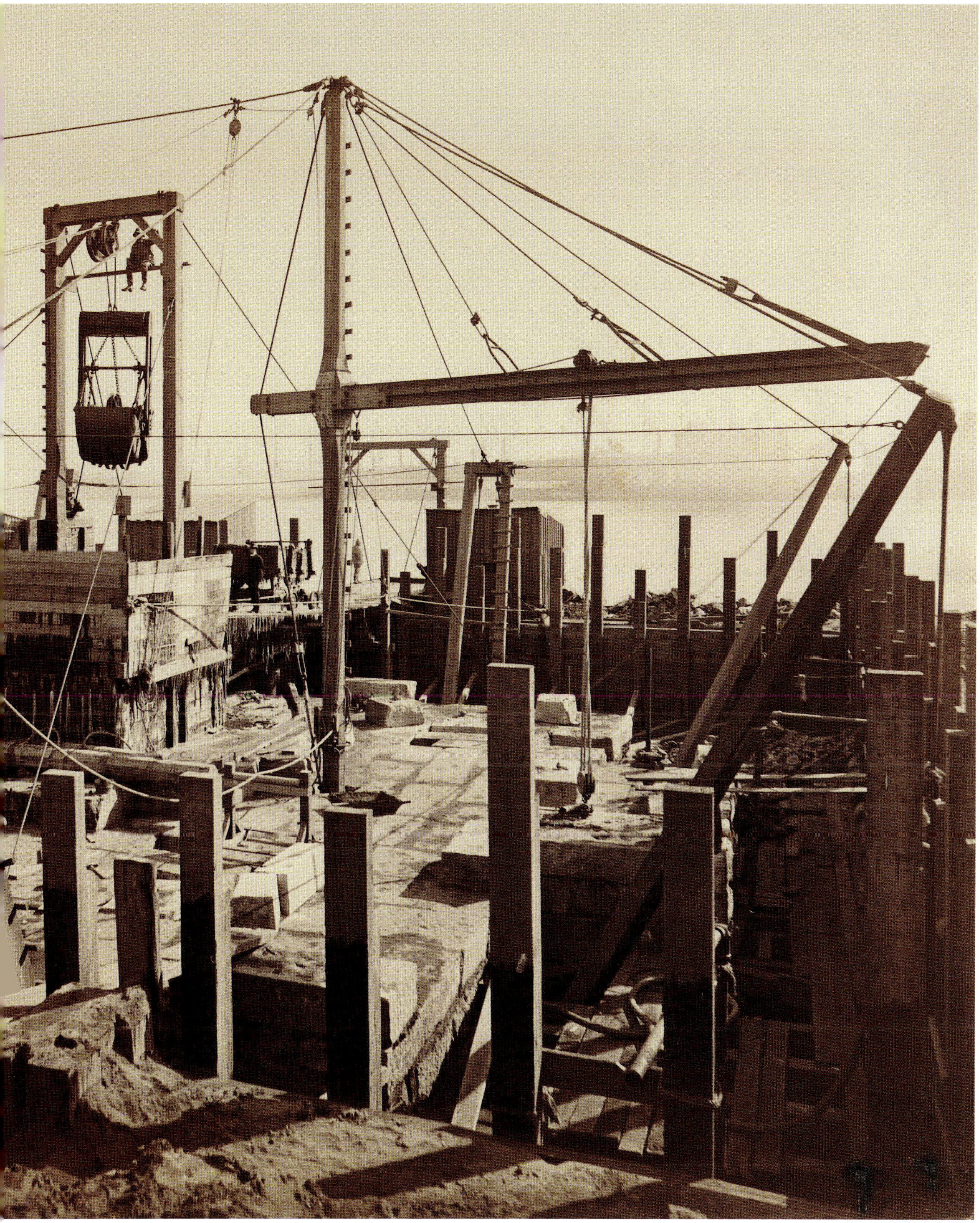

Many patterns of teeth were tested before the most advantageous layout was arrived at. A tooth that answers well for scooping up mud, will not last a day for grappling stones. A supply of five buckets was required to keep two in working order." Ultimately, the clamshell buckets worked well, performing their essential task for both the Brooklyn and New York caissons.

Given the tremendous square footage of the caissons, different ground conditions were often found from one work chamber to another. Men in one might be digging through clay, while in another, workers had to deal with clay and gravel among large rocks that had to be broken up. Inside the caissons, workers first dug directly below their feet, removing all debris before directing their efforts to clearing the area along the perimeter to allow the iron shoe to inch downward. This left the partition walls "hanging," unsupported in the air. As the work proceeded, more weight was being stacked atop the caisson, and it became necessary to find a means of supporting the interior partition walls. Washington wrote that they initially tried "small pillars of earth," about three feet square. But the immense downward pressure of the growing stack of limestone blocks atop the caisson caused the pillars of earth to collapse. The workers then tried wooden blocks and oak wedges to support the partitions. This worked quite well, and the blocks could easily be hammered or split out when the time came to lower the partitions to be level with the iron shoe. Washington commented: "The noise made by splitting blocks and posts was rather ominous and inclined to make the reflecting mind nervous in view of the impending mass of thirty thousand tons overhead."

Washington had designed the caisson's shoe as an iron cutting edge, so that in addition to protecting the wooden edges of the caisson, it would make the digging easier by cutting into the ground. However, it soon became apparent that this cutting edge was not doing any cutting. The resistance against the shoe was just too great. Further complicating matters, rocks got wedged underneath it, and workers had to remove them one by one. The men used heavy hydraulic jacks to pull the largest boulders out from under the shoe. When some of these proved too large to be lifted out by the clamshell buckets, the workers broke them into manageable chunks using sledgehammers and iron bars. Given the large volume of boulders, the engineers assigned an overnight crew to do nothing but pull and smash them.

Day and night, workers had to probe blindly for rocks and boulders in pools of water at their feet, formed by river water that continually seeped in under the shoe and into the work area, despite the compressed air. However, once the Brooklyn caisson was sunk ten feet below the riverbed, the water seepage stopped, making the work a little easier. That is, until the caisson reached twenty feet down, where the boulders were as large as fourteen feet by five feet and presented great challenges for the men.

Washington considered blasting the huge rocks. He was reluctant, though, not knowing the consequences of explosions inside a compressed air caisson. So, as an engineer, he did what engineers are trained to do—proceed carefully and experiment. Inside the caisson, the chief engineer conducted tests, systematically firing a gun with ever-larger black powder charges. Observing no negative result, Washington ordered the workers to begin blasting the boulders. This proved very effective. As Master Mechanic E. F. Farrington wrote, "Blasting now became the order of the day, or more correctly, of the night, for most of it was done then." This new technique sped up the descent of the caisson.

TOP: "Workmen Sawing Timber for Wedges," *Frank Leslie's Illustrated Newspaper*, October 15, 1870. This dramatically lit woodcut depicts workers sawing wood into wedges to be placed beneath the caissons' interior partition walls to temporarily support them. Initially, mounds of soil were used as supports, but they collapsed under the massive weight pressing down on them, leading to the use of sturdy oak wedges. Once diggers cleared a caisson's outer perimeter—its iron shoe—and the exterior walls descended, the men knocked away the wooden wedges, efficiently allowing the interior partitions to drop down to the same level as the shoe. Courtesy of Richard Haw.

BOTTOM: "Drilling the Rock at the Shoe of the Caisson," a woodcut from *Frank Leslie's Illustrated Newspaper*, October 15, 1870. When the workmen digging in the caisson encountered boulders, they had to break them apart—either by splitting them with sledgehammers and iron drills or bars or, later, by blasting them with black powder. Courtesy of Richard Haw.

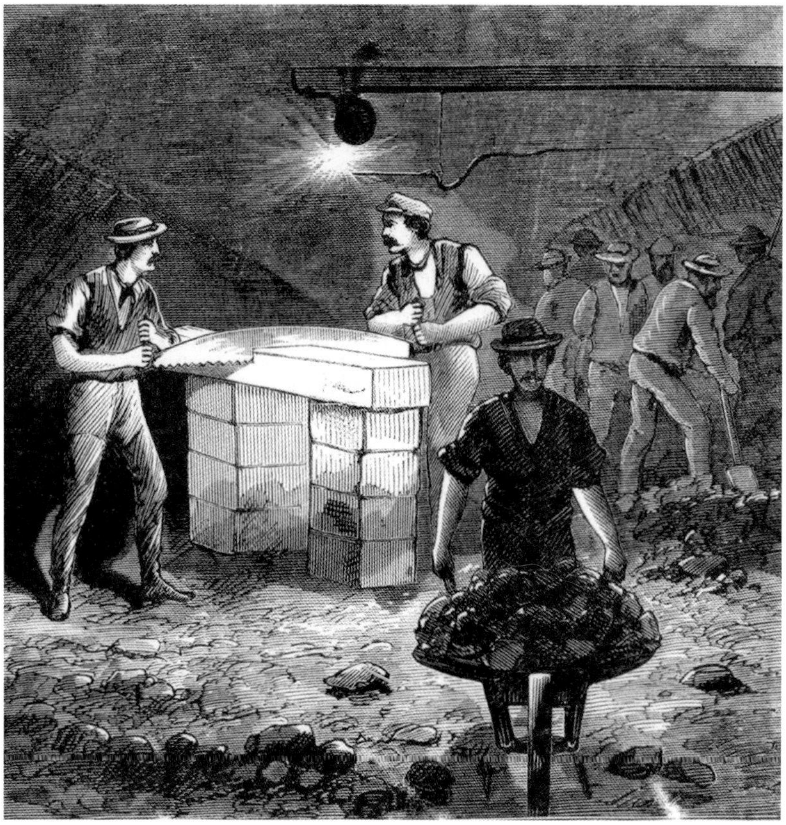

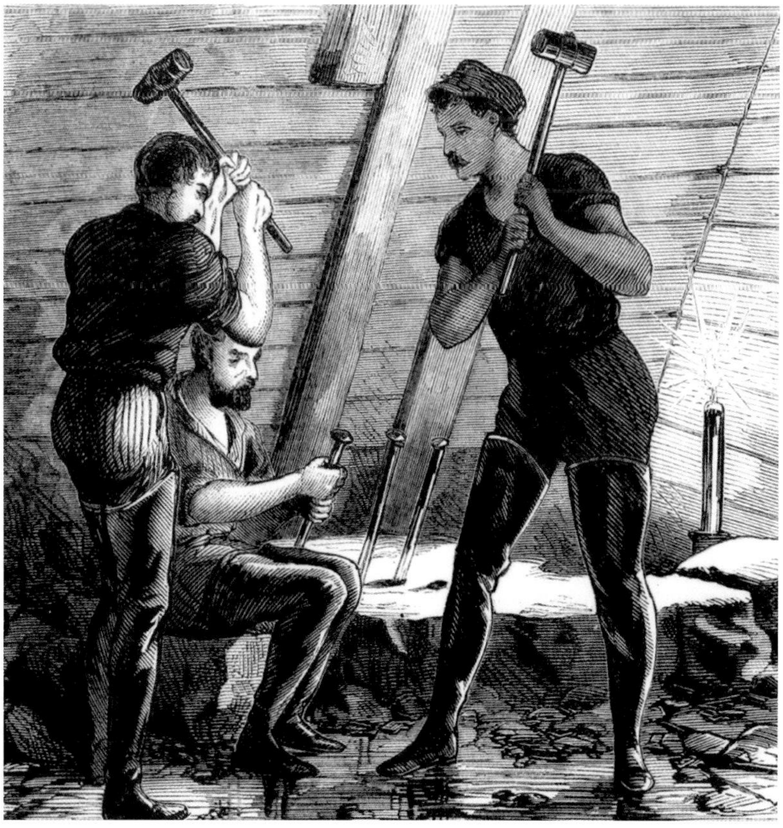

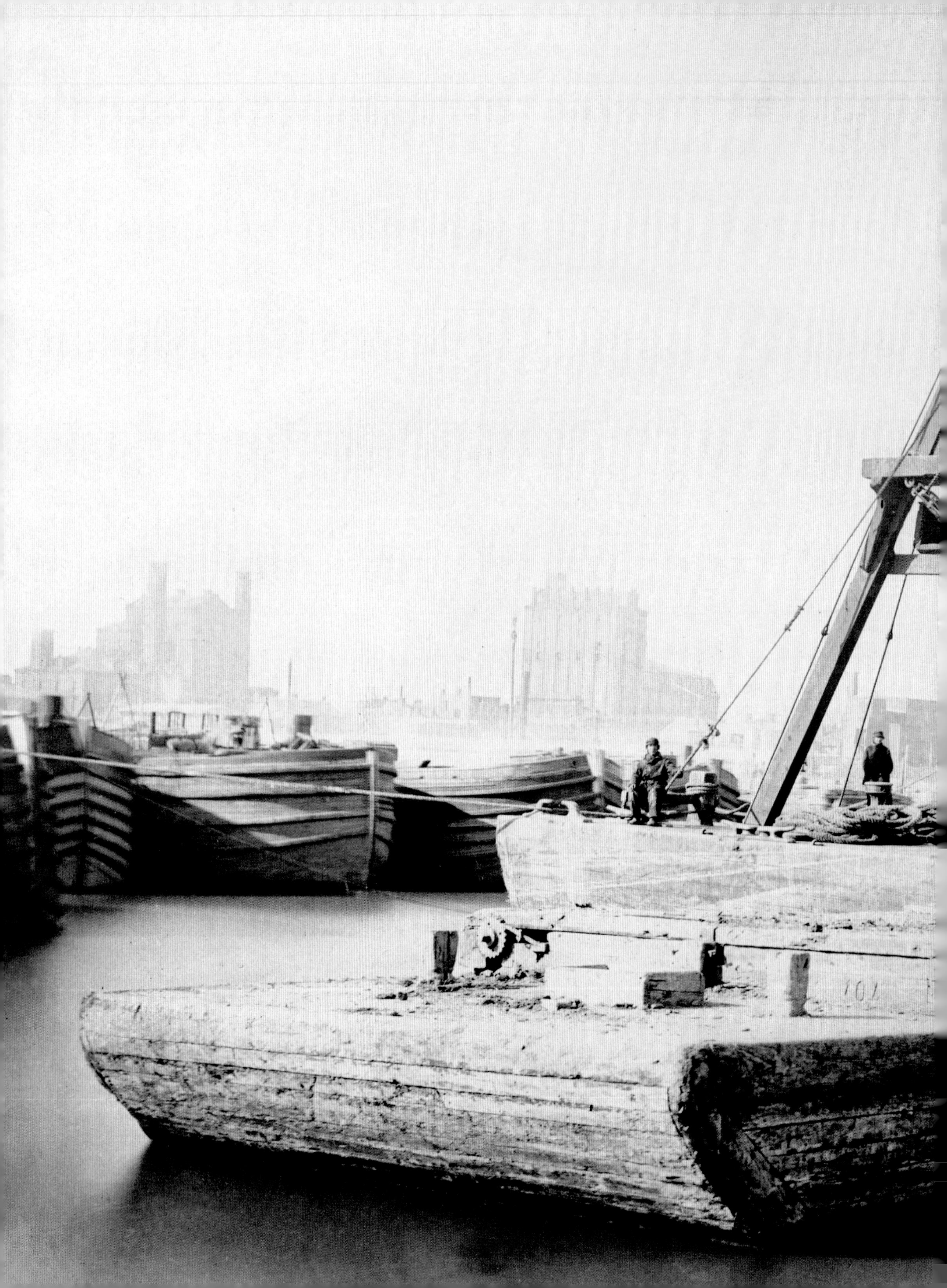

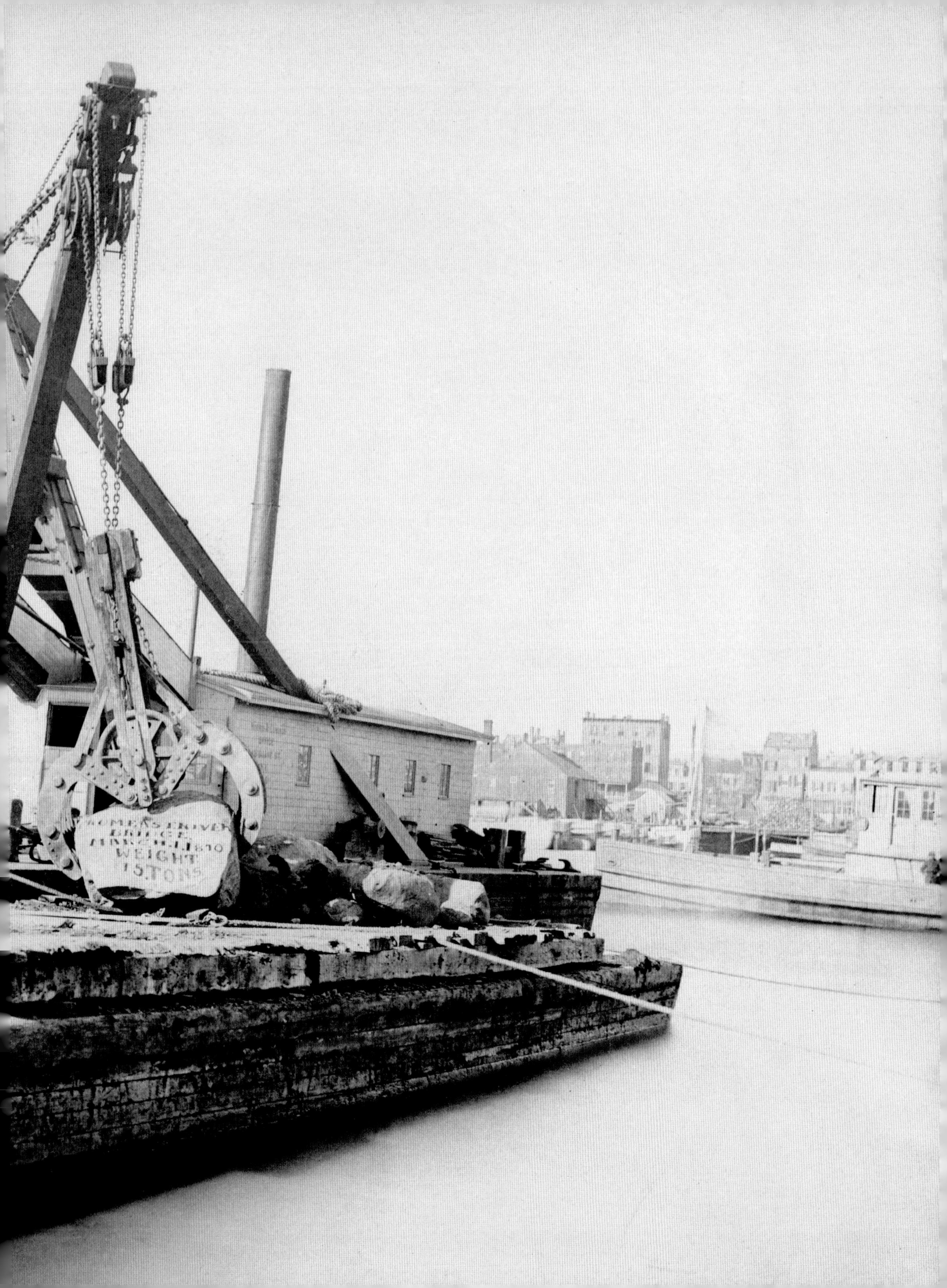
WEIGHT
15 TONS

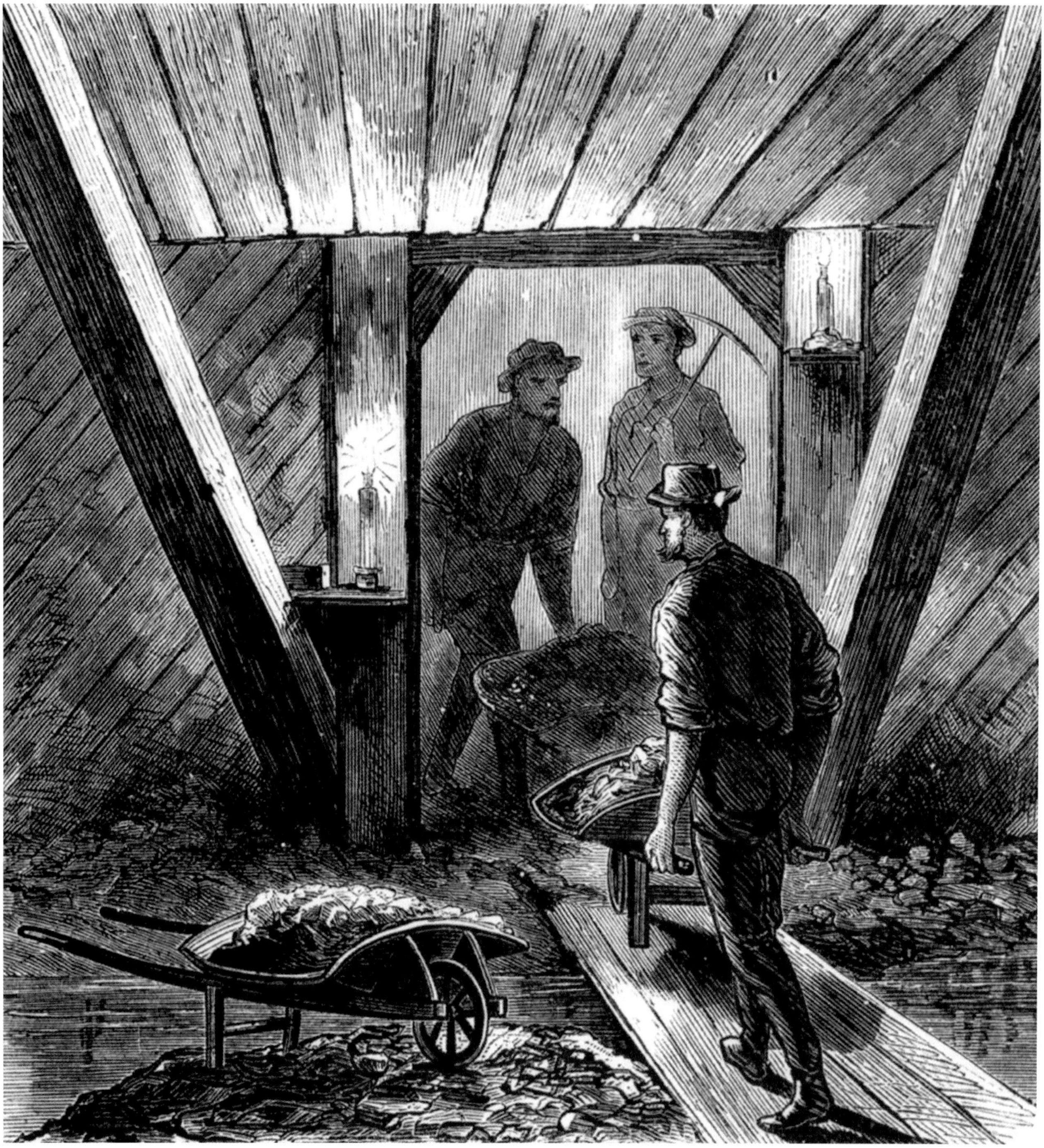

"Door Through Partition, Showing Different Apartments in the Caisson," *Frank Leslie's Illustrated Newspaper*, October 15, 1870. Partition walls, one of Washington Roebling's modifications to his father's plans, divided the large open space within each caisson into separate chambers. These walls helped bear the weight of the caisson and the growing downward pressure as stone blocks were piled on top of it to form the bridge tower. Courtesy of Richard Haw.

PREVIOUS SPREAD: A large boulder photographed on a wooden barge, likely in New Jersey, after being removed during the excavation around the Brooklyn caisson. Workers painted large black lettering across the face of the stone: "From East River Bridge. March 1. 1870. Weight 15. tons." Captured for posterity in a photograph by F. L. Huff of Newark, New Jersey. RPI.

Daily progress jumped from a modest six inches per week to twelve or even eighteen inches in the weeks after.

Master Mechanic E. F. Farrington described what it was like working down below:

> Inside the caisson everything wore an unreal, weird appearance. There was a confused sensation in the head, like "the rush of many waters." The pulse was at first accelerated, then sometimes fell below the normal rate. The voice sounded faint, unnatural, and it became a great effort to speak. What with the flaming lights, the deep shadows, the confusing noise of the hammers, drills, and chains, the half-naked forms flitting about, with here and there a Sisyphus rolling his stone, one might, if of a poetic temperament, get a realizing sense of Dante's inferno. One thing to me was noticeable—time passed quickly in the caisson.

The compressed air in the caisson also diminished the workers' sense of smell, which Washington considered a benefit: "This is a wise provision of nature, because foul odors certainly have their home in a caisson."

Given that the workers spent their entire shift down in the caisson, it was necessary to supply toilets there. In *The Great Bridge*, David McCullough noted the rather unusual one they used:

> It was described in one of Roebling's official reports as a pneumatic water closet and consisted of a wooden box with a lid and a large iron pipe that passed up through the timber roof. The box was kept about half full of water, and whenever its contents were to be discharged, a valve was opened and the pressure from within the caisson would blast everything instantly overhead in the form of a fine mist.

The walls and ceilings of the work chambers were invariably coated with mud. The workers wore rubber boots and moved about on wooden planks to stay above the mud as much as possible. The natural light that came into the caisson through six bull's-eye glass skylights was supplemented by gas lighting, candles, and calcium lights. As Washington reported, "The lighting is done by gas principally, although a double set of pipes has been put in for sixteen calcium lights. Sixty double gas burners have been provided." It was naturally hot and humid inside the caisson, and with all the lighting, temperatures that started at eighty degrees only got hotter as the workday progressed.

Also, given the prevalence of open flames in the chambers, the danger of fire was ever-present. After all, the caissons were made primarily of timber, with highly flammable oakum used as caulking. The compressed air increased the danger—if a fire reached the timber ceiling, the pressure from the compressed air would quickly push the flames up through the many wood courses of the roof.

Early on, there had been several small, inconsequential fires in the Brooklyn caisson. Then, on December 2, 1870, with the caisson almost at bedrock, a large fire, caused by a candle held up near the timbers too long, swept through the caisson roof. Washington Roebling and Foreman Charles Young rushed in to lead the efforts to subdue the fire. They emerged from the caisson before it was contained and became two more victims of the bends, or caisson sickness, as a result of their too-rapid decompression. Suffering from paralysis, they both had to be helped out of the air lock.

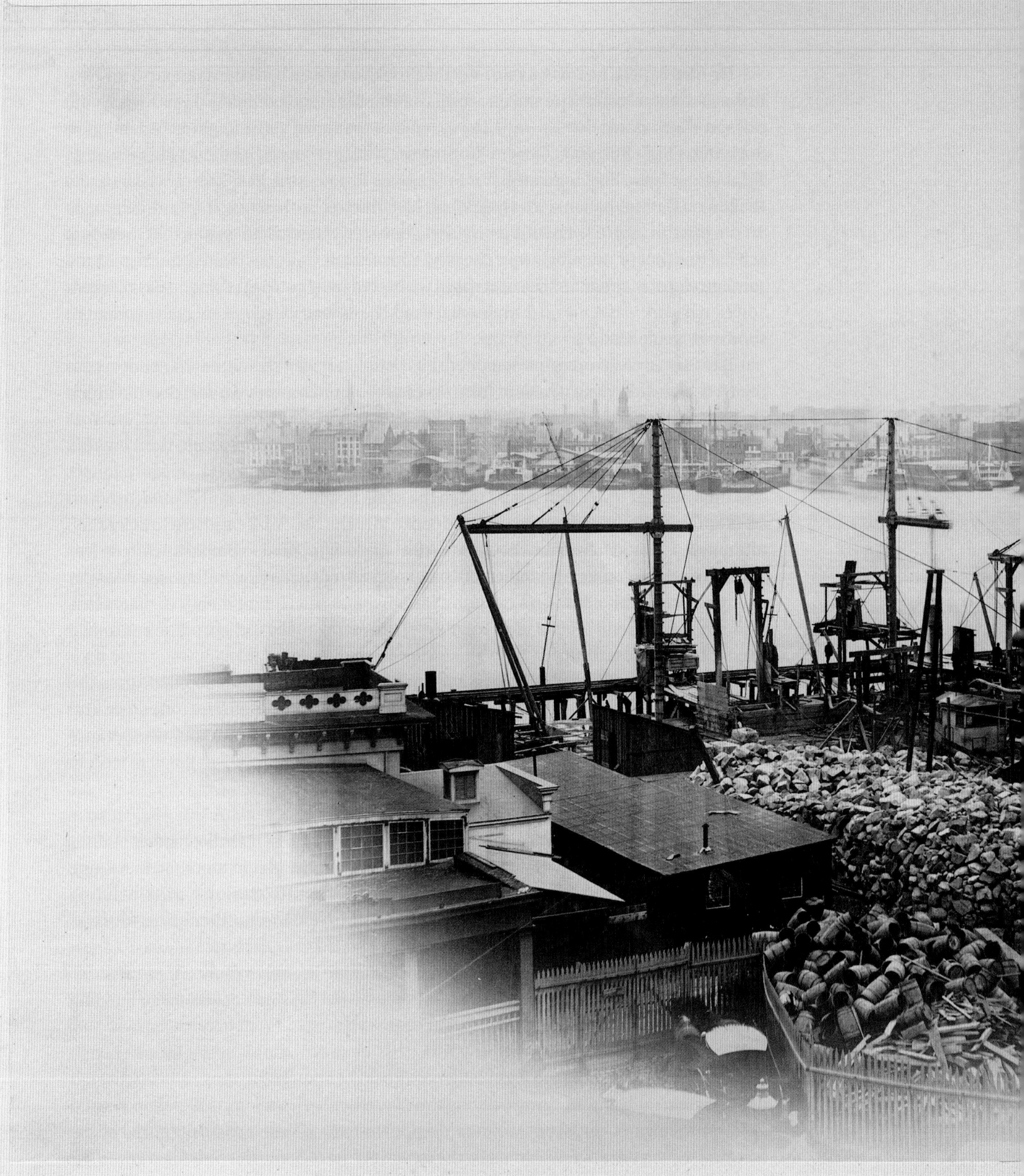

Talfor, Photo.

WORK IN PROGRESS ON THE BROO

East River B

Dec. 1870.

W. A. ROEBLING, ENG

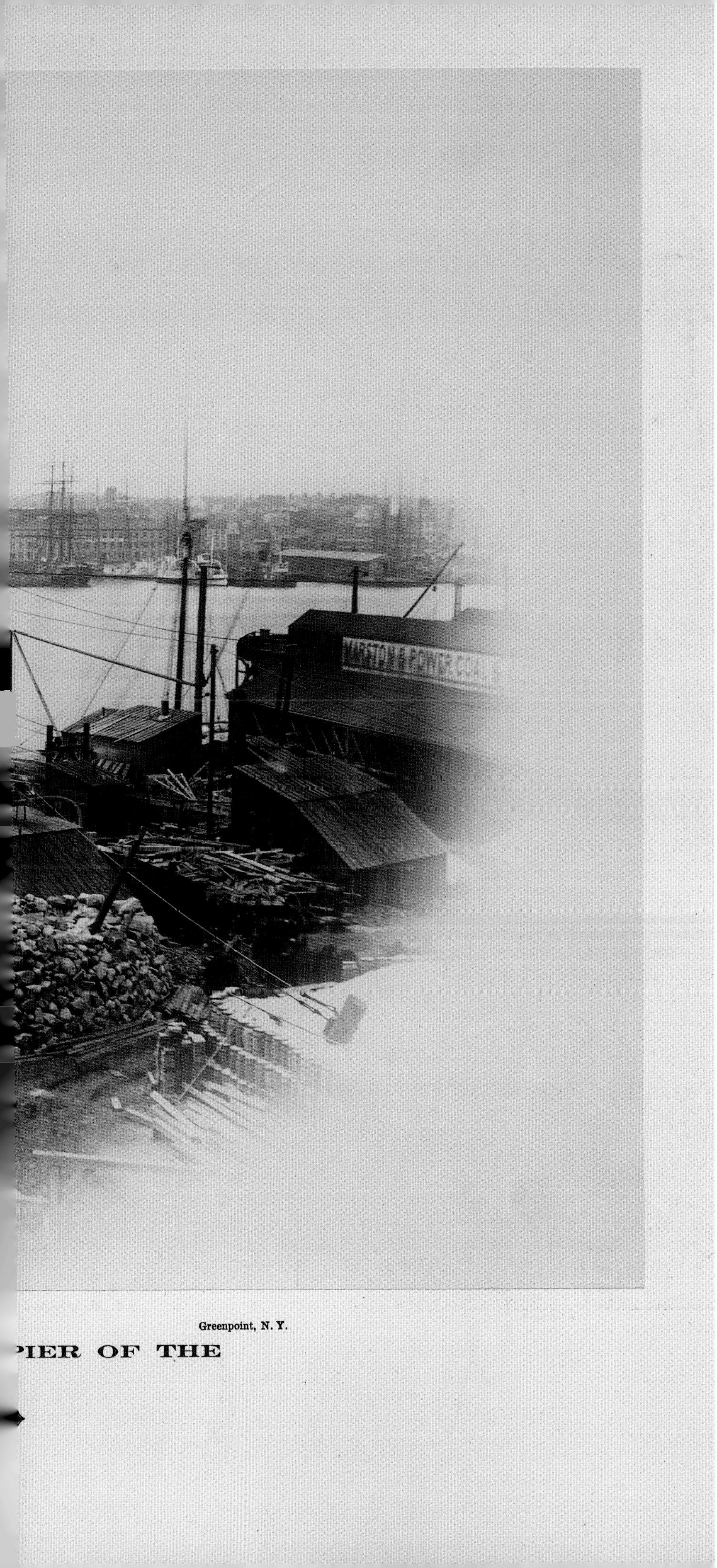

"Talfor, Photo./Greenpoint, N.Y./ Work in Progress on the Brooklyn Pier of the/East River Bridge/Dec. 1870/W. A. Roebling, Eng'r." This photograph shows mast derricks and the two rectangular wooden frames (each has a pulley hanging from it) for the clamshell buckets. While the clamshell bucket is visible in the frame at right, the cables dangling from the frame at left indicate that a missing bucket is likely being repaired—a near-constant occurrence with the clamshell scoops. Seen at the top of the photograph is the East River and the Manhattan shoreline, where both large-masted ships and ferries are docked. At lower left are empty supply barrels stacked next to rock rubble that has been pulled from the caisson. The rubble is either awaiting removal or being stockpiled until the work inside the caisson is complete, at which point it will be used to fill the caisson's interior chambers. RPI.

Progress ensued at a great rate. Typically, the descent toward bedrock was two feet a week, compared to the average half foot on the Brooklyn side (before boulder blasting had been used there). For several weeks, the New York caisson moved down a half foot to almost a foot *each day.* The caisson, the height of a four-story building, continued to descend into the earth, with the tower taking form as the limestone blocks continued to pile up. Washington reported enthusiastically, "The downward movement of the caisson has been under perfect control."

A reporter described the busy scene at the New York caisson:

> At the foot of Roosevelt Street, where the New York tower is being erected, one of the busiest scenes in the city is met with and has been for months—dozens of workmen hurrying hither and thither with wheelbarrows and hods and spikes and shovels; engines puffing away, lifting huge blocks of stone with huge derricks from the barge at the side of the dock, drawing lumber from the foot of the pier, driving the piles of the cofferdam, and condensing and compressing the air to be used by the submarine workmen; men chopping and planing and sawing the immense timbers used in constructing the enormous derricks; others shoveling gravel and sifting sand for the cement; little knots of threes driving immense piles through the heavy timbers of the caisson with their sledges and kept steadily at work by an overseer who evidently enjoys his employment; some wheeling cement for others to lay between the large granite blocks, boring and hammering and cutting stone and carrying iron rails, everything indicating that the work is being pushed rapidly forward.

As the New York caisson was dug down deeper, the pressure of the compressed air was increased to keep the ever-greater weight of water out of the work space. As the difference in pressure between the air outside and the pressurized chamber became greater, the men increasingly suffered when they left the caisson, ignorant of the importance of decompressing slowly. One worker described the condition as "a sort of convulsive fit that twisted one's body like a knot and often made you an invalid for life."

Frank Harris, a sixteen-year-old Irish immigrant, began working on the bridge just days after arriving in New York. In his sensational autobiography, *My Lives and Loves*, published decades later, in 1922, Harris described what it was like working inside the caisson. David McCullough quotes his account in *The Great Bridge*:

> My headaches soon became acute. The six of us were working naked to the waist in a small iron chamber with a temperature of about 80 degrees Fahrenheit: in five minutes the sweat was pouring from us, and all the while we were standing in icy water that was only kept from rising by the terrific pressure.... [One of the workers] assured me that I could stay as long as I liked, but he advised me to leave at the end of the month: it was too unhealthy....

McCullough questioned the accuracy of Harris's account: the young Irishman had incorrectly described the caisson as made of iron, not timber, and had gotten other facts wrong (based on comparing his claims to public records), including the hours men worked, the pay rate, and the time they spent in the caisson. Further, Harris's account was published more than fifty years after his experience, so his recollections may well have been clouded by the passage of time. Nevertheless, with firsthand reports by bridge workers so rare, it is worth considering.

In a single day of working on the bridge, Harris claimed to have made the equivalent of two weeks' wages at another job. But after five or six days of bridge work, as per Harris's account, he began to have awful shooting pains in his ears and feared he might lose his hearing. In order to treat him, the woman with whom he was boarding cut a roasted onion in half and tied each half to one of his ears. Harris reported that the pain quickly dissipated. But just a few days later, while he was in the caisson, a worker nearby collapsed and lay twisting on the ground as blood spurted from his mouth and nose. That was enough for young Harris; he quit.

In January 1872, three months after work inside the New York caisson had begun and with the structure already at a depth of fifty-one feet, Dr. Andrew H. Smith was hired to study the "caisson disease" that was increasingly inflicting the workers. From late January through the end of May, he experimented with various practices and treatments in the hopes of reducing the condition's debilitating impact on the workers. Dr. Smith had the men follow various procedures in an attempt to limit their agony—and the number of worker deaths—but nothing he tried succeeded.

As the men continued to fall ill, Washington Roebling carefully monitored the situation. He now faced one of the key decisions of the entire Brooklyn Bridge project: should he stop the digging and leave the New York caisson where it was, without having reached bedrock, or risk further deaths?

Based on the descent from sixty-eight to seventy-eight feet down, which had been "very tedious and slow," Washington knew that digging the caisson any deeper would be neither easy nor quick. More workers would no doubt die from caisson disease, and it would cost another half a million dollars and take another year. Or he could stop the digging and let the caisson rest on the compact sand and gravel where it was. This was not bedrock, but it was a solid footing.

On May 17, two more workers suffered severe leg pains after emerging from the caisson. One of them, a workman named Reardon, was in agony and retching. He was taken to a nearby hospital, where he died the next morning. Later that day, May 18, Washington ordered a stop to the digging. The New York caisson was seventy-eight feet, six inches below the high-water level of the East River. Washington concluded that the sand and gravel upon which the caisson rested, having not moved for thousands of years, was "good enough to found upon, or at any rate nearly as good as concrete that could be put in place of it." The height of the four corners of the stone courses already stacked atop the caisson was measured. There was only a three-quarter inch difference in their heights, and the caisson was deemed level and in its final position.

Two months later, on July 12, 1872, the work of filling the New York caisson with concrete and crushed boulders was finished. Unlike on the Brooklyn side, the New York caisson had taken many workers' lives. Despite this tragic loss, the two caissons, upon which "the stability of the entire work" rested, were firmly in place. It was now time to finish the towers, which by their unique, unmistakable, and sturdy design would come to define the Brooklyn Bridge in the eyes of everyone who saw them.

"New York Caisson, East River Bridge," photograph by Robert B. Talfor. This caisson, being built in Manhattan along the East River at East Sixth Street, is nearing completion. Printed on the mat, at lower left, are the dimensions of the caisson, and at lower right are the materials used to build it and their weights, as well as the launch date, "May, 1871." The photographer's name, Talfor, is directly below the image, at left, and opposite that, at right, is printed "Greenpoint, B'klyn., E.D.," indicating the location of Talfor's studio. That section of Brooklyn was then known as the Eastern District—thus, "E.D." For more on Robert B. Talfor, photographer, see the appendix. MCNY.

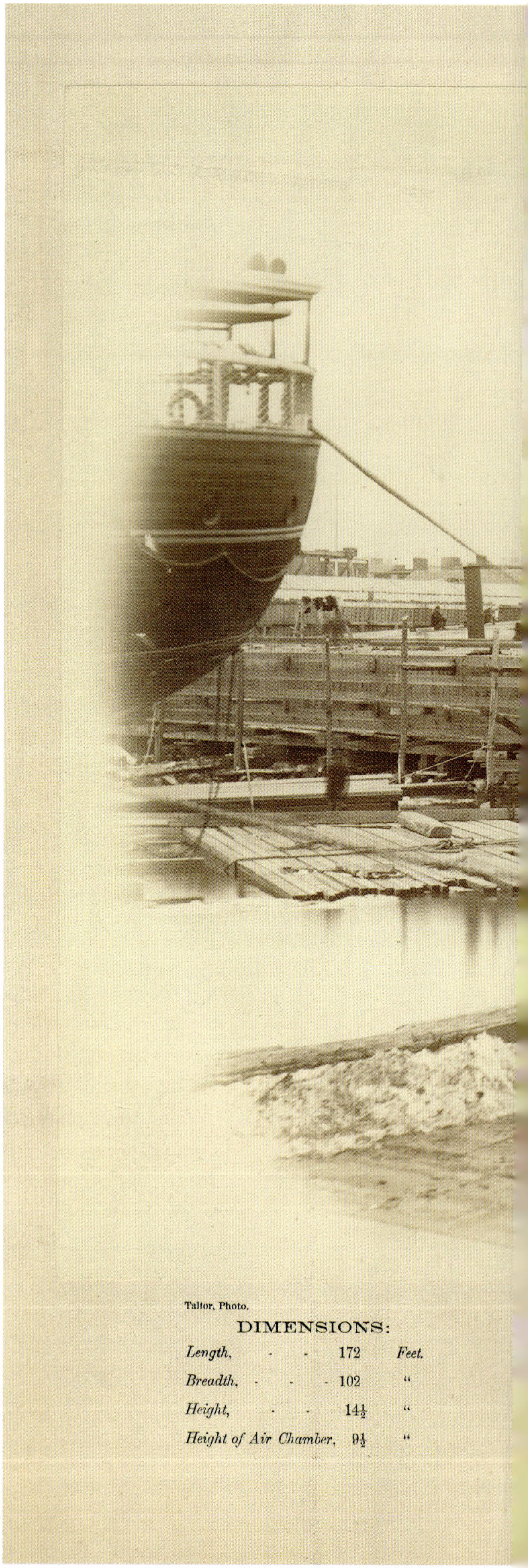

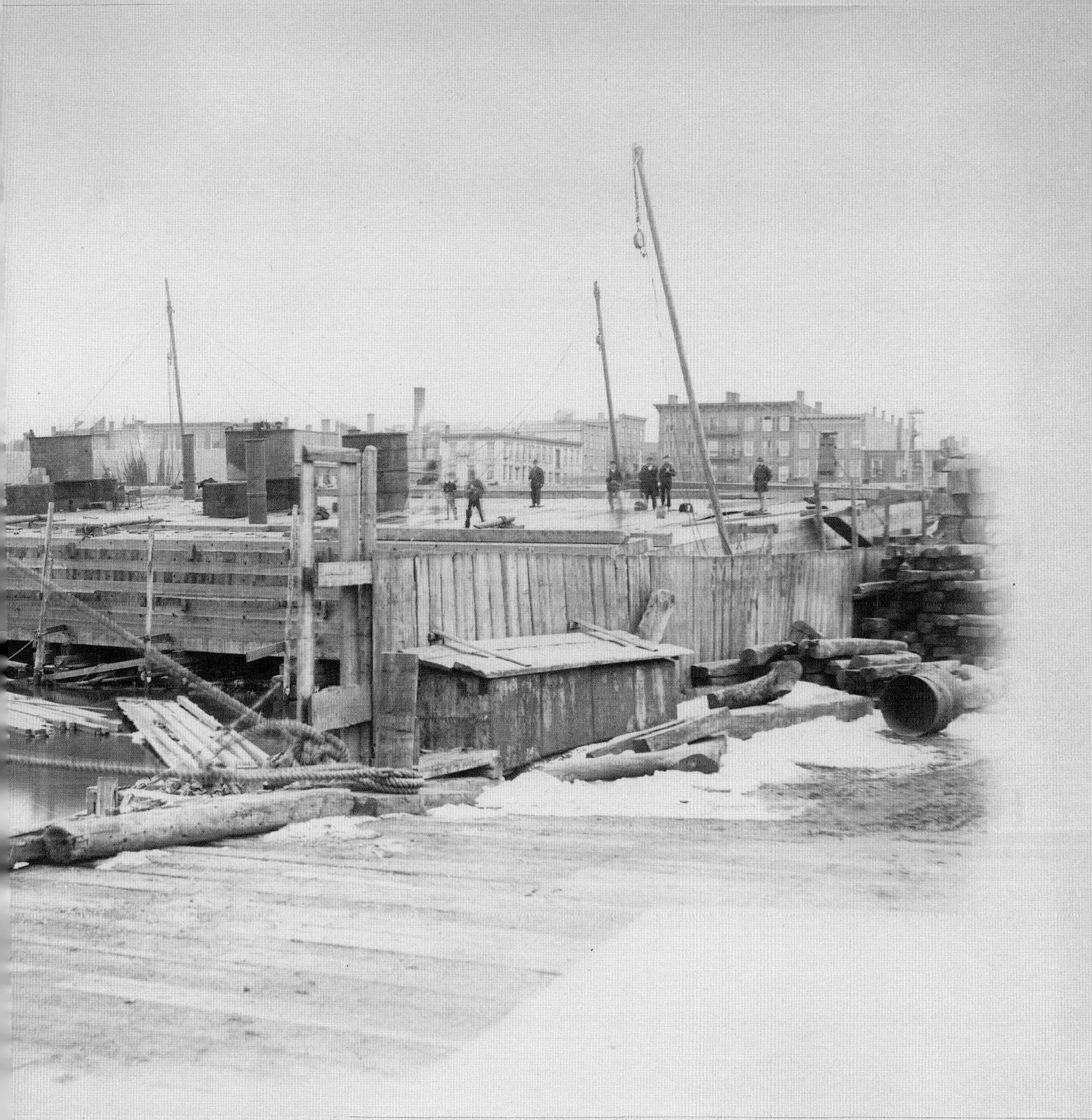

EW YORK CAISSON,

East River Bridge.

Eng'r. Built by WEBB & BELL.

Iron Work, by JOHN ROACH & SON, Morgan Iron Works.

1870

Greenpoint, B'klyn., E. D.

CONTENTS:

Cubic Feet Timber, - - -	125,000
Tons Iron, - - - -	320
Launching Weight, (Tons)	3,600
Launched, May, 1871.	

"Its most conspicuous features, the great towers, will serve as landmarks to the adjoining cities, and they will be entitled to be ranked as national monuments."

—CHIEF ENGINEER JOHN A. ROEBLING

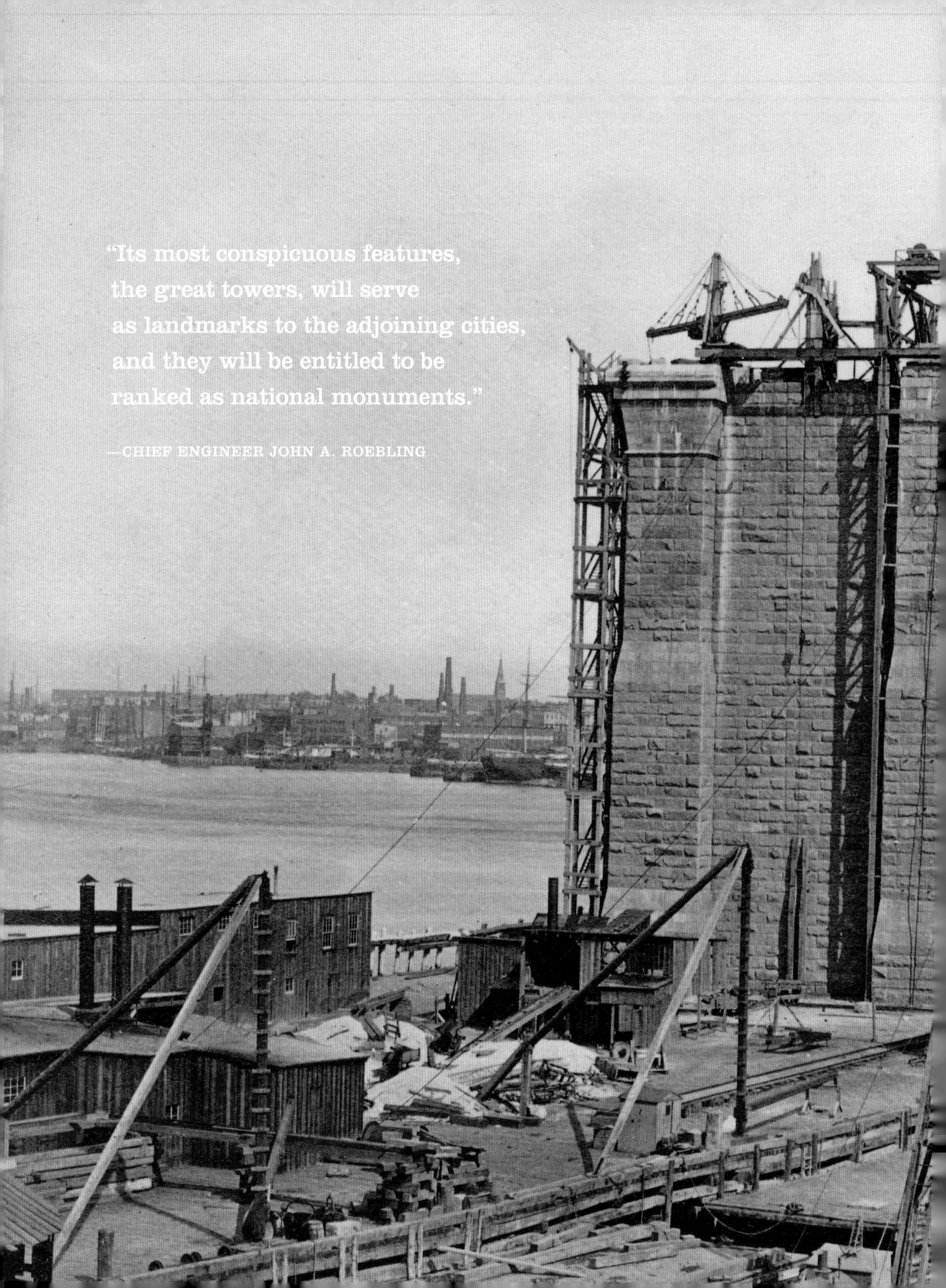

CHAPTER 4

Towers

1870–1876

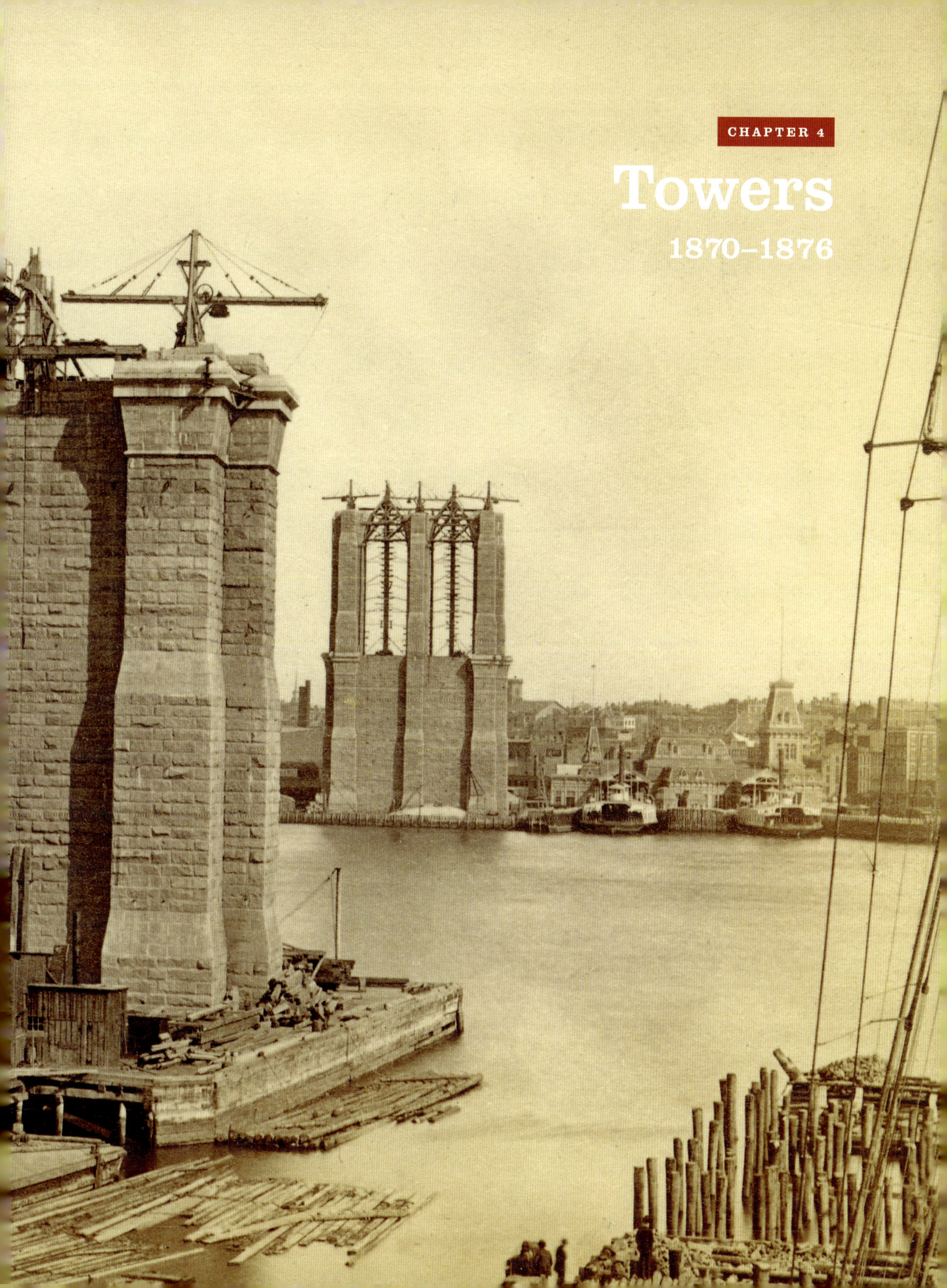

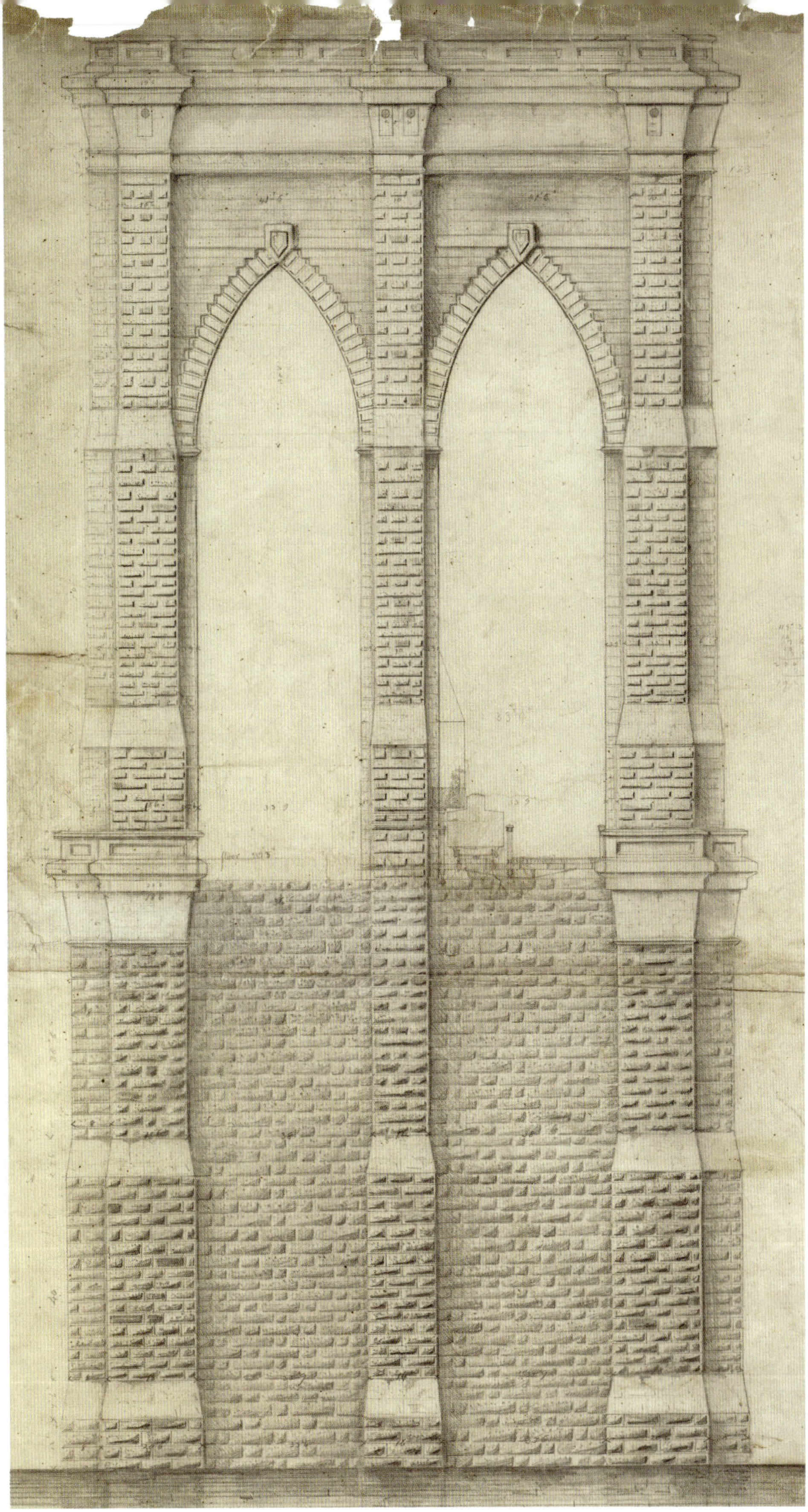

LEFT: Design for the Brooklyn Bridge towers, by John A. Roebling, ca. 1865. These towers, with their impressive Gothic Revival arches through which the roadbed would run, were intended as grand entrances to the cities of New York and Brooklyn. Note the penciled railway car at the bottom of the opening below the right arch. A woman, holding a parasol, stands on the pedestrian promenade, just above the car. NYC Municipal Archives.

PREVIOUS SPREAD: This cabinet card, ca. 1875, shows the bridge's New York tower at about half its ultimate height and, in the distance, the Brooklyn tower approaching completion. The Brooklyn tower was finished thirteen months before the one in New York. Courtesy of the Jeffrey Kraus Collection.

The towers of a suspension bridge are a defining design feature, but they also serve a crucial structural function. The towers support the massive cables from which the roadbed is suspended. To be able to bear the cables' great weight, the towers must be sturdy and strong, their dimensions dictated by the length of the bridge span, the width of the roadbed, the weight of the superstructure, and the anticipated force of local winds and traffic that would cross the bridge. The towers also play a critical part in determining how high a bridge span can hang. When a suspension bridge stretches over a navigable waterway, as does the Brooklyn Bridge, it needs to be high enough that it does not interfere with the river traffic below.

John Roebling wrote with foresight of the bridge he was designing (but sadly would not live to see): "The impression of the whole will be that of massiveness and strength." He designed twin Gothic Revival arches through which the bridge roadways would pass. He wrote, "In a work of such magnitude and located as it is between two great cities, good architectural proportions should be observed." Indeed, his towers brought majesty to the structure, solidity and elegance, creating a grand entrance to two of America's largest municipalities.

Thomas Kinsella, the editor of the *Brooklyn Eagle*, wrote early in the construction of the size and grandeur of the Brooklyn Bridge towers:

> Think of Trinity Church as big at the top of the steeple as at the ground and one solid mass all the way up, and we get some idea of what this great Brooklyn tower is to be. . . . The fame of the Roeblings and the boast of Brooklyn forever will be that, where Nature gave no facilities for a suspension bridge, and seemed indeed to place a veto upon the idea in these low and shelving shores, the genius of the father designed, and the consummate inherited and acquired ability of the son executed, in spite of all obstacles, this most novel and unparalleled masterpiece.

Kinsella continued, voicing his pride in the towers: "Even Europe has no structure of such magnitude as this will be. The most famous cathedrals and castles of the historic Old World are but pygmies by the side of this great Brooklyn tower. And it is our city which is to be forever famous for possessing this greatest architectural and engineering work of the continent, and of the age."

At the time that the bridge was under construction, the tallest office building in America was the Western Union Telegraph Building in Manhattan, on lower Broadway at Dey Street. It reached 10 stories into the sky, rising to a height of 230 feet. The towers of the Brooklyn Bridge would climb nearly 50 feet higher, dwarfing everything around them. They would eventually rise 276.5 feet above the East River, with their bases firmly grounded below the water along its shores.

The towers would elevate the bridge's four massive steel wire cables, from which the longest bridge span yet built anywhere in the world would hang. Each tower would, by John Roebling's estimate, weigh 67,850 tons. Adding the roadway and its traffic and its iron superstructure would bring the total load per tower to 72,603 tons, an immense burden for the towers to shoulder. They needed to be strong enough to withstand that weight.

Limestone blocks were used to build the towers until each reached the low-water line of the East River. From that height on up, limestone blocks were stacked to form the core of the towers, while blocks of granite were used as the facing. Each of the stone courses was approximately 26 inches high; the rectangular blocks of stone varied in size from 30 to 100 cubic feet. Each block was trimmed so that the joints between courses would not exceed four inches. In order to maintain quality control, all the masonry work—including the mortaring of the joints—was done in daylight. The stones of the New York tower, once completed, weighed a total of 93,000 tons.

The stone blocks with which the towers were built came from various quarries and arrived by ship. Unloaded at Red Hook in Brooklyn, the blocks were then transported, as needed, up the East River on scows. Derricks at the bridge site lifted the blocks, one at a time, off the scow and lowered them onto flatcars that moved along temporary rails to position the blocks for raising. A steel wire rope was then attached to each block, and one of two monumental iron pulleys, powered by steam, lifted it up, maneuvered it above the rising tower, then lowered it onto another flatcar, which ran along elevated wooden rails to the stone's destination. The block was finally lowered into position by a derrick mounted atop the tower.

By June 1, 1872, the Brooklyn tower had risen 100 feet above the high-tide level of the East River, about a third of its ultimate height. The tower's base had taken shape: 140 feet long (north to south) and 59 feet wide (east to west).The unfinished top was 19 feet below the level where the roadbed would be. Just a few months later, in October, Washington Roebling reported that more than two-thirds of the final weight of the Brooklyn tower was already atop the caisson, and, despite that fact, the total "settlement" of the caisson's timbers was less than an inch—five-eighths of an inch, to be precise. This indicated that the foundation of the tower, its caisson, was standing firm.

On October 23, the first bridge workers to die from an accident on the job site met their fate. As one of the derricks was hoisting an eight-ton granite block, the socket holding a guy-wire (used to help hold the derrick in place) broke, and three men were hit by falling debris, two of whom died almost instantly. One was Thomas Daugherty, pinned beneath the collapsed derrick. Jacks were rushed to the scene to lift the debris off his body, but it was too late; Daugherty took a few breaths and passed away. The bridge's executive committee made payments of $250—about three months' wages—to the families of the two dead workers. The third, Thomas Douglass, was a mason who was greatly respected for his work in Prospect Park. He was the foreman in charge of the masons at the Brooklyn tower. He had been pinned beneath Daugherty, and the latter, in his death struggle, had repeatedly kicked Douglass in the kidneys, severely injuring him. Douglass lived almost another three years but never recovered. Knowing he would not be returning to work, he prayed that he would at least live long enough to see his tower completed. In June 1875, Douglass was told that the work on the Brooklyn tower—his tower—was done; he died soon thereafter. Assistant Engineer C. C. Martin recalled the foreman with admiration: "He was a splendid man."

By late November, the Brooklyn tower had risen 145 feet above the East River and was above the level of the future bridge deck. The distinctive arched openings in the tower were beginning to take shape. Washington Roebling described the structure of the bridge's two Gothic Revival arches: "The towers consist essentially of three main shafts, united below the floor line by a connecting wall, the latter enclosing two square, hollow spaces for the purpose of saving masonry. Above the floor, these shafts extend singly for a height of eighty feet; here they are united by two Gothic arches of thirty-six feet rise, covering the roadways."

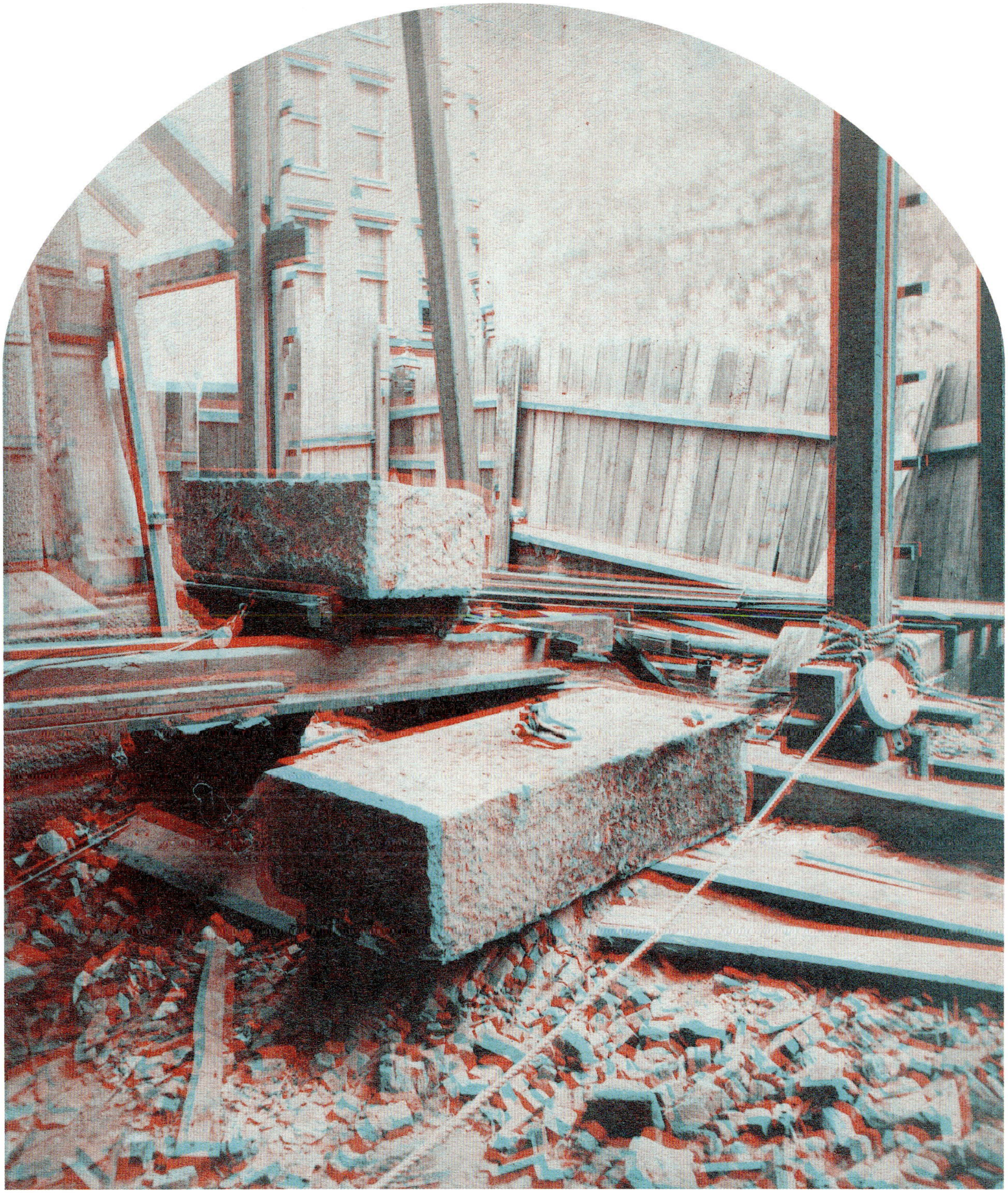

3D Photograph of two granite blocks, dressed and ready to be placed, for a tower, anchorage, or approach. Note the lewis, or iron ring, in the block in the foreground that would allow a derrick to hook and hoist it into place. At top is a block on a wheeled wooden cart; the pulley and rope connected to the cart would allow it to move along rails to the spot where the stone was to be unloaded and placed. Note the rock chips in the foreground, likely broken off the granite blocks as their shapes were refined. Michael E. Ach Collection.

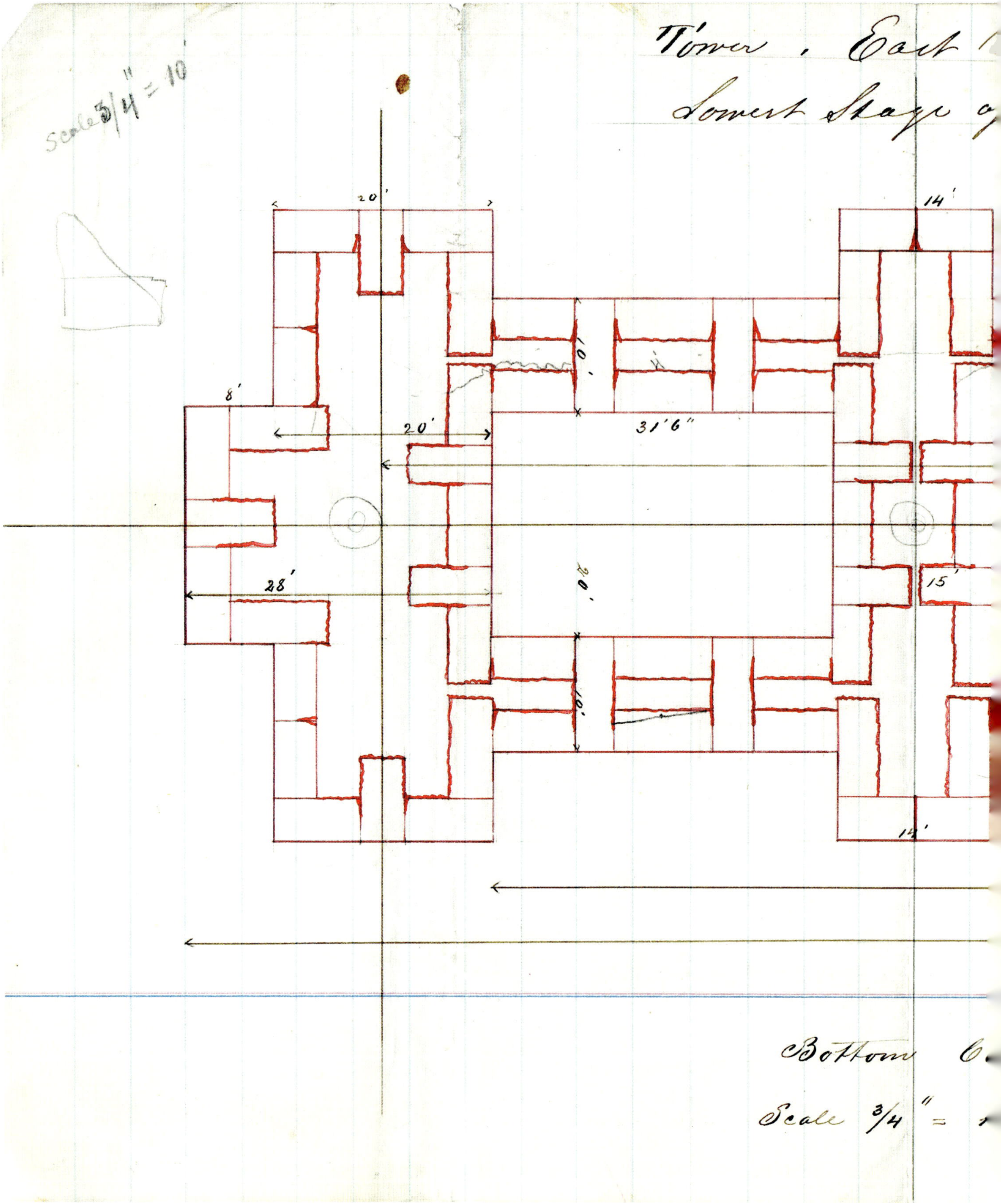
Tower . East
Lowest Stage of
Scale 3/4" = 10
20'
14'
10'
8'
20'
31'6"
28'
20'
15'
10'
14'
Bottom C
Scale 3/4" =

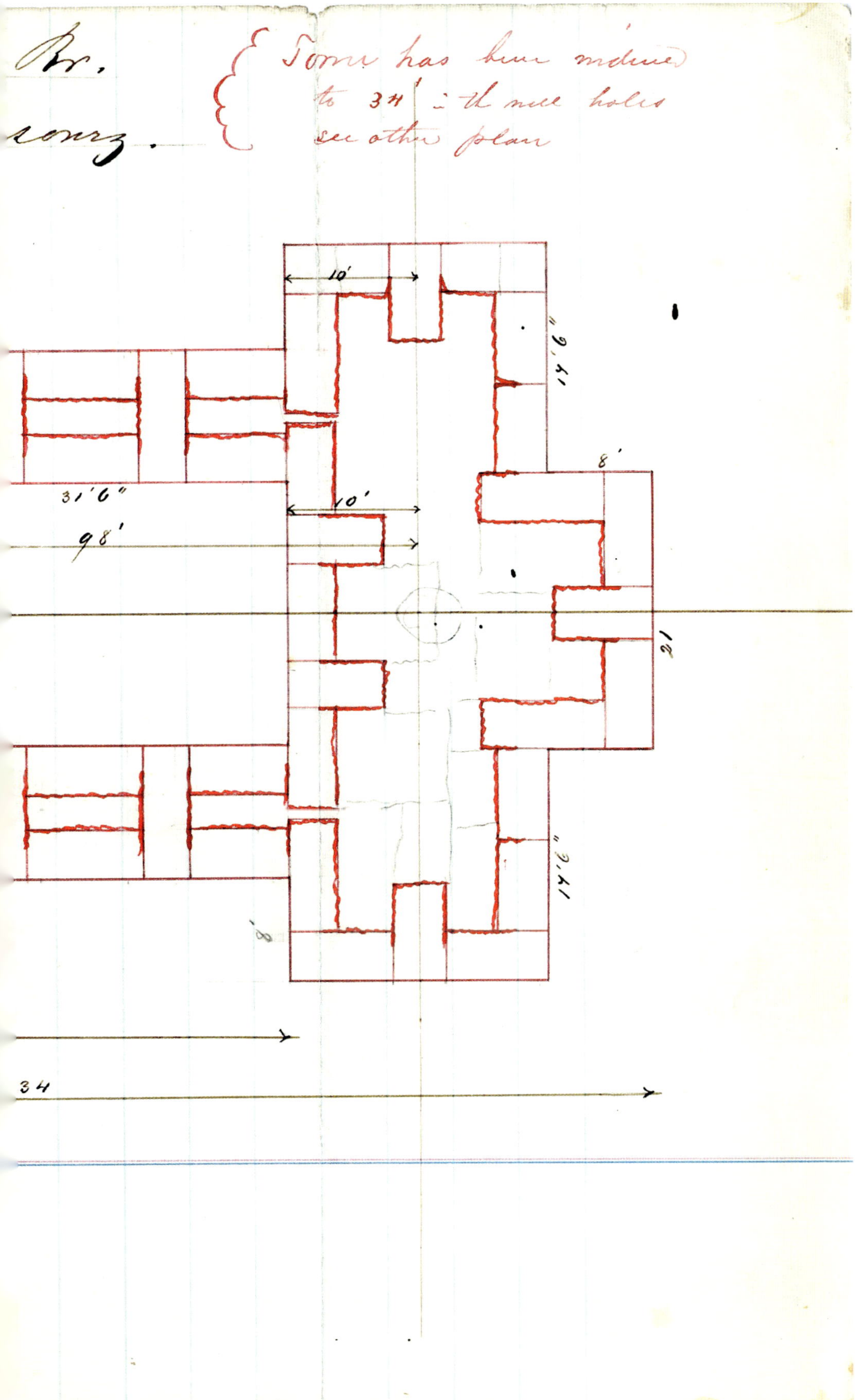

"Tower, East River Br[idge]./Lowest Stage of Masonry." This drawing by Washington Roebling, ca. 1870, shows the shape of the towers from above. Note the two rectangles in the middle; those represent the hollows inside the towers. A similar earlier drawing, signed by John Roebling, is also in the Roebling Collection at Rensselaer Polytechnic Institute. RPI.

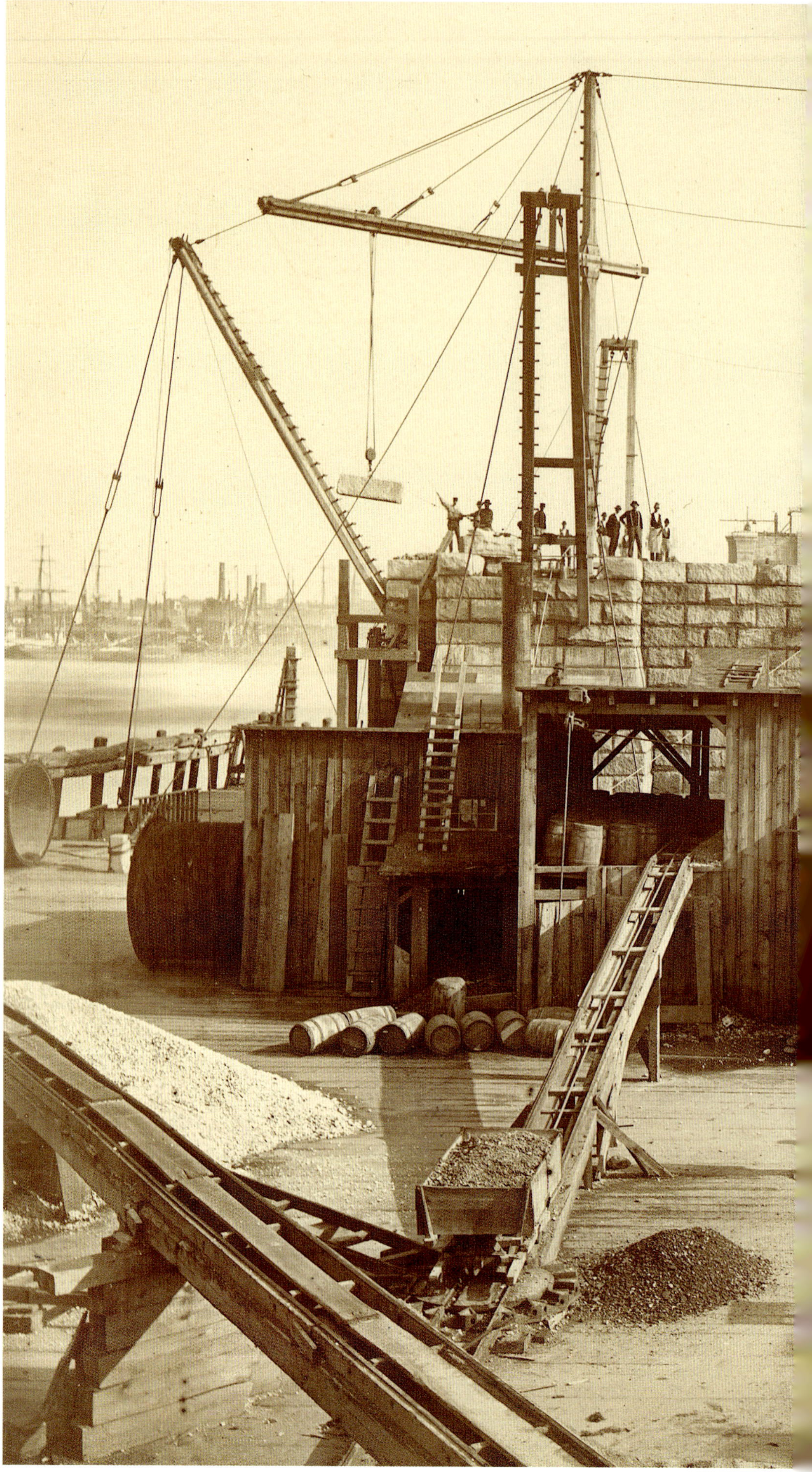

"New York Tower/East River Bridge/in September 1872/with Brooklyn Tower in the distance." A photograph by Silas A. Holmes of the bridge's New York tower being built. Holmes's insider status at the construction site is apparent in this photograph. Note the two granite blocks that hang, poised in midair, from the mast derricks at either end, increasing the drama of the shot—an accommodation to the photographer. Eight boom derricks frame the tower. The gents in the foreground and the workers on the tower stand still for the long exposure, looking at the camera, posing for Holmes. The steam-driven hoisting engines used coal, which is visible at the lower left of the photograph, both in the transport car and on the ground to its right. RPI.

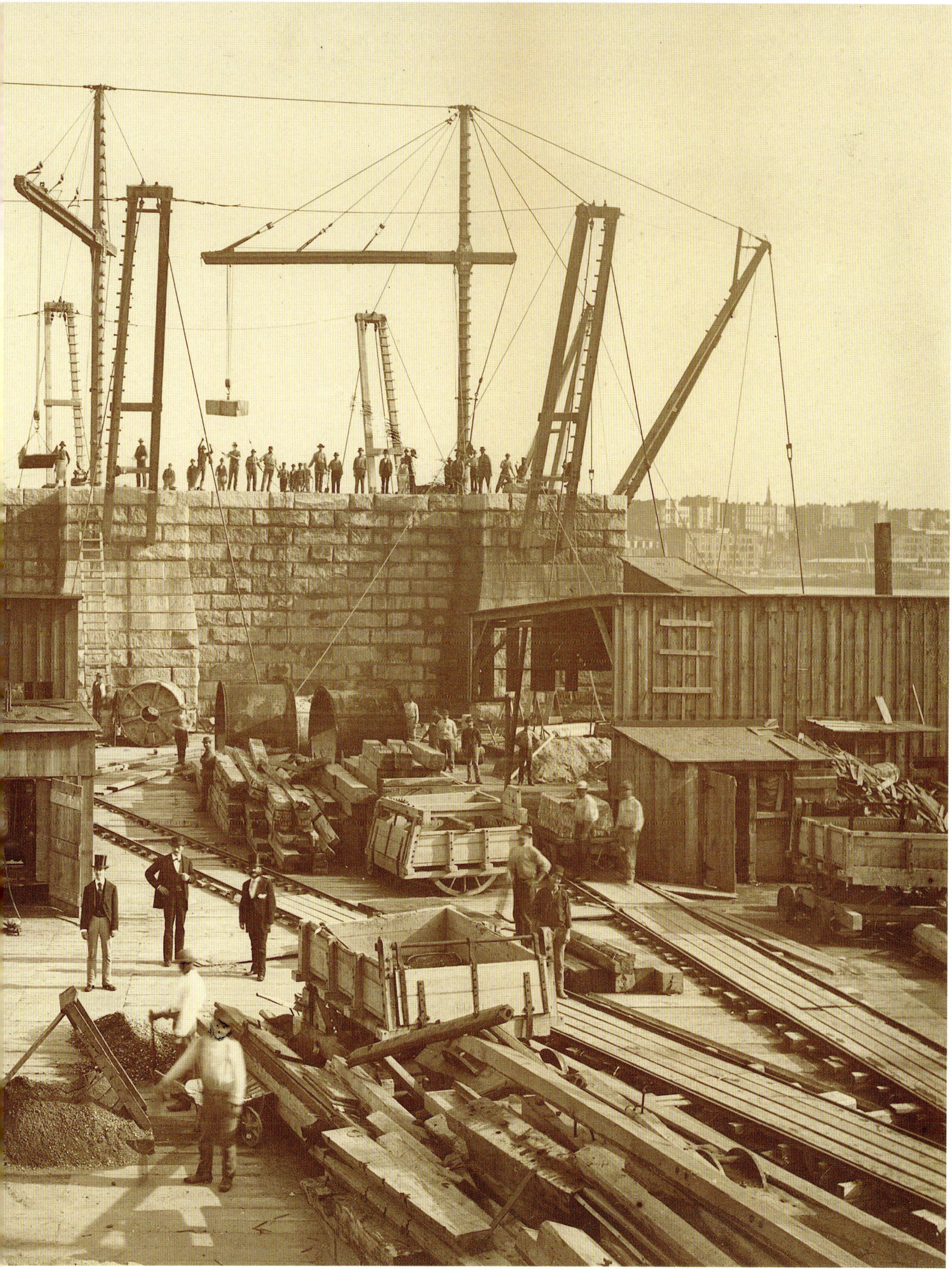

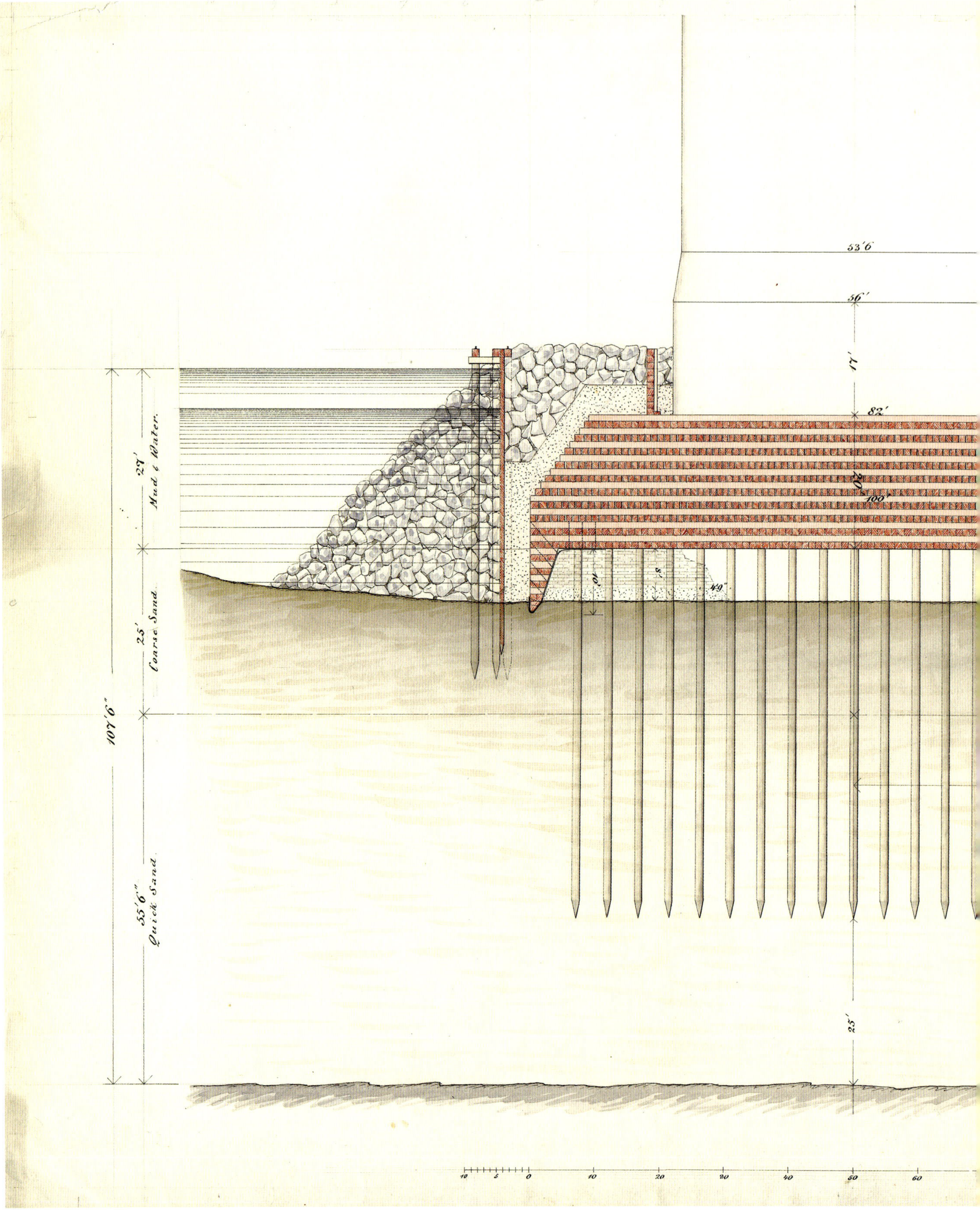

53'6"
56'
17'
82'
20'
100'
27'
Mud & Water.
25'
Coarse Sand.
107'6"
55'6"
Quick Sand.
25'
49'
10
5
0
10
20
30
40
50
60

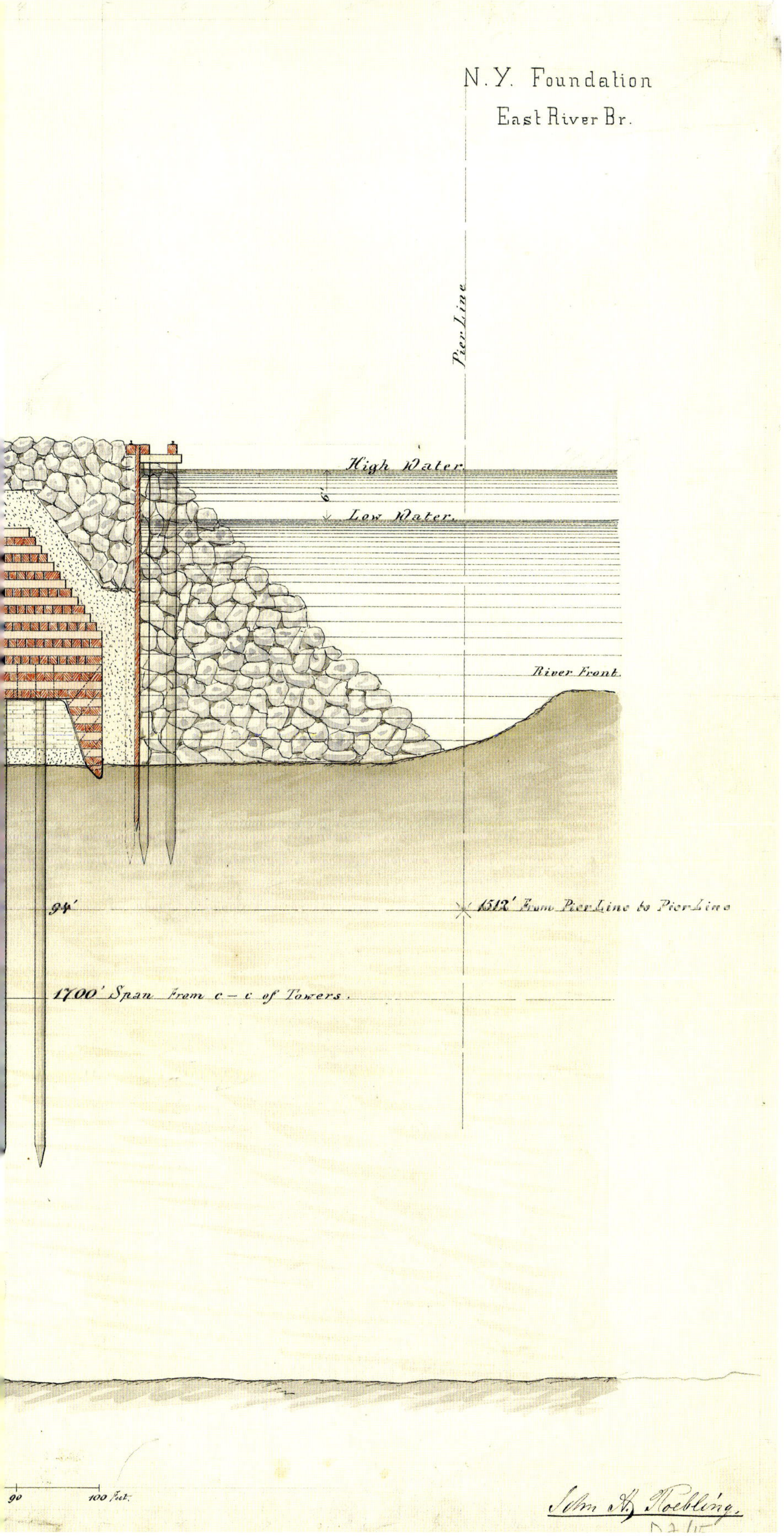

"N.Y. Foundation/East River Br[idge]." A drawing by John A. Roebling, signed at lower right, depicting the New York caisson, with the base of the tower outlined above it, ca. 1865. Note the timber courses, drawn in red. The tower is shown from its side. What look like spikes underneath the caisson are actually directional arrows meant to indicate its downward movement as the workers dug farther into the ground. RPI.

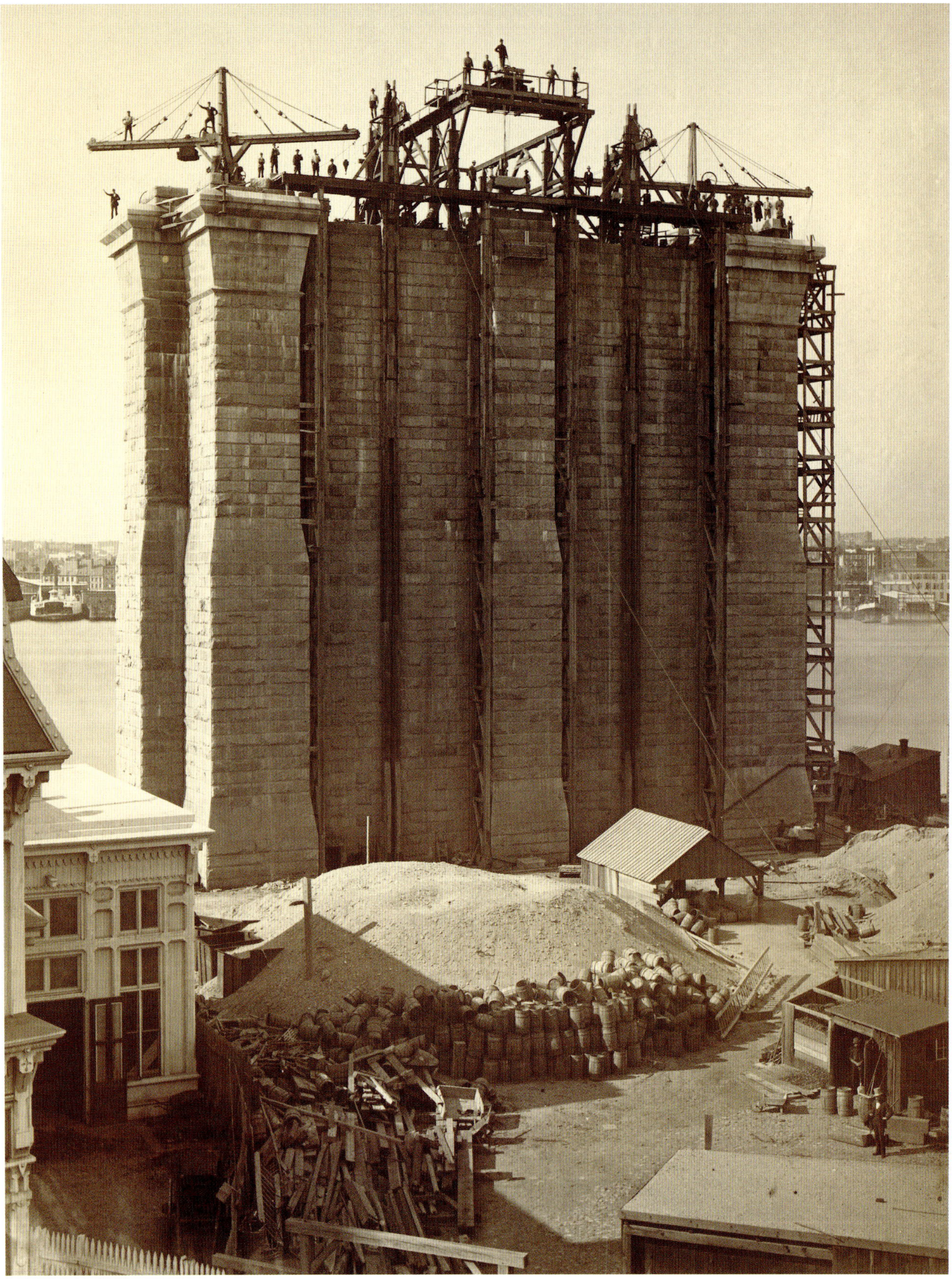

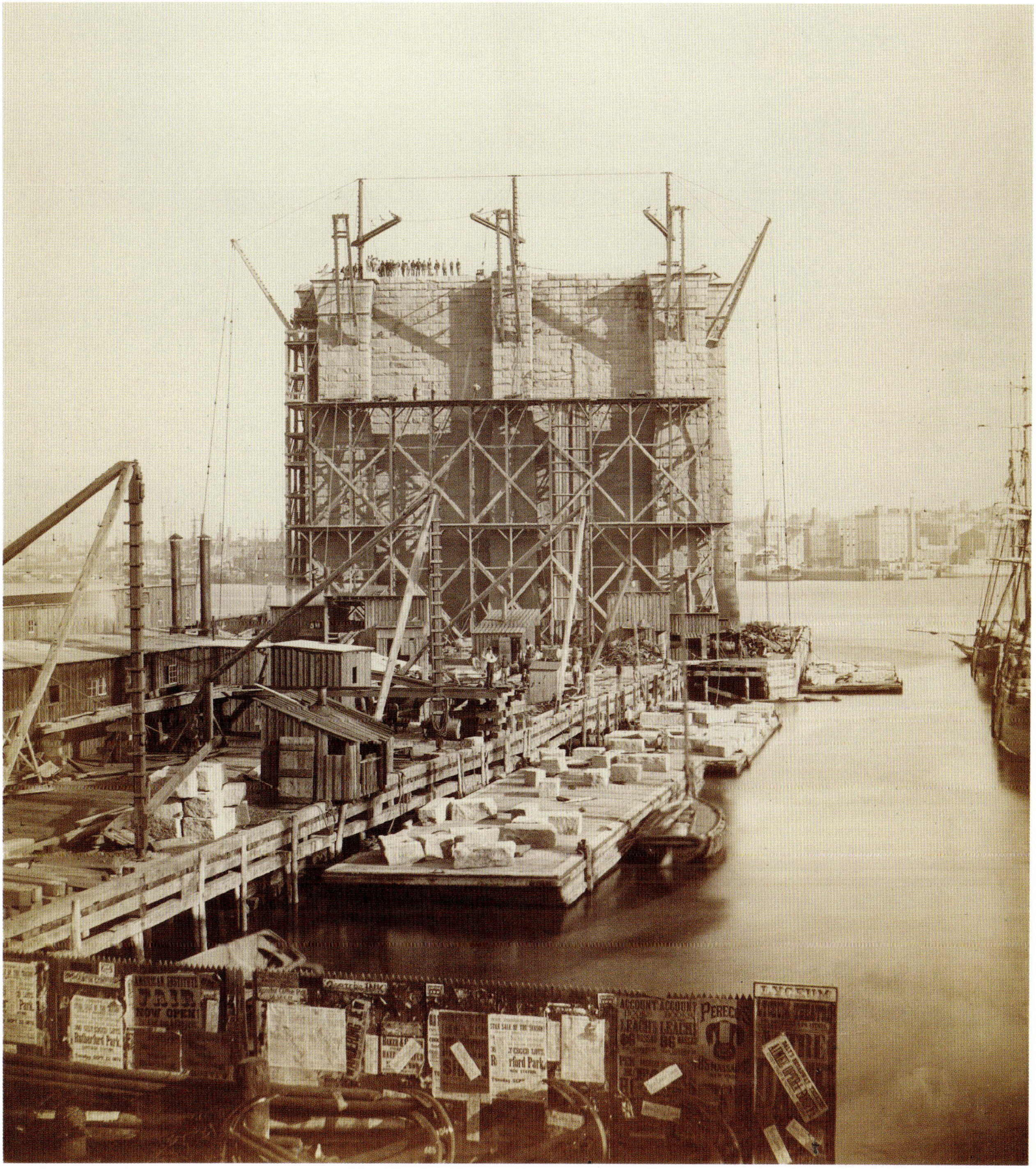

OPPOSITE: "Brooklyn Tower/East River Bridge/in September 1872," photograph by Silas A. Holmes. The sand in the foreground was used in the mortar that joined the granite blocks. The edge of the Fulton Ferry Terminal is visible in the left foreground. With its distinctive mansard roof, the terminal is a valuable landmark in determining which side of the river is in view. This photograph, showing workers neatly posed atop the rising structure, may have been commissioned to mark the occasion of the Brooklyn tower reaching the roadbed level. By the following month, the structure had risen to about seventy feet. RPI.

ABOVE: "New York Pier/about 118 ft. high/ Sept. 1873." Note the stone blocks scattered on the barges, and stacked on the pier, about to be hauled to the top of the tower. Commerce continues, with sailing ships docked along South Street partially visible at far right. MCNY.

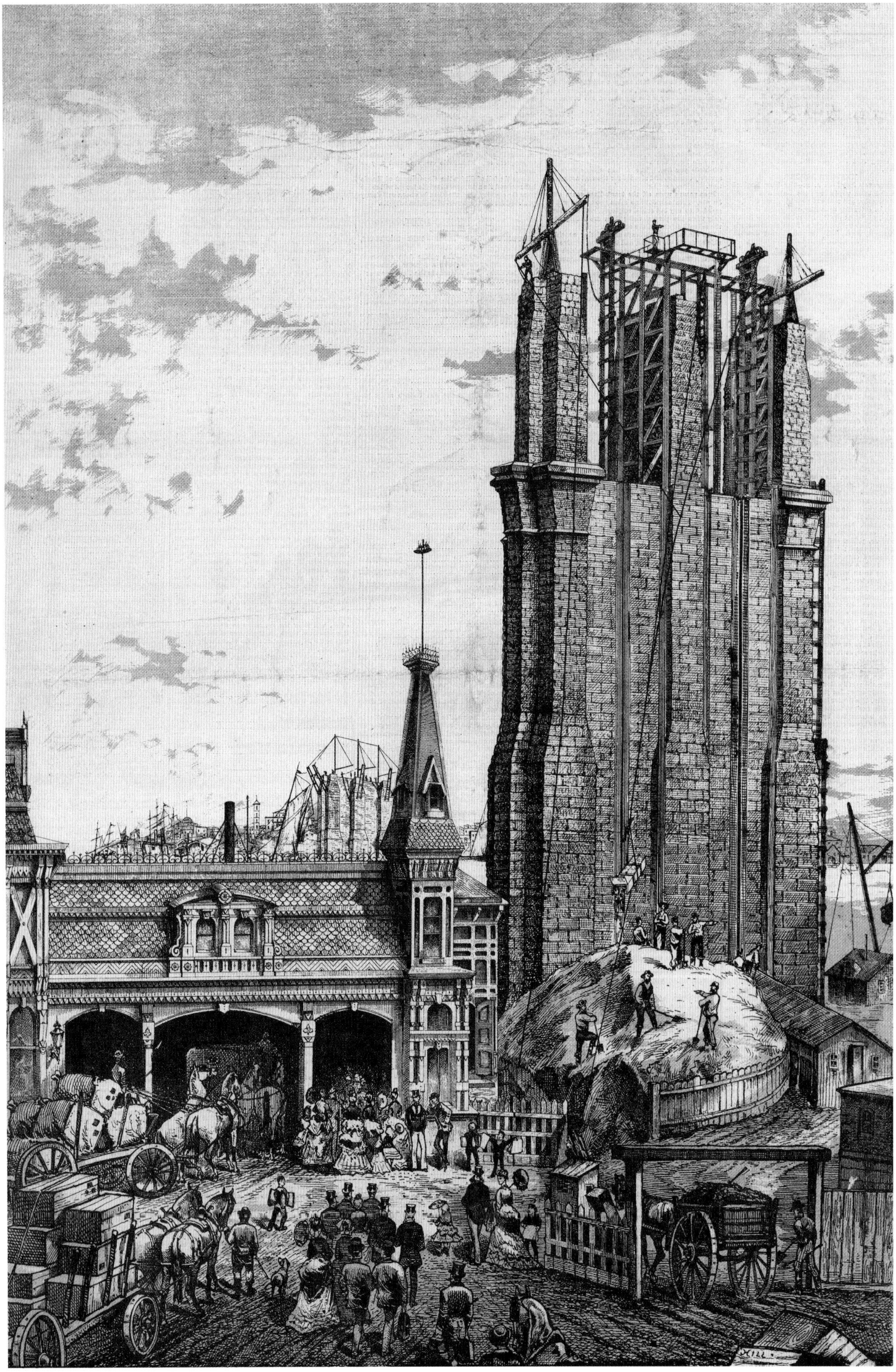

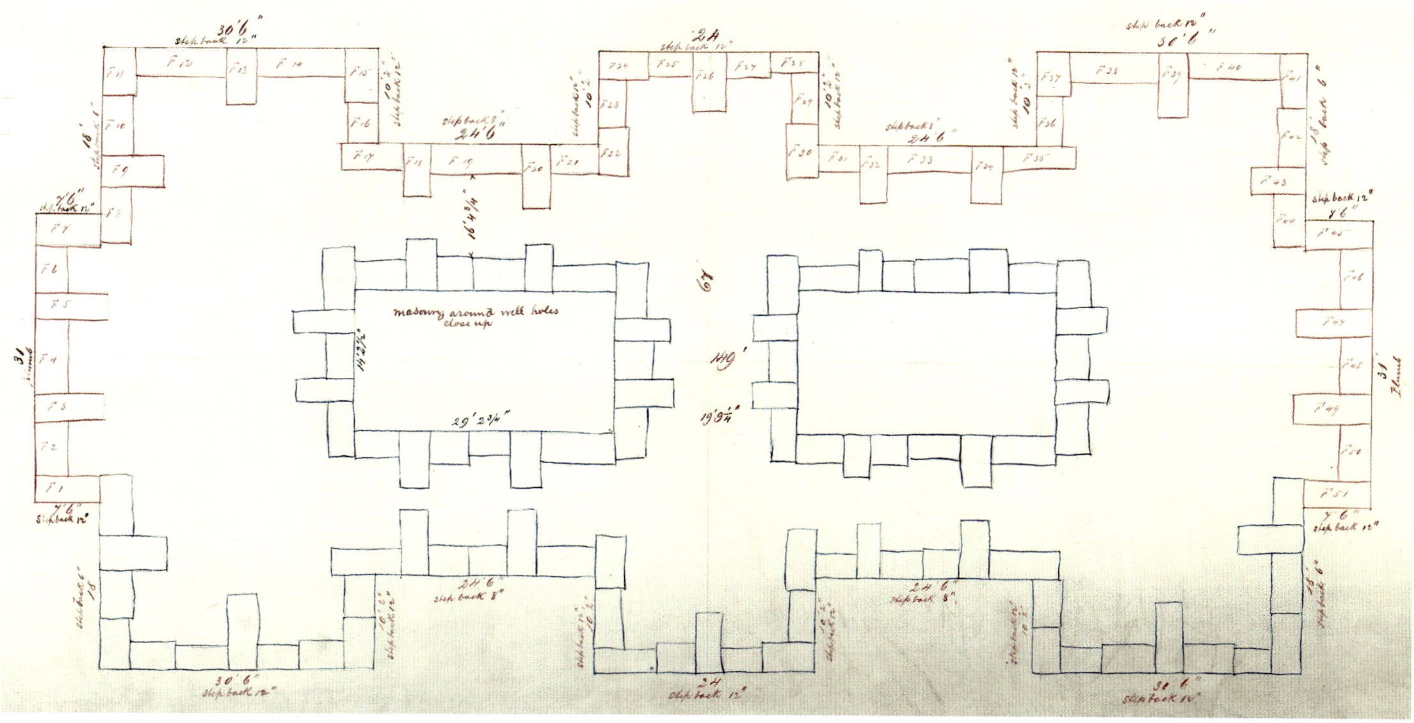

ABOVE: "6th Course Masonry, 20 [inches] Br[idge] Tower, Granite on River Face, River Side." Washington Roebling's drawing bears his initials and the date August 24, 1870. He also included a note with very specific instructions: "The granite facestone colored red is to be cut to ½-inch joints on the builds. The Limestone are colored blue and again no cutting. This course is being cut at Red Hook by numbers." The stone yard cutting the blocks to order was in Red Hook, Brooklyn; the stone itself came from several quarries around the country. NYC Municipal Archives.

OPPOSITE: "Brooklyn Side of the Tower at Fulton Ferry, with the New York Tower in the Distance. The Brooklyn Bridge," *The Daily Graphic*, July 1, 1883. This woodcut shows the tower on the Brooklyn side nearing completion. To its left stands the glorious Fulton Ferry Terminal, where ferry boats arriving in Brooklyn docked until 1924. Note the New York tower beginning to rise in the distance, just above the terminal building. Courtesy of Richard Haw.

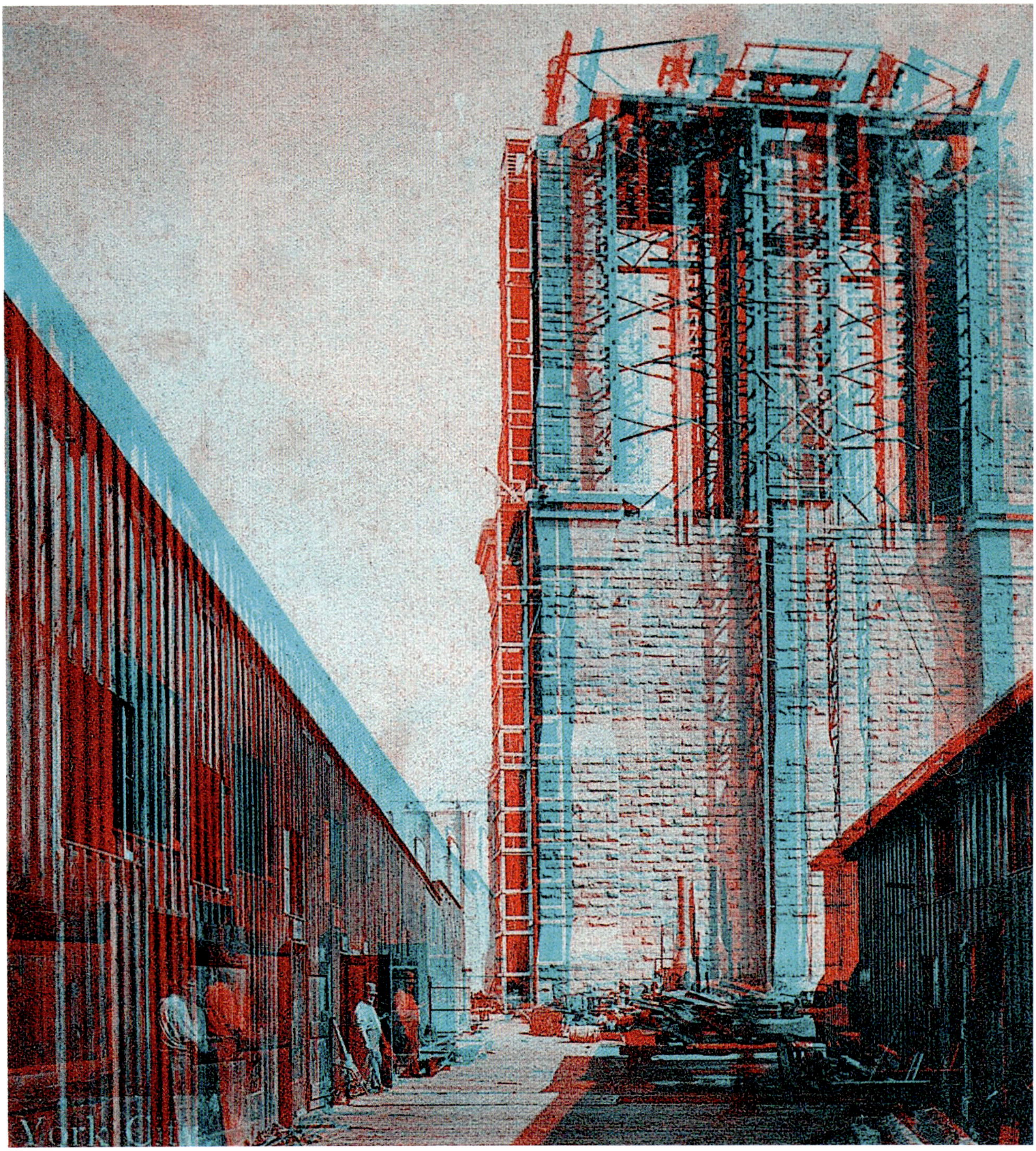

3D Anaglyph of the New York tower, almost at its full height, ca. 1875. The Brooklyn tower is visible in the distance, just above the far end of the building at left. The building at right is the New York caisson's compressor house (see pages 66–67 for an overview of the site). Courtesy of The Green-Wood Historic Fund, Jeffrey I. Richman Stereoview Collection.

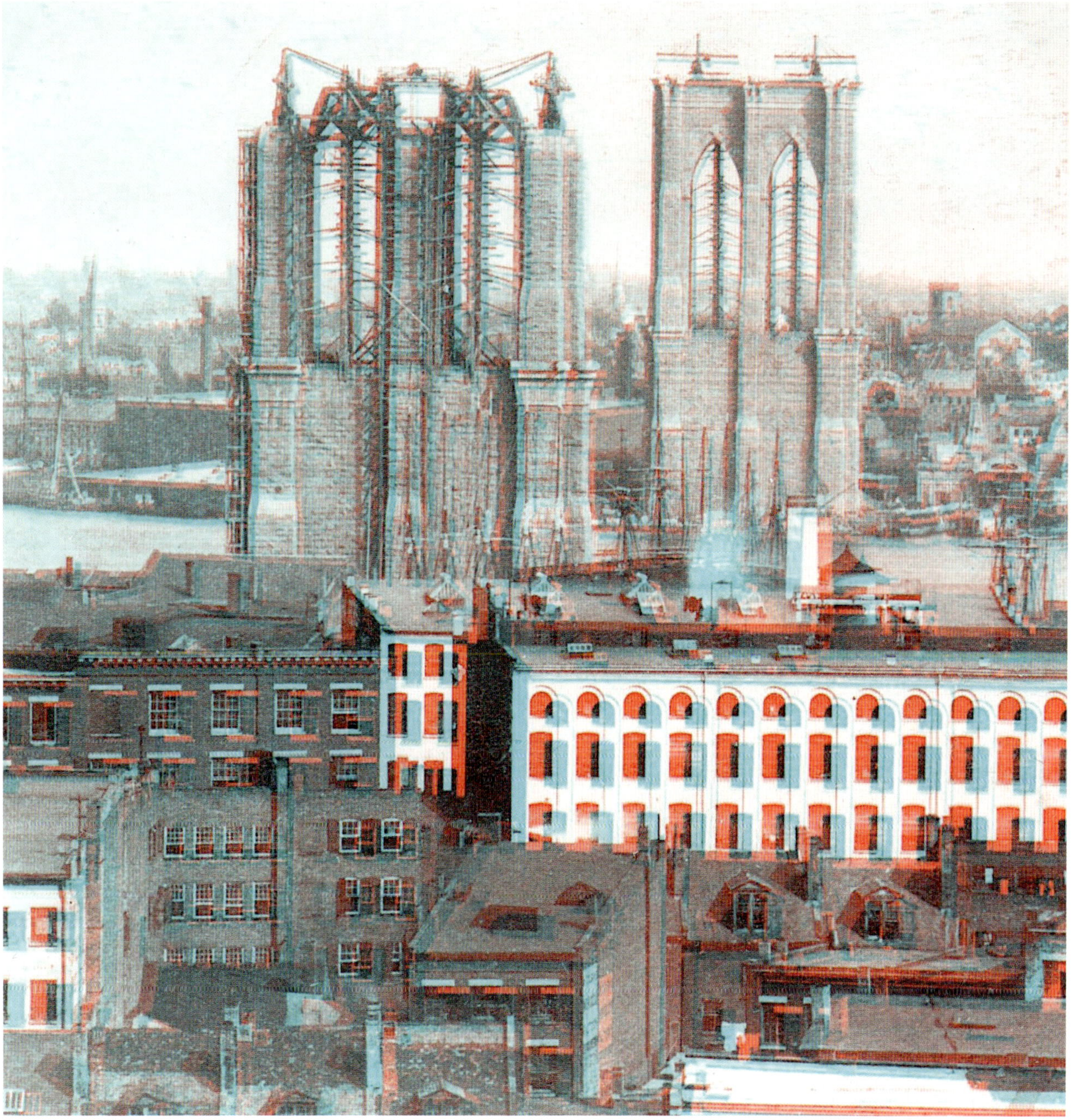

3D Anaglyph of a stereoview taken in Manhattan, looking toward Brooklyn. The photographer used a long lens, which created the effect of shortening the distance between the bridge's two towers. Nevertheless, the photograph gives a good sense of how much lower the buildings in Manhattan were than the towers of the Brooklyn Bridge. By looking at the wisp of smoke above the white building in the foreground, one can understand how the stereoview was shot: when the image is viewed only through one's left eye, using the red lens, there is no puff of smoke seen. However, it is visible when viewed through the right eye, using the cyan lens. That indicates that the two photographs that created this stereoview were not taken in the same moment with one exposure, using a two-lens camera, to achieve a three-dimensional effect. Rather, both images were taken with a single lens at two different times, with the photographer moving the lens horizontally after capturing the first image, and that gap in time explains why the puff of smoke appears only in one image but not in the other. Courtesy of The Green-Wood Historic Fund, Jeffrey I. Richman Stereoview Collection.

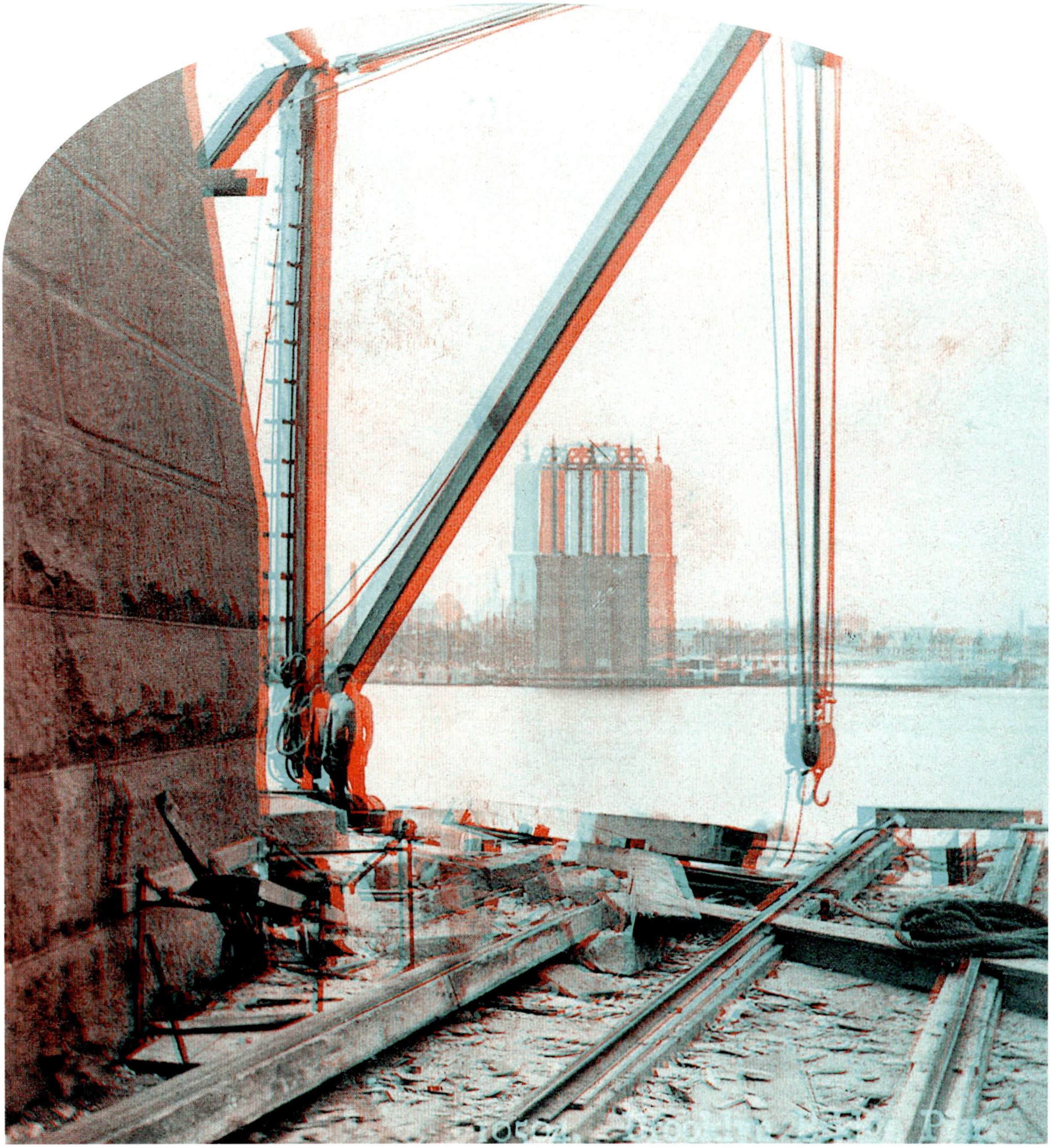

3D "Brooklyn Bridge Pier." Looking across the East River toward the New York tower and into the distance, with the base of the Brooklyn tower at left. This photographer understood how to compose a stereoscopic image; he included objects in both the foreground and the distance to maximize the three-dimensional effect and frame the composition. Courtesy of the Jeffrey Kraus Collection.

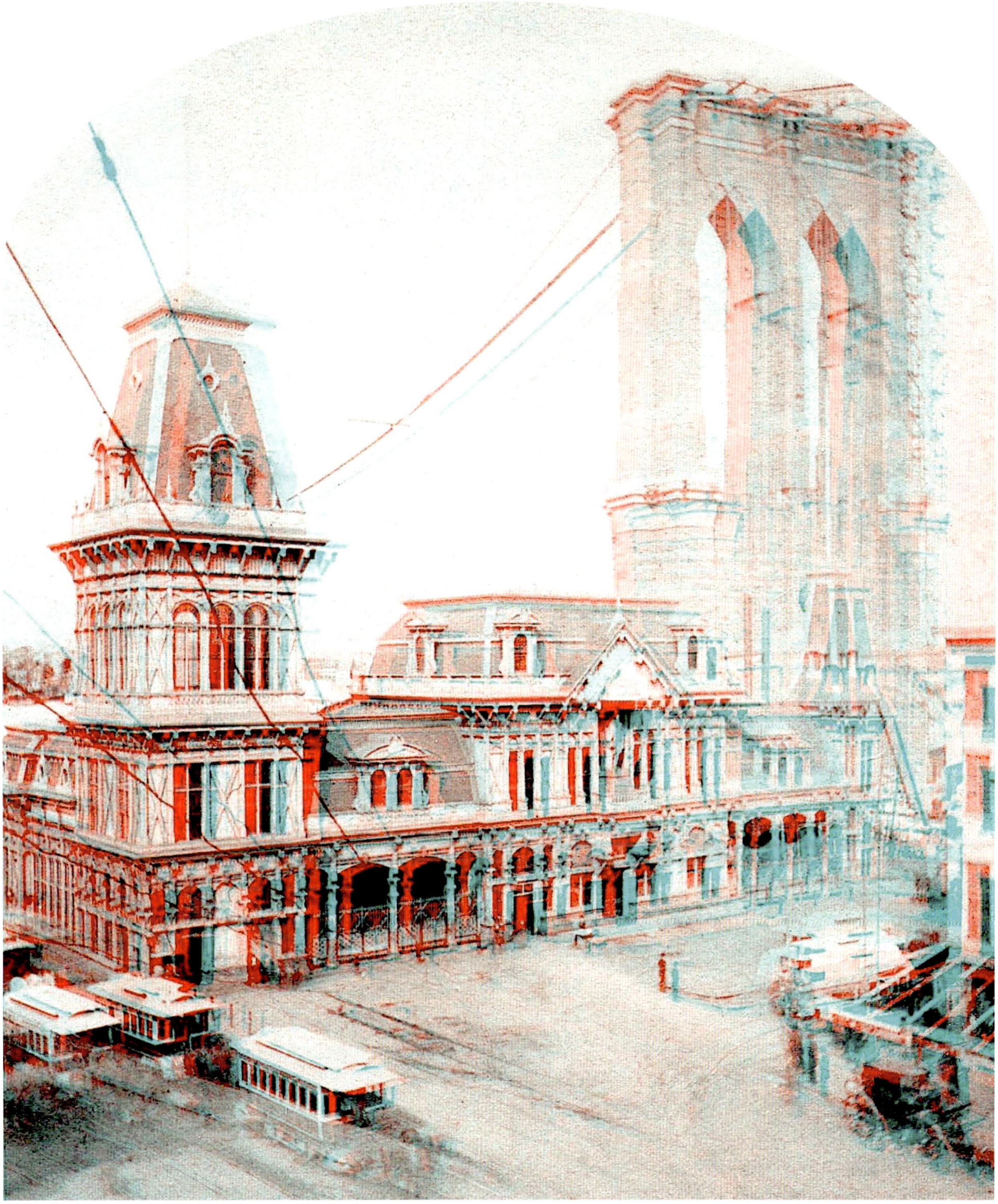

3D Photograph of the Fulton Ferry Terminal on the Brooklyn side of the East River, with the Brooklyn Bridge tower rising just to the right of it, ca. 1875. The ferry building, completed in 1871, when the bridge was already under construction, survived until 1926, two years after its ferries ceased crossing the river. This image gives a good idea of the scale of the bridge towers. Courtesy of The Green-Wood Historic Fund, Jeffrey I. Richman Stereoview Collection.

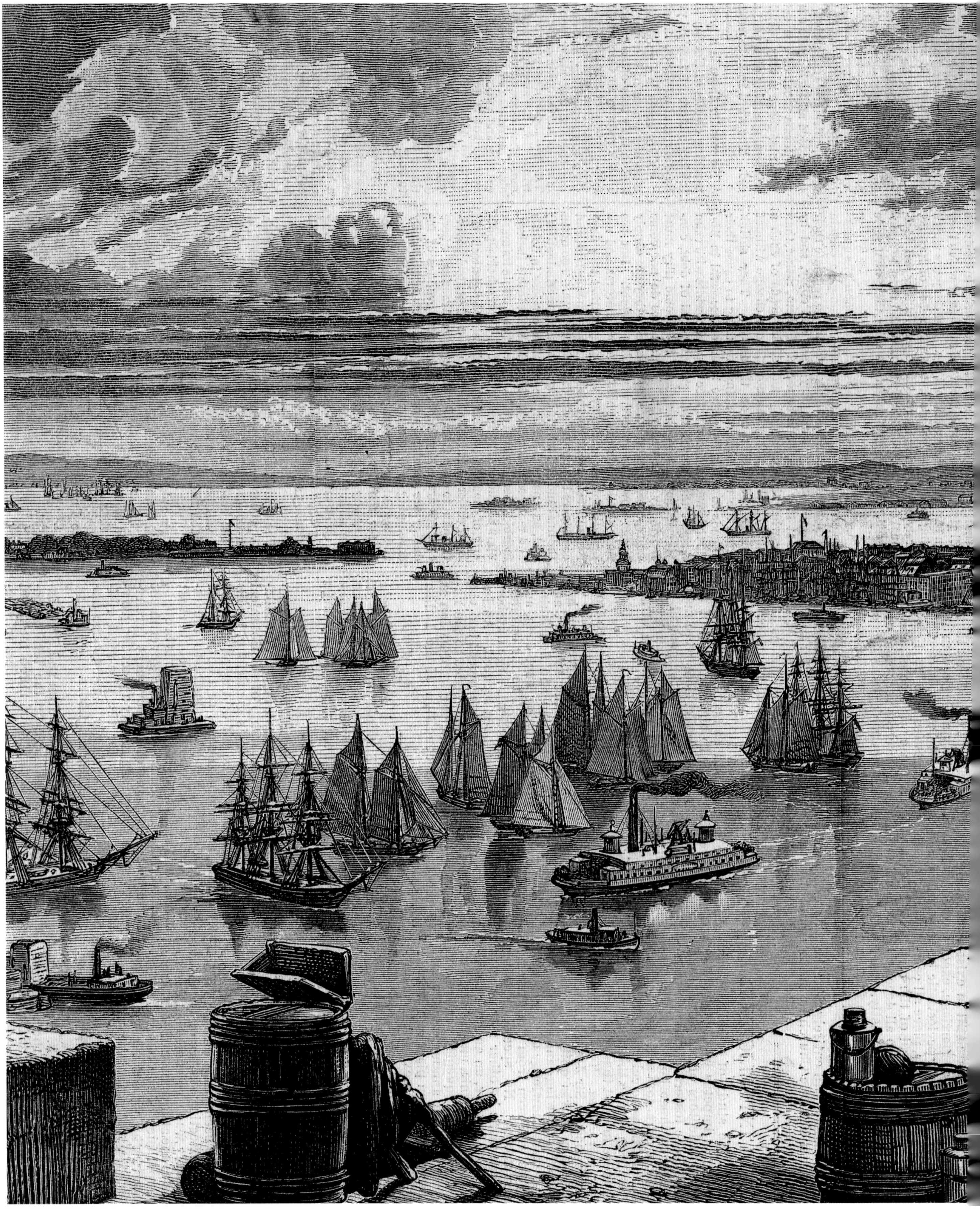

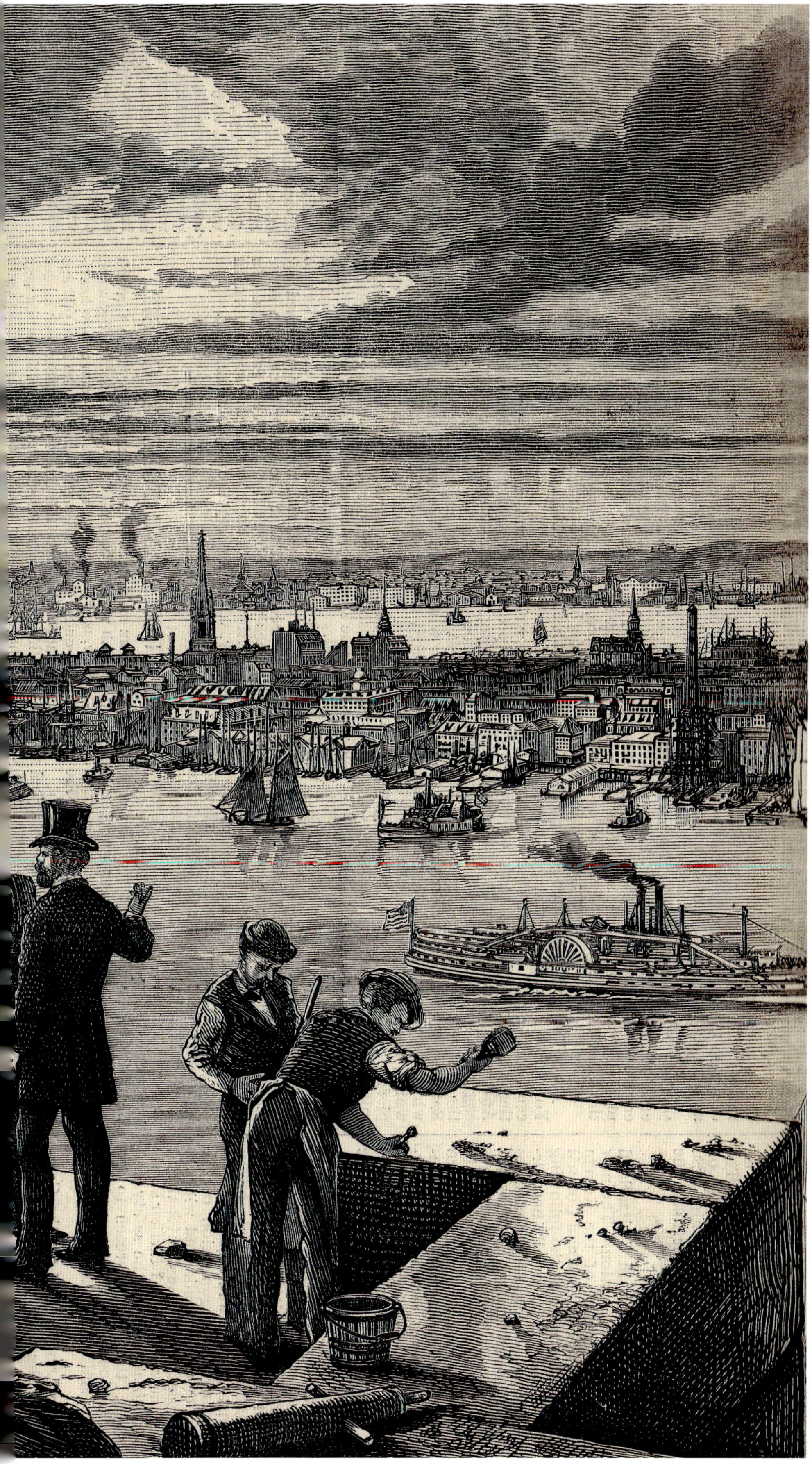

"New York Harbor, As Seen from the Brooklyn Tower of the Suspension Bridge—From a Sketch by C.E.H. Bonwill," *Harper's Weekly*, November 1, 1873. Harper's sent one of its artists to the top of the tower to draw this scene, positing that precious few people would want to climb that high to see this "magnificent panorama." At the time, the spires of Trinity Church and St. Paul's Chapel, as well as the tops of the *New-York Tribune* and the Western Union Telegraph buildings, rose above all the other structures in Manhattan, yet they were dwarfed by the Brooklyn tower. Author's collection.

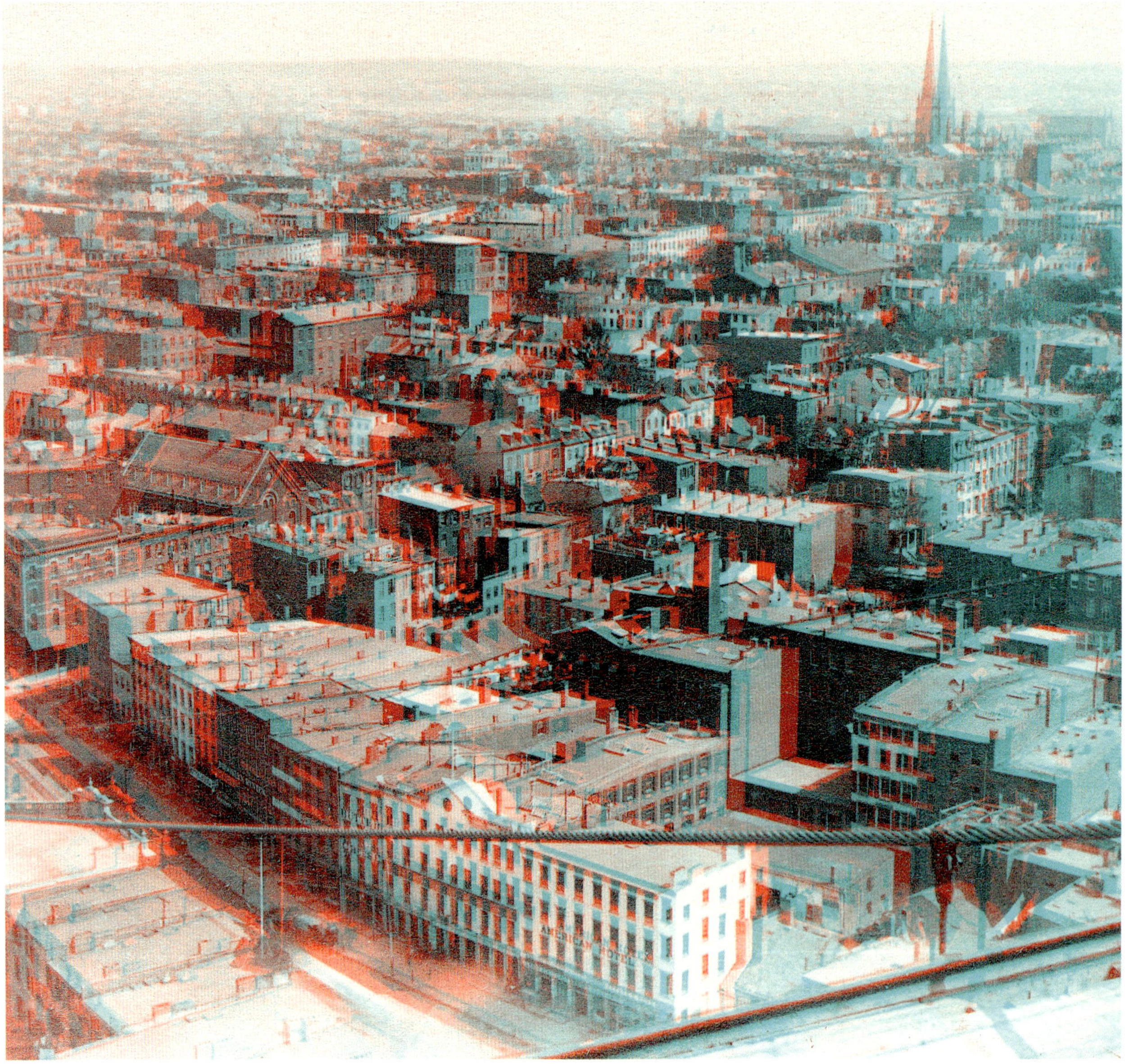

3D View from the top of the bridge's Brooklyn tower, looking southeast. At top right, the spire of Holy Trinity Church, a Brooklyn landmark at the corner of Clinton and Montague Streets, rises above its neighbors. Courtesy of The Green-Wood Historic Fund, Jeffrey I. Richman Stereoview Collection.

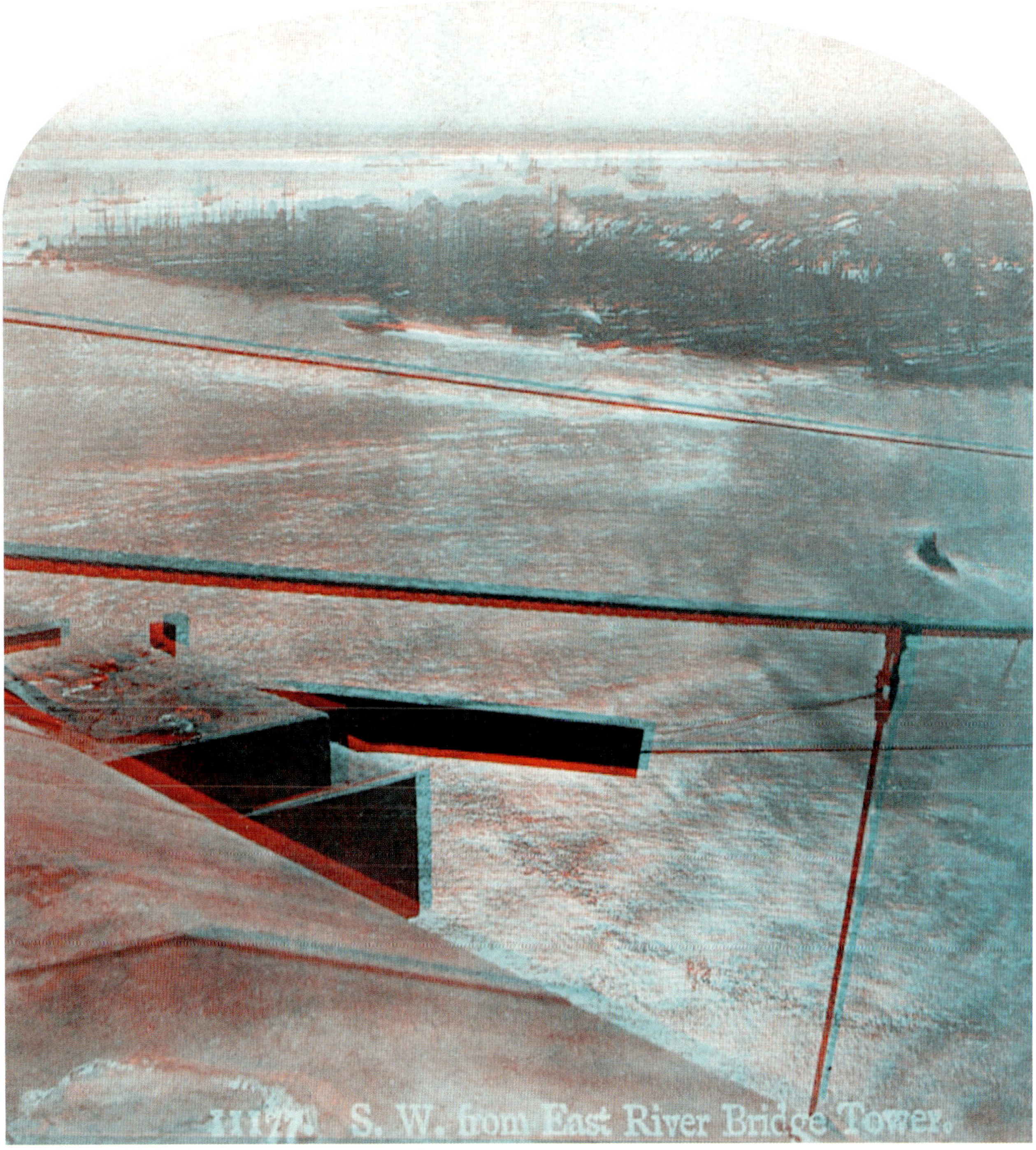

3D View from atop the bridge's Brooklyn tower, looking across the East River toward the southern tip of Manhattan. The stereoview from which this anaglyph was made was published by E. and H. T. Anthony & Co., the period's largest manufacturer and purveyor of photographic equipment, supplies, and images in the world. Note the many masts of the ships docked along the Manhattan side of the river. Courtesy of the Jeffrey Kraus Collection.

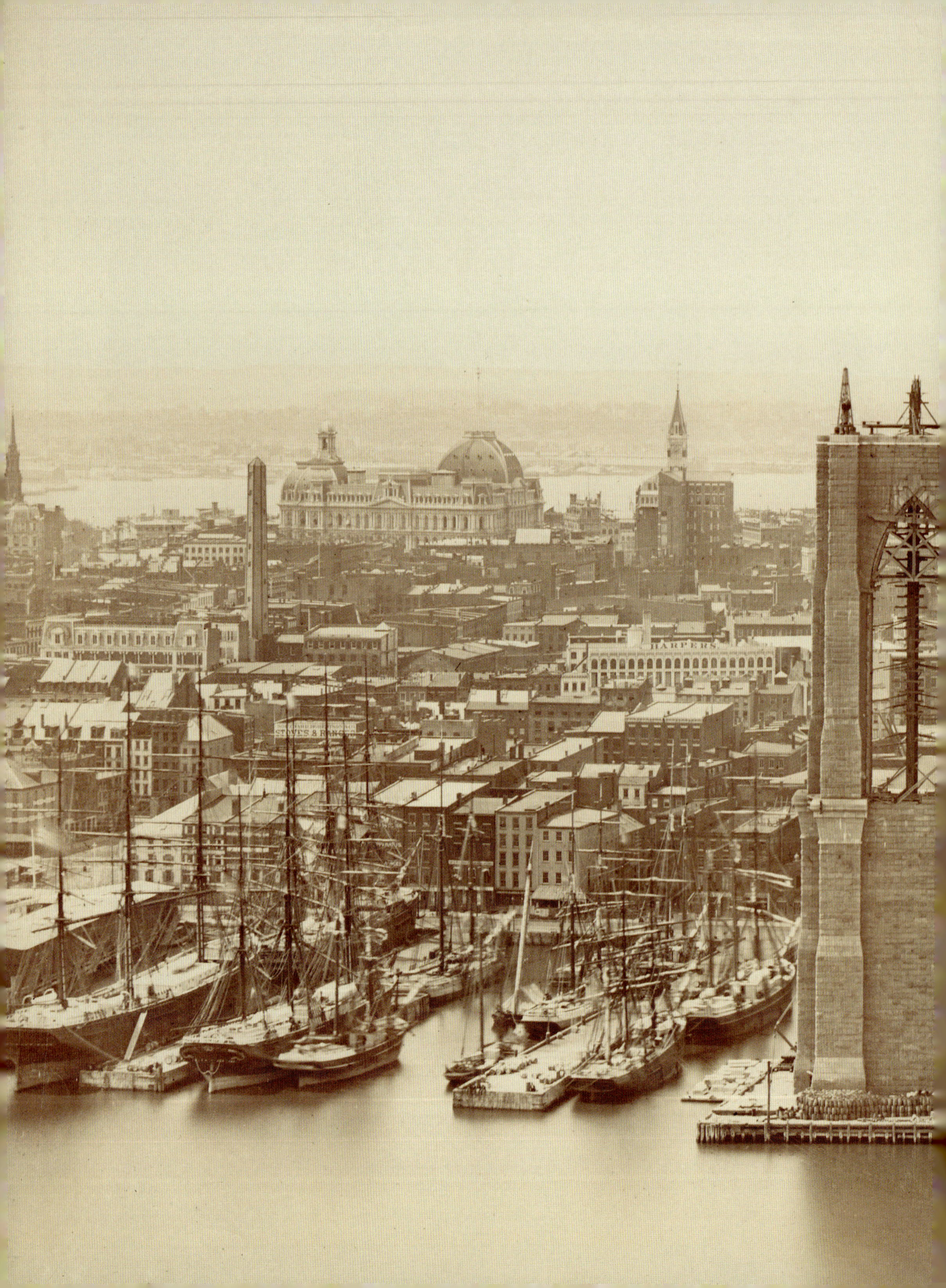
HARPERS

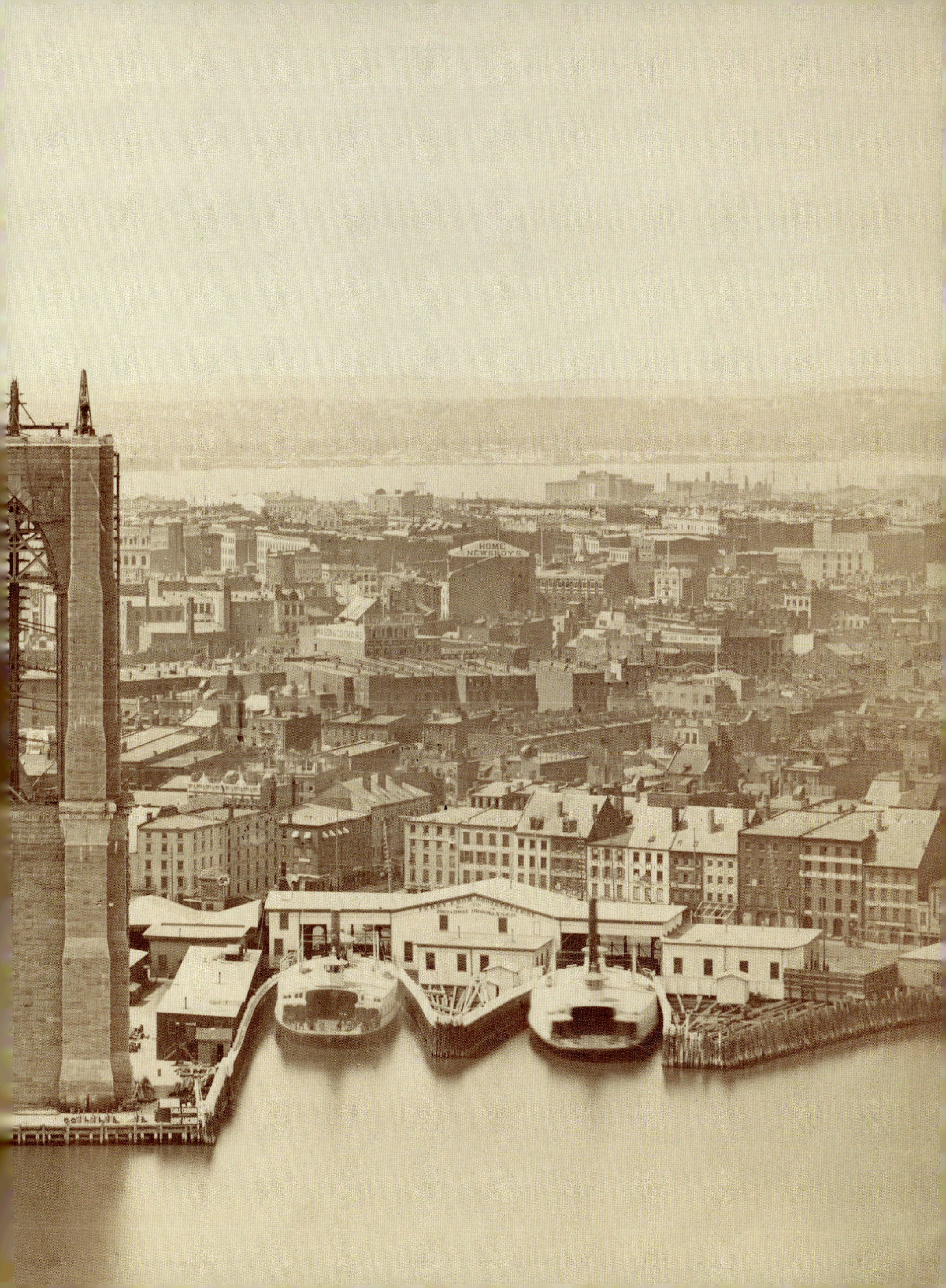
HOME
NEWSBOYS

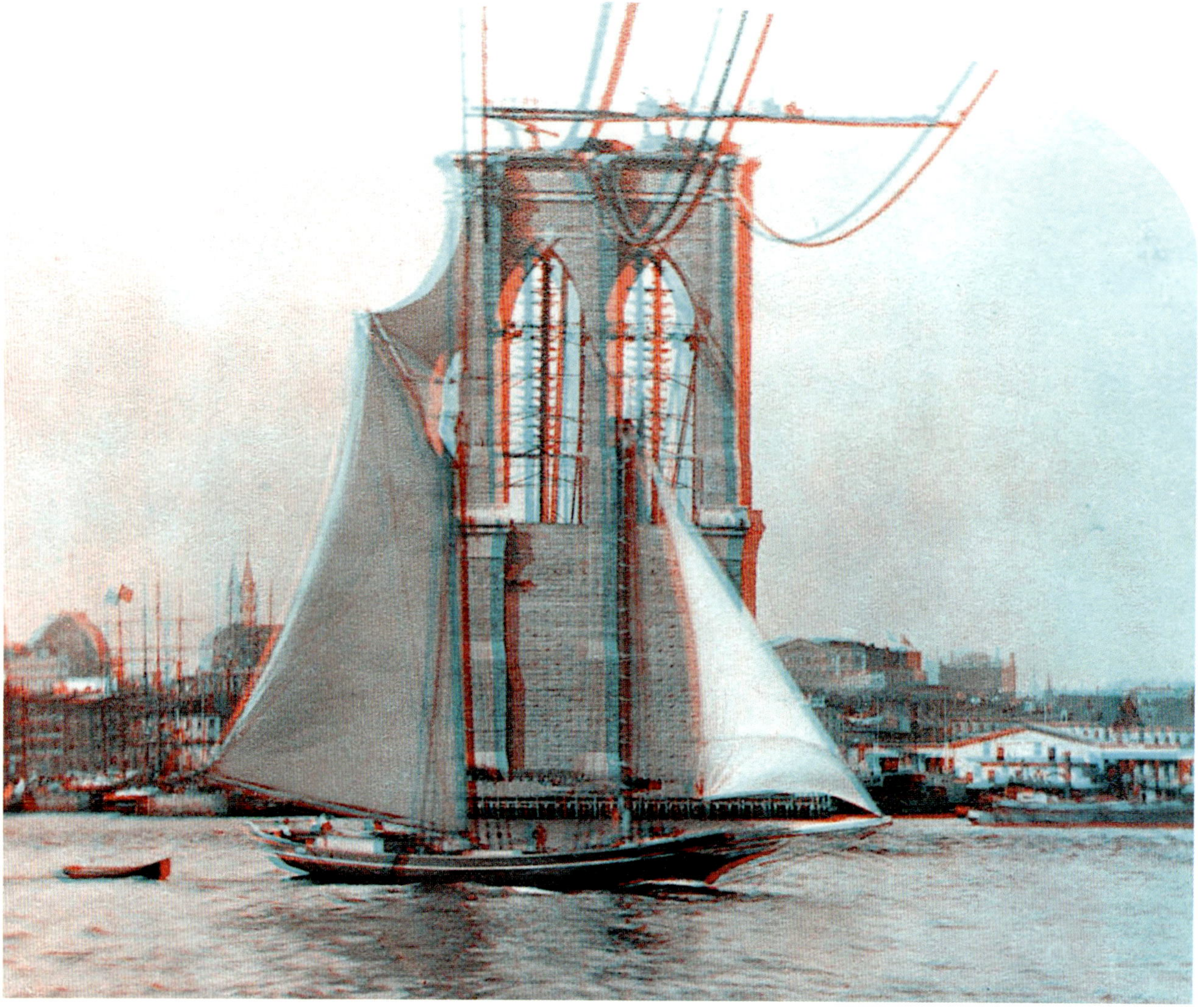

3D Anaglyph generated from stereoview no. 11298, "Bridge Tower," published by E. & H. T. Anthony, 591 Broadway, New York City. This view is ca. 1875, looking west from the Brooklyn shore toward the New York tower and Manhattan beyond. Courtesy of the Jeffrey Kraus Collection.

PREVIOUS SPREAD: Cabinet card photograph of the New York tower of the Brooklyn Bridge, almost completed, with wooden falsework still in place, and lower Manhattan beyond it, ca. 1876. In the distance, west of Manhattan Island, is the Hudson River. Just to the left of the bridge tower is a white building with a sign for Harper Brothers ("HARPERS"), the largest publishing house in the world at the time. Above it is the pointed tower of the New York Tribune Building, which stood on Park Place just east of City Hall Park. To the left of that is the massive and just-completed City Hall Post Office and Courthouse at the intersection of Park Row and Broadway. Note the shot tower (the thin, pencil-like shape) just to the left of the post office. A shot tower was used to make bullets: lead was heated until molten, then dropped through a sieve high in the tower, and by the time the lead reached a water basin at ground level, it was round and ready to be used in guns. The steeple of St. Paul's Chapel, which still stands at Fulton Street and Broadway, is at the far left. MCNY.

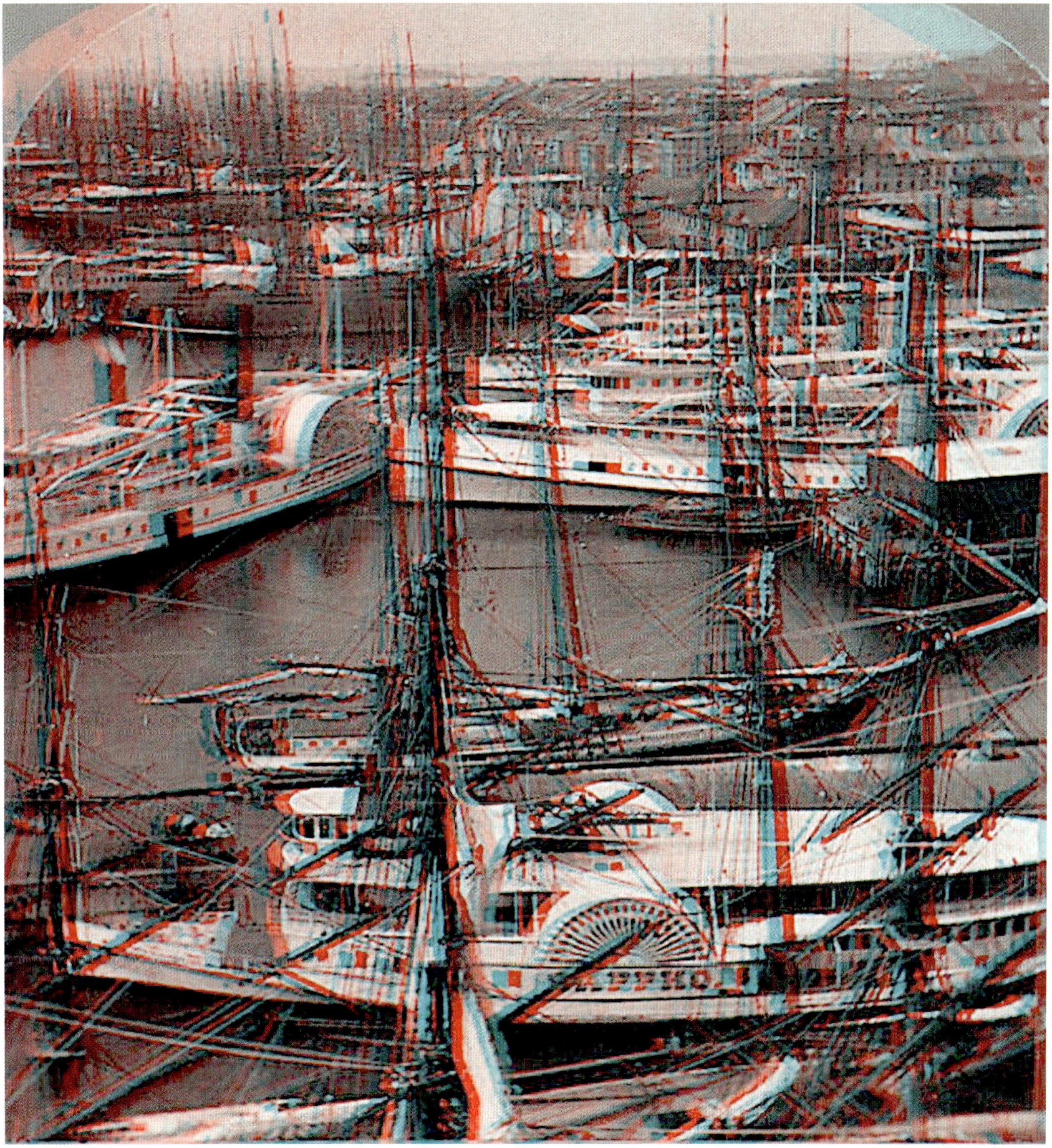

3D View from the New York tower of the Brooklyn Bridge, looking south along the crowded docks just east of South Street. Courtesy of The Green-Wood Historic Fund, Jeffrey I. Richman Stereoview Collection.

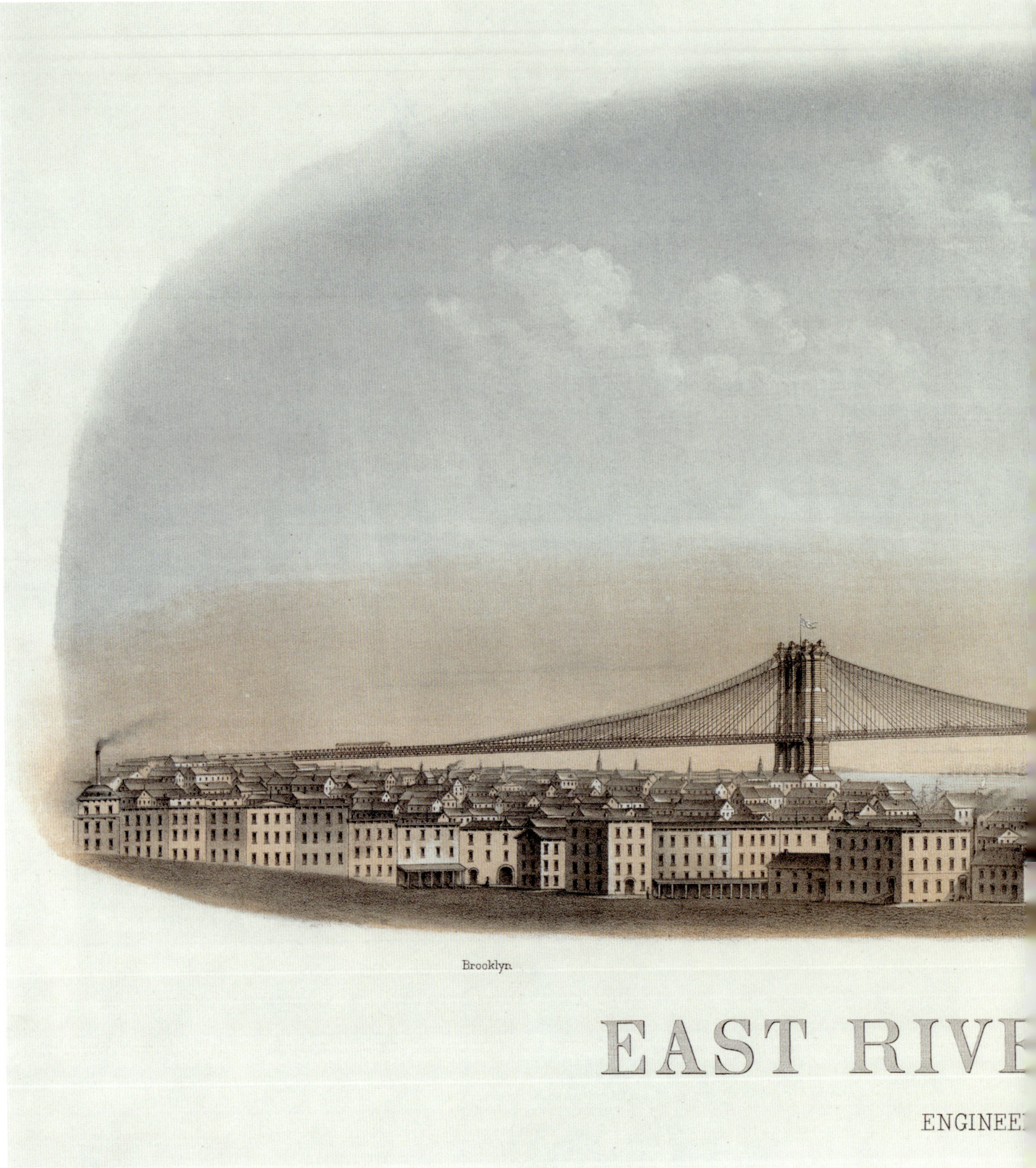
Brooklyn
EAST RIVE
ENGINEE

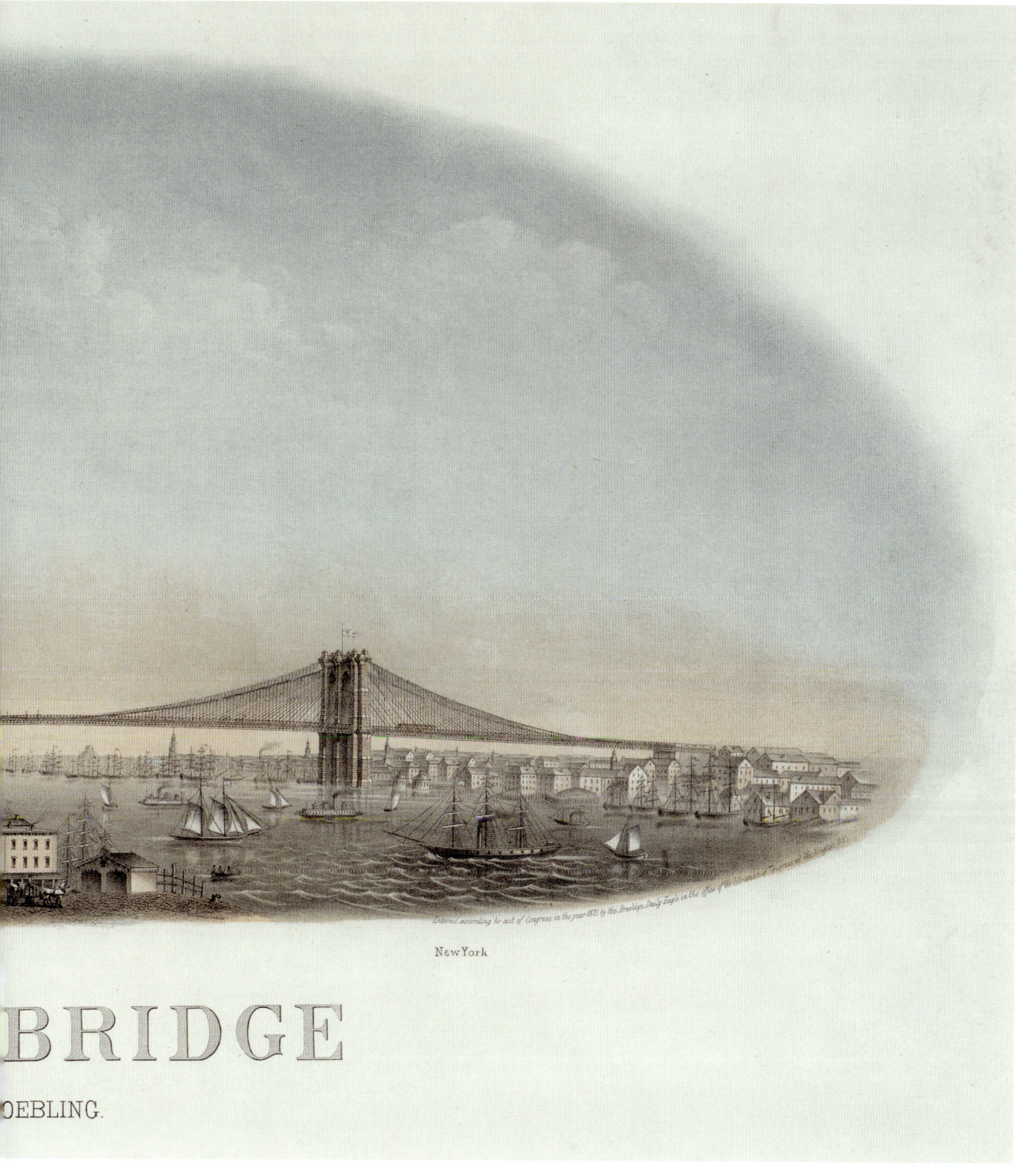

Print of the Brooklyn Bridge, here identified as the "East River Bridge," dated 1873 and published by the *Brooklyn Daily Eagle*. Note the name and title below this rendering of what would become world-famous as the Brooklyn Bridge: "Engineer, John A. Roebling."

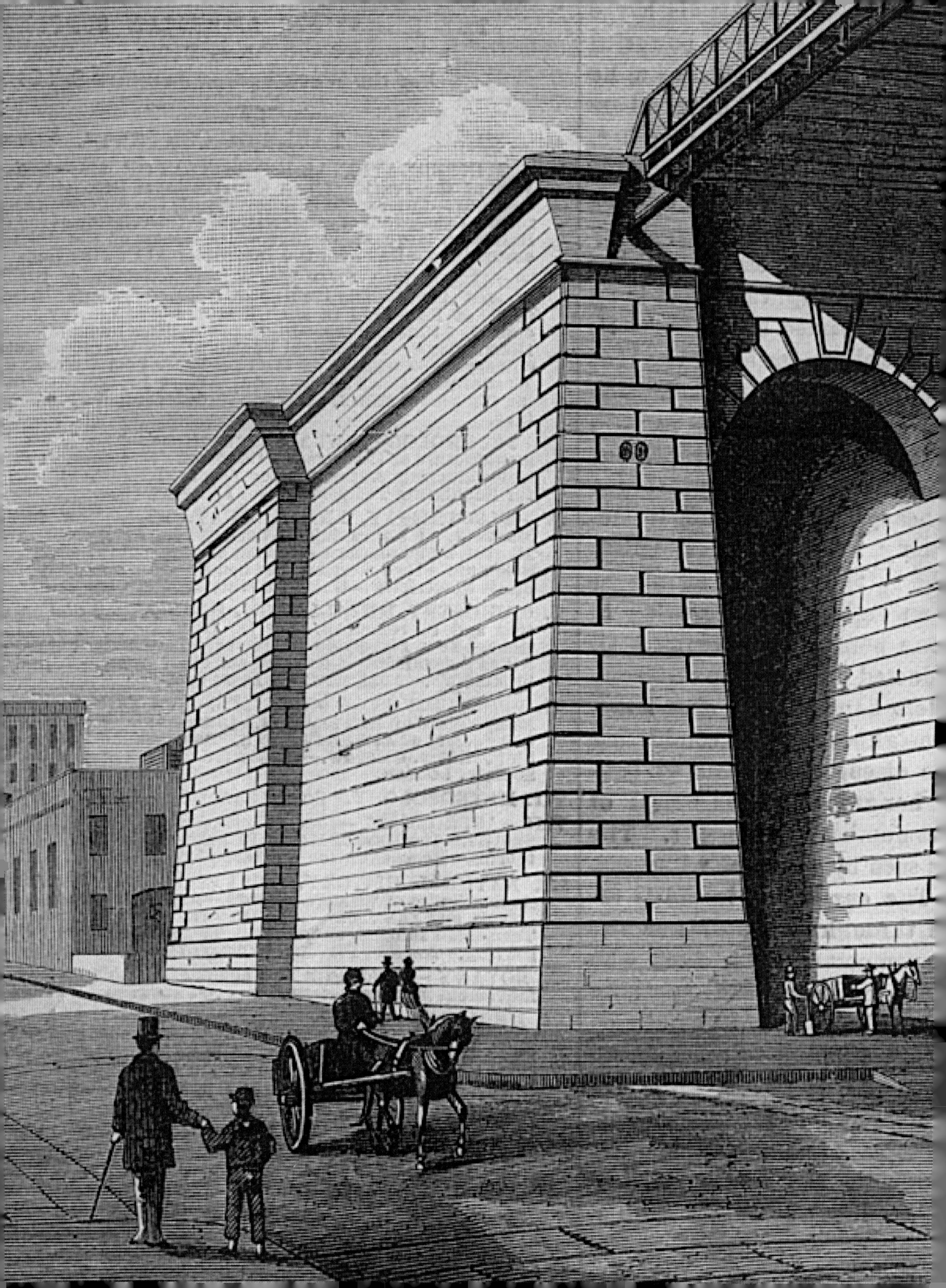

CHAPTER 5

Anchorages

1872–1876

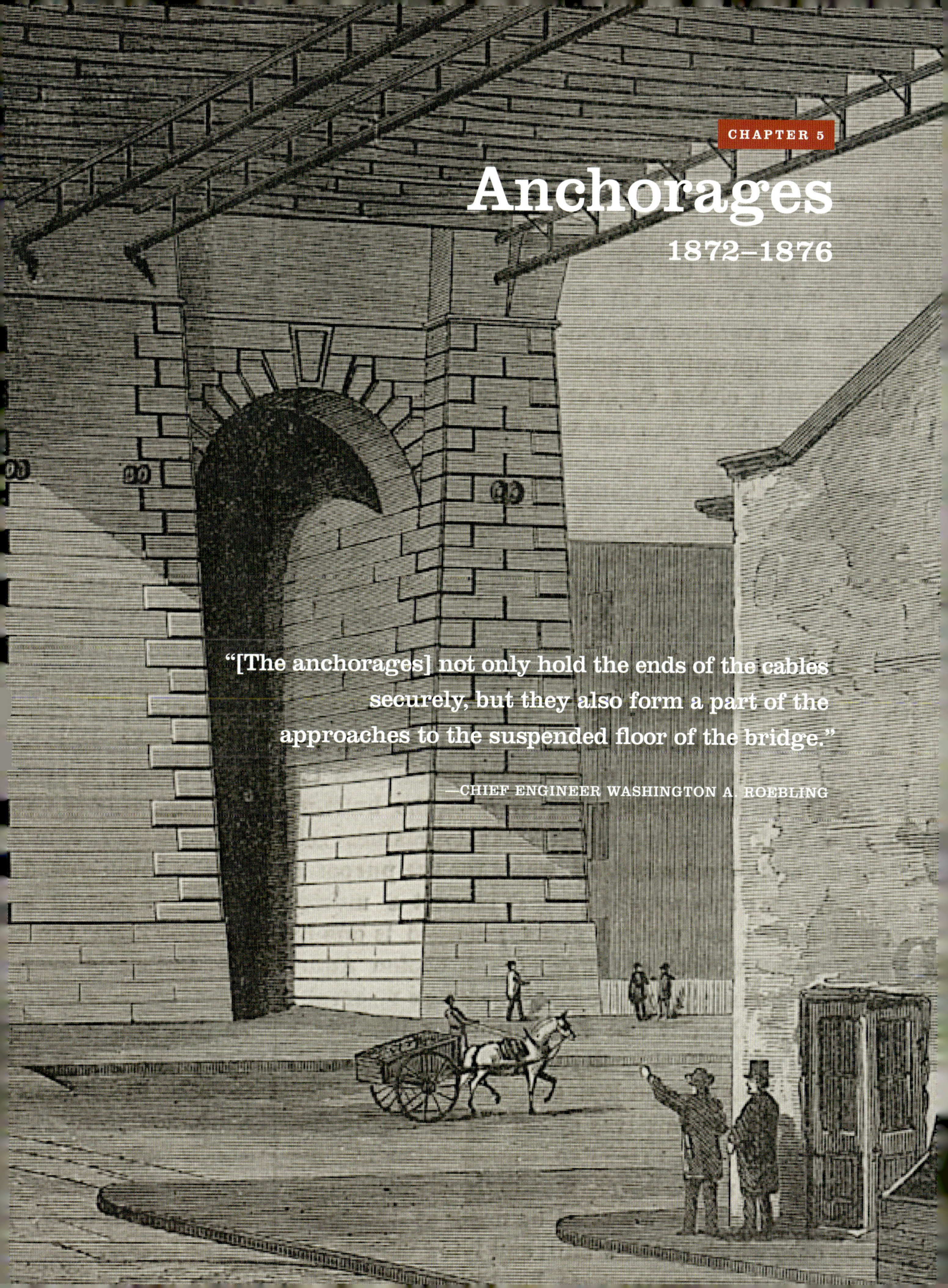

"[The anchorages] not only hold the ends of the cables securely, but they also form a part of the approaches to the suspended floor of the bridge."

—CHIEF ENGINEER WASHINGTON A. ROEBLING

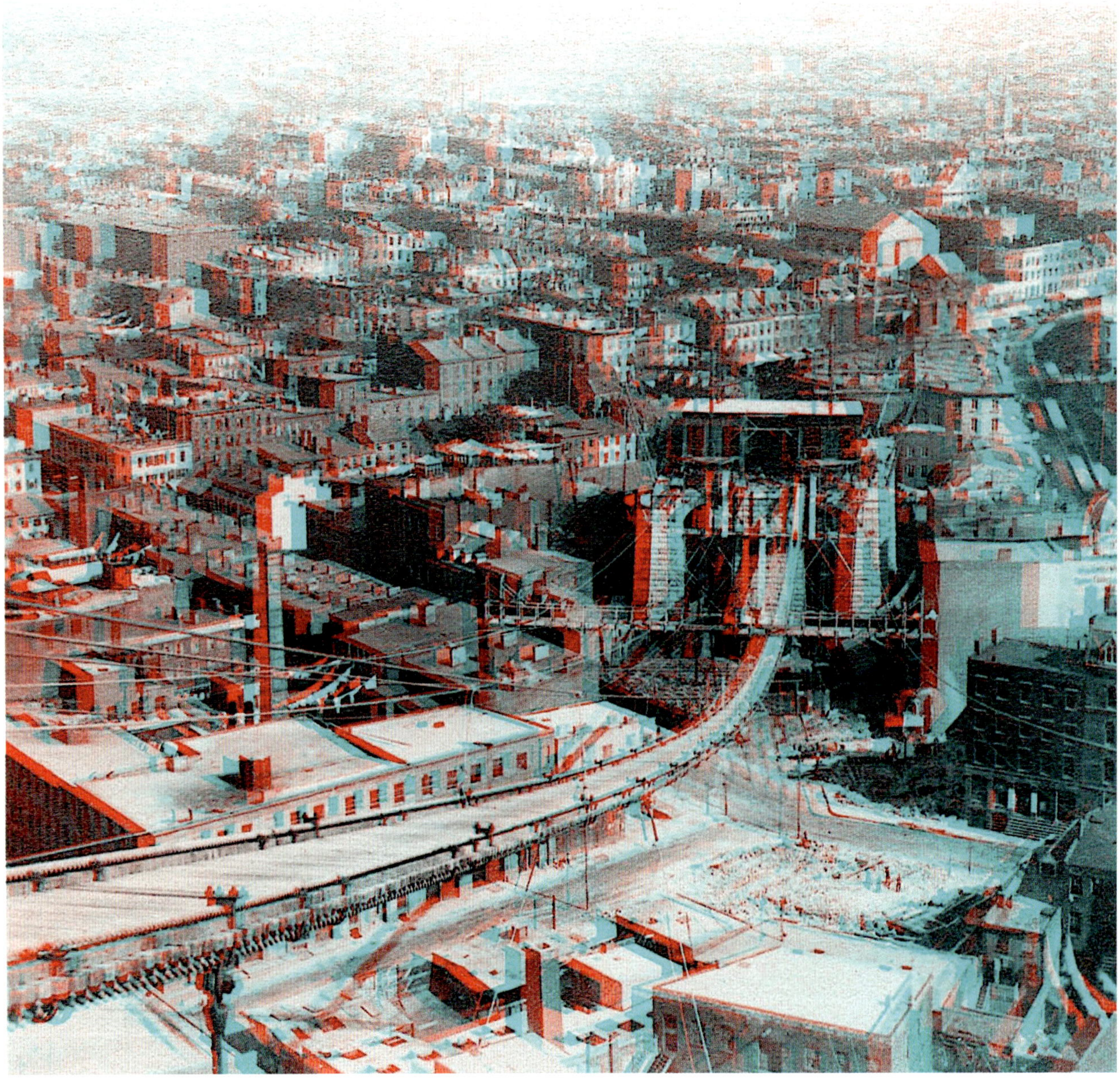

3D View from the bridge's Brooklyn tower, looking down toward the Brooklyn anchorage, with Fulton Street and downtown Brooklyn at right, ca. 1877. The construction of this anchorage is complete, and the footbridge is in place. Eight drums of steel wire for the cable making sit inside the wooden shed atop the anchorage, ready to be spun out toward the New York anchorage. A few of the cable wires are visible, running parallel to and at left of the footbridge. Courtesy of the Jeffrey Kraus Collection.

PREVIOUS SPREAD: "The New York Anchorage of the East River Bridge," *Scientific American*, January 8, 1876. The steel beams of the bridge's deck are visible at top center and right. This woodcut shows how the roadbed of the bridge passes across the top of the anchorage. Courtesy of Richard Haw.

Bridge anchorages, like caissons, provide a largely unseen element that is indispensable to the whole and critical to the success of a suspension bridge. Their main function is to firmly hold the two ends of each of the bridge's cables. Each cable end is attached to a chain of anchor bars. The bars, approximately thirteen feet long, are flat and made of iron. They are linked to other anchor bars to form an anchor chain. At the top, each anchor chain is connected to the strands of a cable; at its bottom, the anchor chains are connected to a huge anchor plate within the anchorage. Both the chains and plates are buried deep within the anchorage, kept in place by multiple courses of heavy stone blocks.

Charles Barnard, in his 1883 *St. Nicholas Magazine* article, explained that if the cables had been simply attached to the towers, the pull of the cables against the masonry would have been so great that they would have pulled the towers down. Instead, it was necessary to run the cables over the tops of the towers and connect the ends of the cables to each of the two massive stone structures, or anchorages, beyond the towers.

Washington Roebling described the anchorages of the Brooklyn Bridge as "artificial, for the reason that there was no accessible rock formation for a natural anchorage." Given the geography at the site, with no natural rock formations in sight, the anchorages had to be constructed from scratch. The two anchorages, as built, are huge stackings of stone blocks. Together, their stones weigh 60,000 tons. Just as in the tower construction, limestone blocks make up the interior, while granite was used for the facings. Both anchorages measure 119 feet by 129 feet at their bases and taper slightly toward their tops. Each is nearly 90 feet high—roughly the height of a seven-story building.

Washington explained that the anchorages "are not solid masses of masonry, but they have arched passageways running through them lengthways to economize material and labor." Washington further explained that the arrangement of the walls of the anchorage created "two large spaces in each, eighty feet long by twenty feet wide, and divided into six stories." These large empty spaces inside the anchorages could be used for storage or commercial purposes.

While the stacked stones of the anchorages provided the flat base and support needed for the approach roadways that ran across their tops, their main function was to firmly secure the ends of the bridge's four enormous cables, which run from one anchorage, up and over each of the two towers, and back down to the anchorage on the opposite side.

The stresses placed on the cables by the suspended bridge spans—and thus on the anchorages where the cables are held down—are immense. Washington noted that there are two key factors to the anchorages' success: granite and gravity. Granite was a hard and durable stone, "a material whose very existence is a defiance to the 'gnawing tooth of time.'" And gravity was a constant, used by engineers to their advantage across the centuries. As Washington wrote, "When I place a certain amount of dead weight, in the shape of granite, on the anchor plates . . . it will remain there beyond all contingencies." According to Washington, "The intention of the engineer is not merely to equal the strength of the cables with that of their anchorage, but also to give the anchorage a solidity to be absolutely unaffected in the slightest degree by the incessant pull of loads and tug of storms for a hundred years, so that no loosening or vibration can ever be initiated."

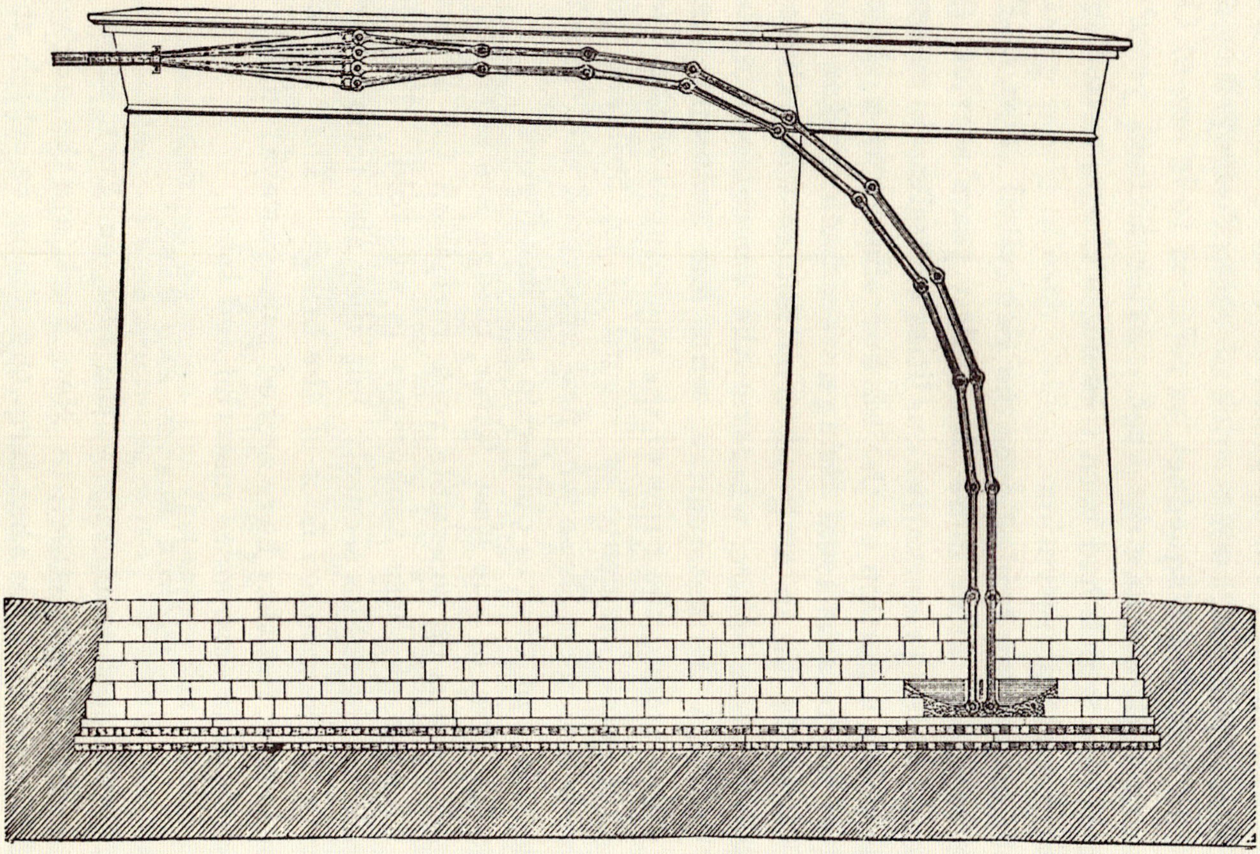

This drawing, which is reproduced in *Concise Description of the East River Bridge* by Master Mechanic E. F. Farrington, is by John A. Roebling. Published in 1881, Farrington's lectures on the construction of the Brooklyn Bridge include a side view of an anchorage and the anchor chain that was buried inside it. At the lower right is an anchor plate with a set of anchor bars attached to it. These anchor bars form the anchor chain that arches to the left, toward the cables. At top left, cable strands are shown attached to the top anchor bars of the chain. The anchor plates, anchor bars, and their connections to the cables were not visible once the anchorage was complete: they were surrounded by the large stone blocks that were piled up as the chain was built, forming the anchorage. Washington Roebling referred to the wall of the anchorage closest to the tower—here, the wall at left—as the "anchor wall." Note the slight incline of the top—that is the angle of the approach as it ran from ground level up to the nearby tower.

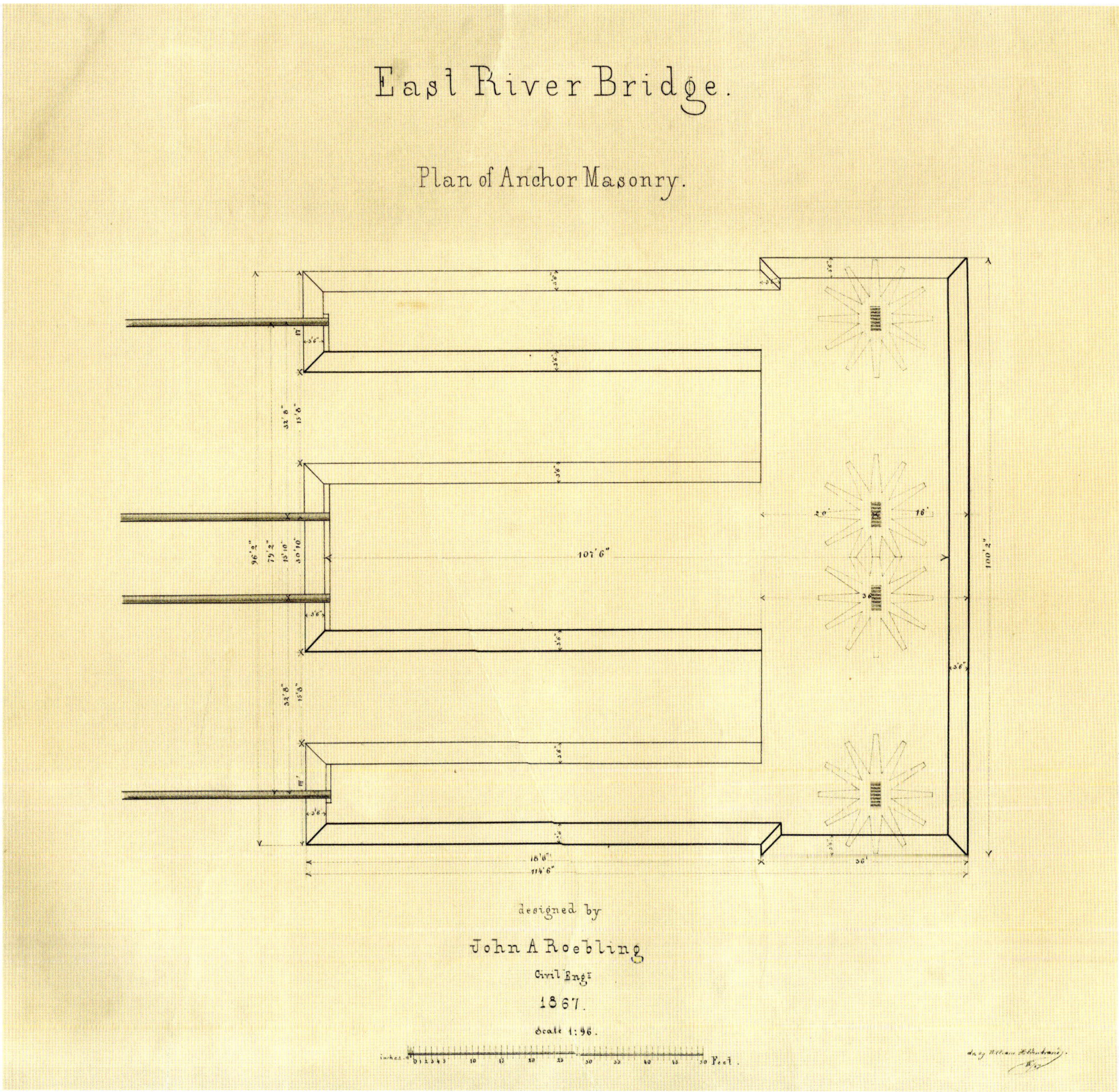

"East River Bridge/Plan of Anchor Masonry/designed by John A. Roebling/Civil Engr./1867." A notation at lower right identifies this as having been drawn by Assistant Engineer Wilhelm Hildenbrand in 1867. It shows one of the anchorages from above, with the bridge's four cables entering from the left. At right are the four anchor plates, shaped like starfish, buried within the walls of the anchorage. This drawing makes it apparent that the anchorages, as large as they were, were not solid throughout. Rather, they had three "fingers" that were separated by empty spaces. RPI.

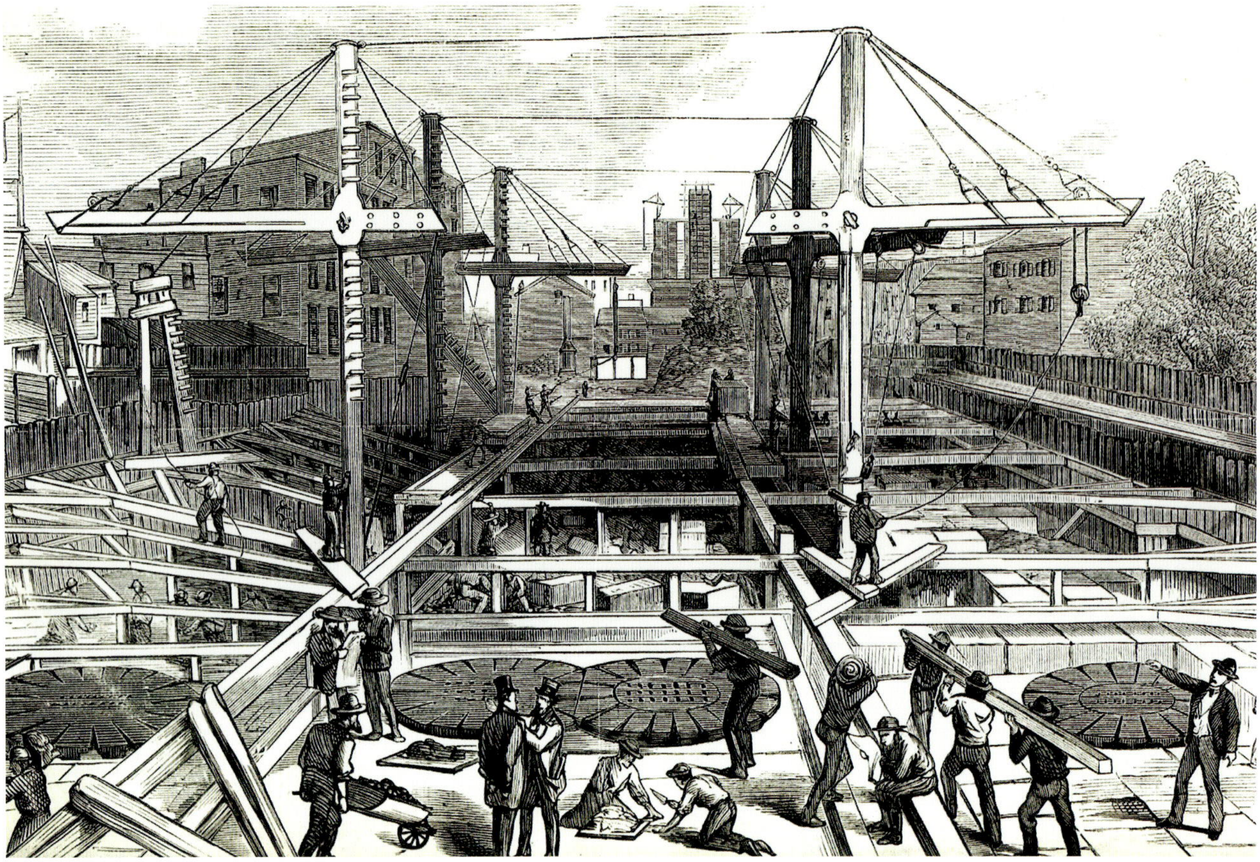

A woodcut showing the four anchor plates positioned at the base of the bridge's Brooklyn anchorage, *Frank Leslie's Illustrated Newspaper,* August 30, 1873. Work on the New York anchorage would not begin until May 1875. Hence, this must be the Brooklyn anchorage, where work began in February 1873—it would take more than two years to complete. Note, in the center distance, the bridge's Brooklyn tower topped by derricks. Courtesy of Richard Haw.

In 1873, soon after the work on the anchorages had begun, Washington Roebling described how the cable ends were held down:

> The four cables enter the masonry at an elevation of eighty-two feet, and after passing into the wall for twenty-five feet, they form a connection with the chains. The latter consist of cast steel bars, thirteen feet long and ten inches wide, by one and a half inches thick, arranged in ten sections, each containing nineteen bars and forming in all four curved lines extending from the cables to the anchor-plates. The connections are made by six-inch steel pins.

As indicated in his above description, Washington had considered using steel for the anchor bars. But Krupp Works, the manufacturer in Germany that Washington wanted for the job, could not assure him that their steel bars would have the strength needed. So instead he chose iron. The bars were ultimately made in the United States by three companies and were described by one observer as "splendid specimens of what American Manufacturers can do."

Every anchor bar widens toward both ends to allow for an "eye hole;" thus they are sometimes referred to as "eyebars." These openings accommodate a pin that connects an anchor bar to one above and one below, thereby creating the chain. Each chain consists of ten bars linked together. Nineteen of these chains are attached to each of the four anchor plates. The anchor bars vary slightly in size, depending on their position in the chain; the bars higher up in the chain have to bear more strain and, therefore, are somewhat larger.

While in Europe, learning about the latest in suspension bridge technology, Washington had discovered that engineers there typically placed their anchor chains in tunnels to afford access for maintenance. He decided that this method was "contrary to the genius of the American people, with whom everything has to look out for itself; hence, in the arrangement of this anchorage, the chains are inaccessibly preserved and are not entrusted to the neglect of posterity." Washington's plan was to encase the chain, as it rose, within courses of large limestone blocks. This would not only aid in holding the chains securely in place as the enormous weight of the cables pulled on them, but, since any sort of maintenance would be impossible, it made them "self-reliant," a treasured American quality.

Holding down each anchor chain is a huge anchor plate embedded at the base of each anchorage, on the side farthest away from the towers and river. Each anchorage has four identical anchor plates, one for each cable. The anchor plates each weigh 46,000 pounds and measure 16.5 by 17.5 feet on their face, with a thickness of 2.5 feet at their center.

The center of each anchor plate has a large oblong opening, through which the bottom set of nineteen anchor bars was slid and then held in place using a sturdy wrought iron pin. After this lowest group of bars was secured to the plate, massive stone blocks were placed atop the anchor plate and around the bars, fixing them in place. Additional layers of stone were then stacked until only the top ends of those first anchor bars were visible.

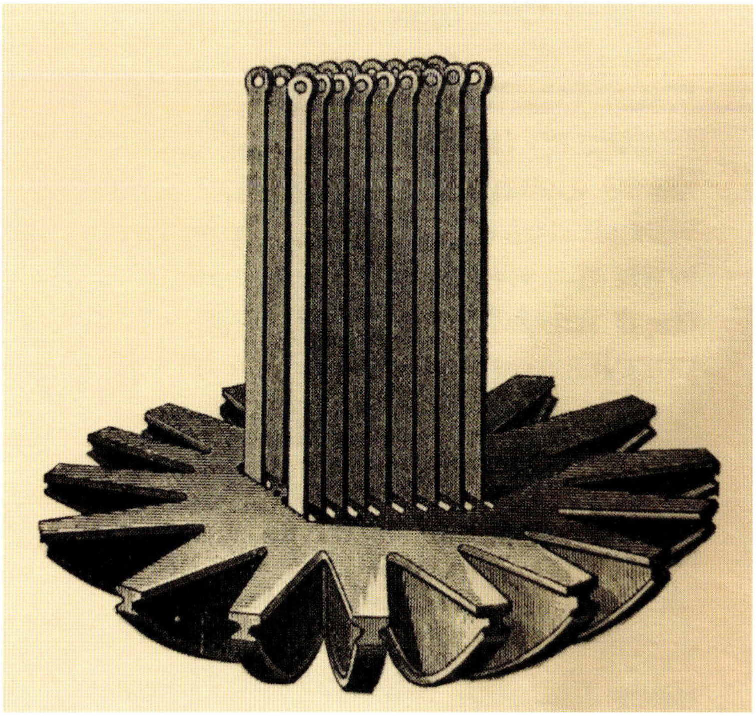

An anchor plate, with the first tier of anchor bars threaded into its slots and pinned below, *Scientific American,* January 8, 1876. Author's collection.

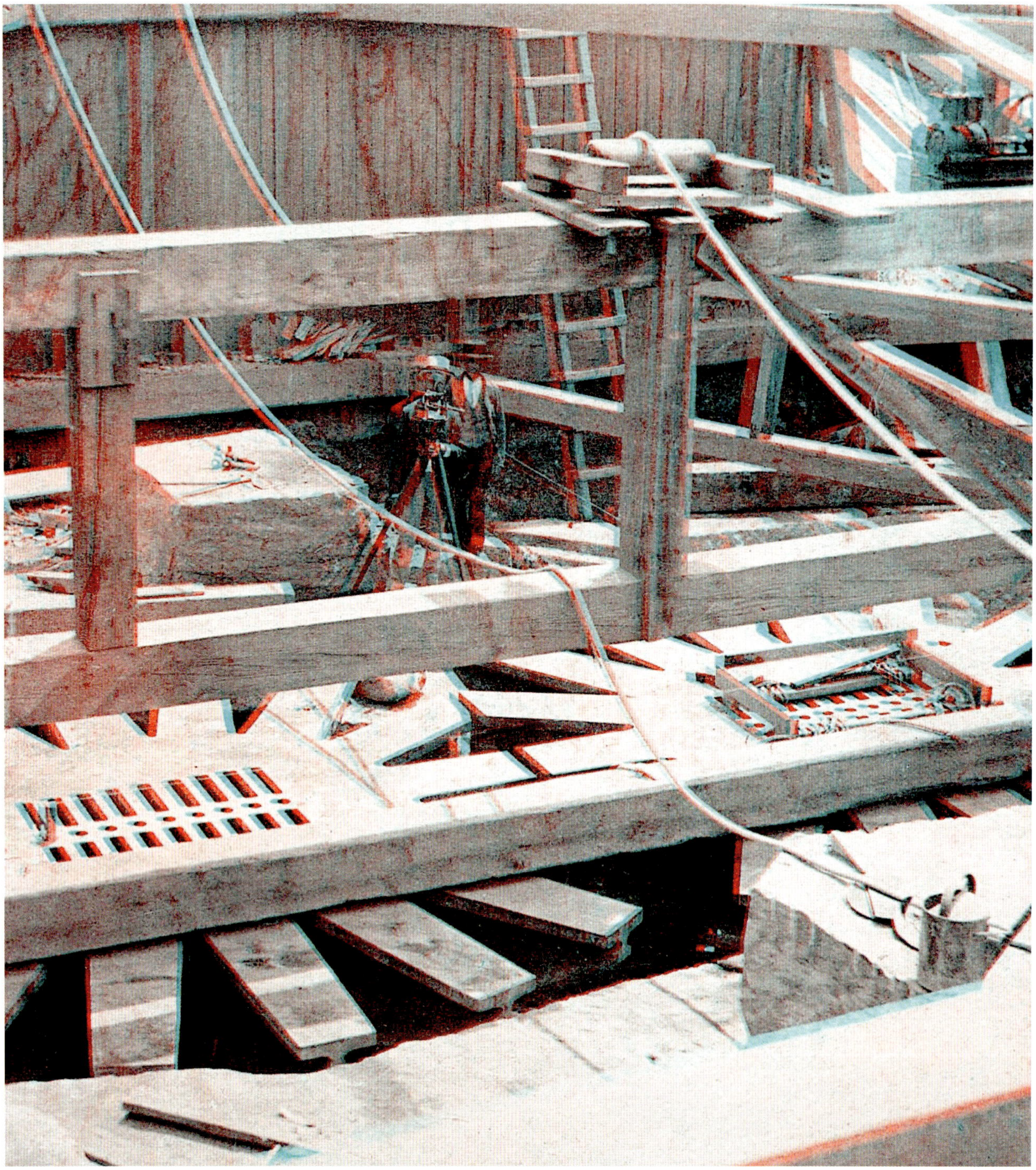

3D A label on the stereoview from which this anaglyph was created identifies this as the bridge's Brooklyn anchorage. Two of the enormous iron anchor plates—each with multiple radiating arms—are visible here. At the center of each plate are two rows of nine rectangular slots, through which eighteen anchor bars will be threaded. Their ends, once inserted into the slots, will be secured by inserting heavy iron pins through the holes below the plate. Once fastened in place, these anchor bars—the lowest tier of the anchor chain—will ensure that the chain is secured at the bottom of the anchorage. Courtesy of the Jeffrey Kraus Collection.

Six-inch-long pins, ranging in diameter from five to seven inches, were then used to connect the next set of nineteen anchor bars. As each set of bars was added to one of the four anchor chains, the anchorage's masonry rose with them; courses of huge limestone blocks were placed around the most-recently-added anchor bars, firmly holding them in place. The workers continued building the anchor chain until ten sets of bars were connected, forming an incredibly strong chain. Once the stone courses had been piled up around the highest and last set of anchor bars, it was time to connect the cable ends (one strand at a time) to the anchor chain.

This photograph, ca. 1876, shows the New York anchorage nearing its full height. The anchorage and its derricks soar above the nearby buildings, some of which will have to be demolished to build the approach. In the distance, at far right just below the horizon and across the East River in Brooklyn, is the distinctive tower of the Fulton Ferry Terminal. Author's collection.

At the base of the anchor chains the bars are vertical. But, as the chains rose, they began to arc toward the towers, finally resulting in a fully horizontal set of bars at the top. This arc was engineered so that the top bars of the anchor chain are about twenty-five feet from the wall of the anchorage closest to the nearby tower. Once the nineteen strands that make up each of the four large cables were connected to the last set of bars of each anchor chain, additional stone blocks were stacked on top, burying the chains within the anchorage. As a result, the cables do not emerge from the top of the anchorage, but rather from its wall.

By summer 1876, the anchor chains were complete and the stone courses of the anchorage were built up to the height of the top of the last set of anchor bars, which were now ready to be attached to the cables. Just one small matter: the cables had to be made. But before that could happen, a footbridge had to be built and "cradles," or platforms, positioned, allowing the workers to monitor the spinning of the cables.

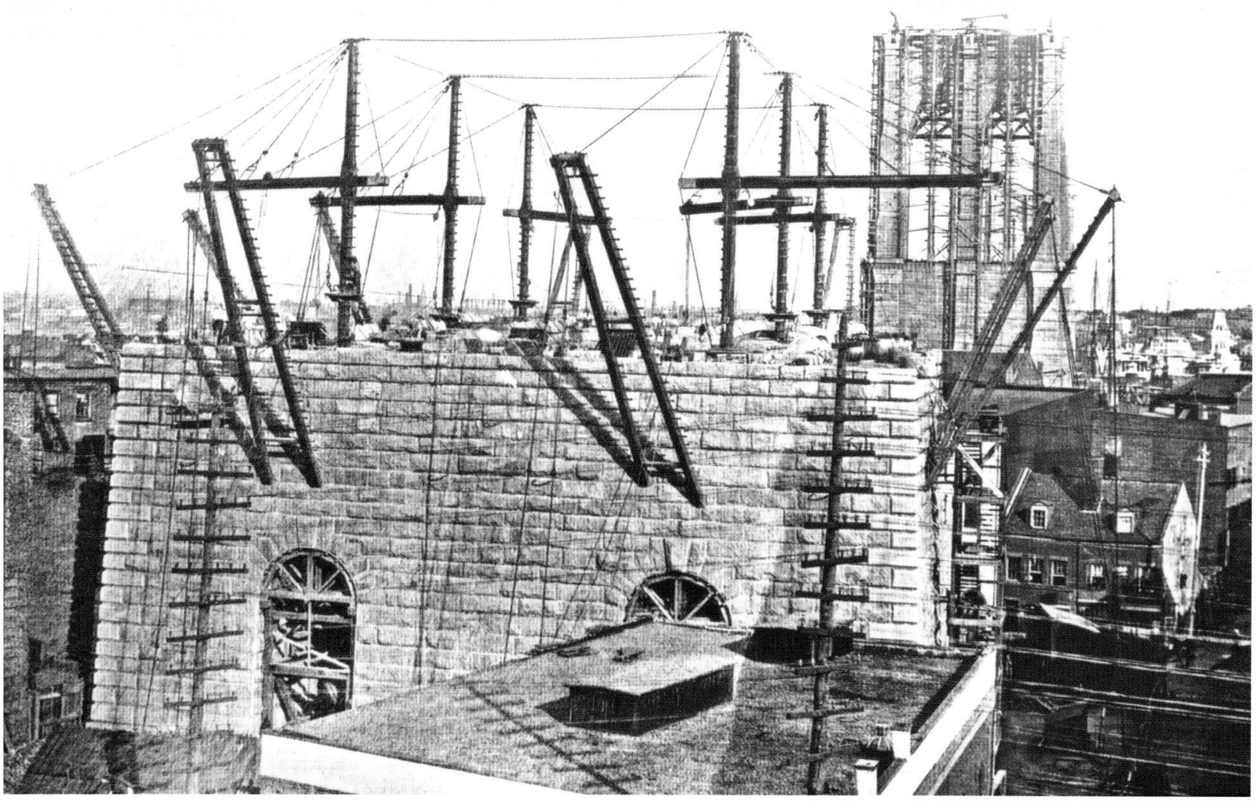

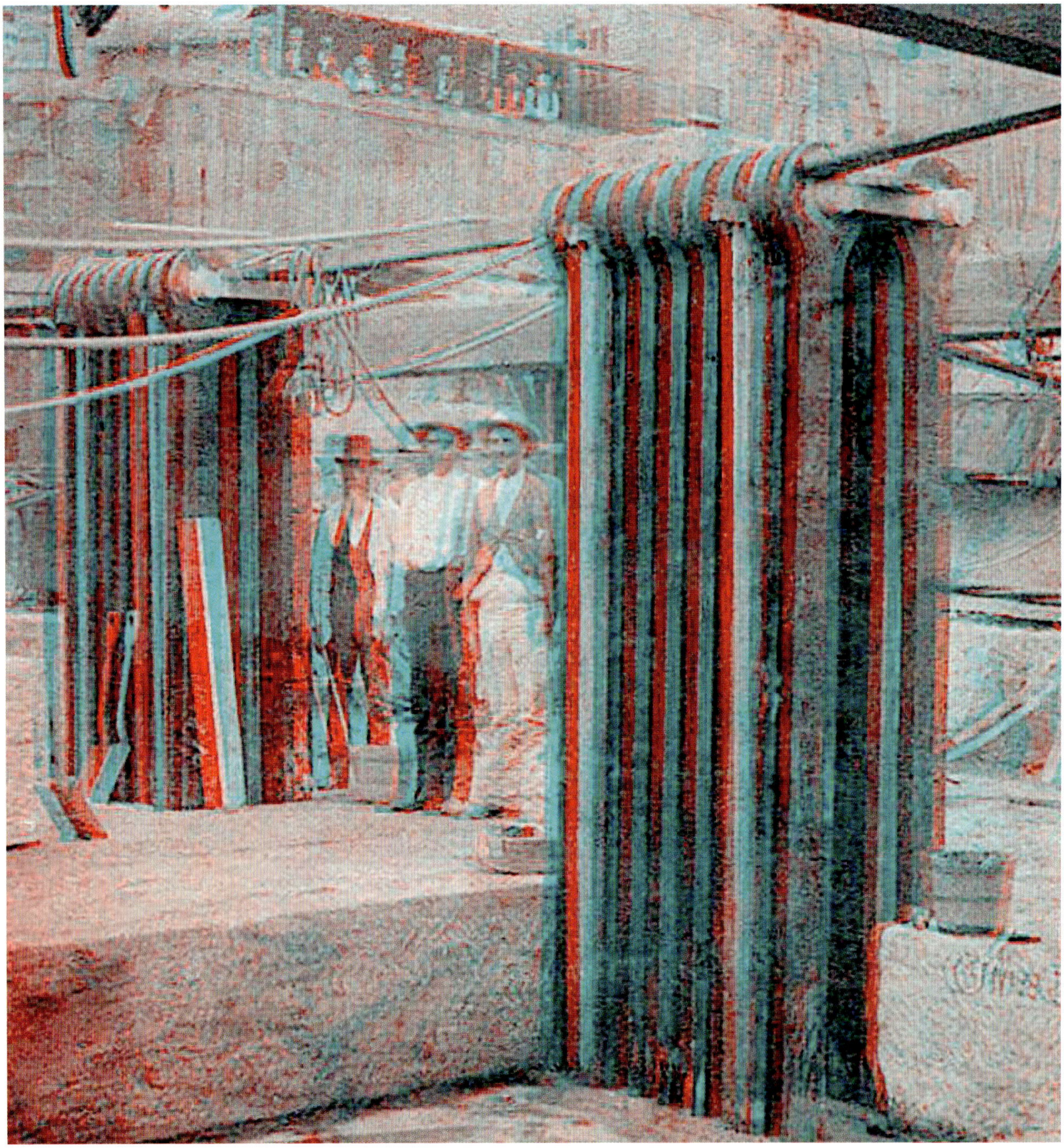

3D An anaglyph showing two sets of bars forming two of the four anchor chains in an anchorage. These bars are already connected to the chain below them, and the workers have begun to surround the cluster of bars with stone blocks that will be piled up just below the "eyes," or openings, at their top (see next photograph). Here, pins have been passed temporarily through the eyes at the top of the bars to hold them in place. Wooden props, or spacers, press against the pins to keep the eyes aligned. Courtesy of the Jeffrey Kraus Collection.

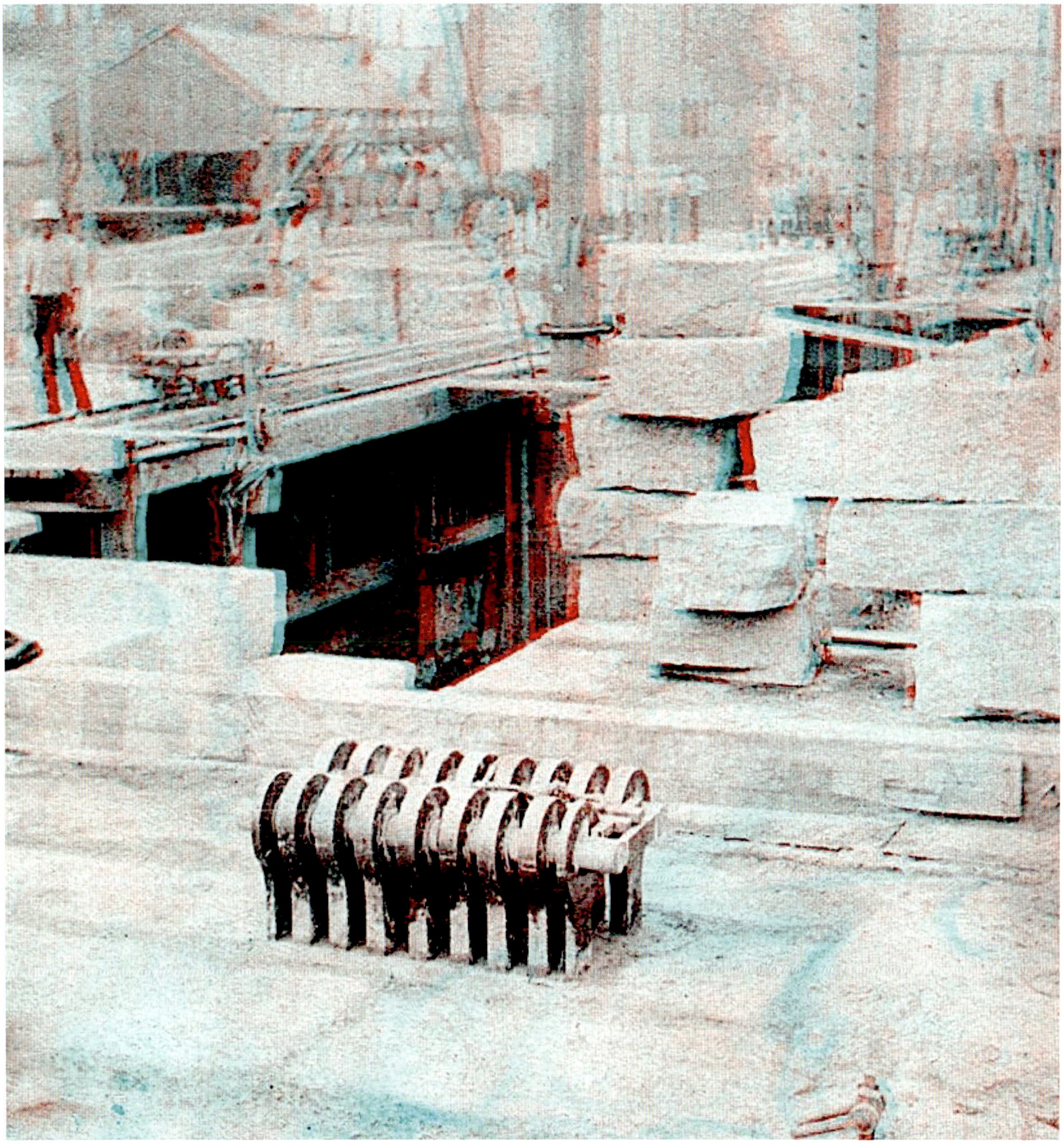

3D

Anaglyph showing the upper ends of a set of anchor bars, now surrounded by stone, awaiting the attachment of another set of anchor bars to them to continue the chain upward. This process of encasing and connecting was repeated until the anchor chain, two rows of nine bars each, was completed. These anchor bars are vertical, indicating that they belong to one of the two lowest sets of bars in the anchor chain. Starting with the third set, the bars and, thus, the chain, begin to arch toward the bridge towers, with the tenth and last set fully horizontal. This last set of bars would then be connected to one of the strands that makes up one of the cables. Courtesy of the Jeffrey Kraus Collection.

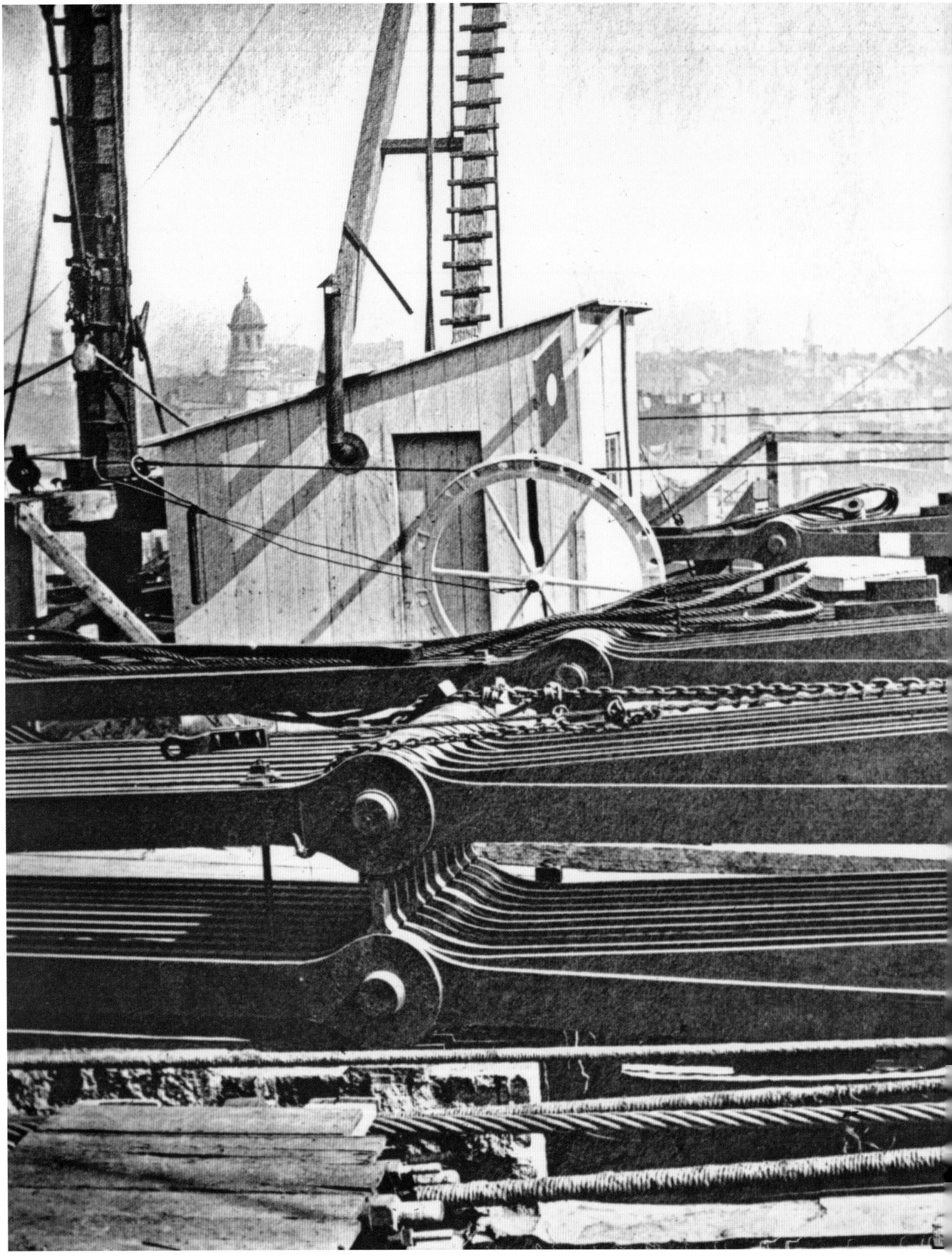

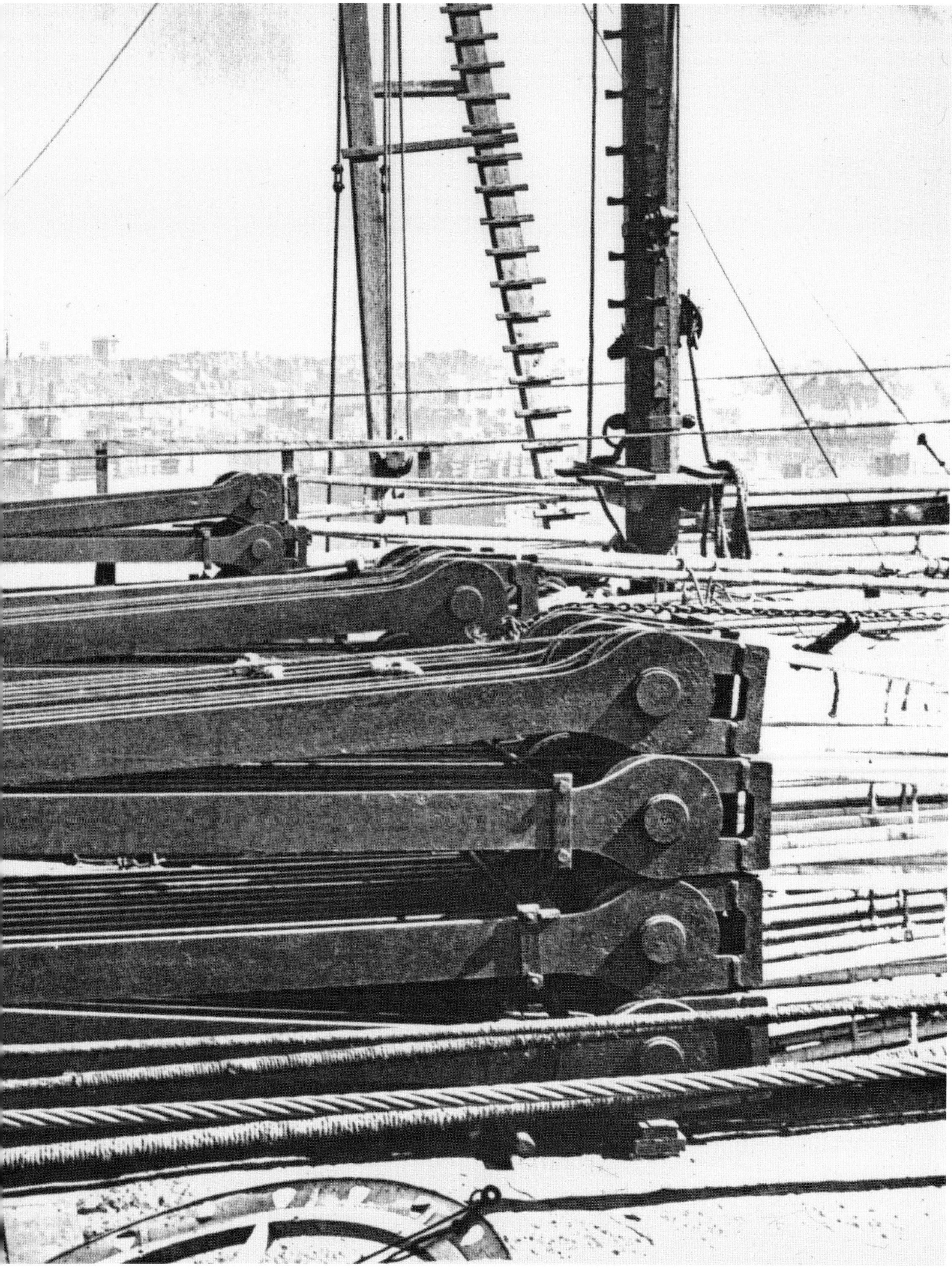

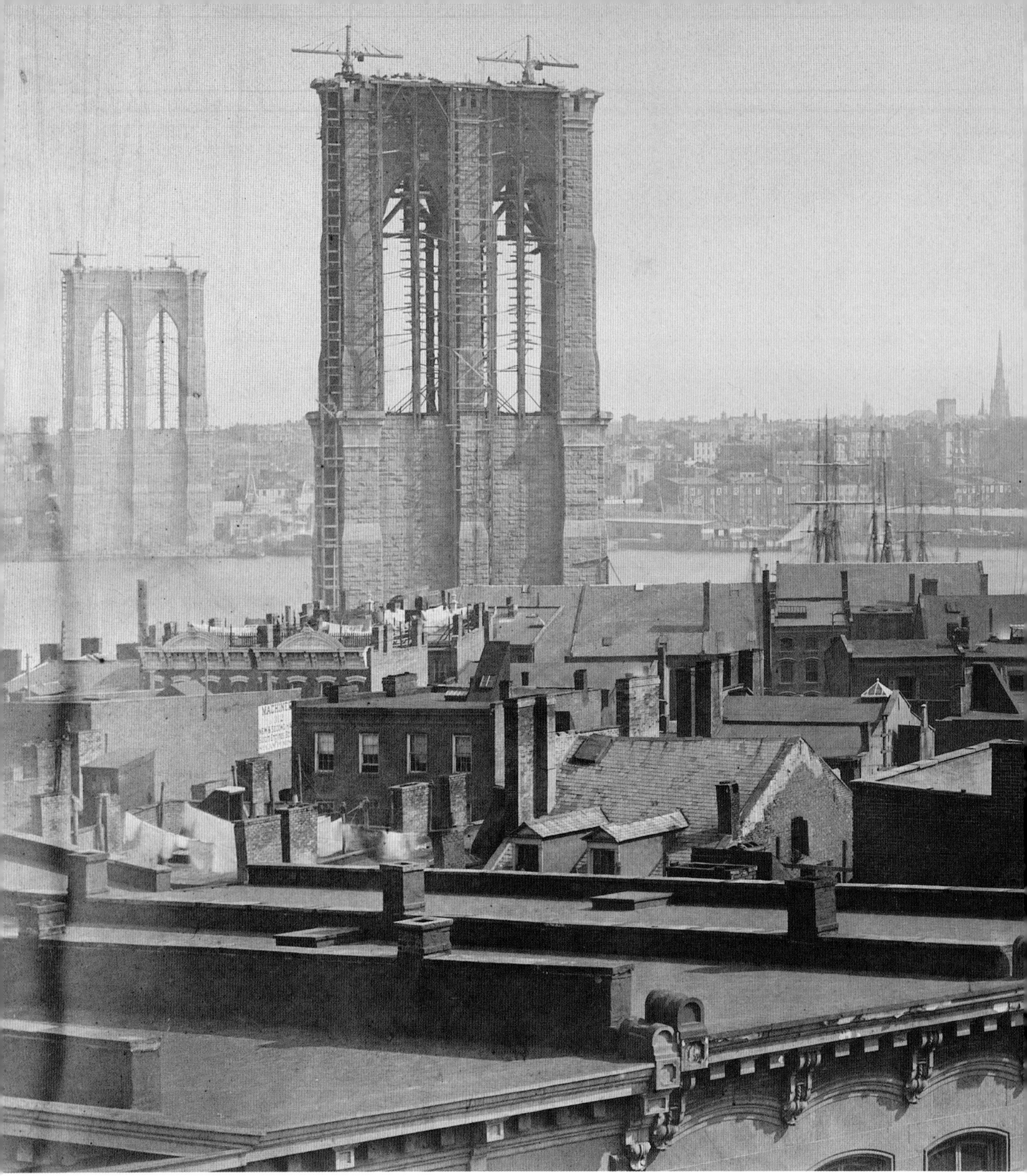

PREVIOUS SPREAD: This photograph, printed from a lantern slide, shows the set of anchor bars that forms the end of the anchor chain. At right, the cable strands (white) are pinned to the anchor chain. The photographer also shot the image on page 186; here, the image shows more of the anchor bars. Author's collection.

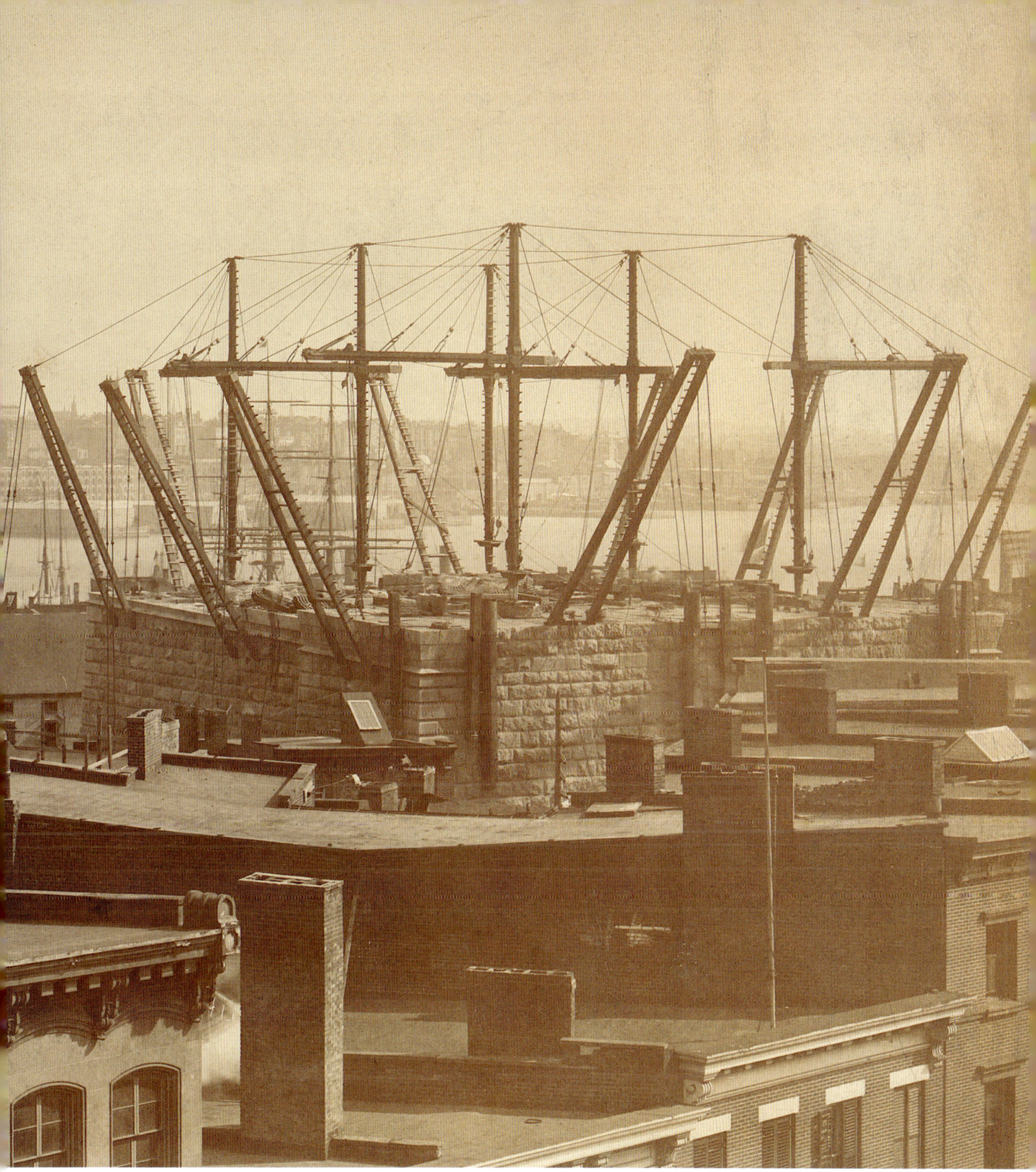

ABOVE: "View of work from New York. July 1876." Penciled on the back of this photograph: "New York Anchorage previous to Cable-Making." The New York anchorage, at right, is close to reaching its full height; it would be completed by the end of that month. A profusion of derricks are in position near the anchorage's top to lift additional stone blocks up onto it. The anchorage, though not yet at its full height, already rises above the nearby buildings. MCNY.

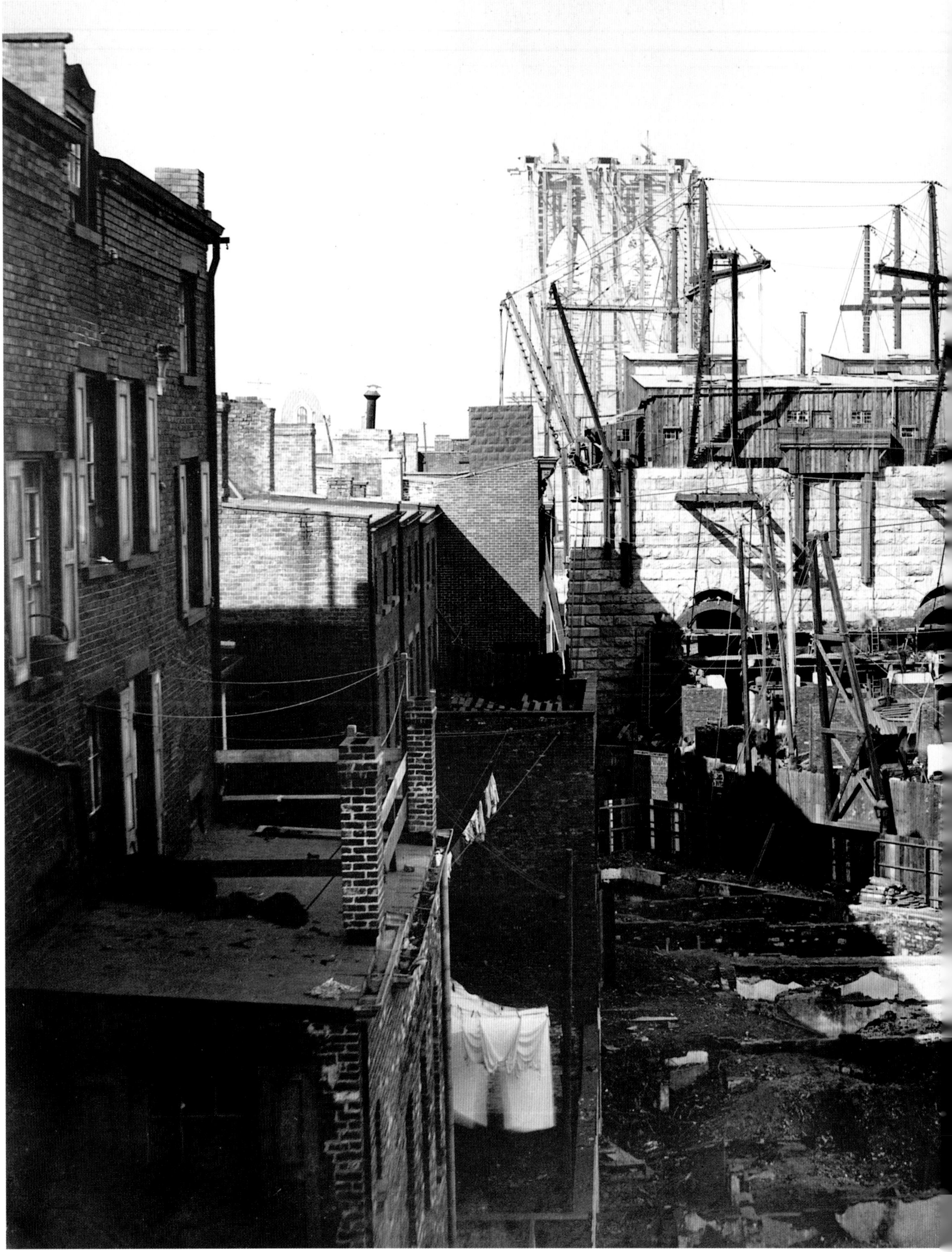

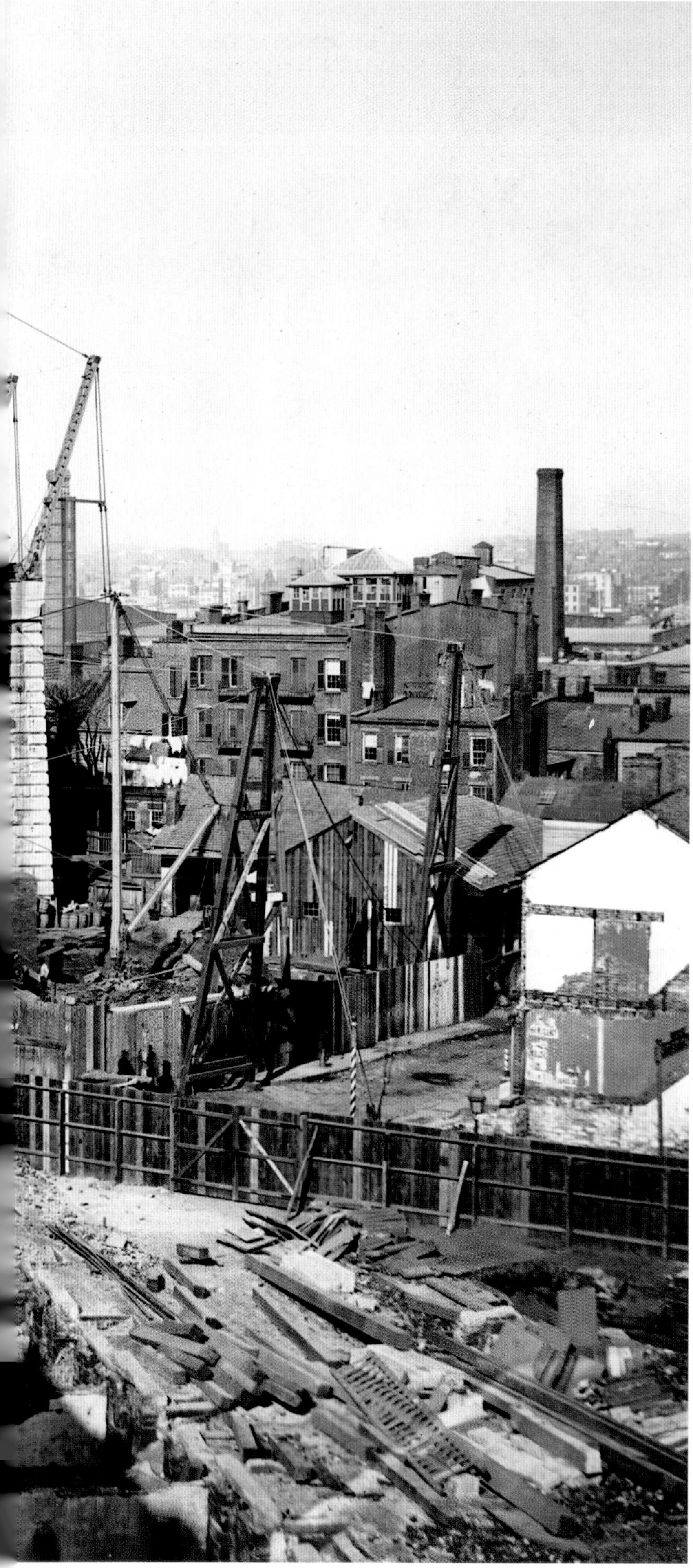

This photograph of the bridge's Brooklyn anchorage and tower, ca. 1877, gives some perspective on their relative positions. Each anchorage was placed on the land side of the nearest tower and held the ends of the massive steel wire cables that draped down from the tower top. The area in the foreground would soon be cleared for the Brooklyn approach. Note the wooden shed atop the anchorage; it held the large spools of steel wire rope for cable making. There was no such shed atop the New York anchorage; all of the wire was fed from the Brooklyn side. RPI.

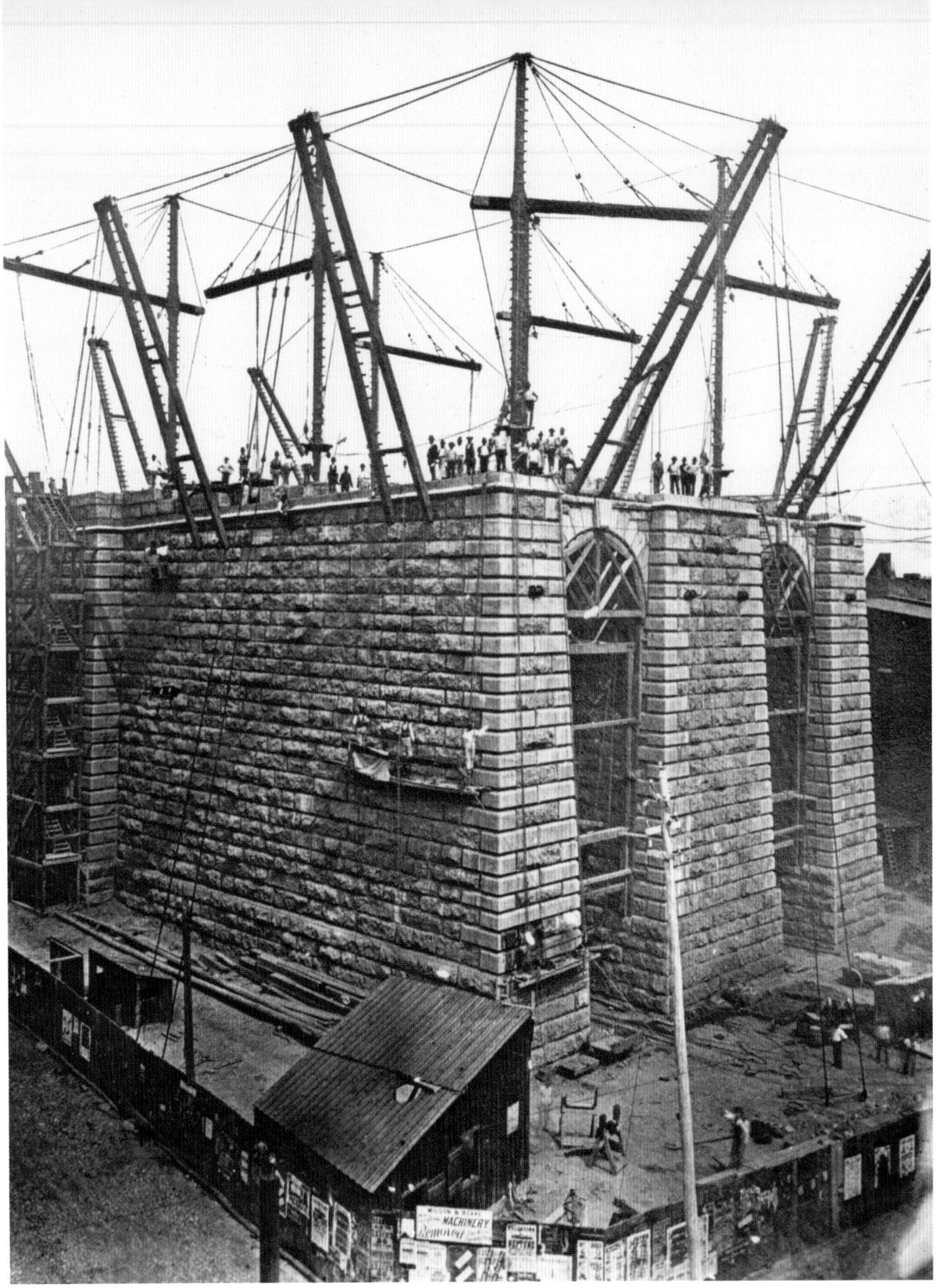

ABOVE: A lantern slide photograph of one of the anchorages under construction, rising to just above its arches, just short of its full height. Note the falsework—the wood framing—underneath the arches, keeping the stones in place until the mortar sets. This photograph was likely taken by Silas A. Holmes. The mount for a print of the same image in the Museum of the City of New York, while unidentified, is similar to the mounts of the museum's other Brooklyn Bridge photographs that are stamped as the work of Holmes. Also, given that the men atop the masonry have stopped work to pose for the camera, Holmes was likely the photographer: his images are the only identified ones in which the workers stand still while his camera captures their image. Note, however, that three masons on the scaffold on the left side of the anchorage continued to fill the joints between the granite blocks, their backs to the camera. Author's collection.

OPPOSITE: A photograph of the New York anchorage, ca. 1876. The ends of all four cables are fully encased by granite, at top right, where the slats of the footbridge are also visible. Temporary wooden stairs, at left, lead up to the roadbed of the bridge approach, allowing workers access to the area just beyond the anchorage. MCNY.

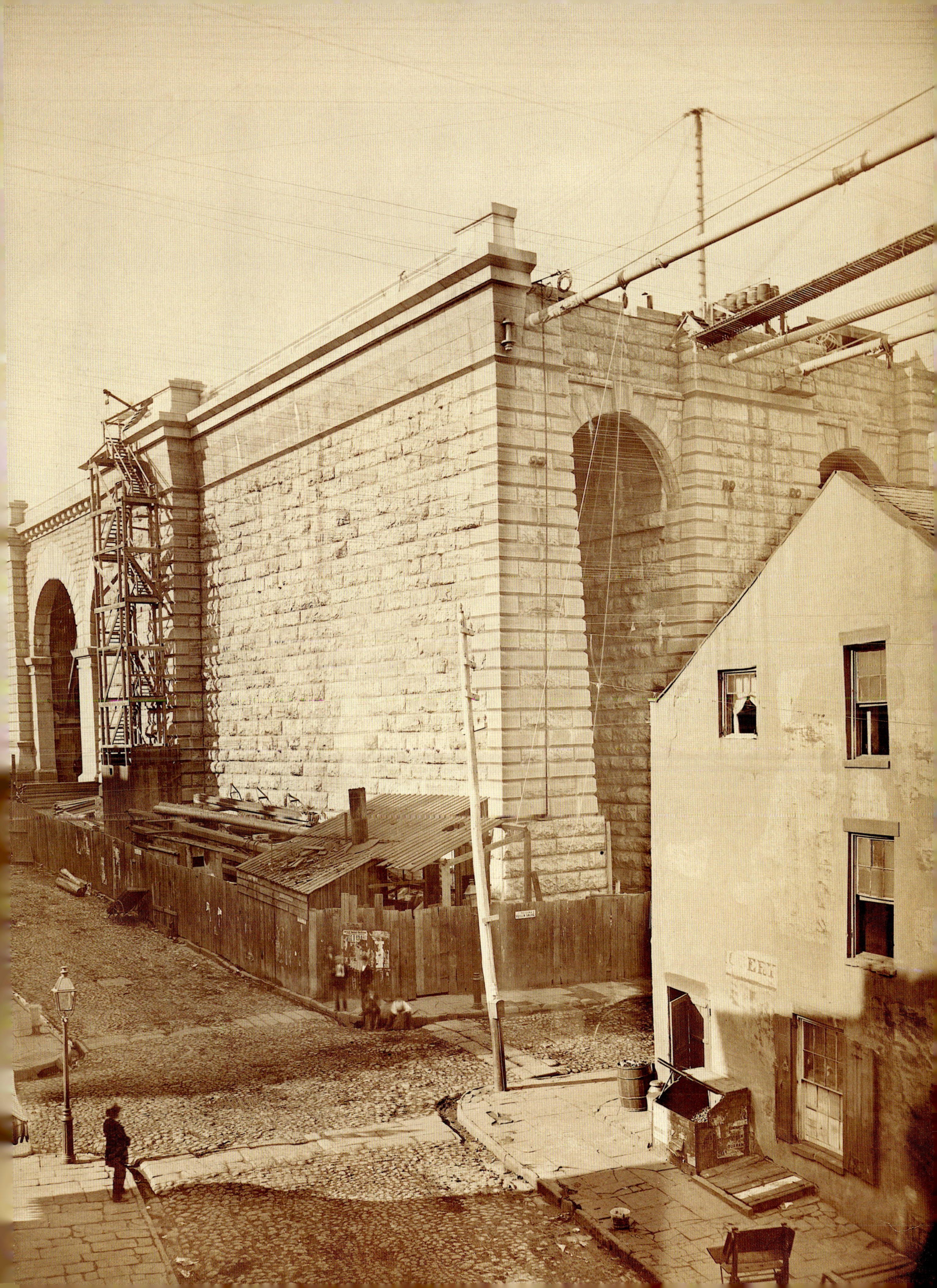

This photograph, printed from a lantern slide, shows the work on one of the anchorages almost complete, ca. 1875. The cables have already been attached to the anchor chains, but those attachments are no longer visible, buried in the courses of granite. At the center of the image, a ladder allows workers to climb up to the footbridge (its slats are clearly visible at top center). Note also the wooden shanty, at lower left, likely for the storage of equipment or for the men to use to change into and out of their work clothes. Of the sand in the foreground, which may have been dug from inside the caisson, Master Mechanic E. F. Farrington wrote: "The sand proved to be admirably adapted to building purposes, and it was transported in boxes and cars over our railway to the yard at the foot of the tower, and there gathered in an enormous heap, which was afterwards used in constructing the masonry." Author's collection.

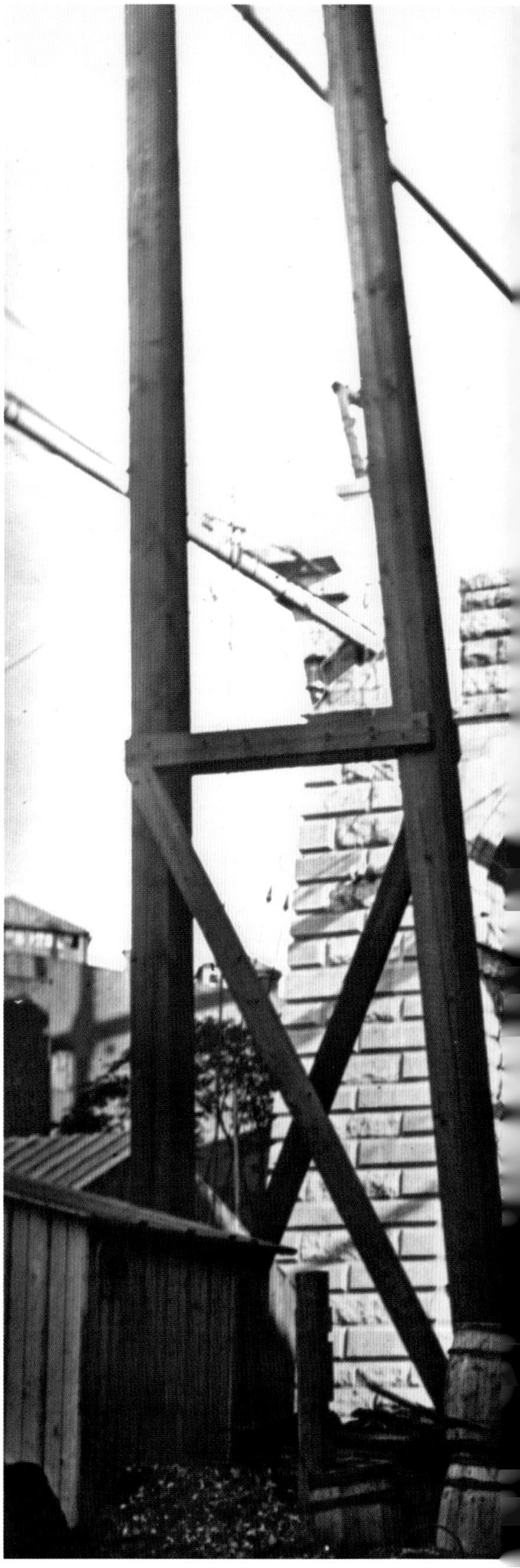

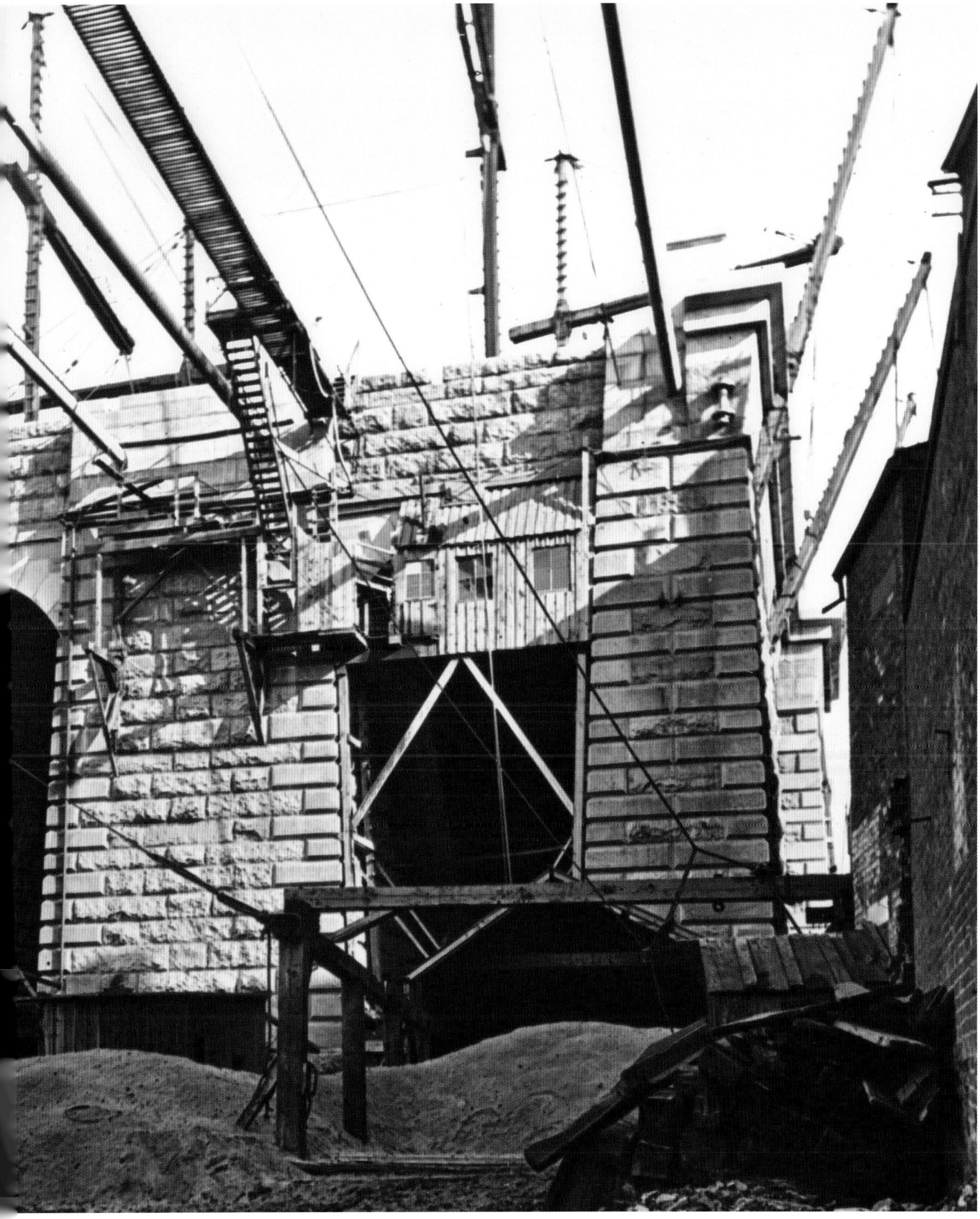

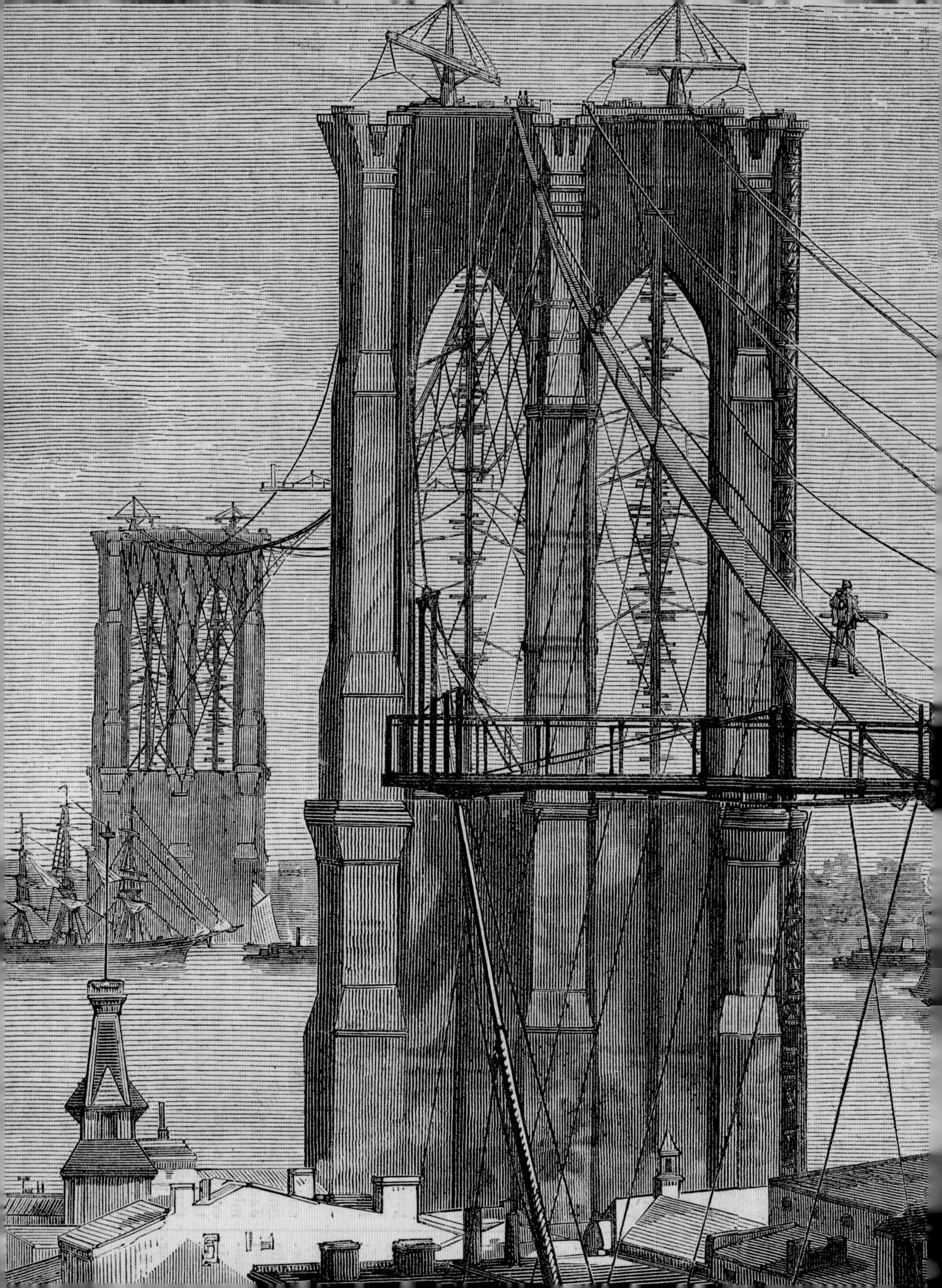

CHAPTER 6

Footbridge and Cradles

1876–1882

"…'this pathway in the sky,' as some have called it."

—MASTER MECHANIC E. F. FARRINGTON

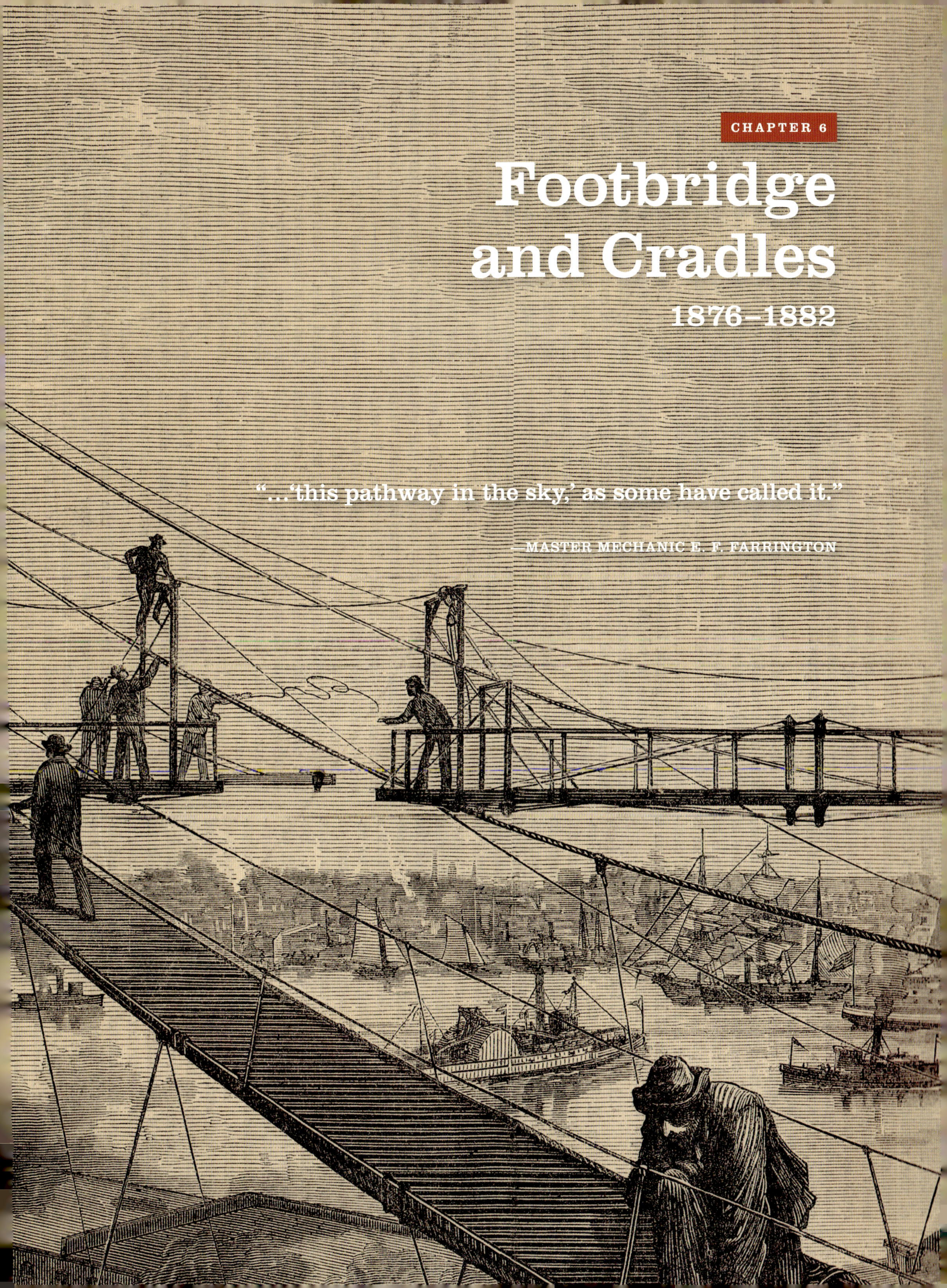

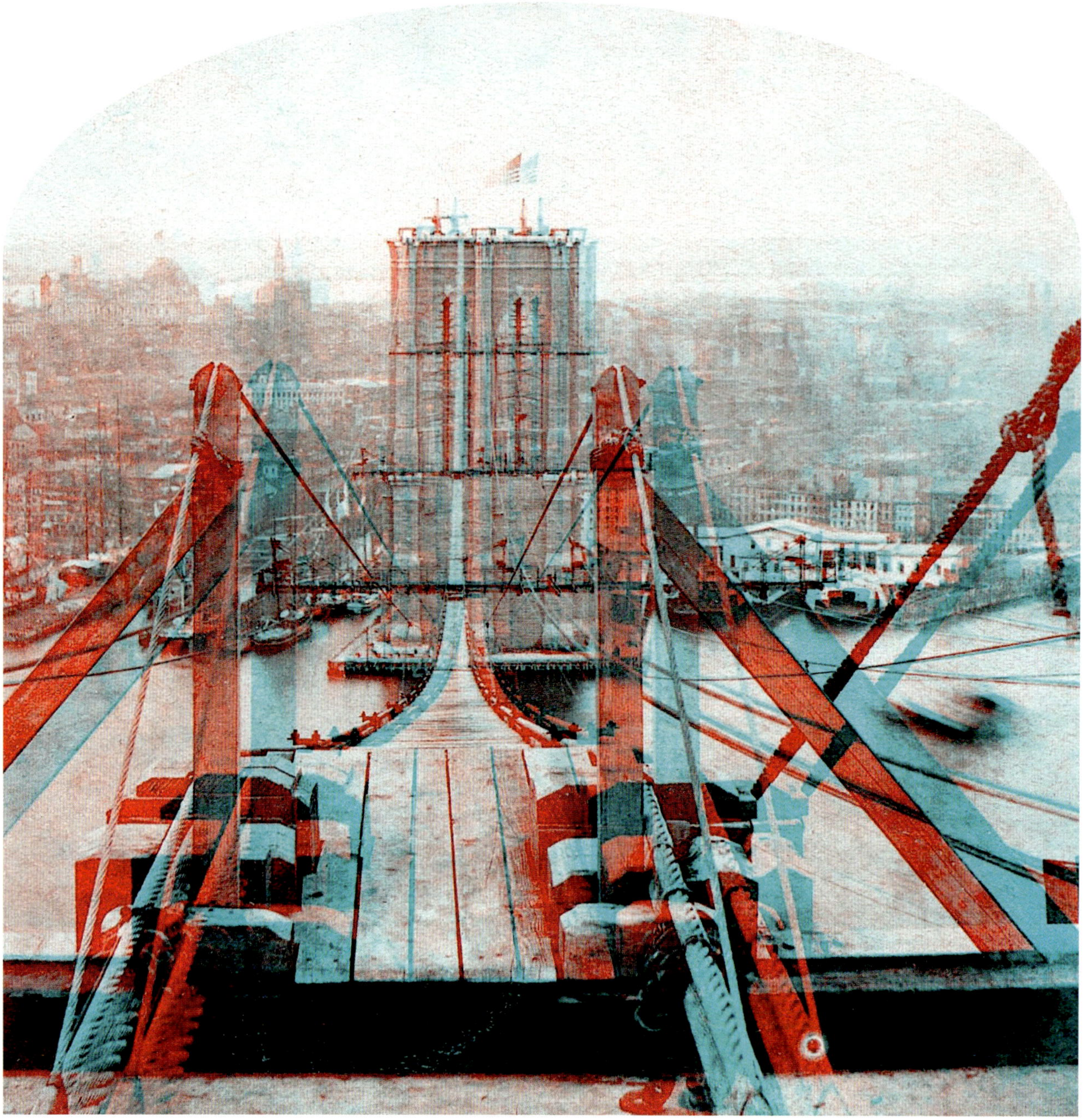

3D Anaglyph of the footbridge taken from atop the bridge's Brooklyn tower, looking toward Manhattan. The 3D effect is outstanding here. Courtesy of The Green-Wood Historic Fund, Jeffrey I. Richman Stereoview Collection.

PREVIOUS SPREAD: "A Promenade in Mid-Air—The Brooklyn Ascent to the Bridge Tower—Drawn by Schell and Hogan," *Harper's Weekly*, March 31, 1877. The primary purpose of the footbridge was to allow workers to get to the cradles—the horizontal scaffolds at right angles, and to either side, of the footbridge—to monitor the cable wires and keep their arcs uniform as they were spun from anchorage to anchorage. The cradles were in pairs and joined together to create a continuous catwalk across the width of the bridge, so the men could reach the wires of all four cables. Author's collection.

Unlike other features of the bridge that had already been completed—the caissons, towers, and anchorages—the footbridge was not intended to be a permanent part of it. Rather, it was a means to an end, put in place temporarily to aid in the construction. The footbridge allowed workers to move conveniently about the site, from anchorage to anchorage. Indeed, Master Mechanic E. F. Farrington described the footbridge as "indispensable as a means of ready communication with all parts of the work." But the footbridge's primary purpose was to place the workers within easy reach of the full length of the cable wires, from Brooklyn to New York, so that they could adjust those wires as needed.

From the footbridge, the men were able to gain access to the "cradles," temporary platforms placed high up across the length of the bridge. These cradles allowed the workers to monitor the steel wires as they were being spun into cables, ensuring that every wire was arranged as uniformly as possible before being bound into strands and eventually cables. Standing in the cradles, the men could adjust the position and tension of a wire as necessary. After the cables were completed, the footbridge and cradles had served their purpose, and they were taken down in 1882.

The footbridge could not be built until 1876, when the towers and anchorages were completed. Yet Washington Roebling had begun to think about its design as early as 1874, as indicated in his papers concerning the "High Foot Bridge" dated June–July of that year, which have survived in the Roebling Collection at Rensselaer Polytechnic Institute. In those documents, Washington outlined the dimensions, strains, deflections, and "[d]escriptions of mode of getting the ropes across and elevated in their place."

Washington's "Specifications for the Wire Ropes for the East River Bridge," issued early in 1876, included instructions that "proposals will be received until the first day of February 1876, for the manufacture and delivery of the following steel & iron wire Ropes required for the erection of the temporary foot-bridge, cradle cables, and other appliances necessary for the construction of the main cables of the East River Bridge." The footbridge itself was made up of oak slats held aloft by wire ropes slightly more than 3,600 feet long. Each of these slats was 4 feet long and 3 inches wide by 1.5 inches thick. The slats were laid crossways on the suspended wire ropes, leaving 2-inch gaps between them "to permit a free passage of the wind and thus reduce oscillation."

The footbridge was completed in early February 1877. As described succinctly by David McCullough in *The Great Bridge*, it was "a sort of hanging catwalk strung from city to city, draped above the river at an elevation sixty feet higher than the actual roadway would be." The footbridge hung down near the center line of the bridge, two hundred feet in the air—not a good height for the faint of heart. However, with guy wires and storm cables attached to it, the footbridge was pretty stable. Contemporary descriptions noted that it typically rocked from side to side but only a bit, and rarely did the footbridge move up and down.

Addressing the matter of working at such heights, Master Mechanic E. F. Farrington wrote: "No man can be a bridge-builder who must educate his nerves. It must be a constitutional gift. He cannot when 200 feet in the air, use his brain to keep his hand steady. He needs it all to make his delicate and difficult work secure. They must plant their feet by instinct . . . and be able to look sheer down hundreds of feet without a muscle trembling."

He further noted that his workers quickly took to the job: "[N]o one but myself had any previous experience in this kind of work, but the men being for the most part sailors and riggers, readily learned the business, and became very expert in it."

As news of the completion of the footbridge spread, newspapermen were dispatched to report on it. The Bridge Company welcomed them to walk across it, but only when accompanied by Master Mechanic E. F. Farrington, upon whom the duties of chaperone fell. One reporter did not have fond recollections of his stroll along the footbridge, and his description suggests that the master mechanic had embraced this duty as an entertaining diversion:

> The undulating of the bridge caused by the wind, which was blowing a gale, the gradually increasing distance between the apparently frail support and the ground, the houses beneath bristling all over with chimneys, looking small enough to impale a falling man, the necessity of holding securely to the handrail, to prevent being blown off, produced sensations in the reporter's head—and stomach—never experienced before. In vain he glanced furtively into his companion's face to detect any signs of flinching on his part. Stolidly the master mechanic kept on, and the reporter fancied once that he caught a backward glance of enjoyment at his discomposure.

Strangely enough, the footbridge was open for a time to the general public. No gates blocked access to it. No security guards barred entry. In a single day, soon after a notice signed by "WA Roebling, Chief Engineer" was posted stating that it was "safe for only 25 men at one time," about one hundred people climbed the workers' ladder up to the top of the Brooklyn anchorage, walked across it to the footbridge entrance, and headed out over the East River. *The Brooklyn Eagle* questioned the sanity of those who wanted to do so:

> While Revs. Drs. Storrs and Buddington and several excellent ladies are moving in the matter of providing a new insane asylum for this city, a considerable number of our people are proving the necessity of such an institution and their own fitness to be life occupants of it . . . by crossing the footbridge . . . without call, without necessity, out of no business or artistic impulse, and from sheer foolhardy and peripatetic "cussedness."

Some of those who availed themselves of this once-in-a-lifetime adventure soon regretted their decision. One such man was Brooklyn Bridge President Henry Cruse Murphy, who had championed the bridge in the New York State Legislature, so much so that he was dubbed "the Moses of the Brooklyn Bridge." (Sadly, Murphy died mere months before the bridge opened, just as the "promised land" was within sight.) Like many others, Murphy decided to walk the footbridge, but it did not go well for him: "While I looked upward or ahead I was all right; but I chanced to look down . . . and I determined I couldn't afford to lose the president of the company just then, and so I went back."

Soon after his own adventure, Murphy saw a print in an illustrated newspaper depicting thrill-seekers up on the footbridge who, regretting their choice of how to spend their day, were down on their hands and knees in terror. In reaction, he decreed that access to the footbridge would henceforth be limited to those who applied for, and were granted, passes. For a brief time, that requirement reduced traffic, but soon so many thrill-seekers were applying that Murphy was spending an hour every morning doing nothing but interviewing applicants.

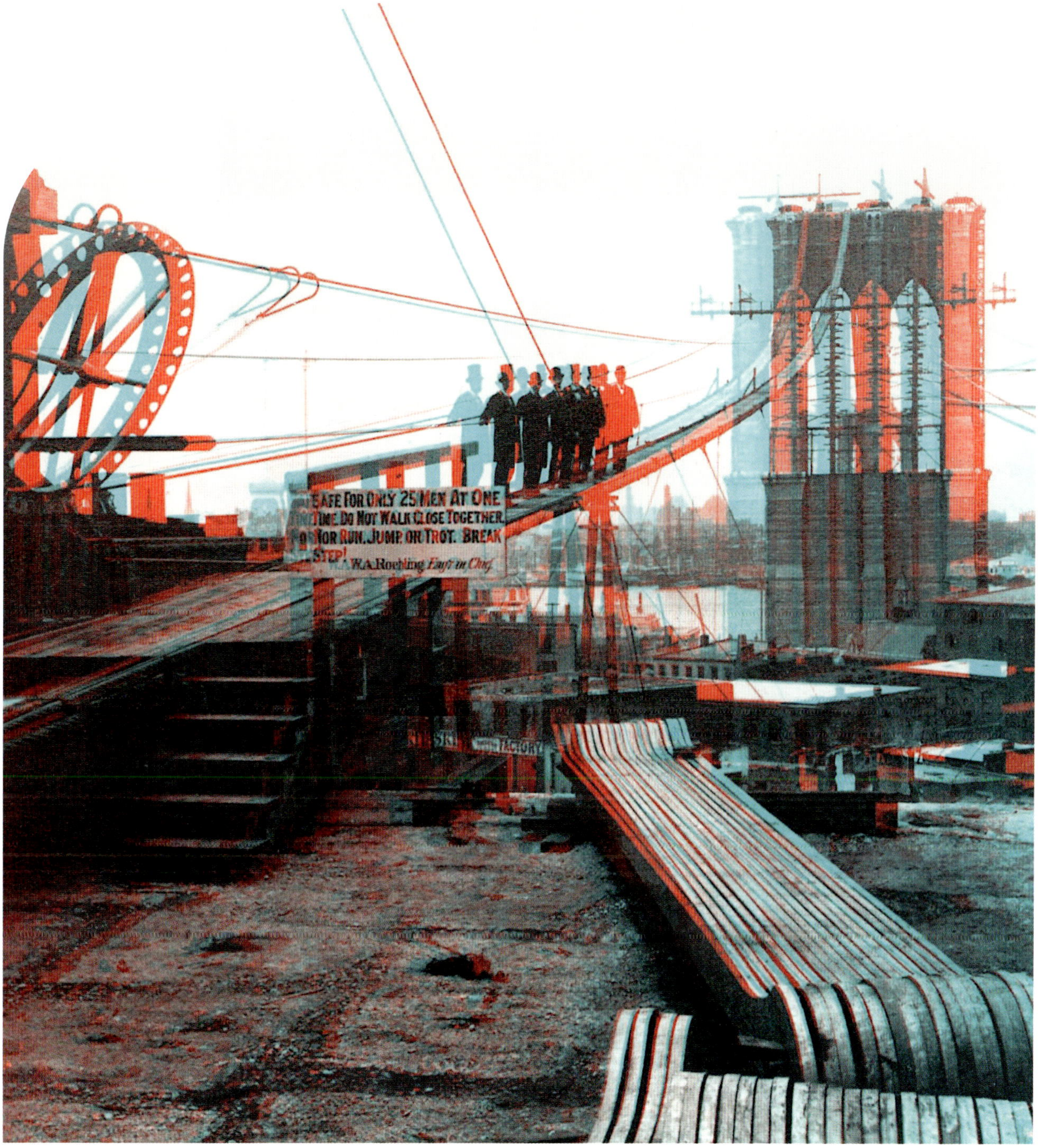

3D The entrance to the footbridge atop the Brooklyn anchorage, ca. 1877. David McCullough reports in *The Great Bridge* that the men in their top hats were bridge trustees posed on the James Street side of the Brooklyn anchorage. Once the footbridge was completed, it was open to the public. Note the sign posted at the steps leading up to it: "Safe For Only 25 Men At One Time. Do Not Walk Together. Nor Run, Jump, or Trot. Break Step! W. A. Roebling. Eng'r in Chief." Anchor bars are arrayed at lower right. Courtesy of the Jeffrey Kraus Collection.

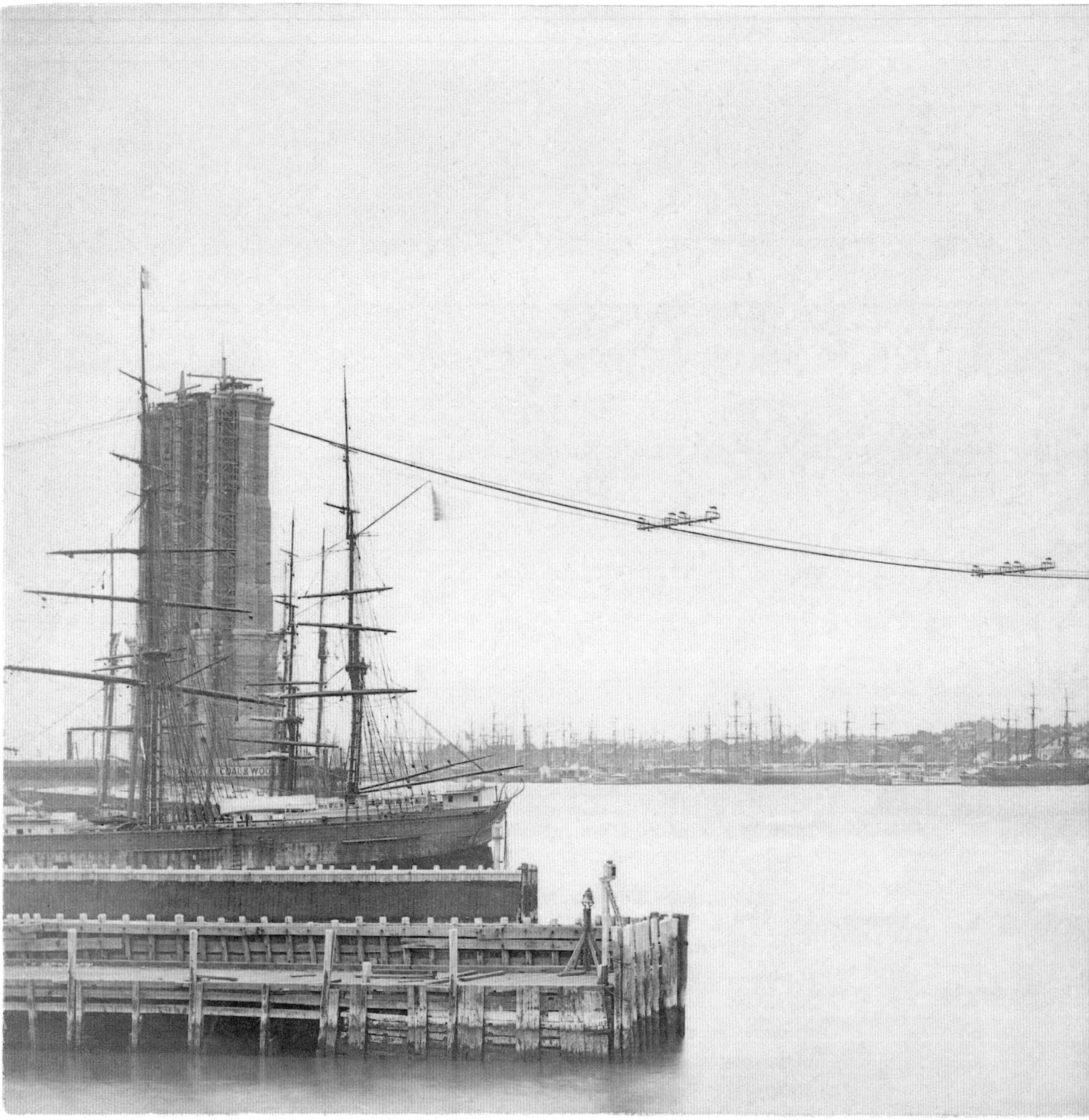

A cabinet card photograph of the Brooklyn Bridge towers, footbridge, and cradles, looking southwest, from Brooklyn toward lower Manhattan, ca. 1877. A worker could easily reach out from his position on a cradle to adjust the wires, keeping them in order. The three pairs of cradles, equidistant between the towers, are visible, while a pair of cradles between the New York tower and anchorage is barely discernible at far right. The pair of cradles on the Brooklyn side, between the anchorage and tower, cannot be seen from this angle. Courtesy of the Jeffrey Kraus Collection.

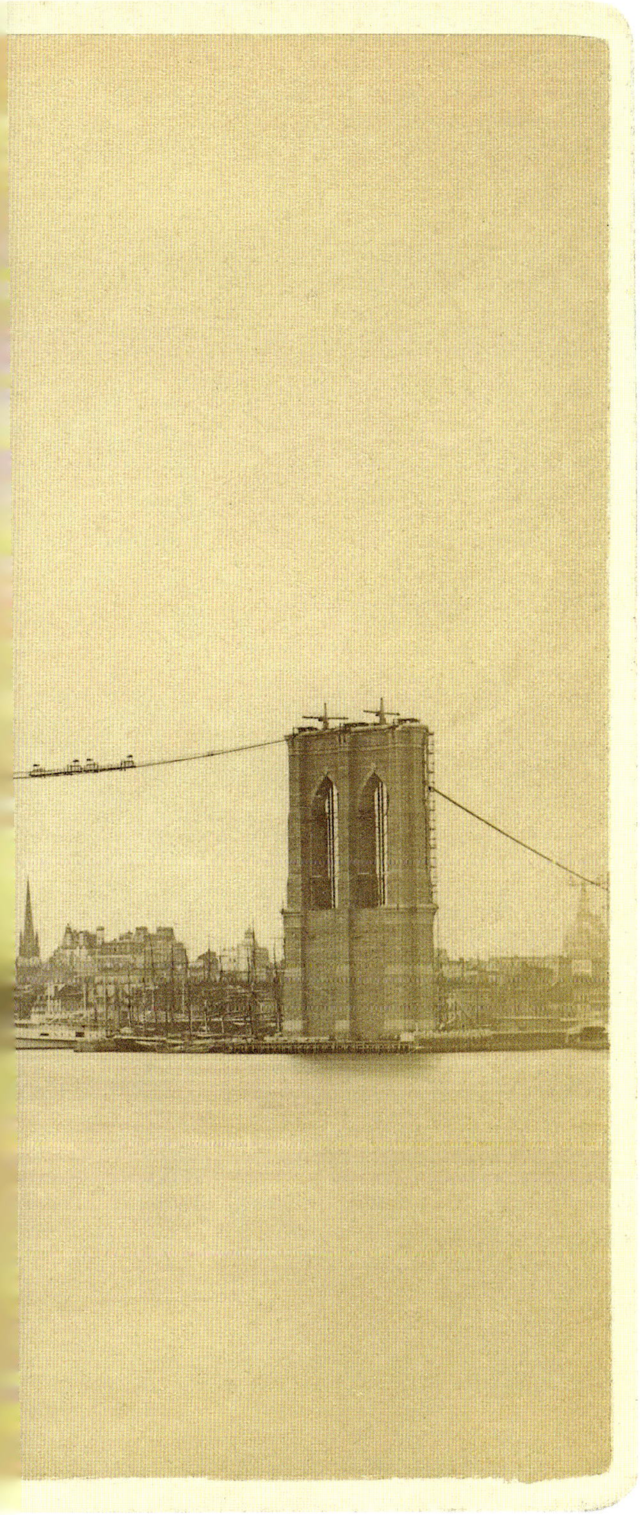

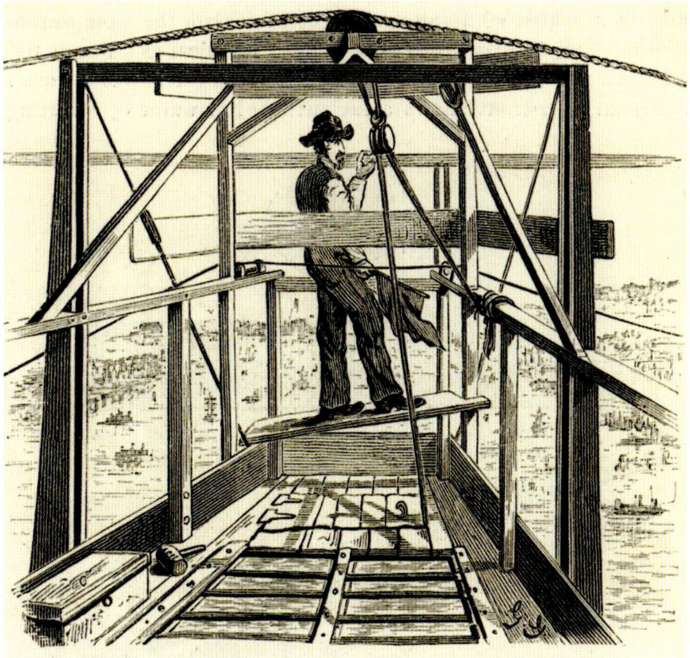

"In the Cradle," *Appleton's Journal*, January 1878. A worker monitors the wires with his left hand. In his right hand, he holds a flag to signal to his co-workers in the distance. Courtesy of Richard Haw.

FOLLOWING SPREAD: "From Tower to Tower—The Suspension-Bridge Over the East River—View from the Brooklyn Tower," *Harper's Weekly*, April 14, 1877. This view was published a year or so after the towers had been completed and while the footbridge was still in place. In the distance, men work on the three pairs of cradles that are spaced between the towers. Note the arrangement of subjects atop the tower, in the foreground: laborers at left and who are likely trustees, at right. Courtesy of Richard Haw.

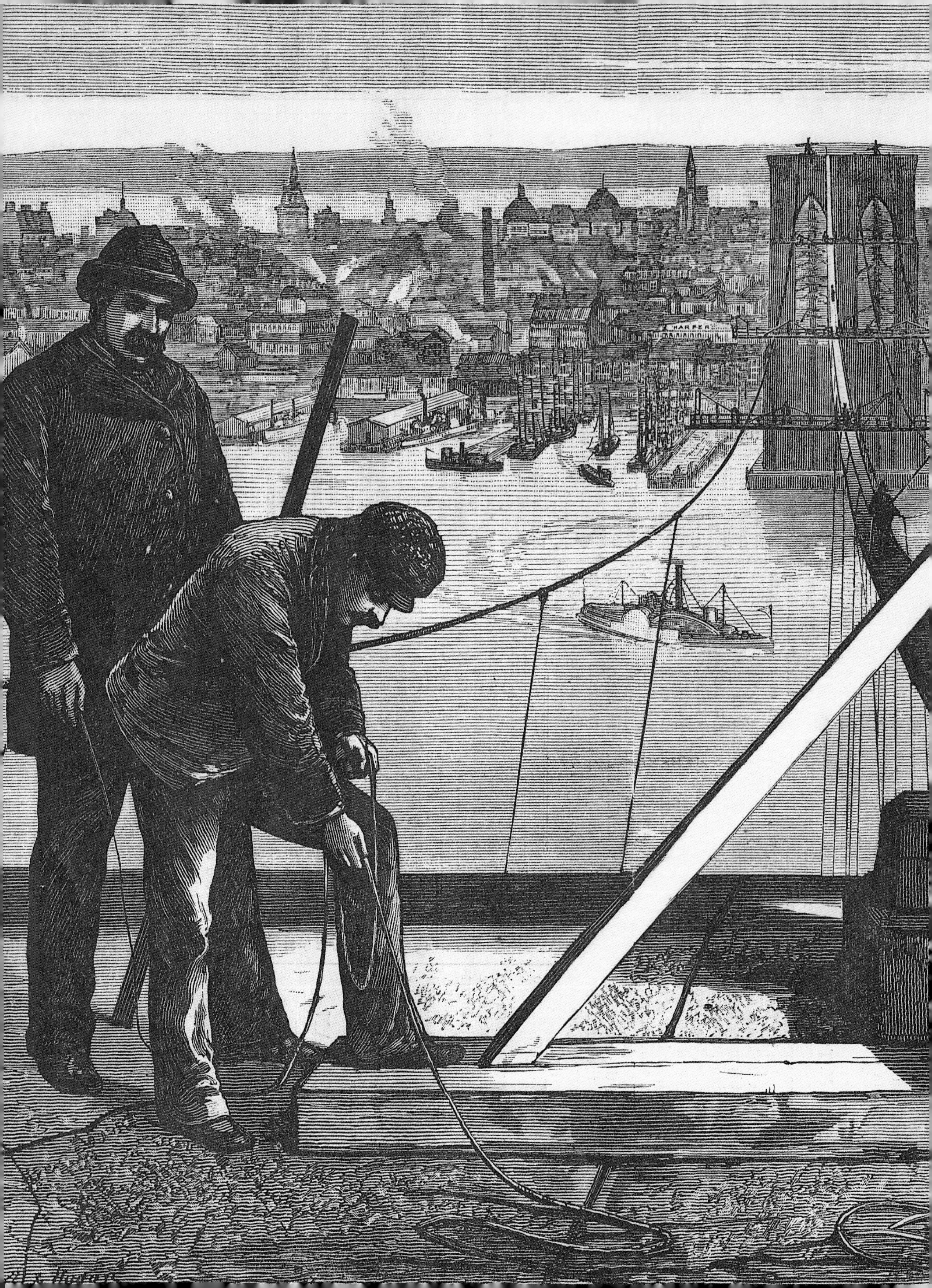
HARPER

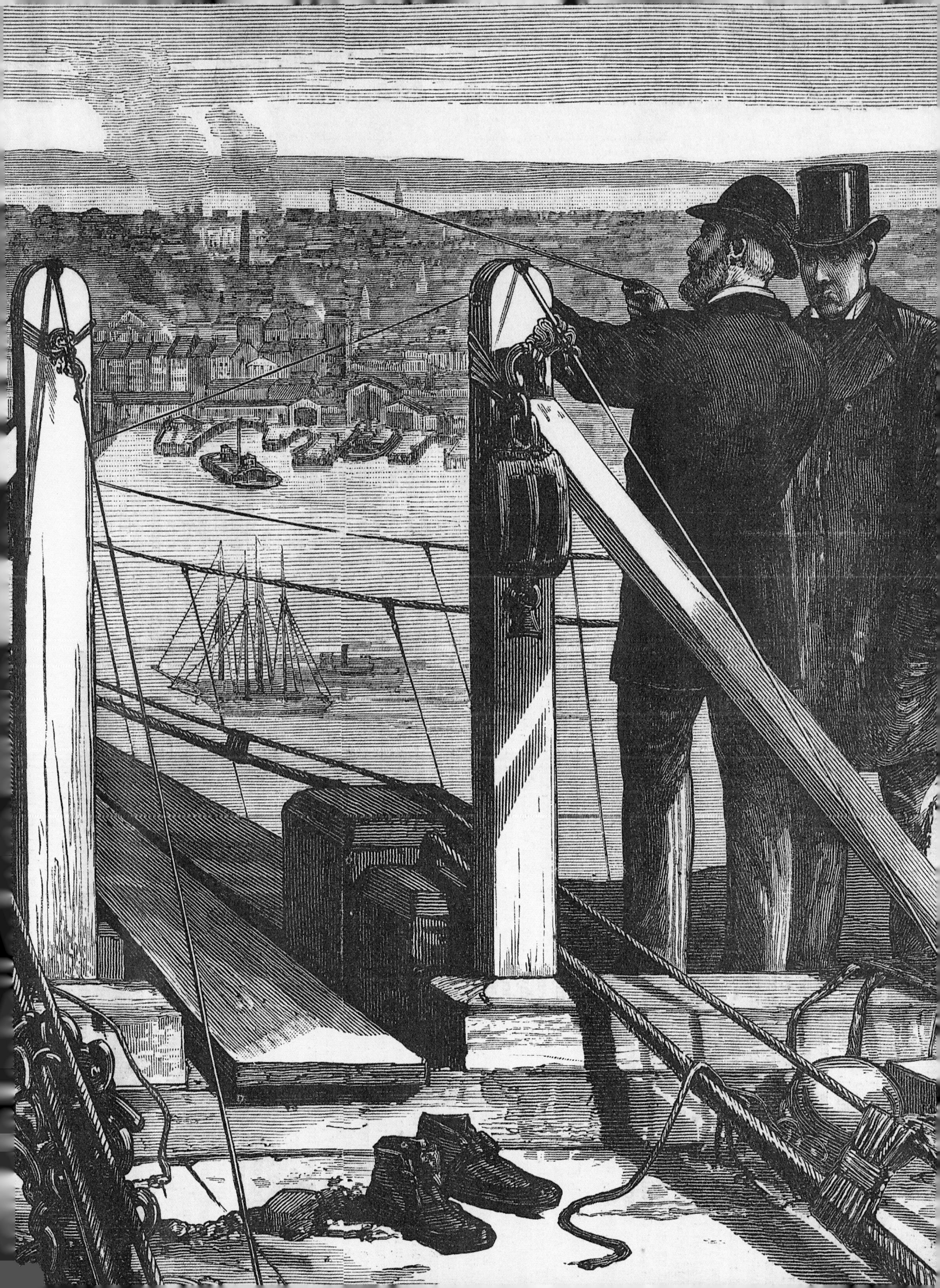

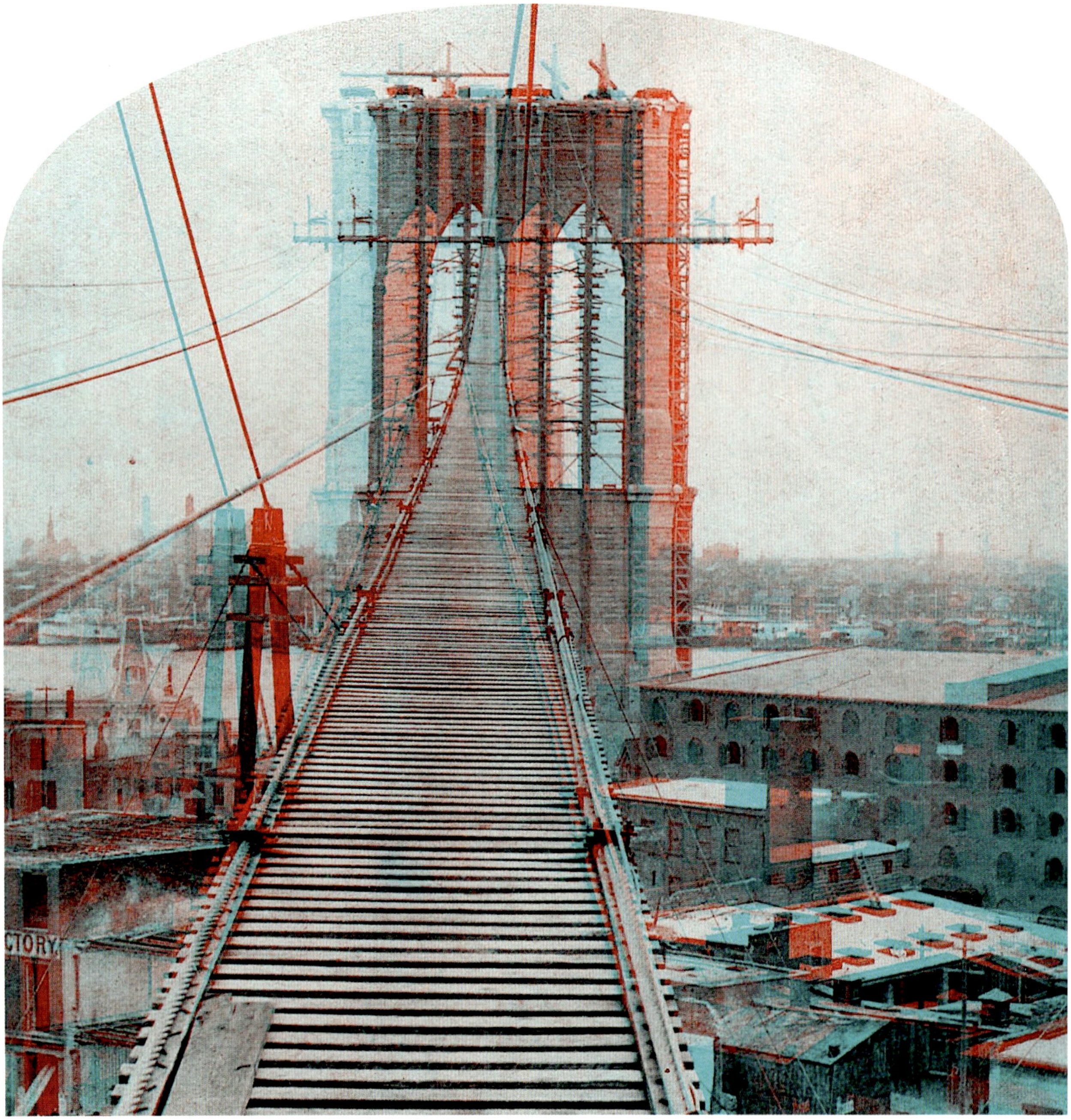

3D An anaglyph of the footbridge between the bridge's Brooklyn anchorage and its tower, showing Manhattan in the distance, across the East River, ca. 1877. The spacing between the oak slats of the footbridge, and the wires that supported them, are clearly visible here. Courtesy of the Jeffrey Kraus Collection.

OPPOSITE: "The Foot Bridge over the East River, New York," *Scientific American*, March 10, 1877. Looking through the arches of the bridge's Brooklyn tower, with New York City in the distance. At right, the three pairs of cradles between the towers—with several workers on them—are visible. The two buildings in Manhattan with the flags flying atop them are, at left, the Western Union Telegraph Building (the tallest building in New York City at the time) and, at right, the City Hall Post Office and Courthouse. Author's collection.

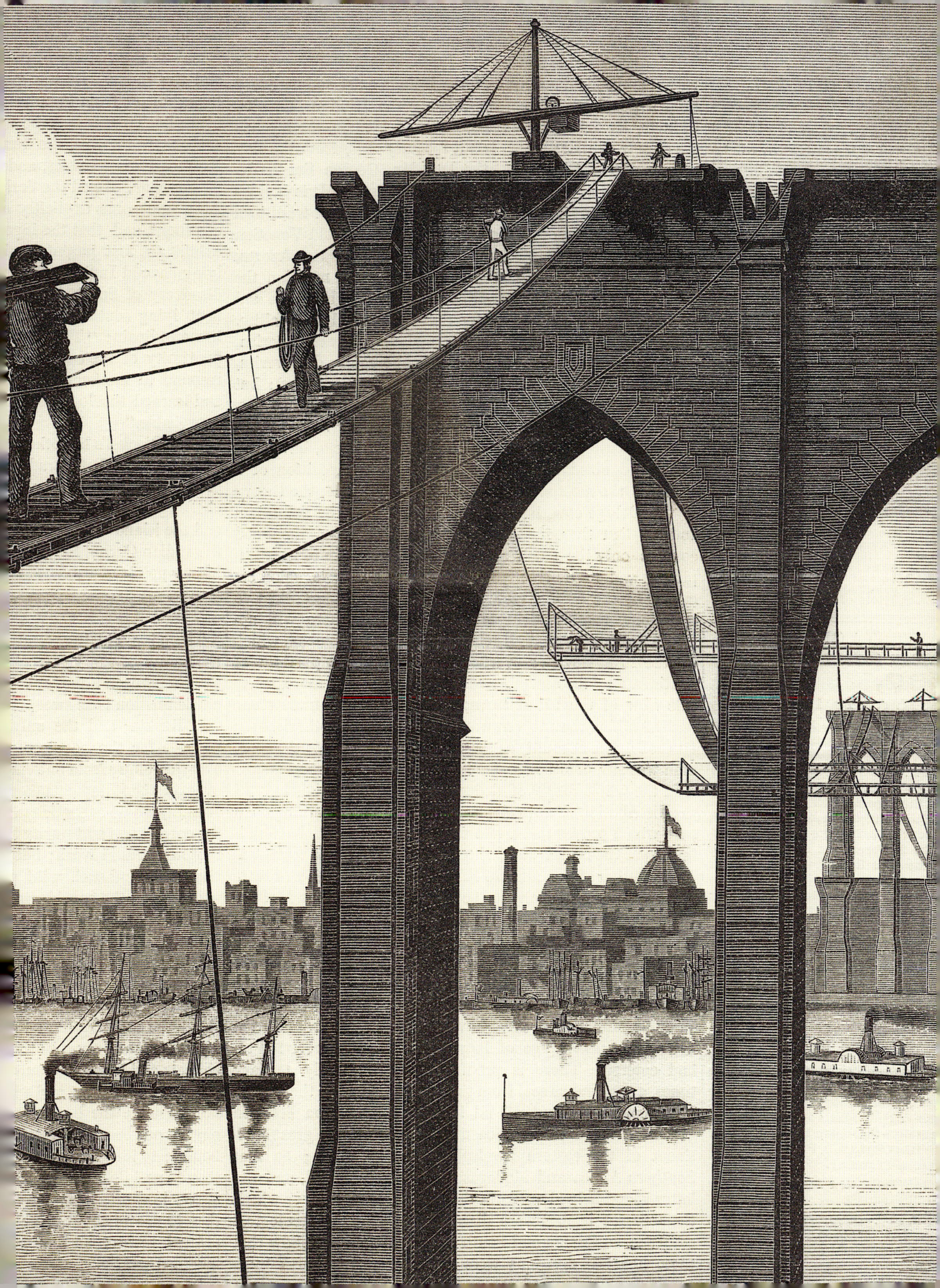

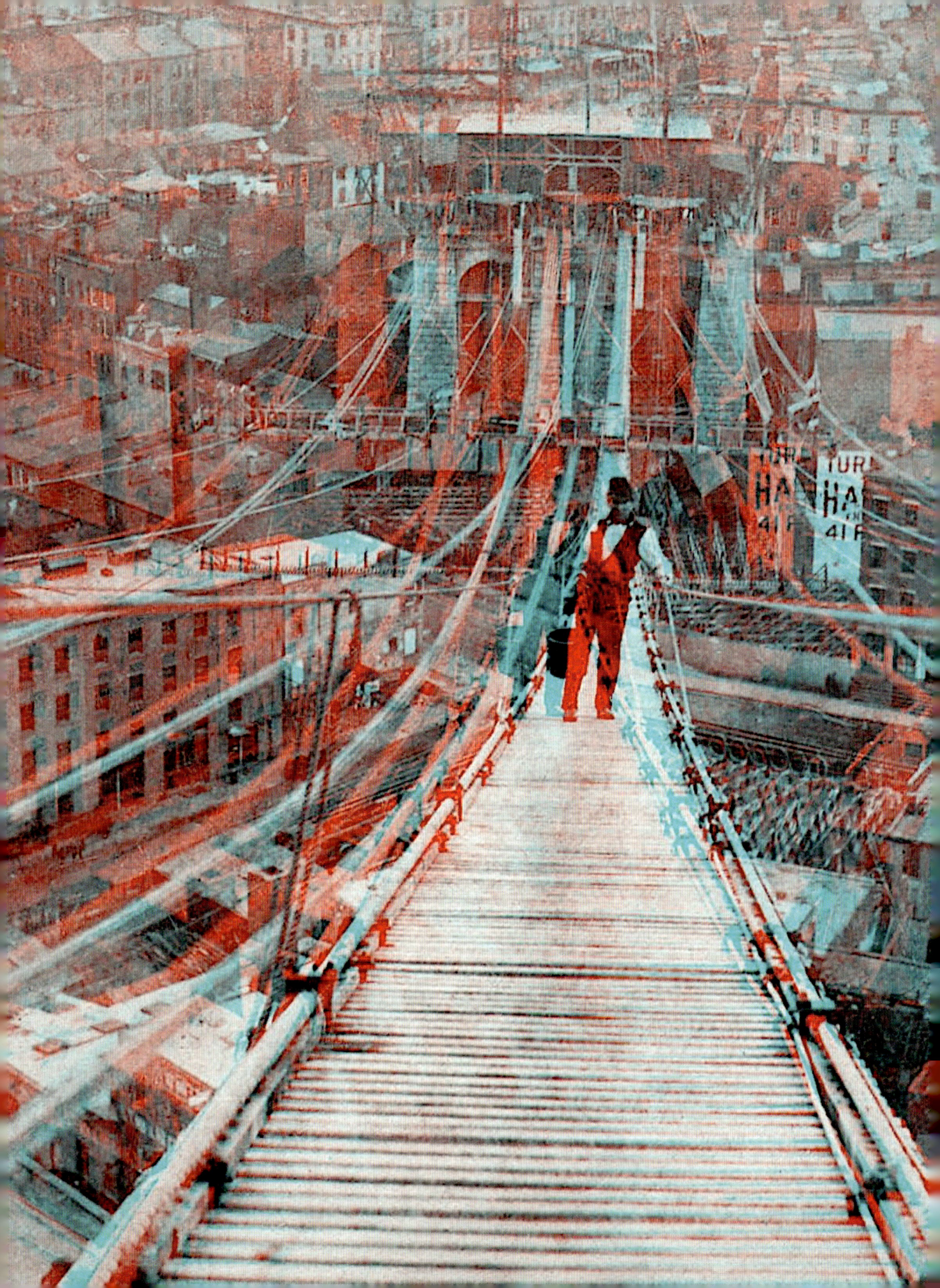

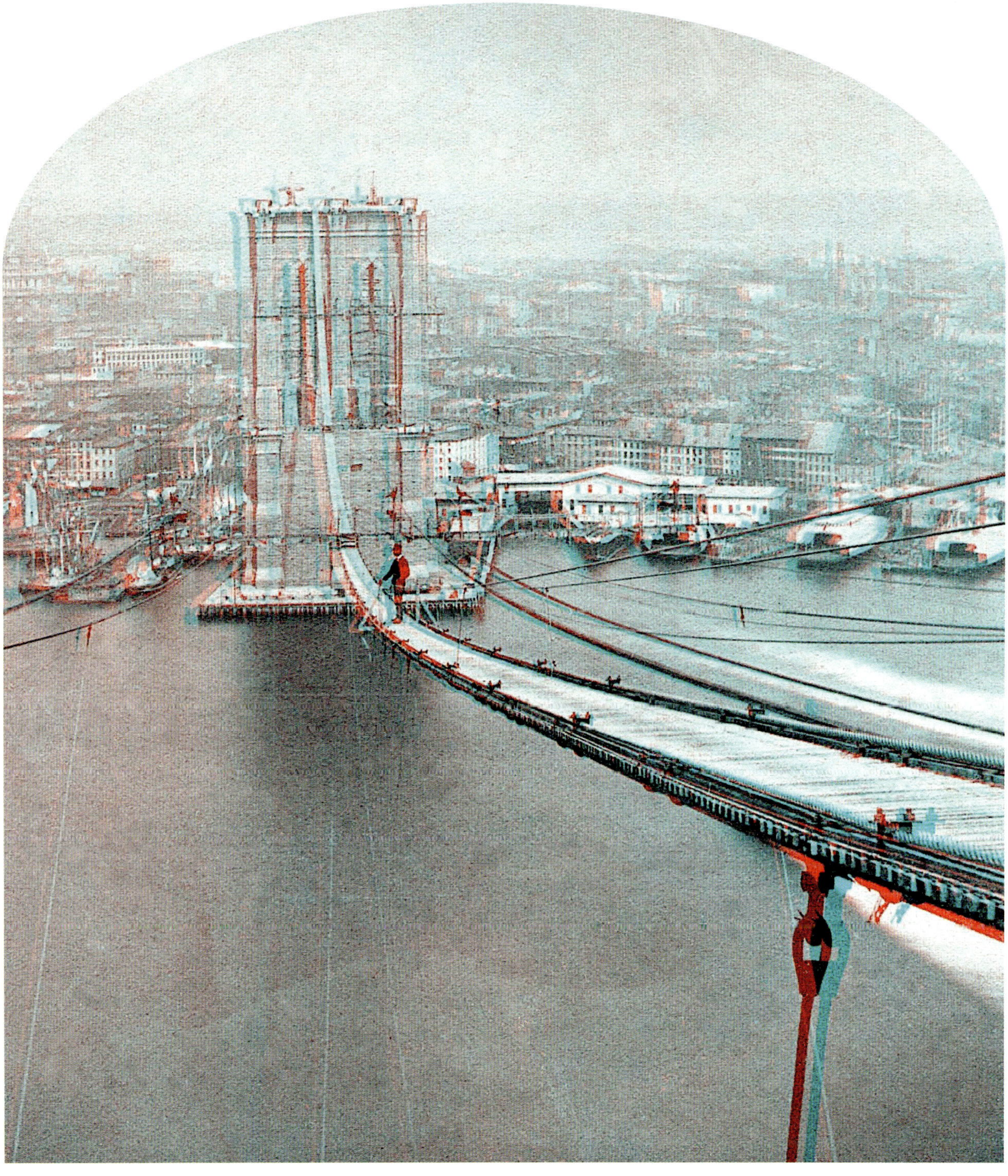

3D Looking west at the footbridge and the bridge's New York tower. Note the guy wires atop and below the footbridge to give it greater stability. Courtesy of The Green-Wood Historic Fund, Jeffrey I. Richman Stereoview Collection.

3D OPPOSITE: Anaglyph of a bridge worker on the footbridge, between the Brooklyn tower and its anchorage. A pair of cradles is visible just above his head in the distance. Courtesy of the Jeffrey Kraus Collection.

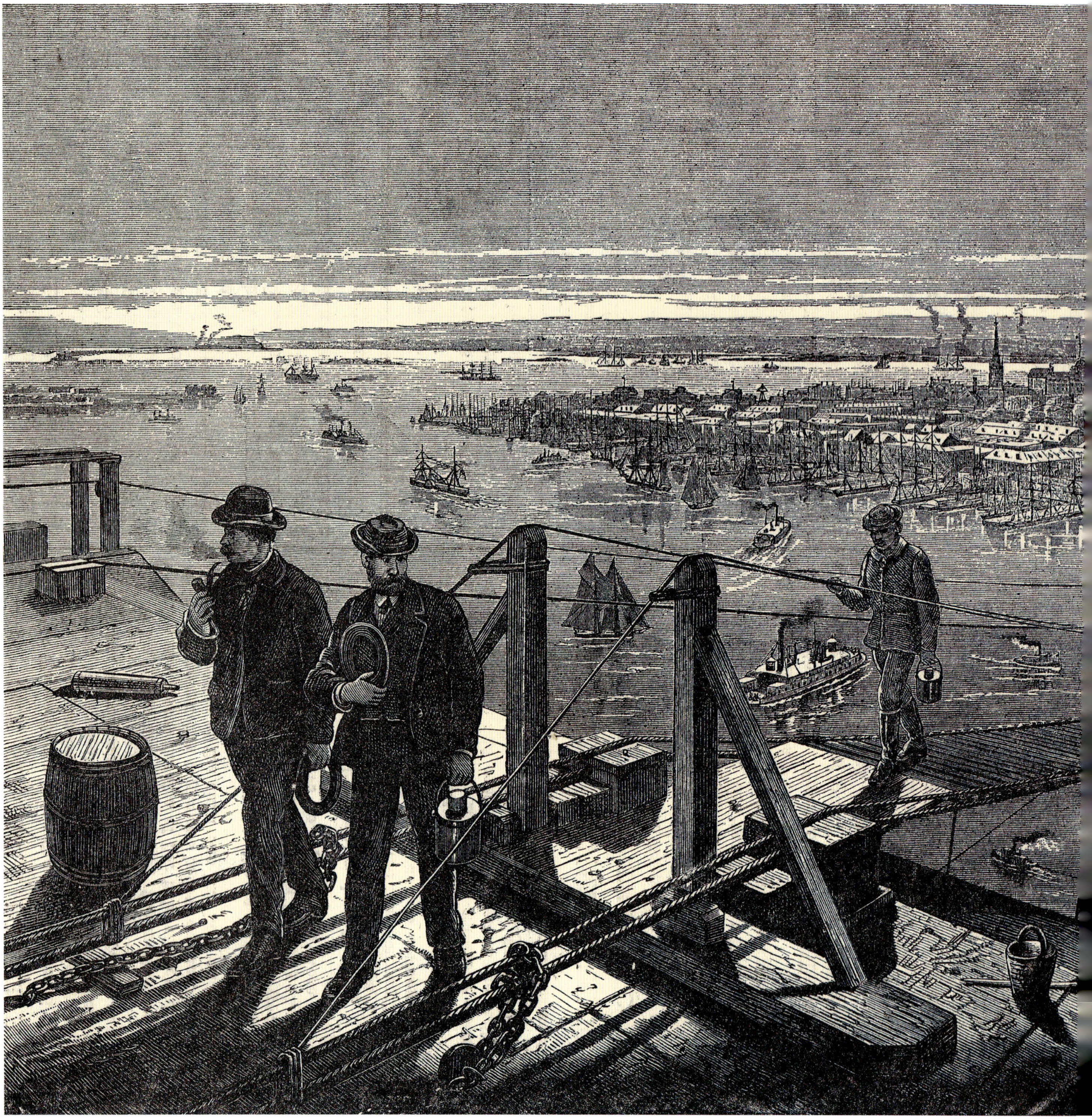

"The Brooklyn Bridge after the Temporary Footpath was Laid," *Frank Leslie's Popular Monthly*, July, 1883. In this woodcut, lower Manhattan and the New York Harbor are visible in the distance. Author's collection.

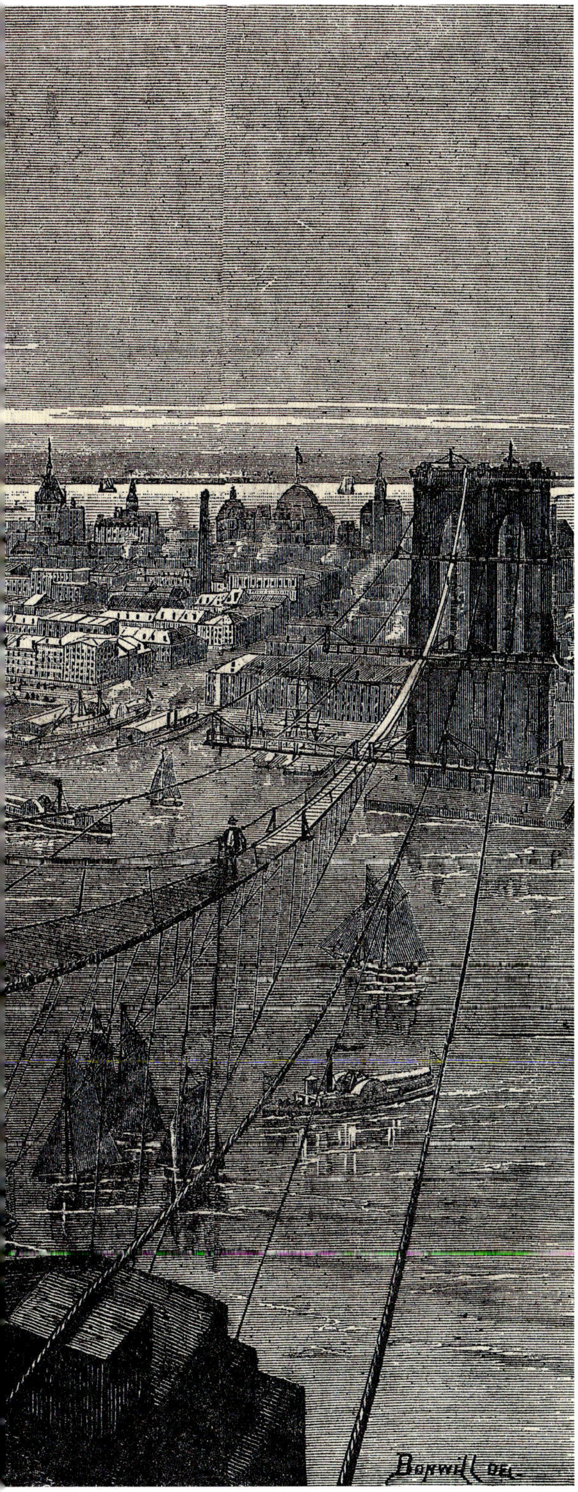

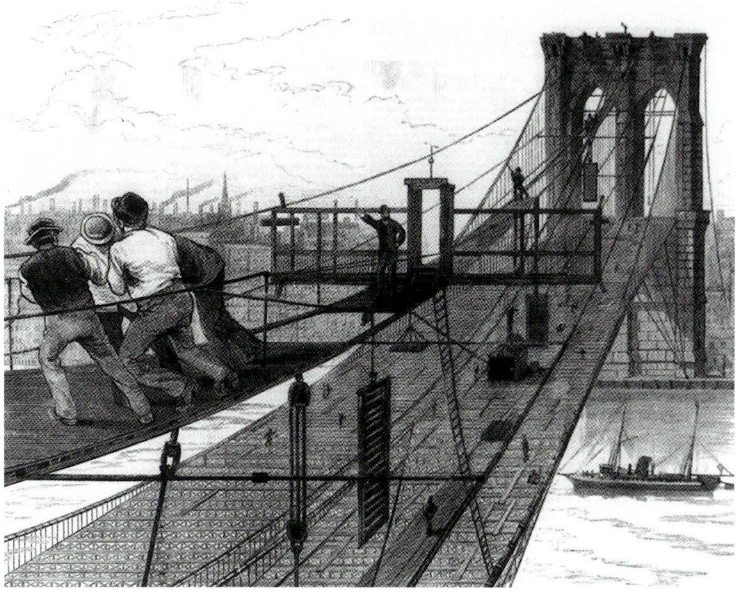

"Removal of the Footbridge," *Frank Leslie's Illustrated Newspaper*, April 1, 1882. Once the cables had been completed, the footbridge was no longer needed. This woodcut depicts gaps where sections of the footbridge have already been removed. Note that one section hangs down on a rope. The artist has depicted the superstructure below, which would support the roadbed, already well advanced. Courtesy of Richard Haw.

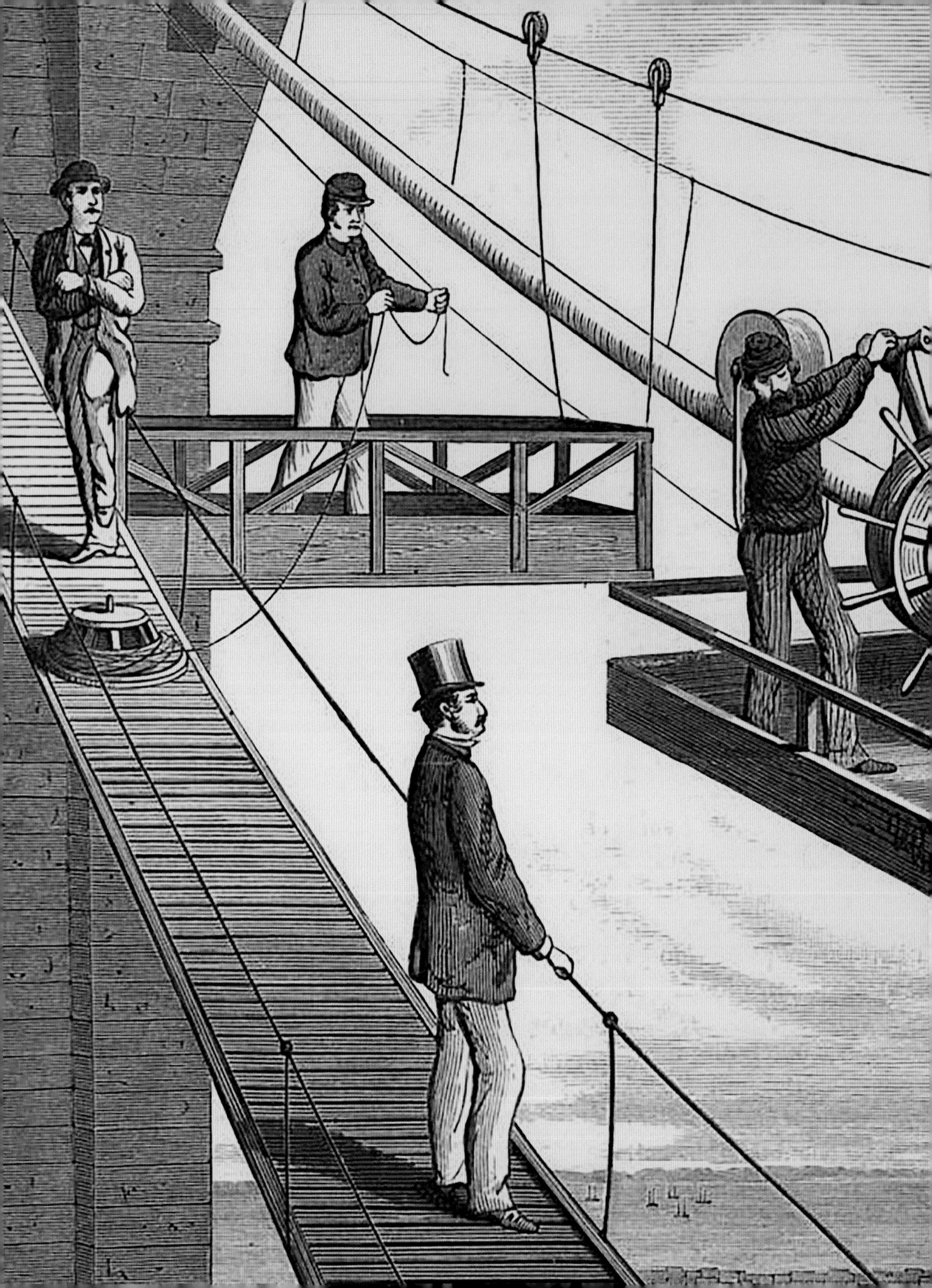

CHAPTER 7

Cables

1876–1879

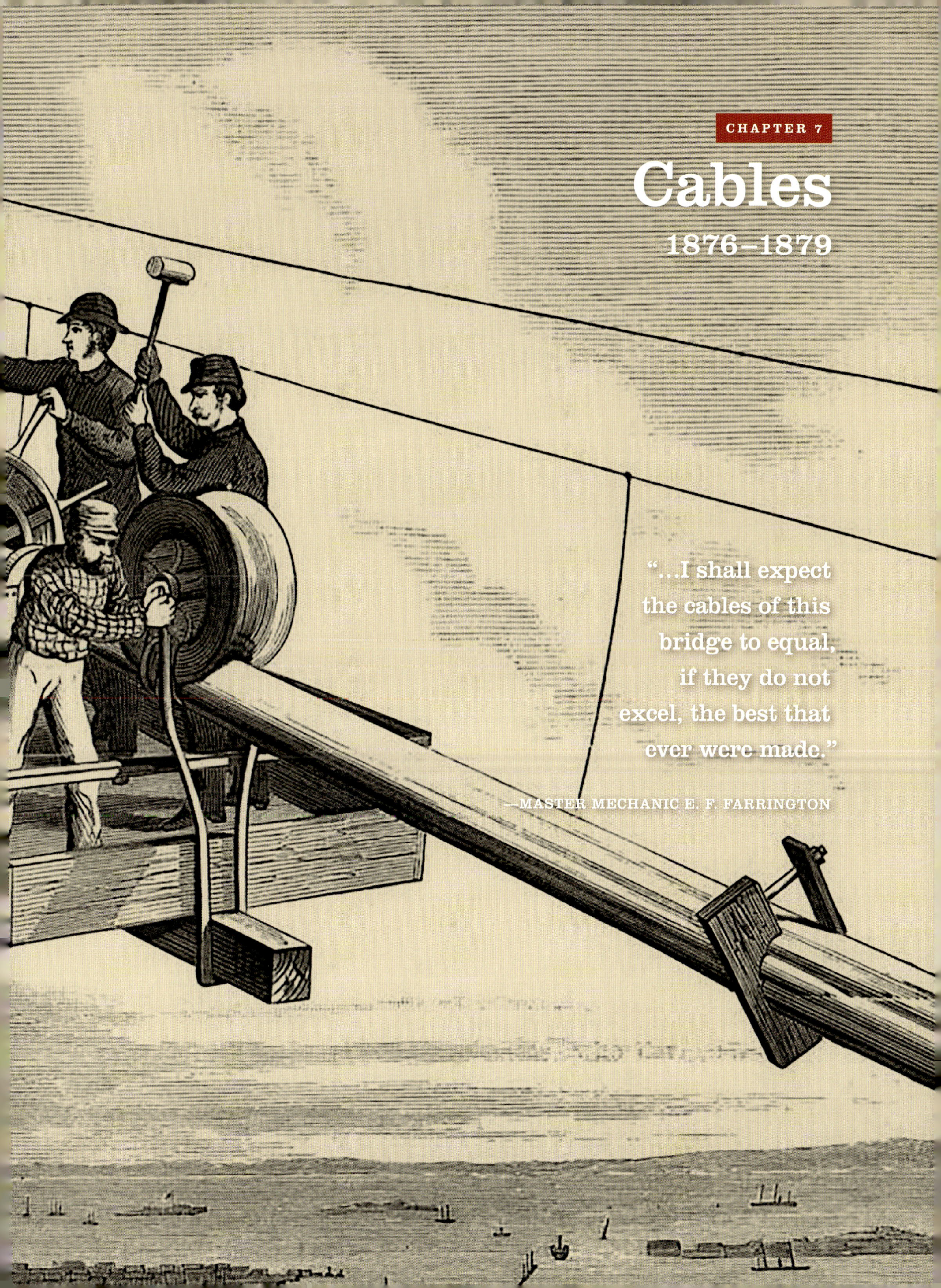

"…I shall expect the cables of this bridge to equal, if they do not excel, the best that ever were made."

—MASTER MECHANIC E. F. FARRINGTON

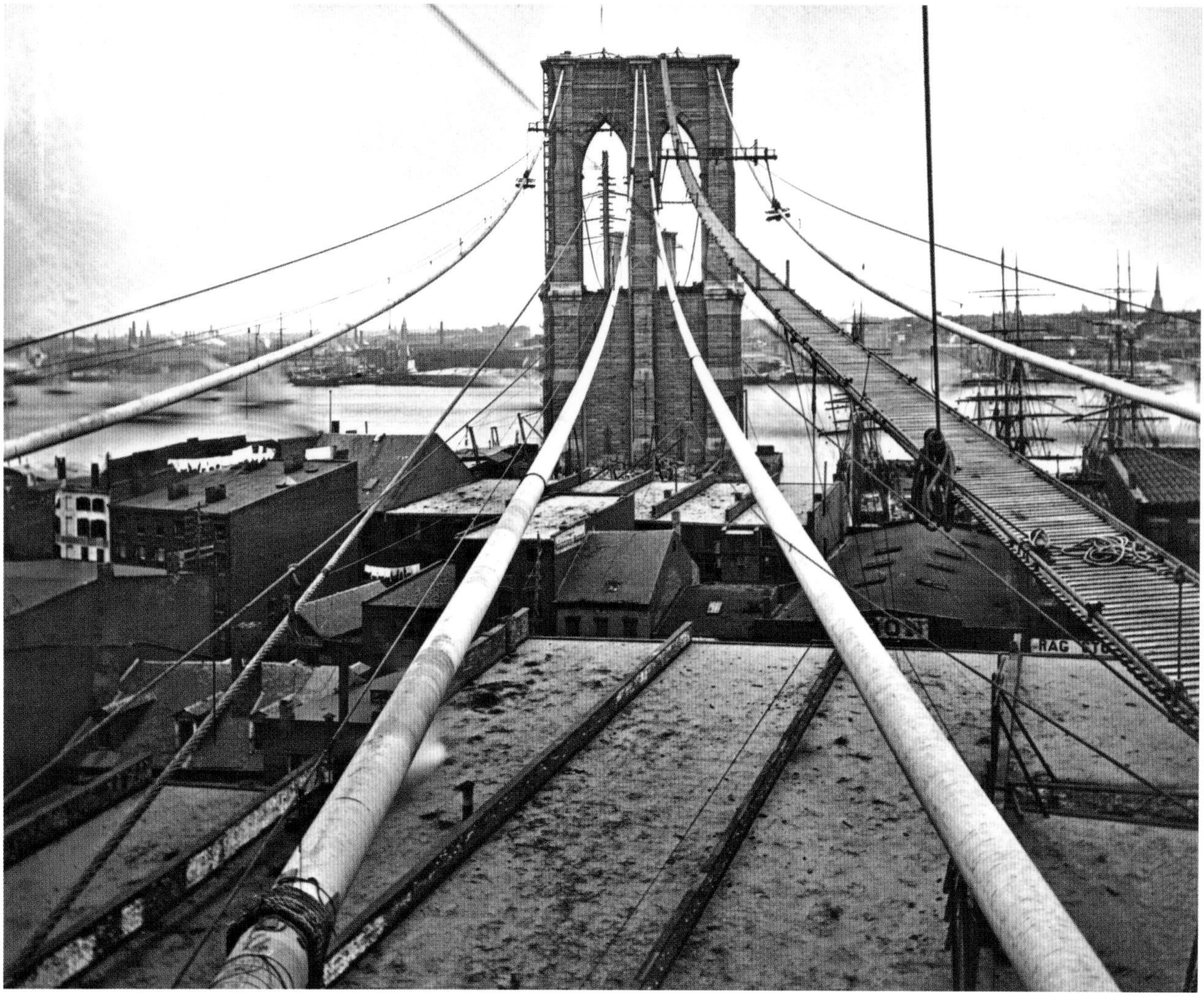

A photograph, printed from a lantern slide, of the New York tower, with the four cables of the Brooklyn Bridge fully wrapped. Author's collection.

PREVIOUS SPREAD: "The Great Suspension Bridge Between New York and Brooklyn," *Scientific American*, November 9, 1878. A team of workers typically wrapped ten feet of cable in a day; sixteen teams of men, four per cable, operated simultaneously. Note the footbridge at left and the cradle to its right. Courtesy of Richard Haw.

Cables are essential elements of a suspension bridge. It is the cables from which the bridge deck, or superstructure, is suspended. Stretching from anchorage to anchorage, the cables arch up from each anchorage and over the towers, then dip in graceful curves, or catenaries, between them.

The Brooklyn Bridge was built with four massive cables, one along the downriver flank of the bridge, two over the center of the towers, and one near the upriver edge. Each cable is 15.5 inches in diameter—an unprecedented size at the time. Attached to these cables are suspender wires that hang down and connect to the deck beams below, upon which the roadbed sits. These huge cables can support a tremendous amount of weight.

In Charles Barnard's article, "The Brooklyn Bridge," which appeared in *St. Nicholas Magazine* in 1883, he explained to his young readers the fundamental nature of the Brooklyn Bridge cables, stating that they "are not chains with links, nor are they twisted, like ropes. They are bundles of straight wires laid side by side, and bound together by wires wound tightly around the outside."

The pros and cons of suspension bridge cables made of twisted wire ropes versus straight wires, and of making cables in place versus off-site and moving them to the bridge site, were detailed by Assistant Engineer Wilhelm Hildenbrand in his 1877 essay, "Cable-Making for Suspension Bridges, With Special Reference to the Cables of the East River Bridge." In it, he lays out the three options for making suspension bridge cables:

> First, the cable is composed of twisted wire ropes.
>
> Secondly, it consists of parallel wire strands [each strand is a group of wires; a cable is made up of strands], previously made on shore and then hoisted or hauled in position, or
>
> Thirdly, it is made right in place; each wire is taken over separately and adjusted; and all of them are by wrapping combined into a cable.

Hildenbrand further explained that there were substantial advantages to the third option—making or spinning the wires in place. This process was called "spinning" because, much as a spider spins its web, one fiber at a time, the steel wires (which would eventually total more than five thousand) would be pulled across one at a time. As Hildenbrand noted, the tensile strength of a straight wire is superior to that of a twisted wire, making the first option unattractive for such a heavy bridge. Moreover, a straight wire takes up less space than a twisted wire, resulting in the benefits of less wind resistance, less surface area subject to corrosion, and less cost. Also, straight wires are easier to connect to an anchor chain than twisted ones.

The second option above was not possible in Brooklyn or Manhattan because there was no open space near the bridge where prefabricated wire strands could be laid out before being lifted over the towers. Moreover, given that each Brooklyn Bridge cable would extend 3,577 feet and weigh tens of thousands of tons, it would be impossible to move them, even one strand at a time, from their place of manufacture. Hildenbrand did note, however, that there was a substantial drawback to the third option, that is, spinning the cables in place: work on the cables would be necessarily delayed until the towers and anchorages were finished. But this concern was clearly outweighed by the other considerations.

Looking west from the New York tower, down toward the New York anchorage. As in the previous photograph, the footbridge and all four cables are visible. But, in contrast, here the photographer is standing on the New York tower looking down at the anchorage, not up from the anchorage. This view also shows the buildings along the bridge's right of way, beyond the anchorage, that have yet to be cleared for the construction of the approach. Author's collection.

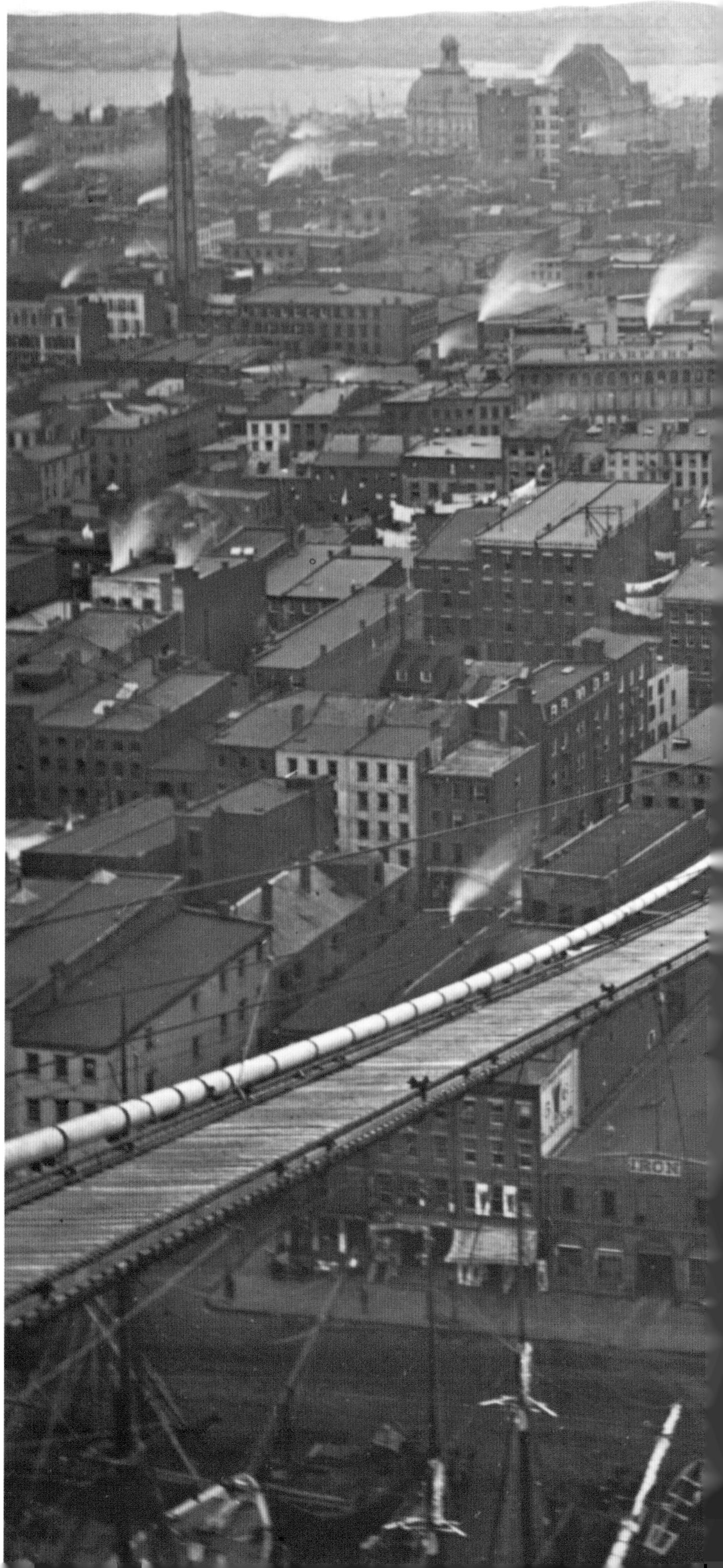

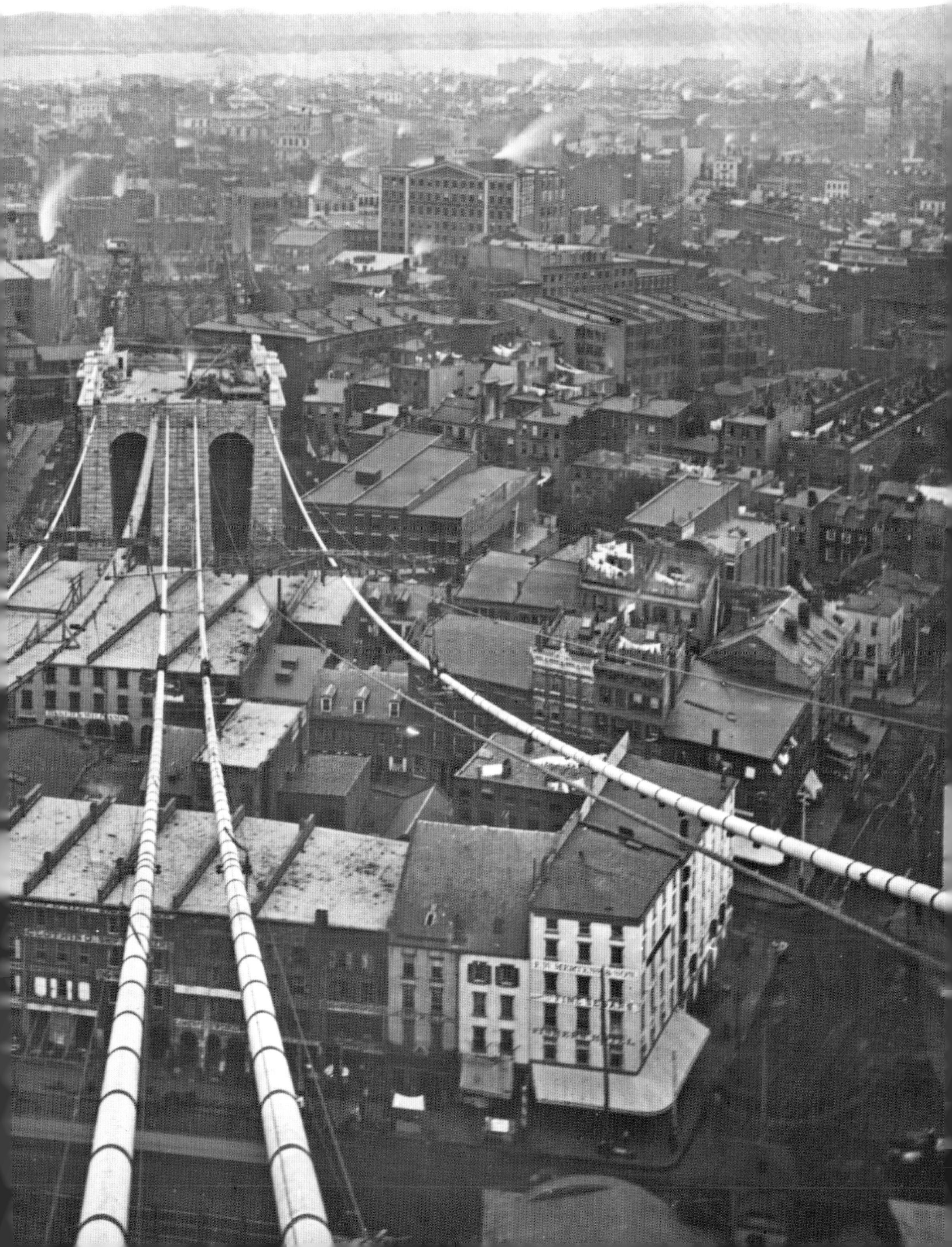

Plus, the Roeblings, and Master Mechanic E. F. Farrington, were experienced with spinning cables in place, which they had done for their earlier suspension bridges.

While the Roeblings and Farrington had used these same spinning techniques elsewhere, the metal used to fabricate the wire was different. John Roebling had used iron wire to make the cables for his earlier suspension bridges, but for the Brooklyn Bridge, he specified steel—"the metal of the future"—because it offered greater strength in tension. Chief Engineer Washington Roebling noted in his specifications for the cables that "the general character of the wire is as follows: It must be made of crucible cast steel; it must be hardened and tempered; and lastly, it must be all galvanized." The cables for the Brooklyn Bridge, he noted, would require 3,400 tons of wire "of superior quality steel."

By early 1876, Washington Roebling had finalized his plans for the cables. They would be gigantic; spinning them in place would be an enormous undertaking. Each of the four cables was to be made up of nineteen strands. However, the exact number of wires per strand is unclear. Assistant Engineer Wilhelm Hildenbrand reported in his "Cable-making for Suspension Bridges" that each cable for the Brooklyn Bridge "consists of nineteen strands of 332 parallel steel-wires and contains therefore altogether 6,308 wires." But Master Mechanic E. F. Farrington recorded that the total number of wires per cable was 5,434—the same number that David McCullough reports in his appendix to *The Great Bridge*. Farrington's and McCullough's number, when divided by the 19 strands per cable, indicates that each strand is composed of 286 wires.

Regardless of the precise number of wires per strand, there were roughly six thousand of them. If all the wires in a single strand were laid out end to end, they would stretch for about 184 miles. If all the wires making up the four cables were laid end to end, they would stretch approximately 14,000 miles, the equivalent of crossing the United States from New York to California and back again, then one more round trip, plus almost another full trip out to California. That is a lot of steel wire.

Before the process of making the cables could begin, it was necessary to create two wire loops, or "travelers." It was these travelers—one on the upstream side of the bridge, the other on the downstream side—that would "spin" the cables, pulling their component wires from the Brooklyn anchorage to the one on the New York side. Once in place, each of the travelers would "travel" in a loop, pulling across each individual wire, one after another, to make up the strands that would be wrapped together to form the bridge's four great cables. The loop functioned like a traditional clothesline, only in the case of the spinning of the cables, the loop was powered, or moved, by a steam engine. It went in only one direction, pulling wires east to west.

And so the preliminaries began. On August 14, 1876, the first wire for the traveler was pulled across the East River on a scow escorted by two tugboats. That wire, which had already been tied to the Brooklyn anchorage and hoisted over the Brooklyn tower, was now lifted into position atop the New York tower before being run down to the New York anchorage, to which it was secured. Then the process was repeated with a second wire. This wire was then spliced, or joined, to the first, completing the loop of this traveler.

Assistant Engineer Wilhelm Hildenbrand was on site watching as the first wire was positioned. It was, he announced triumphantly, "the first connecting link between the cities of New York and Brooklyn, destined never more to be broken." He further described the event:

> Simple as the operation was, it created considerable excitement and interest among the population probably caused by a feeling of historical importance for the day, which practically should unite two cities. Thousands of spectators lined the shores, who greeted with loud cheers the appearance of this little rope, as it rapidly ascended high up in the air. It seemed that all doubts, hitherto entertained as to the erection and subsequent safety of the bridge, had vanished from that moment. The East River Bridge was considered an established fact.

As described by the *New York Herald,* this draping of the first wire rope over the towers was "the engagement ring in the marriage preparations of the two cities." The *Brooklyn Eagle* was more succinct: "Wedded" read its headline.

Once the first traveler was completed, Master Mechanic E. F. Farrington assigned himself the task—indeed, the glory—of being pulled across from anchorage to anchorage:

> Being in charge of this part of the work, I had a desire, natural to any one in the same position, to cross first. Moreover, on all previous work of this kind, I had found some difficulty to venture out to work on the ropes at first, and I thought my passage on the smallest of them would inspire confidence. I intended to cross quietly at a convenient time, but my intention was suspected. Some reporter got information of it, and published it.

Farrington's plan, to "cross quietly," was soon in tatters. The day before his scheduled ride, the *Brooklyn Eagle* broke the story that the first crossing of the bridge would occur the next morning. Excited by the news, enthusiastic crowds gathered early the next day along the docks and around the Brooklyn tower and anchorage, from which Farrington's historic trip would start. By his own estimate, twenty thousand spectators were in attendance, watching expectantly as he was about to become "the first man over the wire."

On August 25, 1876, the day of Farrington's great public triumph, he was decked out in a linen suit and straw hat. The master mechanic was "perfectly cool and collected" as spectators waved and cheered when he, atop the Brooklyn anchorage, seated himself on a wooden board that hung by ropes from the wire traveler. A large steam engine was started up, the traveler was engaged, and the wire loop—with the master mechanic attached—began moving. Pulled upward toward the Brooklyn tower, Farrington was off on the journey of his lifetime. As he rose above the enthralled spectators, he was swept up in the moment. Unfastening the rope that had been tied across his chest as a safety belt, Farrington stood up on his plank seat. Now standing with both hands holding the ropes from which his seat hung, he let go, first with one hand and then the other, as the crowd cheered. Exhilarated, the usually reserved Farrington waved to the spectators below and threw them kisses. It took Farrington twenty-two minutes to complete the journey to the New York anchorage—including two stops when Farrington had to walk his seat and himself over the tops of each of the massive towers, because there was not enough clearance between the traveler wire and the tower top.

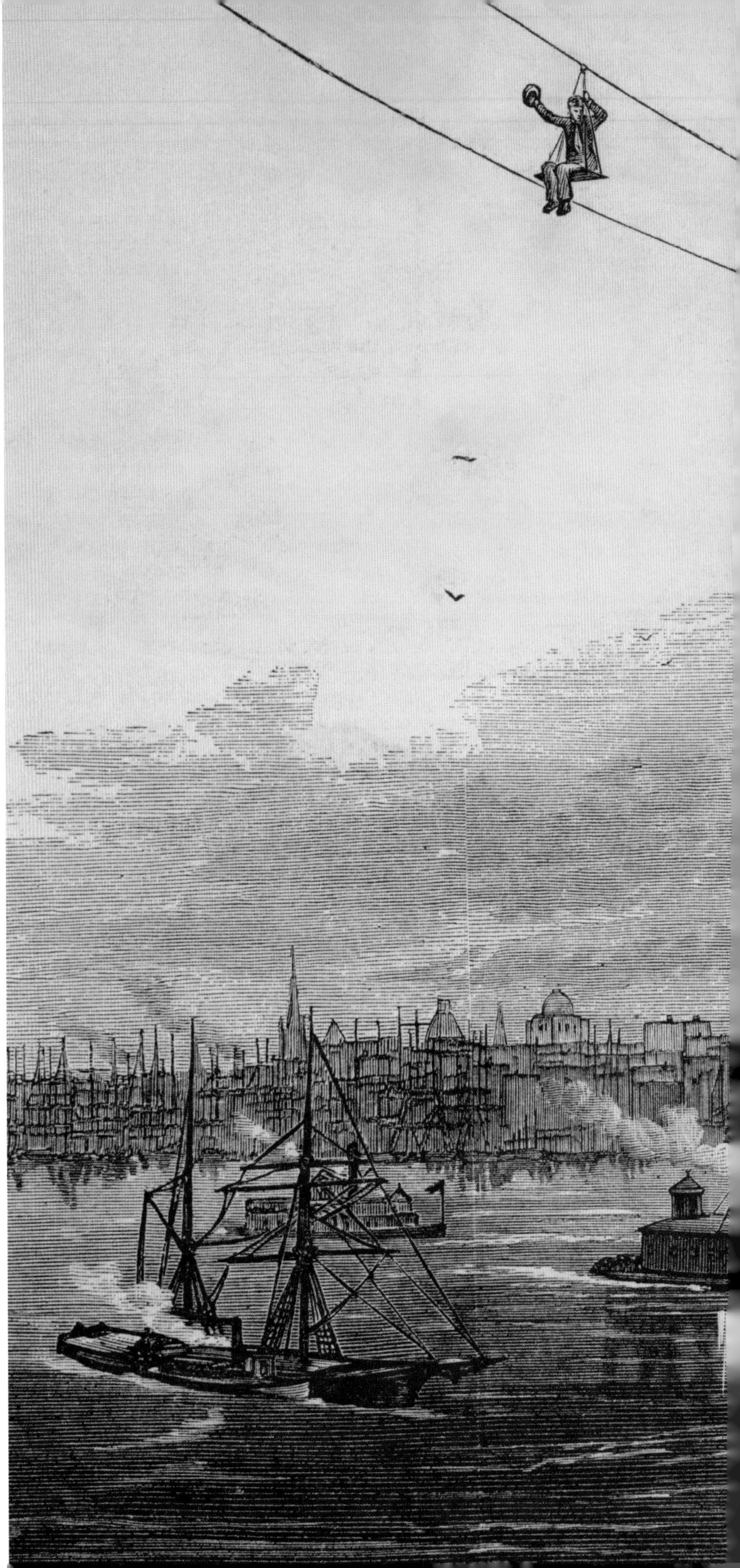

"New York—The First Passenger Across the East River Bridge—Testing the Temporary Cable," *Frank Leslie's Illustrated Newspaper*, September 9, 1876. Master Mechanic E. F. Farrington riding the wire across from Brooklyn to New York City.

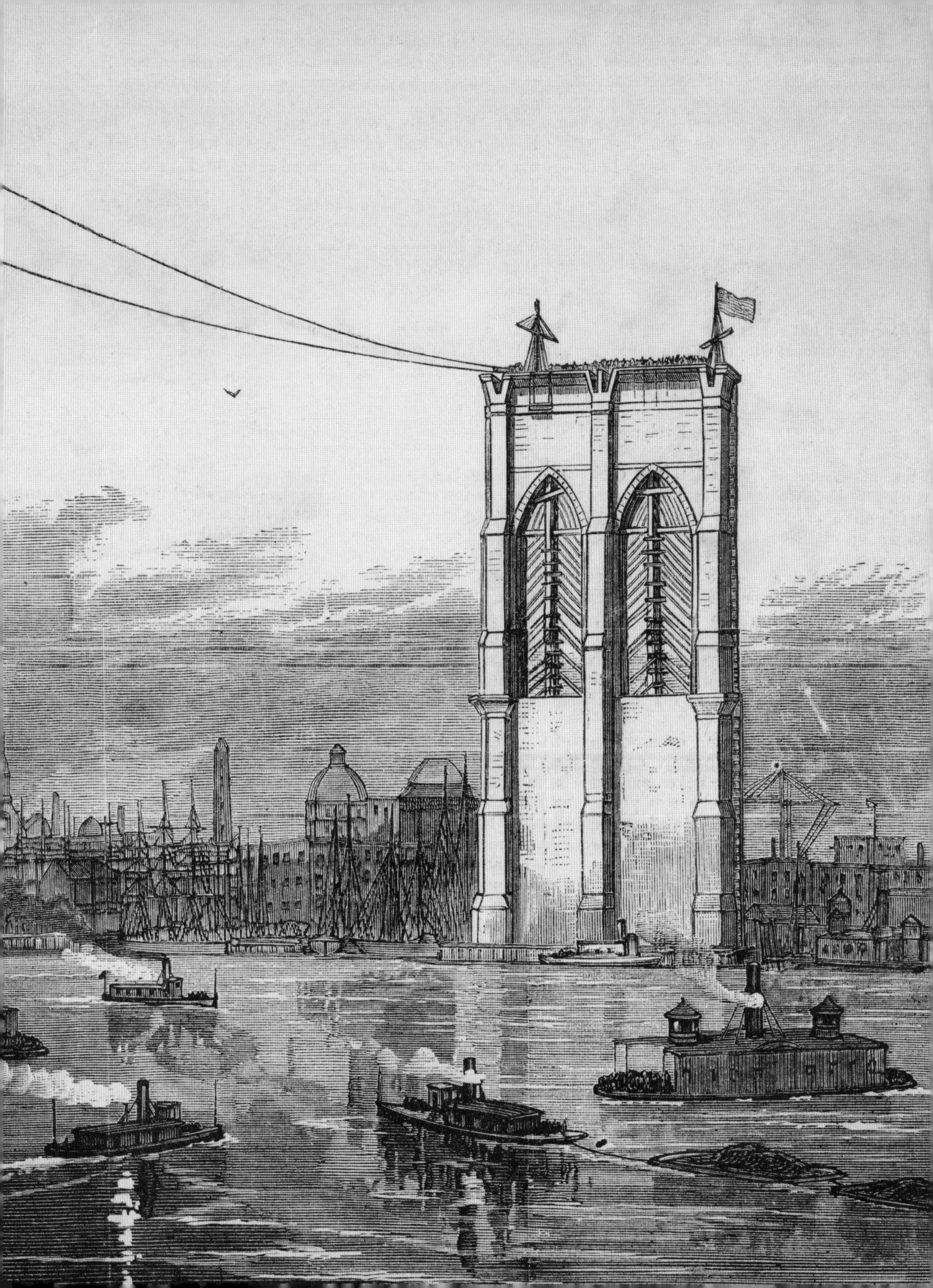

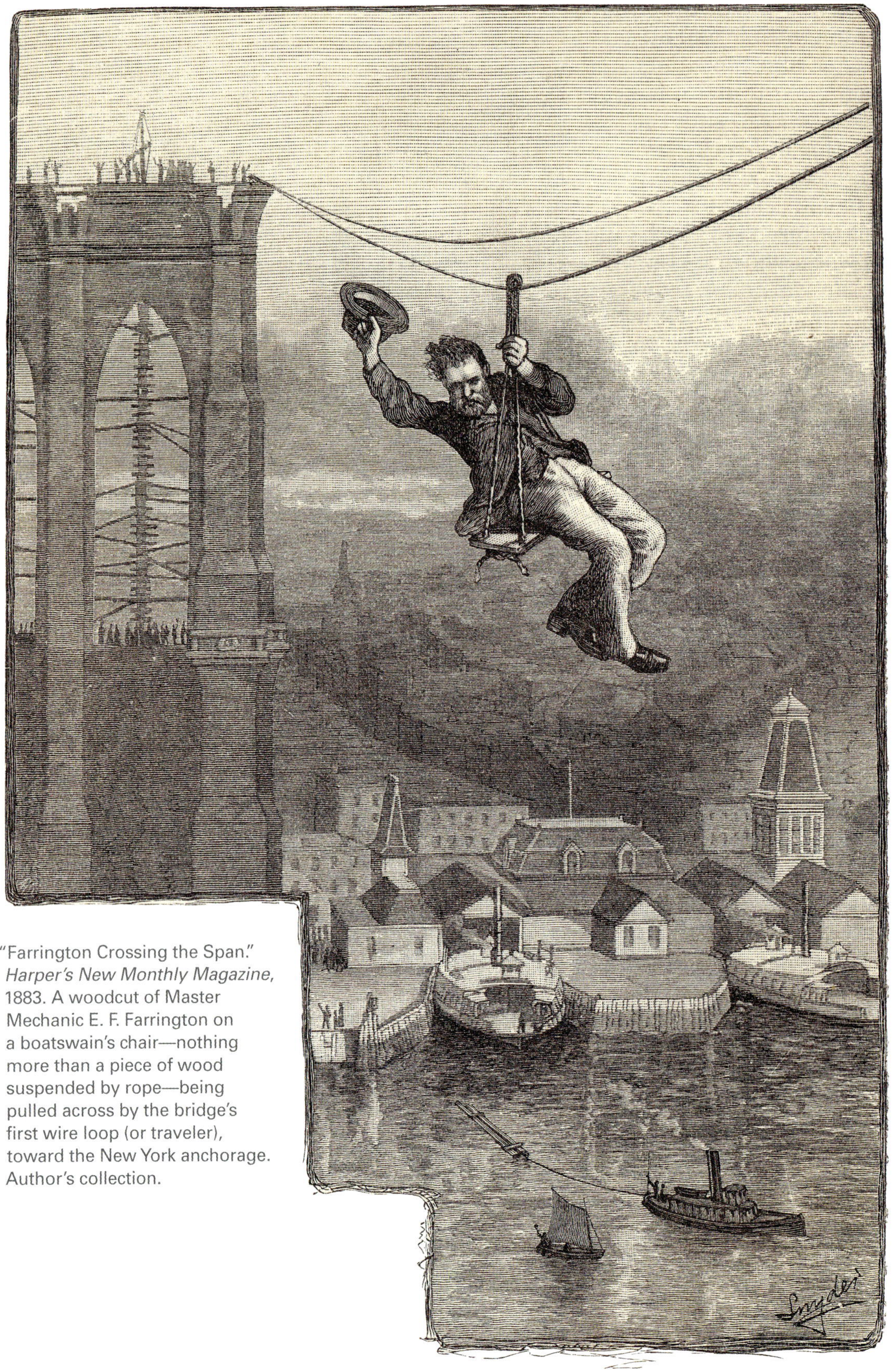

"Farrington Crossing the Span." *Harper's New Monthly Magazine*, 1883. A woodcut of Master Mechanic E. F. Farrington on a boatswain's chair—nothing more than a piece of wood suspended by rope—being pulled across by the bridge's first wire loop (or traveler), toward the New York anchorage. Author's collection.

After Farrington completed his joyous ride, he climbed down off the New York anchorage, intending to quietly and anonymously take a ferry back to work in Brooklyn. But the crowds on the streets, having recognized him, had other ideas. They closed in, hoping to celebrate his triumph with him. Fleeing, Farrington reported that he was "obliged to seek refuge from their attentions in the bridge yard." His trip, he continued, would be hailed "as a wonderful and daring achievement; and so no doubt it appeared to the uninitiated; but to sailors and riggers accustomed to face real dangers, it must have seemed a commonplace affair." Noting the joy of the ride and the downside of celebrity, Farrington recalled: "The ride gave me a magnificent view, and such pleasing sensations as probably I shall never experience again. It also gained for me what was in my position a troublesome and annoying notoriety; and I was well pleased when public attention was attracted to other events of more importance."

Farrington, having had the honor of being the first person to cross the East River using the Brooklyn Bridge—however incomplete—quickly sent the triumphant news off to his boss, Chief Engineer Washington Roebling: "I have carried out your instructions to the letter . . . and from my perfect familiarity with your plans, and my own experience, I shall expect the cables of this bridge to equal, if they do not excel, the best that ever were made."

For the cable work above the East River, Farrington trained his workers in the proper techniques, explaining the steps necessary to spin the cables in place: "The operations in cable making are oiling, splicing, drumming up, running out and regulating the wires, wrapping the strands, and letting off and lowering them, and clamping and wrapping the cables." Washington Roebling predicted that this entire process of making the cables would take about two and a half years. And that is indeed what it took.

On the same day as Farrington's ride, the second traveler was completed. A thick steel wire—three-quarters inch in diameter—was attached to the traveler upon which Farrington had just ridden. That wire was then pulled across as the steam powered "endless rope"—the first traveler—made its rotation. Then, a second wire of the same thickness was attached to the loop and sent across. Like the first wire, it was disconnected from the traveler once it reached the other anchorage, where the workers then spliced the ends of the two newly arrived wires together. This process was both quicker and easier than using a scow and tugboats, especially given the weight of this second wire rope. No doubt the workers were relieved not to have to hoist these up out of the East River and over the towers.

Now the travelers that would do all the cable spinning were complete, ready to begin their long and arduous tasks. As Farrington wrote, splicing the ends of these wires together formed two endless ropes, "capable of being worked back and forth, to draw loads and carry burdens."

These loops would labor for two-plus years, pulling the thousands of wires that would make up the bridge cables. Propelled by the large steam engine located at the Brooklyn anchorage, each traveler pulled two "carrier wheels" that were attached equidistant from one another. Each carrier was a large iron wheel that "carried," or pulled, the lengths of steel wire from the Brooklyn anchorage to the New York one. These steel wires were fed from one of eight large drums, or wheels, of wire positioned inside a wooden shed that had been built atop the Brooklyn anchorage. The steel wire from which the cables were made was coiled around the drums; two drums were dedicated to each of the four cables.

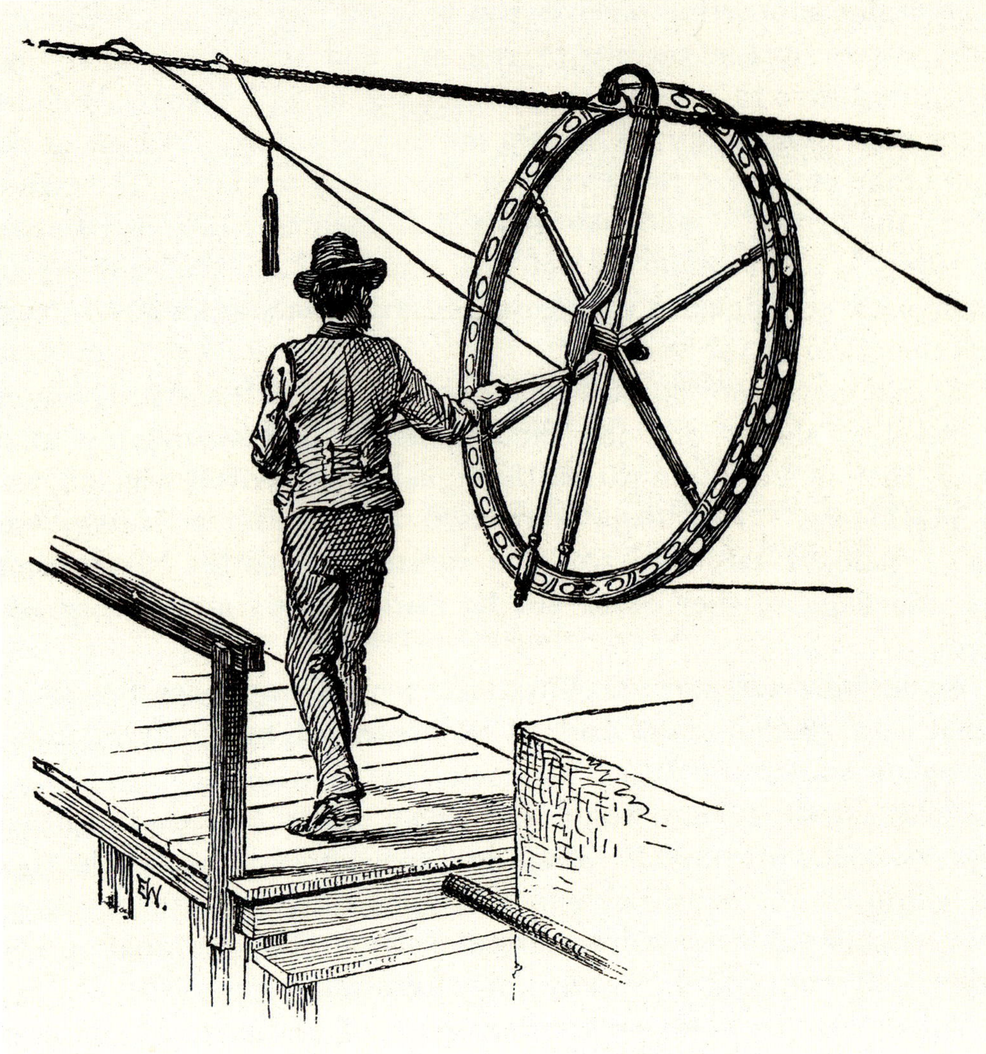

"Carrier-wheel," *Appletons' Journal*, January 1878. This woodcut depicts the mechanism that pulled the steel wires across from the Brooklyn to the New York anchorage in order to make the four cables. The heavy wire rope of the traveler—to which the carrier wheel is attached—is visible at the top. A thinner wire wraps around the wheel. Master Mechanic E. F. Farrington reproduced this image in his *Concise Description of the East River Bridge*. Courtesy of Richard Haw.

The wire end coming off a drum was looped around the carrier wheel at the Brooklyn anchorage, and the traveler then pulled the carrier with the wire trailing behind it across to the New York anchorage. Simultaneously, an empty carrier was heading back to Brooklyn from New York. It took a carrier only about ten minutes to travel from anchorage to anchorage. When the carrier wheel arrived at the New York side, the end of the wire was taken off the carrier and connected to a "shoe," or iron brace, that in turn was firmly attached to a "leg" that had been fastened to the anchorage. It remained there until all the wires that made up a strand had been spun, and the strand was ready to be permanently attached to the anchor bars and, thus, the anchorage.

After the wire was disconnected from the carrier wheel at the New York anchorage, the empty carrier wheel was pulled in the opposite direction by the traveler; that carrier, upon its return to the Brooklyn anchorage, picked up another wire that it then pulled back to the New York anchorage. These wires unspooled from the drums as the carrier wheel, to which they were attached, was pulled across by the traveler. The process was repeated over and over again.

TOP: "Plan and Elevation Showing Cable, Cradles, and Foot Bridge," 1876. This drawing illustrates how the wires were spun higher than the cables' final resting elevation. Once their spinning was completed, the cables were lowered into place. Author's collection.

BOTTOM: "Arrangement of Cable Drums, Foot Bridge, Cradles, and Cables, Brooklyn Anchorage," *Scientific American*, November 25, 1876. This print shows the eight drums, or wire wheels, atop the bridge's Brooklyn anchorage; two drums supply wire for each of the four cables. In actuality, these drums, each about two feet wide and eight feet in diameter, were stored in a wooden shed, but the illustrator eliminated the shed so that the reader could see the positioning of the drums.

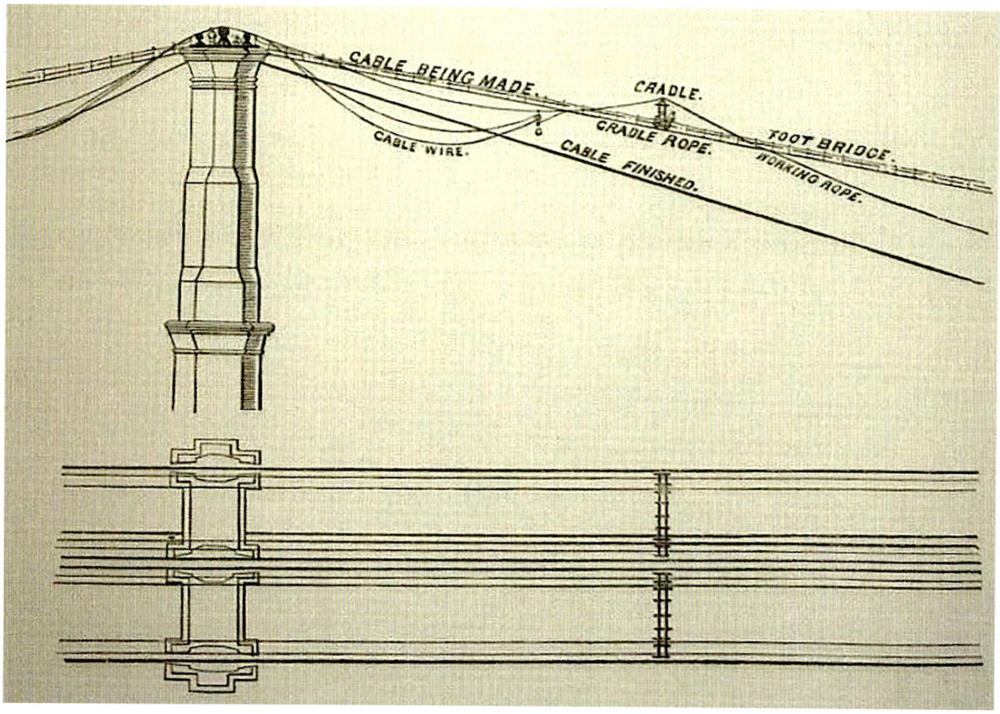

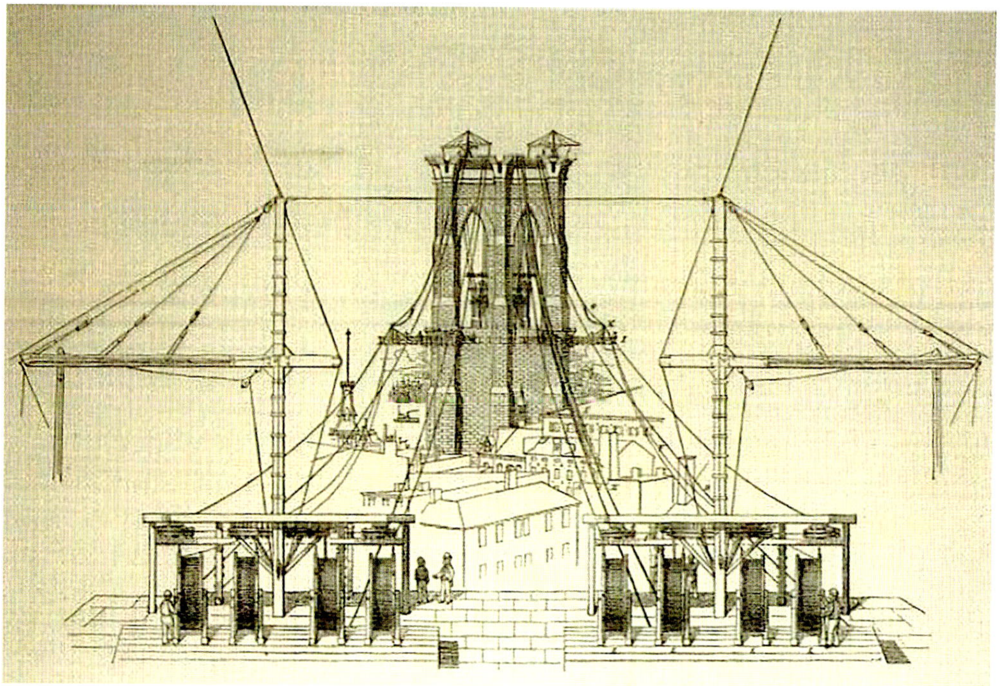

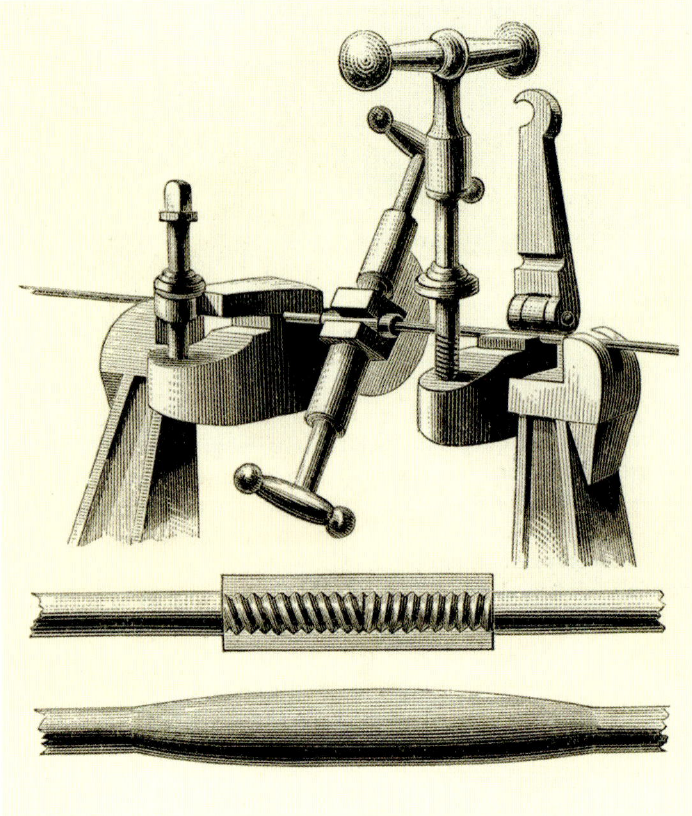

"Machine for Coupling the Wire," *Appletons' Journal*, January 1878. This woodcut shows the contraption that was used to splice the wire ends together. Each of the coils of wire shipped from the factory contained only several hundred feet of wire. Yet the distance from anchorage to anchorage, which a single wire had to stretch across, was almost 3,600 feet. Therefore, the wires from many coils had to be spliced together before being wound around one of the eight large drums used in the cable making. By one count, 200,000 splices were made to connect the wires that made up the four cables of the Brooklyn Bridge. Each cable was spun from the wire of two drums, and each drum contained almost ten miles of wire. Further, wires often broke as they were being spun across the bridge span. These broken wires would then have to be hauled in and spliced back together. It took the bridge engineers two years to create this "coupling" system, including the machine, which they designed specifically for the construction of the Brooklyn Bridge. The machine played a key role, creating splices that were equal in strength and as free from corrosion as the wires themselves. Courtesy of Richard Haw.

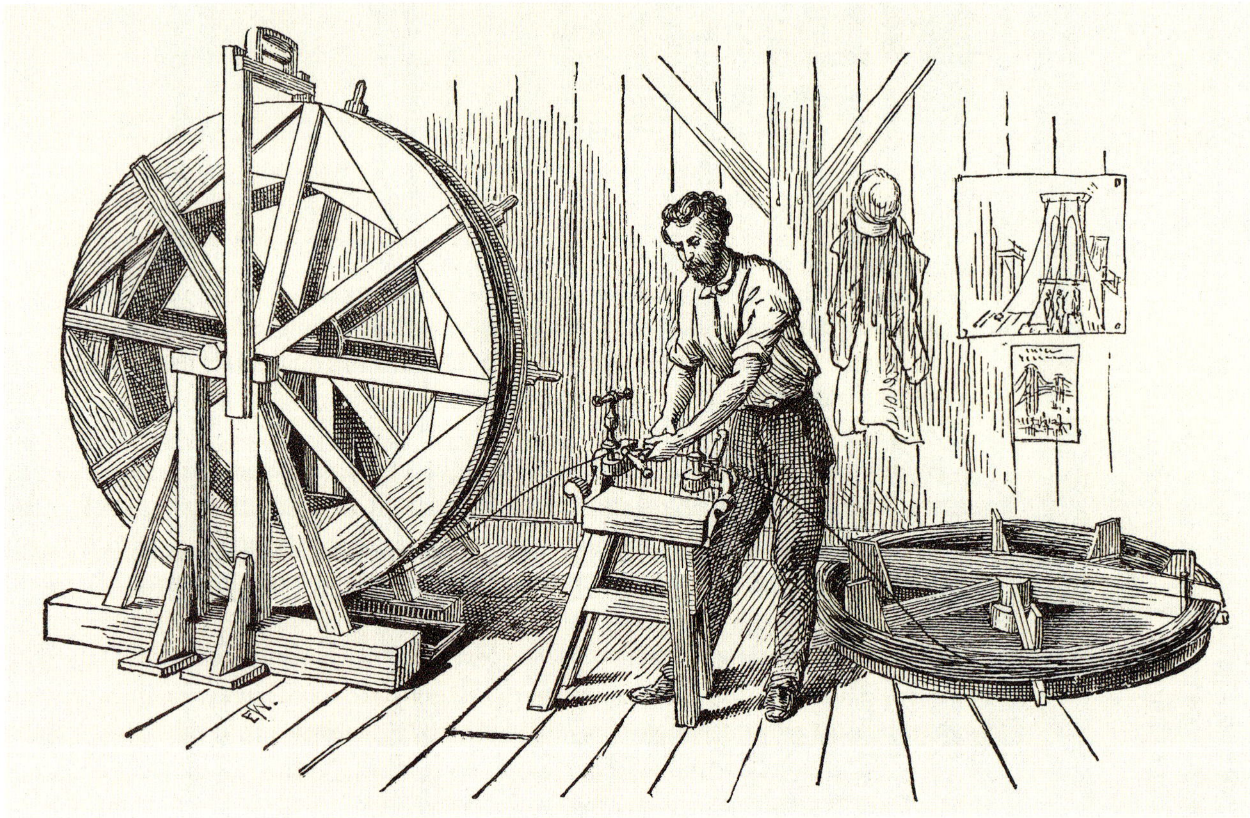

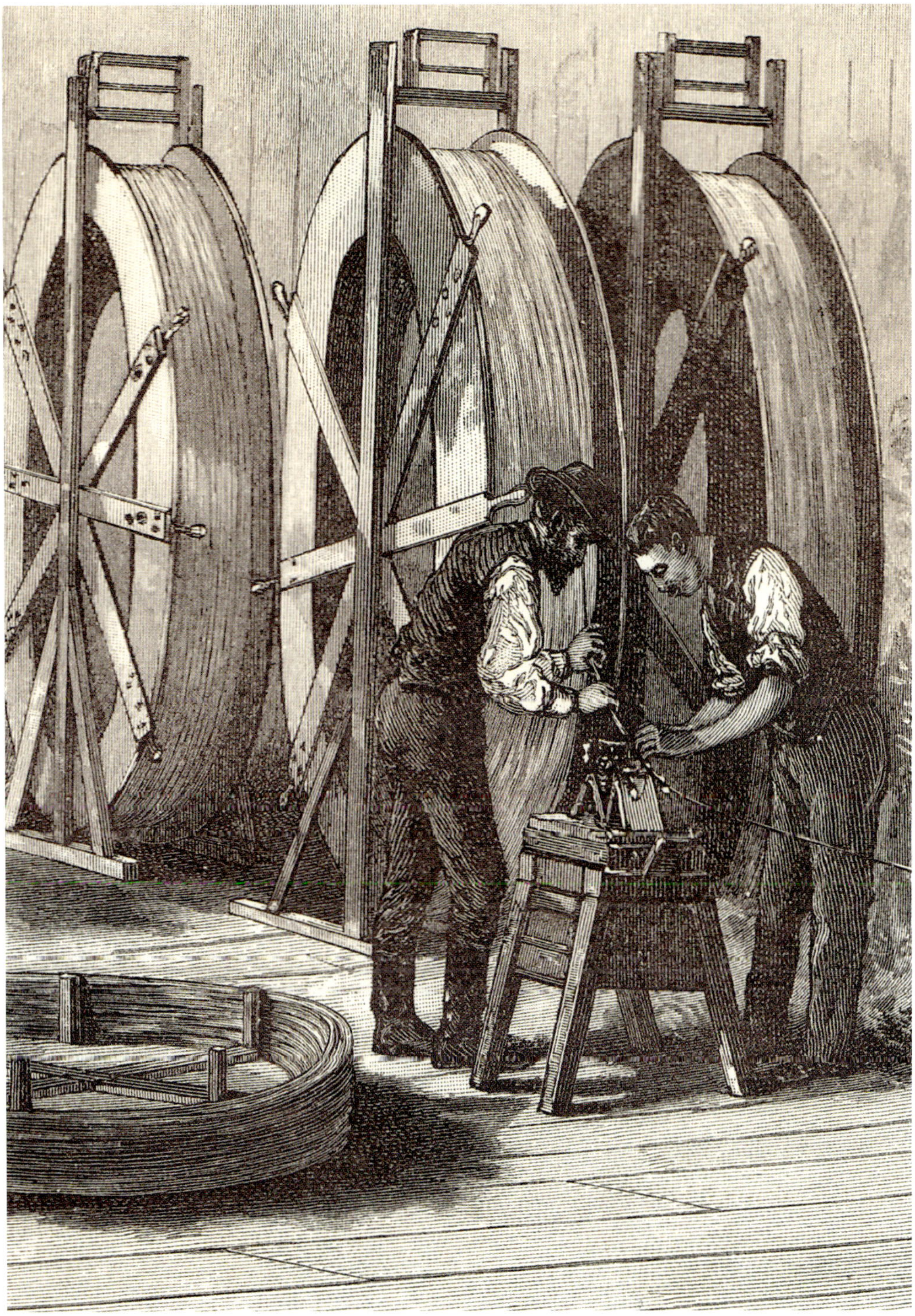

"The Drums—Splicing the Wires," *Harper's New Monthly Magazine*, May 1883.

OPPOSITE: "Machine for Splicing Steel Wire," a print from *Concise Description of the East River Bridge, With Full Details of Construction, Being the Substance of Two Lectures Delivered at the Cooper Union, March 6 and 13, 1880, by E. F. Farrington, Master Mechanic*, published in 1881.

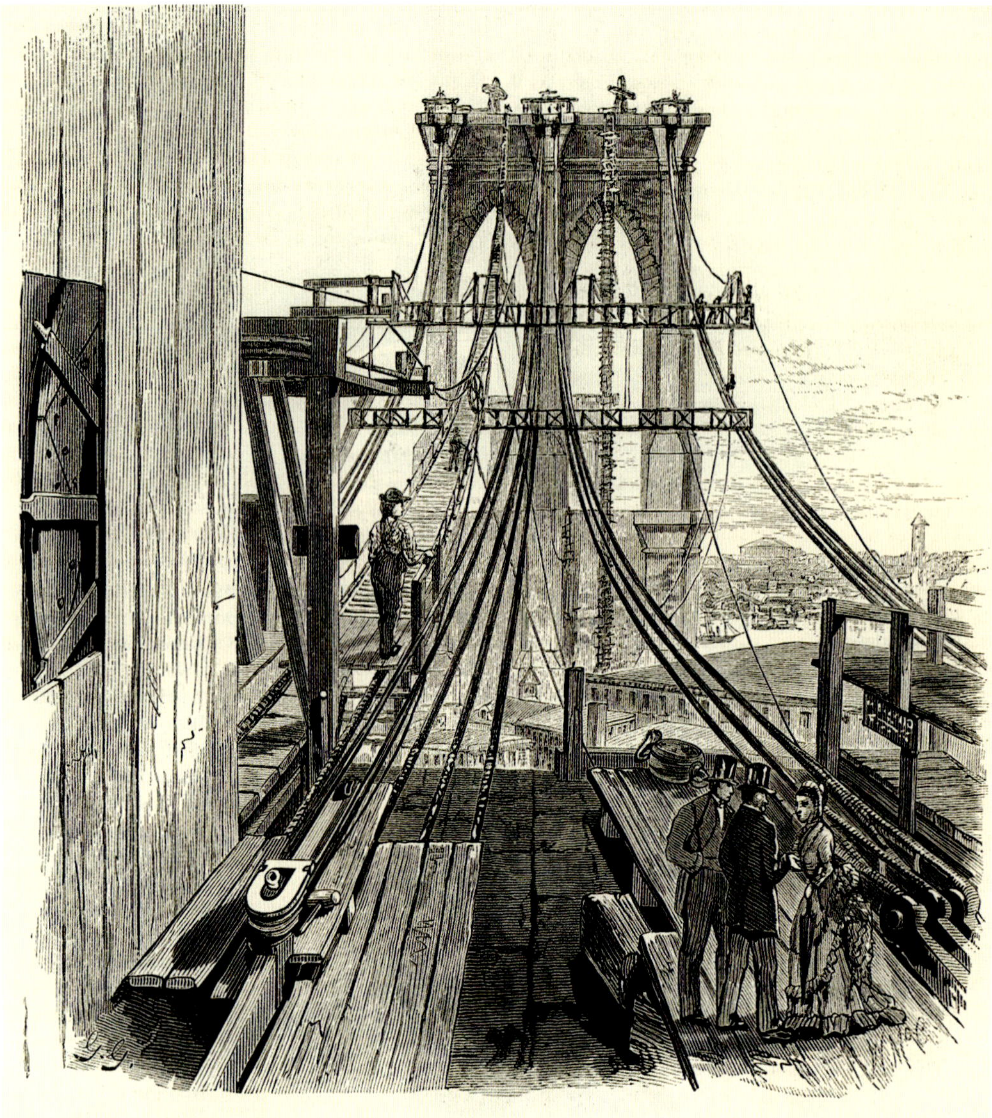

"Looking Toward Tower from Brooklyn Anchorage," *Appletons' Journal*, January 1878. Note the drum of wire in the far left of the woodcut, visible inside the opening at the side of the wooden shed. The wire is unspooling from the top of the drum as it is being pulled across the bridge—this was called the "spinning," or the making, of the cables. Just to the right of the wall of the shed, a "shoe" at the end of a strand is depicted. At lower left, wooden planks cover the anchor bars below them, allowing the workers to walk on top of them. Anchor bars, to which wire strands are attached, are visible to the right of the three people in the foreground. Courtesy of Richard Haw.

OPPOSITE: "The Great Suspension Bridge Between New York and Brooklyn—The Cable Fastenings," *Scientific American*, August 11, 1877. Woodcut showing a pair of anchor bars that have been pivoted up, held in place atop stacked wooden blocks, as one strand of a cable, attached to its U-shaped shoe (seen where a worker stands on top of the anchor bars, with his left arm extended and his right hand on the shoe), is about to be eased off farther and then pinned between two anchor bars at the right end of the bars. Courtesy of Richard Haw.

A newspaper clipping saved by Emily Roebling lauded the early wire work:

> There is something colossal in the look of the East River piers as they show in the morning sunlight; the ropes already connecting the two piers seem like slender threads, and as the vessels pass and repass under them some idea may be formed of what may be the effect when the graceful upper wire structure is completed, with the roadway crowded with passengers and vehicles of all descriptions and the high-masted clippers and coasting traders passing underneath.

Upon completion of the spinning of all the wires that made up one strand, it was necessary to perform an operation known as "letting off the strands." This involved first releasing the "shoe"—to which the ends of the strands are secured—from the "leg" where it had been temporarily held in place, then easing it forward and pinning it between two of the anchor bars at the top of the anchor chain.

David McCullough, in *The Great Bridge*, describes this "letting off" process:

> By July 2 [1877] the first of two strands was completed for the two cables on the downstream side of the bridge. The work of lowering the strands into position then began. At the anchorages the strands were drawn back by a hoisting engine until the shoe was released from its fastening. The shoe and strand were lowered slowly, carefully forward, twelve feet, the hoisting engine and a block and tackle holding the immense pull of the strand. Because there was a twist in the tackle, the shoe turned up on edge as it came forward and slipped neatly in between the eyelets of the anchor bars. The forward motion was stopped then and a seven-inch steel pin passed through the eyelets and the shoe.

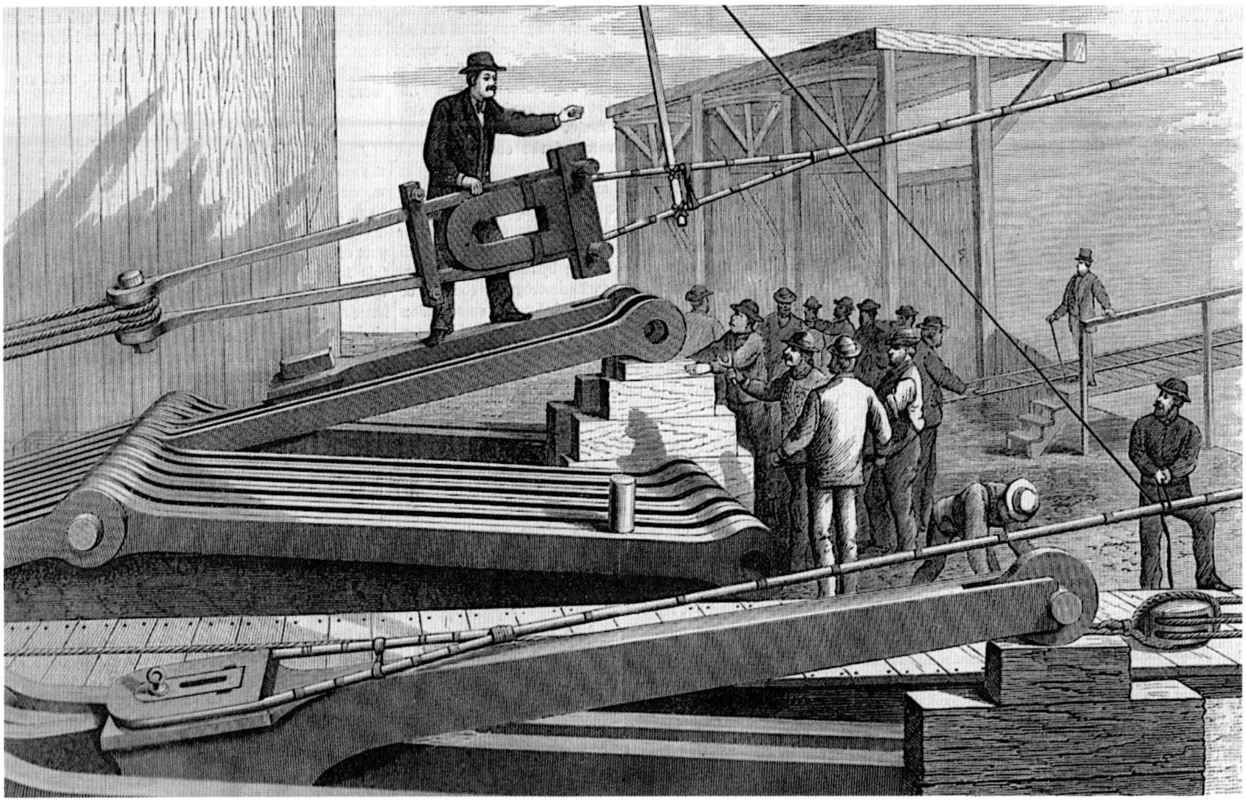

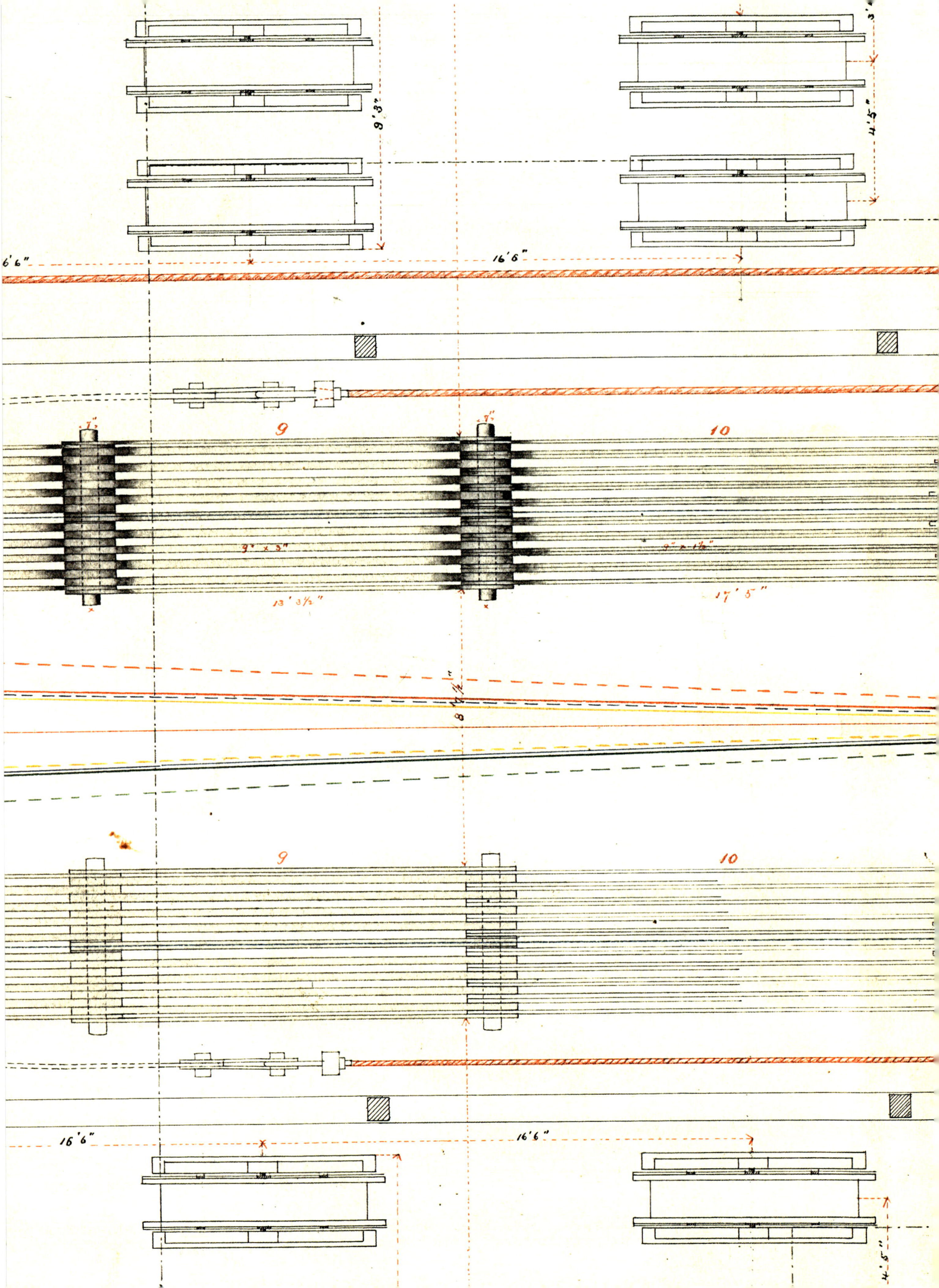

9'3"
3'
4'5"
6'6"
16'6"
9
10
7"
9" x 3"
9" x 1½"
13' 3½"
17' 5"
8' 4½"
9
10
16'6"
16'6"
4'5"

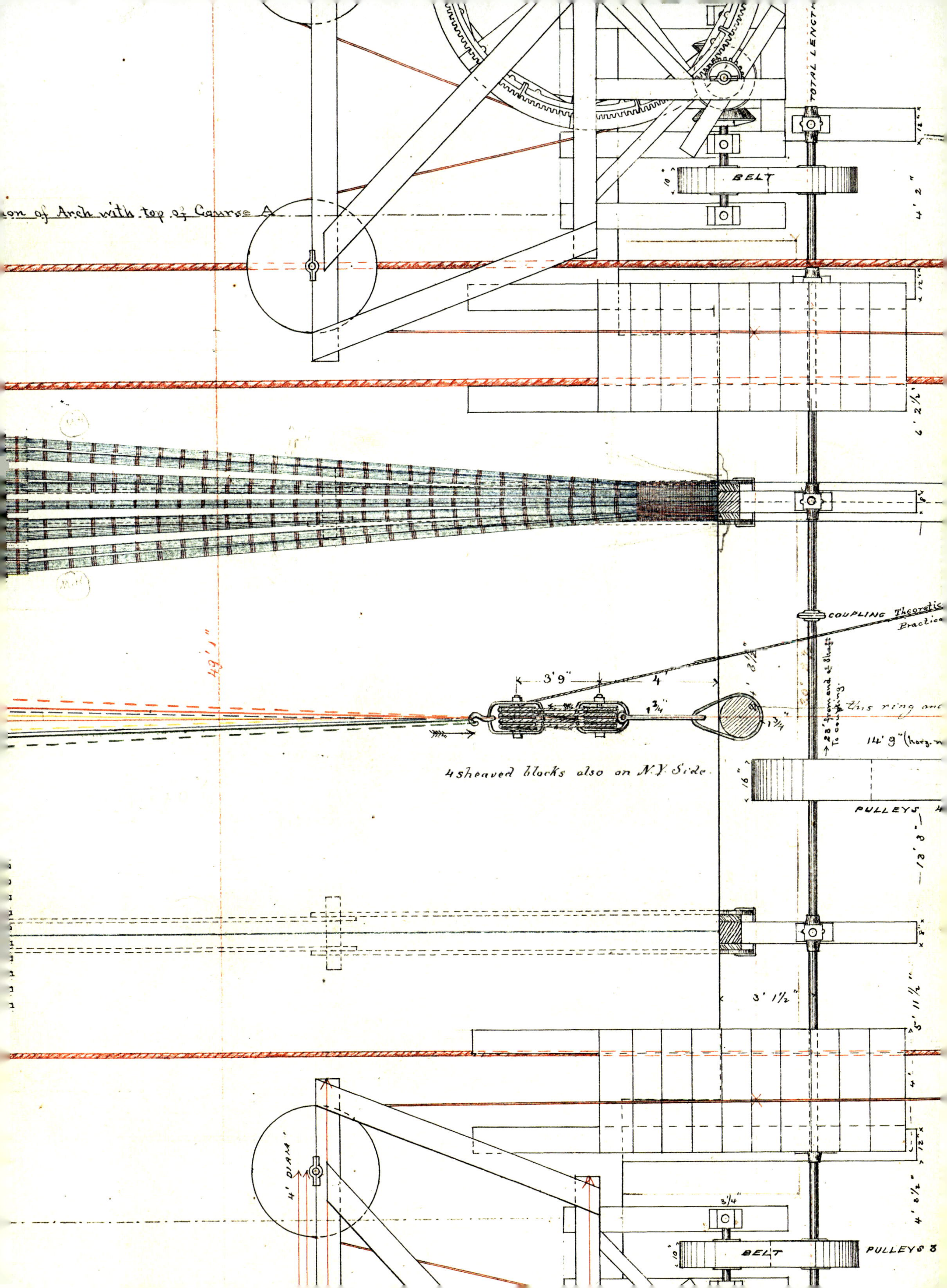

on of Arch with top of Course A
BELT
4sheaved blocks also on N.Y. Side
3'9"
4'
COUPLING Theoretic
Practica
this ring and
14' 9" (horz.
PULLEYS
3' 1½"
4' DIAM.
BELT
PULLEYS 3

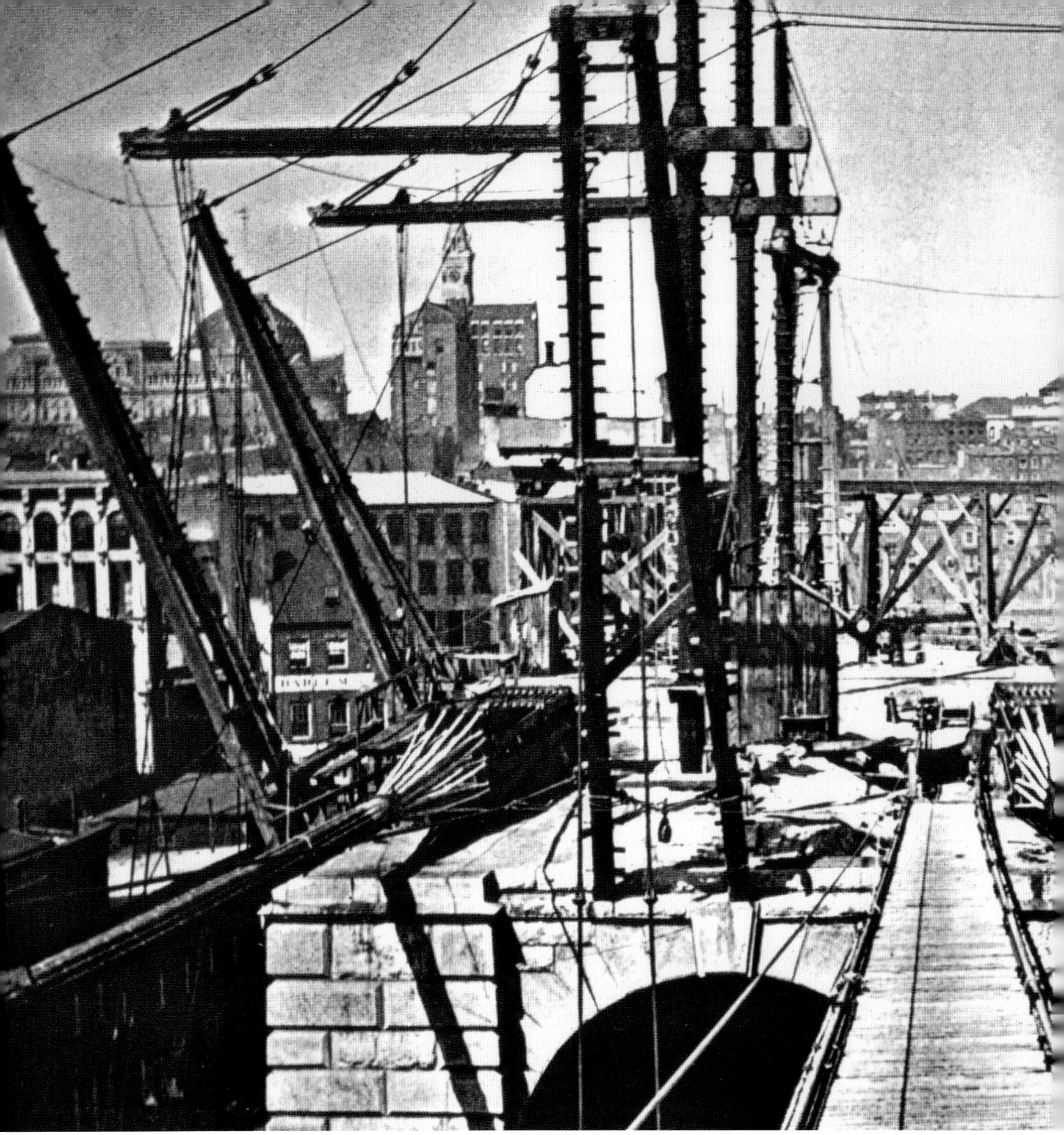

PREVIOUS SPREAD: Drawing by Chief Engineer Washington Roebling, showing the joining of the anchor bars (at left) and the cable strands (at right). RPI.

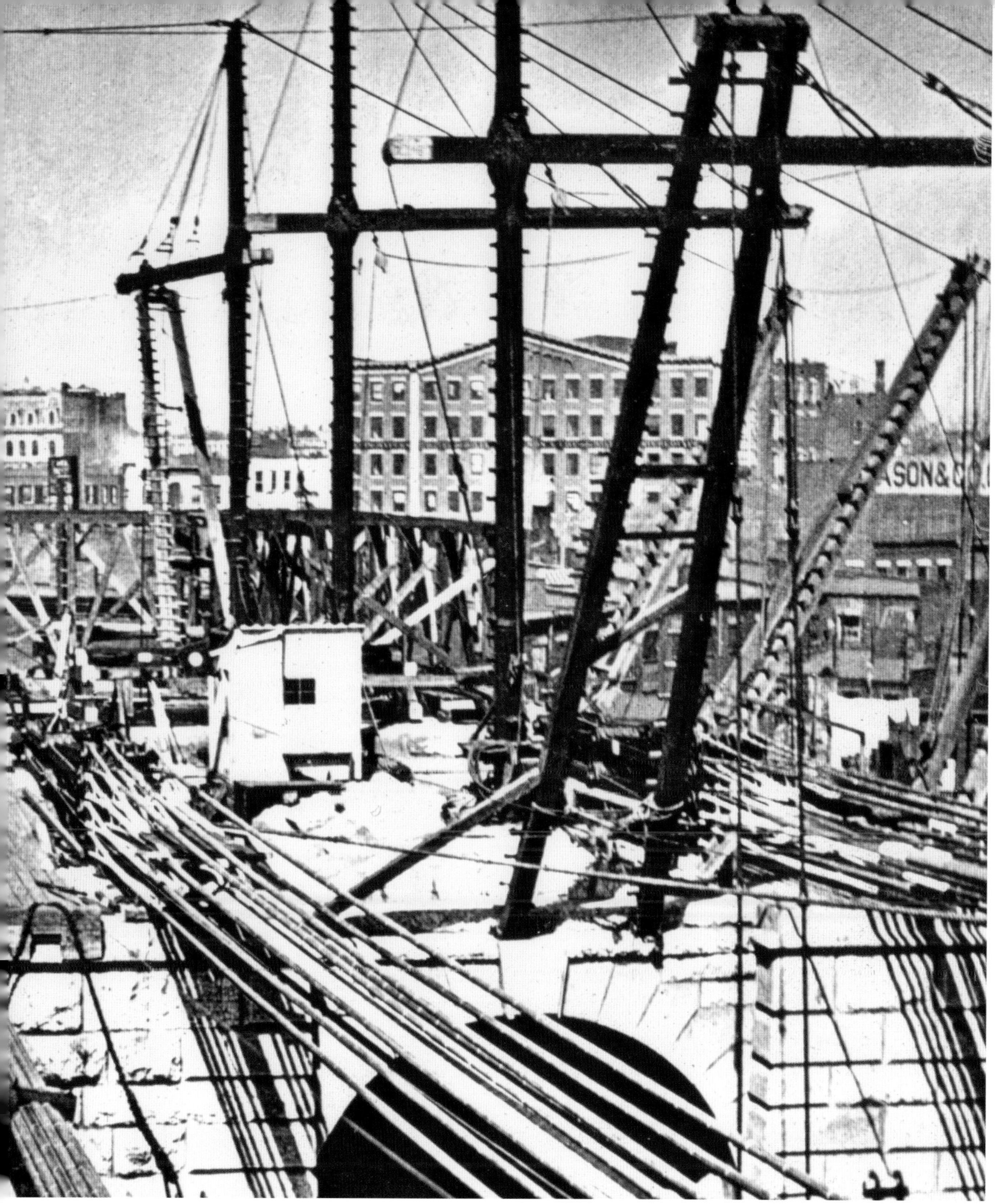

Print from a lantern slide photograph of the top of the New York anchorage, looking west into Manhattan, showing the strands of the four cables, ca. 1879. Note that the two bridge cables to the left have already been arrayed and clamped together, awaiting wrapping, while the two cables on the right have yet to be arrayed and clamped; their nineteen strands remain separate. The footbridge is visible at center, parallel to the cables. The derricks remain in place, ready to lift the last of the stone blocks onto the anchorage. The wooden framing in the center distance is a craneway, an elevated track used to move heavy stone blocks into position along the approach. At left, in the distance, is the distinctive tower of the New York Tribune Building, and, to the far left, at the horizon, is the hulking new City Hall Post Office and Courthouse, which opened in 1876. Author's collection.

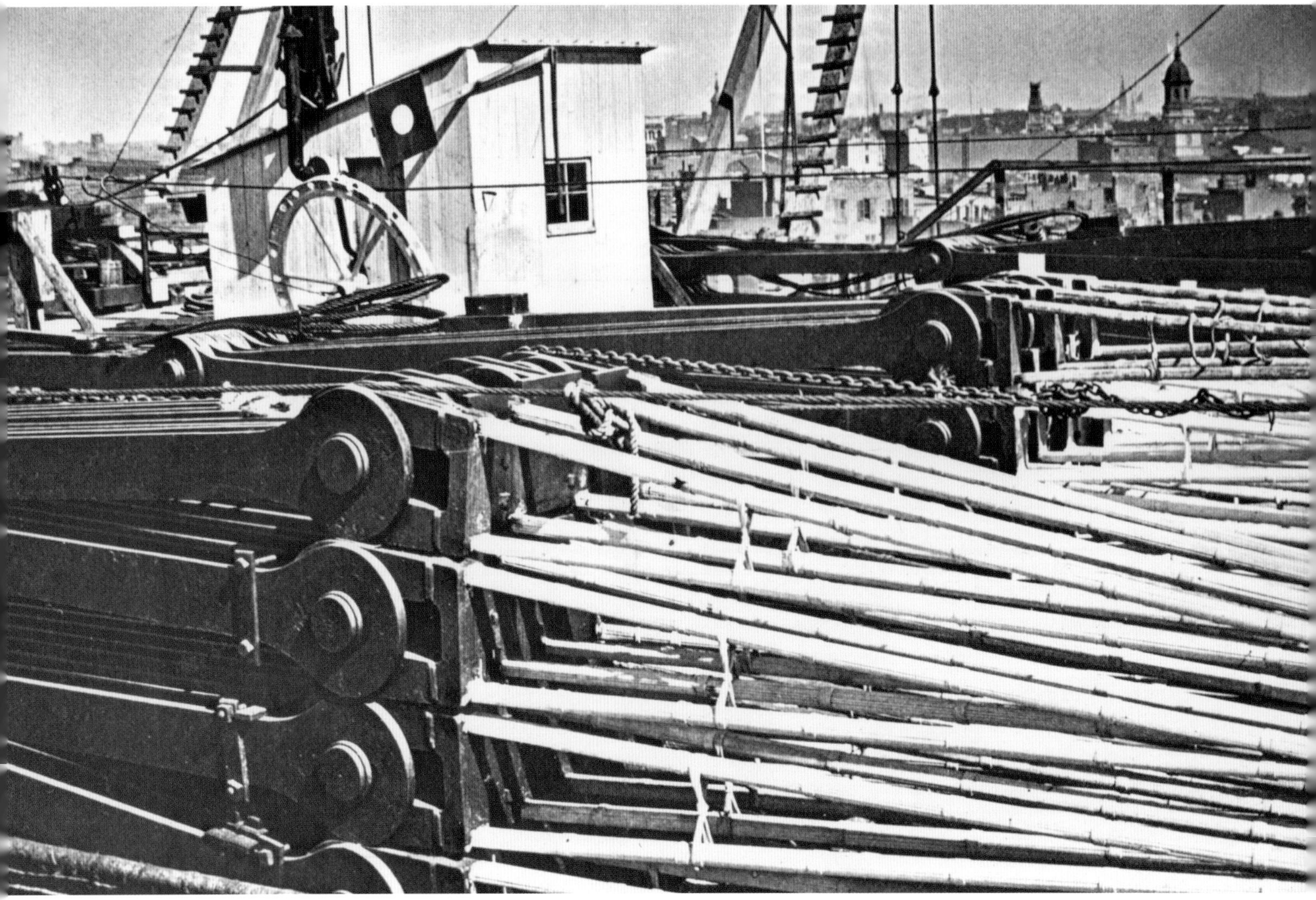

This print, from a lantern slide photograph, ca. 1878, shows anchor bars (at left) and the white cable strands (at right) that have been attached to them. A variant of this image, at the Rensselaer Polytechnic Institute, is titled, "Cable connections with anchorchains, N.Y. anchorage." As each of the nineteen strands that formed a single cable was completed, it was eased (or let) off its temporary holding position by a steam engine, using a very heavy block and tackle, and attached to the end of two anchor bars, using a steel pin to connect the "shoe," holding the end of the strand, to the "eyes" of the anchor bar. This image depicts the completion of that process. Note the carrier wheel behind the anchor bars, on this side of the wooden shack. The church steeple at top right was in Manhattan and confirms that this photograph was taken on top of the New York anchorage. Author's collection.

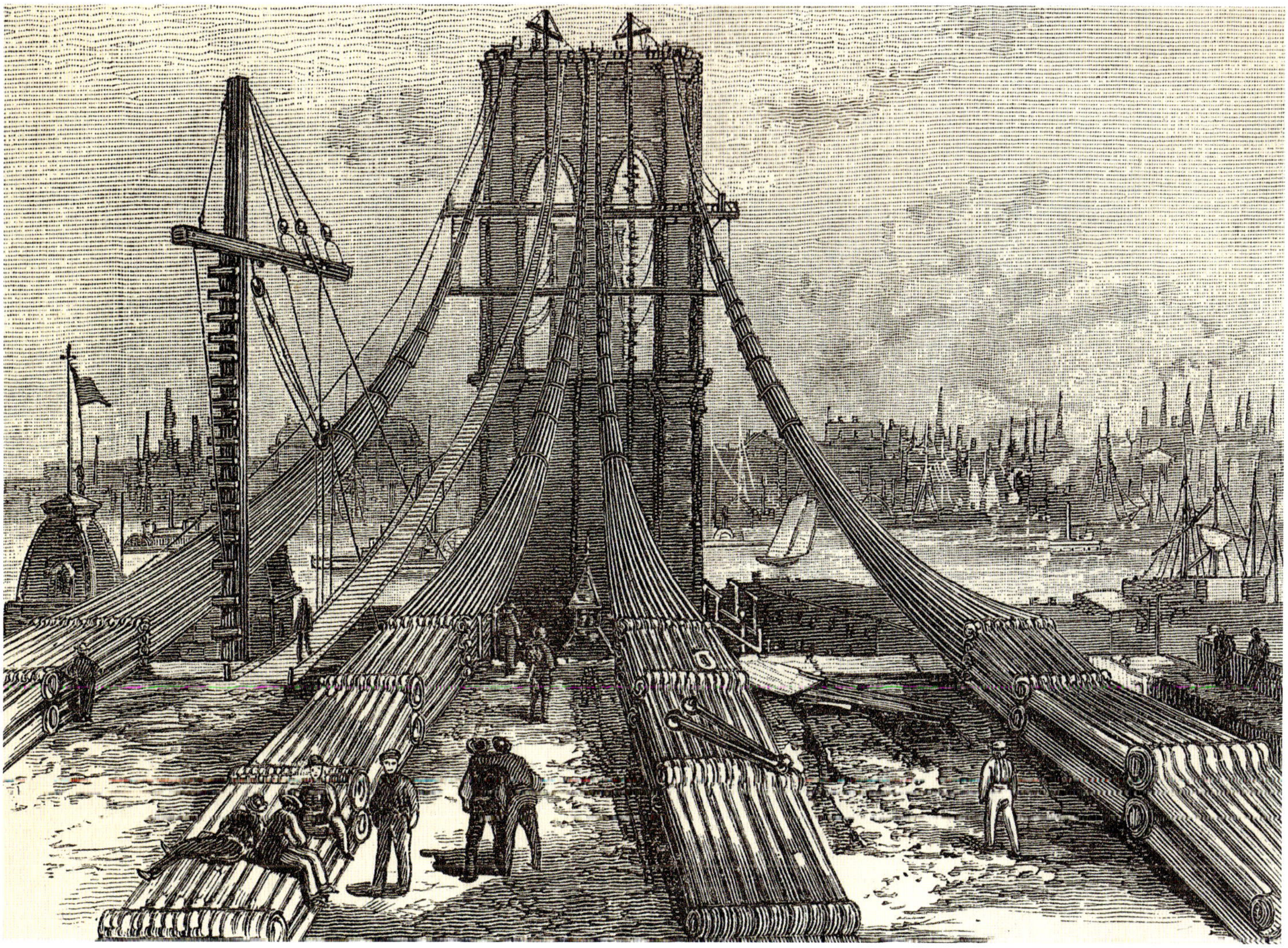

"The Brooklyn Anchorage." *Harper's Monthly*, May 1883. Looking west from the anchorage toward the Brooklyn tower, with Manhattan in the distance. The cable strands have already been attached to the anchor chains; they are clamped and await wrapping. The extent of the work necessary to attach the cables to the anchor bars (and thus the anchorage) is apparent here.

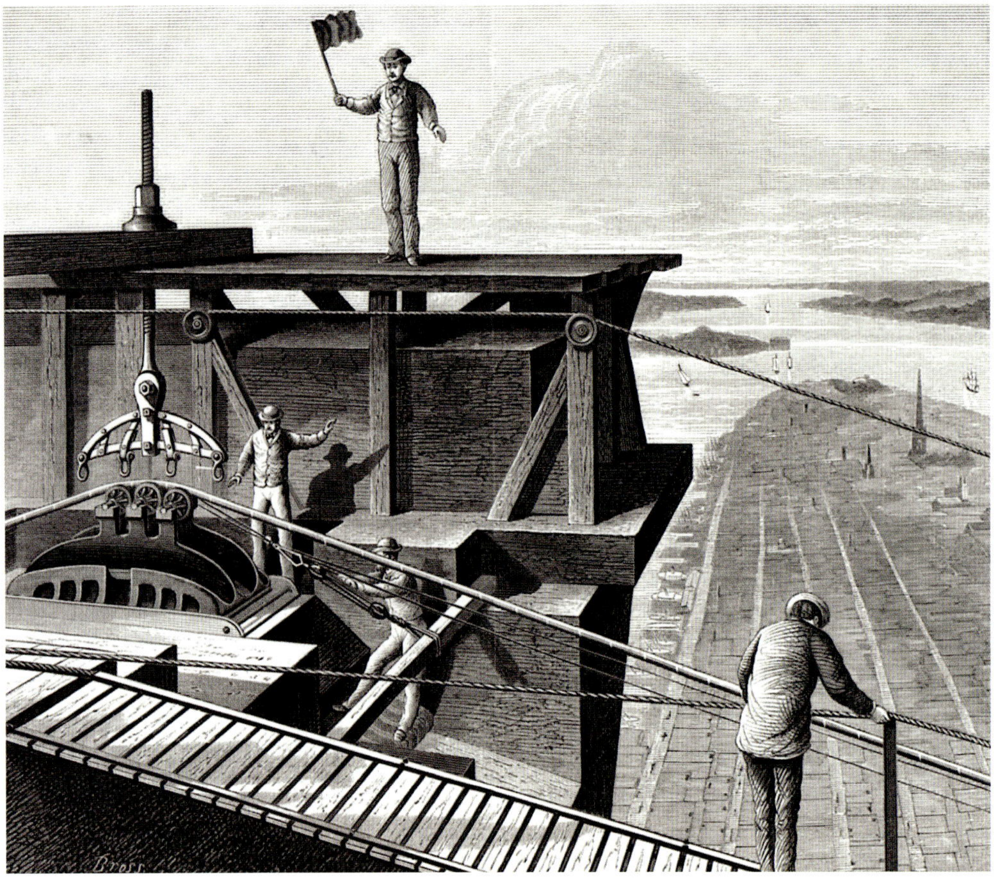

TOP: "Cable Making on the East River Bridge," *Scientific American*, August 4, 1877. This illustration shows workers atop the New York tower, with a saddle at left and the footbridge in the foreground. At right, Manhattan stretches to the south. The spire of Trinity Church appears to be bisected by the top bridge wire. Courtesy of Richard Haw.

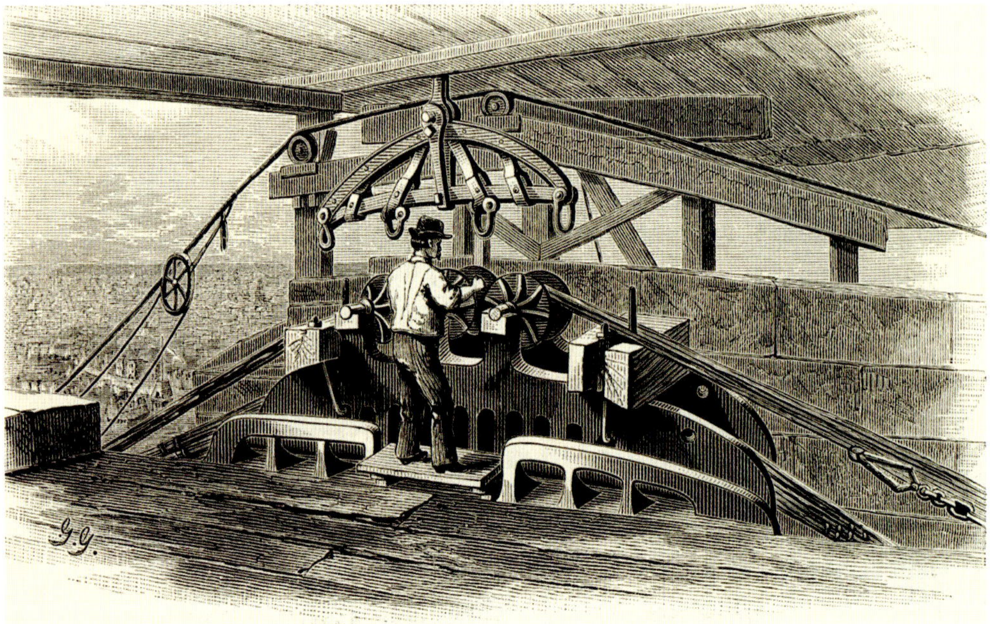

BOTTOM: "The Saddle," *Appletons' Journal*, January 1878. Note the contraption above the man's head. That is the device, with its four clevises hanging down, that was used to grab and lift a strand. Courtesy of Richard Haw.

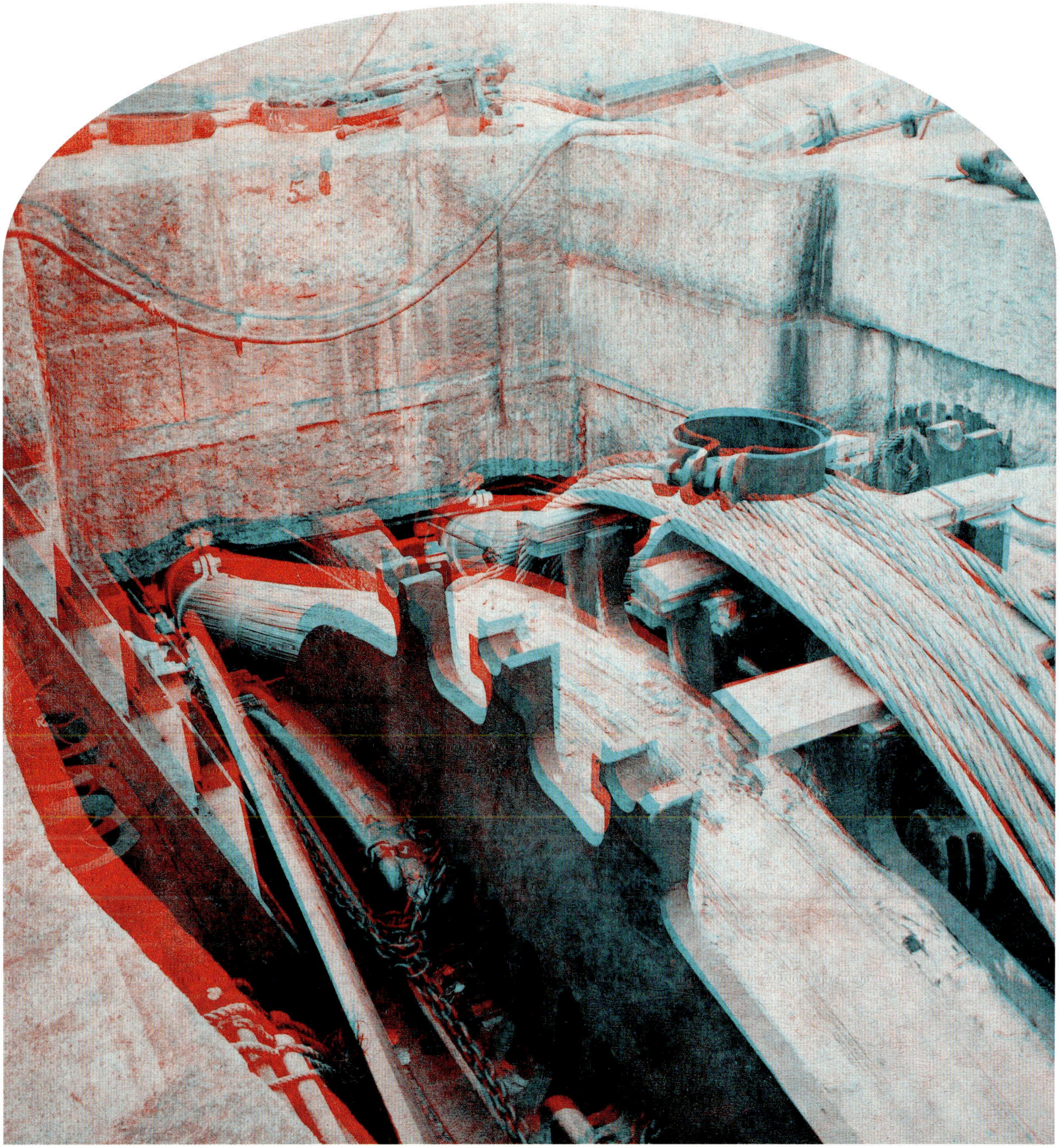

3D This anaglyph shows the process atop one of the towers as two of the four cables are being lowered onto their saddles. Note that the strands of the cable at left have already been arrayed and clamped—the clamp is visible around the strands just before they disappear from sight. The strands of the cable at right, resting temporarily on wooden planks, have yet to be aligned; a clamp sits on the strands, awaiting use. At top left are several unused clamps. The three U-shaped grooves at the center are intended to hold in place the cover that will protect the cables and saddles from the weather. The rollers, upon which the saddles and cables were to move as expansion and contraction occurred, are below the cables and are, therefore, not visible here. Unfortunately, once the weight of the cables was placed on the saddles, that weight drove the rollers into the saddle plates, freezing all movement and making the carefully designed mechanisms unable to perform their intended function. Courtesy of the Michael E. Ach Collection.

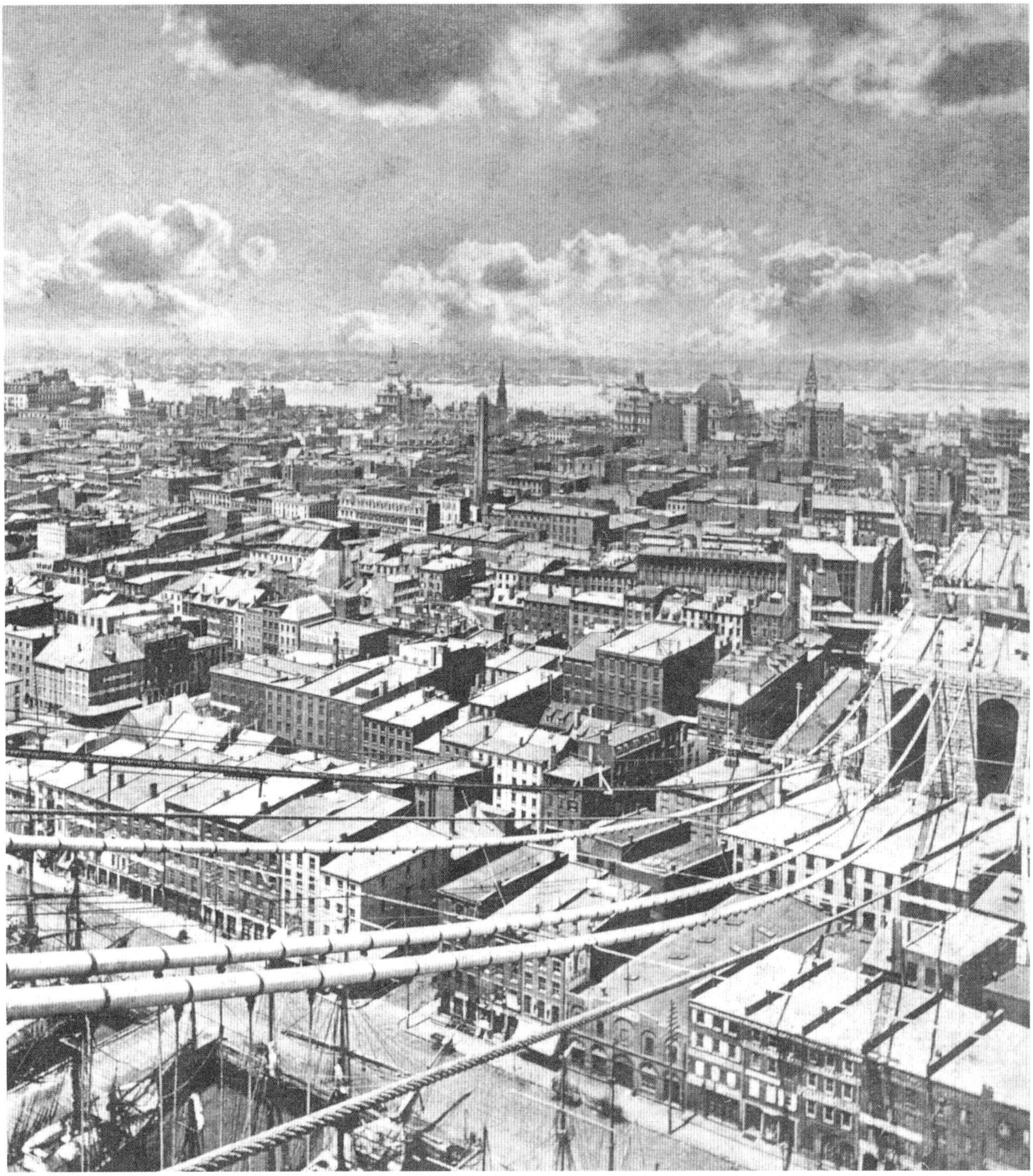

This view of the New York anchorage and lower Manhattan dates from 1881. The tower and anchorage had been completed, the cables had been wrapped and finished, and, at left, the suspenders are being hung from the cables. Courtesy of The Green-Wood Historic Fund, Jeffrey I. Richman Stereoview Collection.

For his young readers of *St. Nicholas Magazine*, Charles Barnard vividly described the cold and dangerous work on these buggies:

> These seem like simple things to do. But just think of it a little while! Think of working in a little wooden cage swinging and swaying two hundred and fifty feet in the air! The days were bleak and cold and the wind blew—oh! How it does blow up there sometimes! Below was the black water, perhaps dotted with ragged ice. A misstep, and—good-bye. No man would ever come back alive. There was nothing between them and death but the wire ropes suspended high over the masts of the ships.

These workers, hanging high above the river on their buggies, bound the cables with wire for their entire length, with the exception of where the cables crossed the saddles; binding them there would have been tremendously difficult both because of the confined work space and each strand's great weight.

The cable wrapping stopped in November 1878, after New York City refused to provide its share of the bridge funding. The work would not resume until the following March, when New York City finally met its commitment. The work on the cables was completed soon thereafter.

On the hundredth anniversary of the opening of the Brooklyn Bridge, Paul Goldberger, then the architectural critic of *The New York Times*, wrote "an appreciation" entitled "Brooklyn Bridge, at 100, Embodies Spirit of the Age." Of the cables, he rhapsodized: "The cables offer a gentle counterpoint, so delicate that they look like harp strings, and though they are, in fact, made of heavy strands of steel bound together, they make us feel that if we plucked them they would respond with beautiful music."

With the cable-making process now completed, the approaches to the bridge had to be built before work on the suspended deck could begin. After all, the bridge would be useless unless people, wagons, carriages, horses, and livestock could get up onto it.

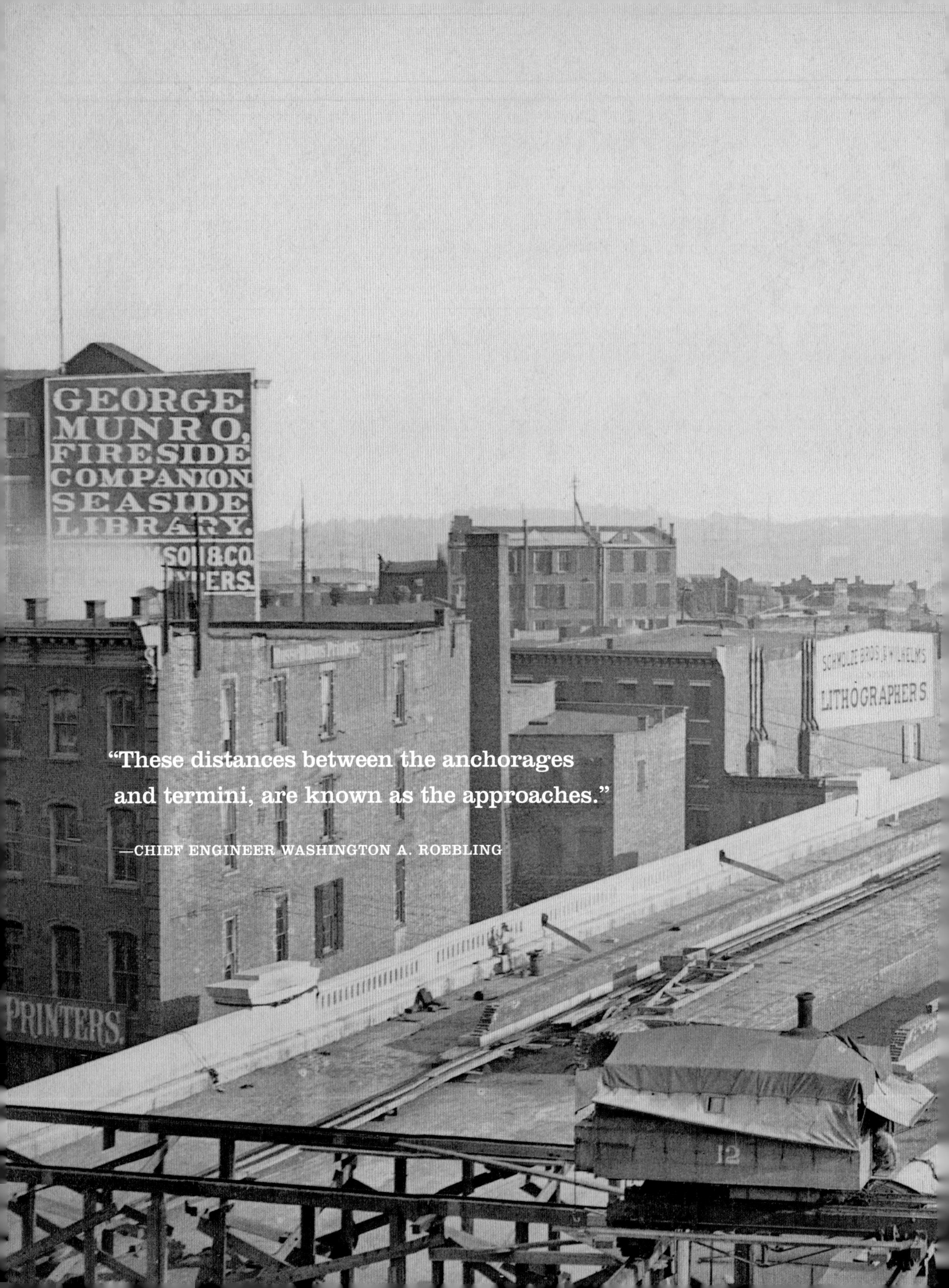

"These distances between the anchorages and termini, are known as the approaches."

—CHIEF ENGINEER WASHINGTON A. ROEBLING

CHAPTER 8

Approaches

1877–1882

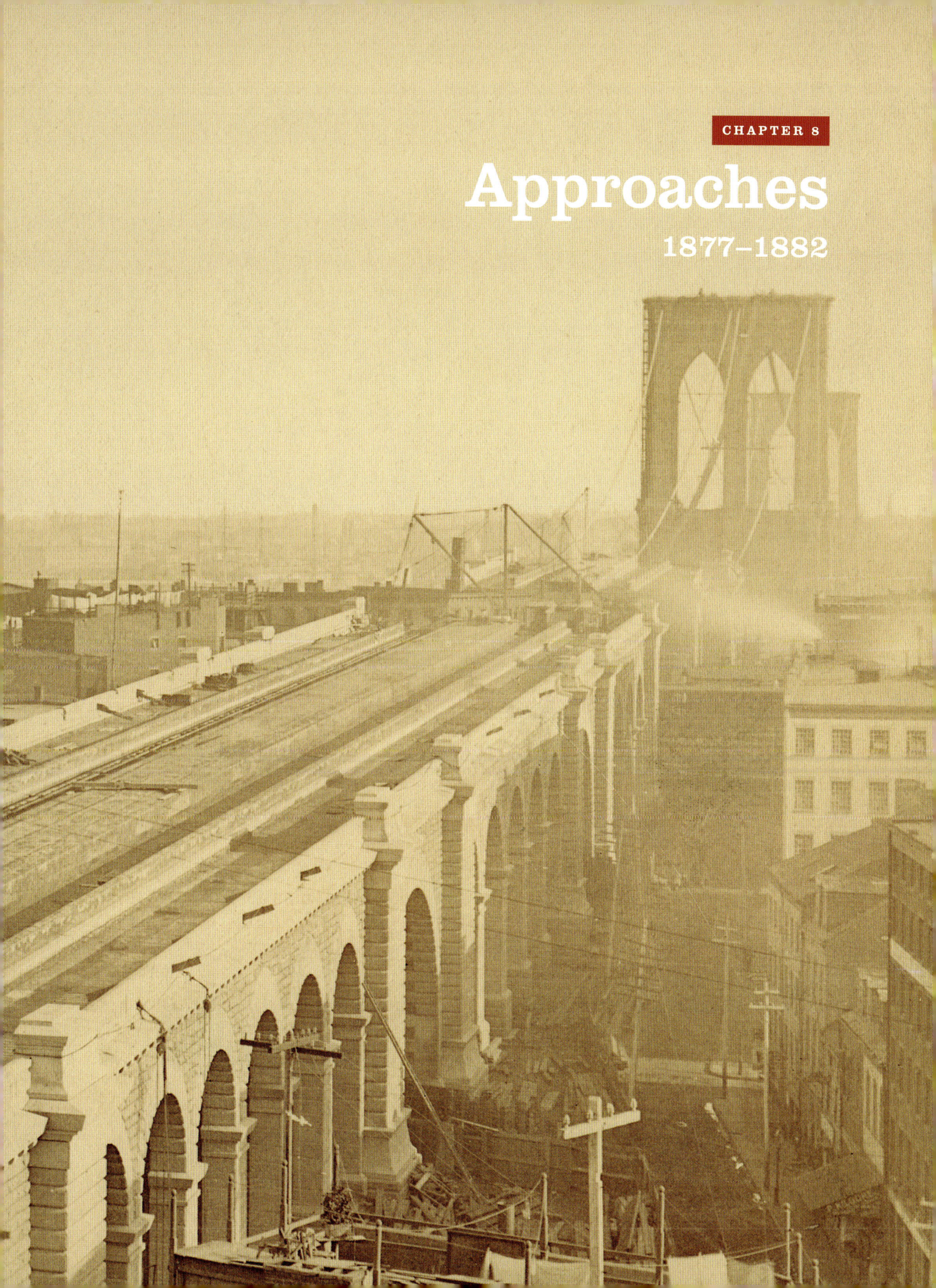

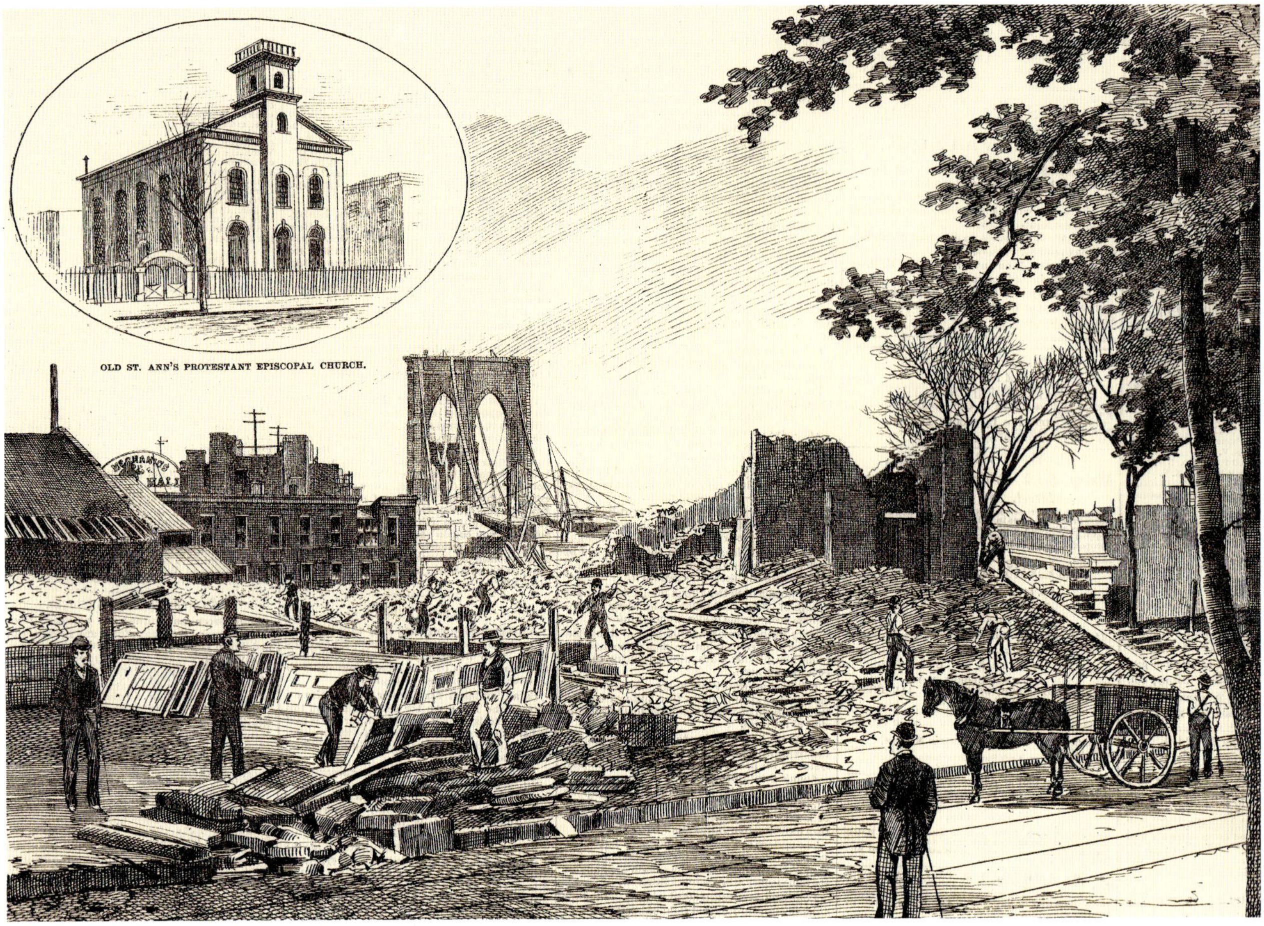

"Making Way for the Brooklyn Bridge—the Demolition of Old St. Ann's Protestant Episcopal Church, Corner of Washington and Prospect Streets, Brooklyn," *Daily Graphic*, May 7, 1880. The church's tower, visible from afar, was used to mark the bridge's centerline in 1869, before any construction had begun. In 1880, the Bridge Company purchased the church building and land; the engineers had the structure demolished later that year. Author's collection.

PREVIOUS SPREAD: A photograph of the New York approach, close to completion, showing the parallel divisions of the roadbed taking shape. The center section would soon be divided into three parts: the pedestrian promenade, elevated in the middle, and railway tracks on either side. The one-way roadways are already defined along the outside of the roadbed. Note car number 12 atop the craneway in the foreground; it makes several other appearances in this book. RPI.

Work on the bridge approaches began in 1877 and continued for five years. This was one of the most time-consuming phases of the construction—not because the work was pioneering or complicated, but rather because it was so massive. Each of the two approaches runs gently downward from its respective anchorage to ground level and to the railway terminals at the ends of the bridge. The approaches support that part of the bridge floor that is not suspended from the cables.

Those using the bridge to cross the East River, whether pedestrians, carriages, railway cars, or livestock, would enter one side of the bridge at its terminus. They would then proceed along the approach as it increased in elevation until crossing over the top of the anchorage, then traverse the suspended spans of the bridge. Arriving at the approach on the other end of the bridge, they would gradually proceed down off the bridge and continue on their way.

Washington Roebling described the span and elevation of each of the two approaches:

> The Brooklyn approach is to be a structure about nine hundred feet long on the centre [*sic*] line, starting from grade at Sands Street, and increasing in height until at the rear of the Brooklyn anchorage it will be about sixty feet above the ground. . . . The New York approach will be fifteen hundred and forty-six feet in length, starting from grade at Chatham Street, and increasing in height to about sixty-eight feet, at the rear of the New York anchorage.

While the approach on the New York side ran in a straight line for its entirety, the shorter approach on the Brooklyn side curved to the south as it moved eastward. As a result, the Brooklyn approach crossed above several streets at odd angles. Master Mechanic E. F. Farrington explained that, as a consequence, "Steel truss or plate girders will be substituted for masonry to carry the bridge floor over these streets."

Before any work could begin on the bridge construction, the engineers had to first survey the land and mark the centerline, or right of way, of the bridge. This would determine its exact path. So, in 1869, Chief Engineer Washington Roebling and Assistant Engineer William H. Paine set out together onto Brooklyn's streets and decided to put the Brooklyn (or eastern) terminus on the land along Washington Street where St. Ann's Church then stood, between Prospect and Sands Streets. Since the church tower, rising above its neighbors, could be seen from a great distance, it would serve as a landmark to identify the bridge's centerline—particularly useful for engineers and workers looking east from the Brooklyn tower, anchorage, or approach. The church would remain in place as the work slowly progressed. In 1880, when the easternmost end of the Brooklyn approach was ready to be built, the church was torn down. As the *Brooklyn Daily Eagle* reported coldly, St. Ann's was "swept away by the march of modern improvement."

In 1877, when construction on the approaches was ready to start, workers began to tear down all the buildings remaining at street level. Much of the bridge construction before this had been undertaken without the need to clear the land purchased by the Bridge Company: the caissons and towers were built on the shores of the East River that were largely free of structures, and other construction work, like the spinning of the cables, occurred above old buildings.

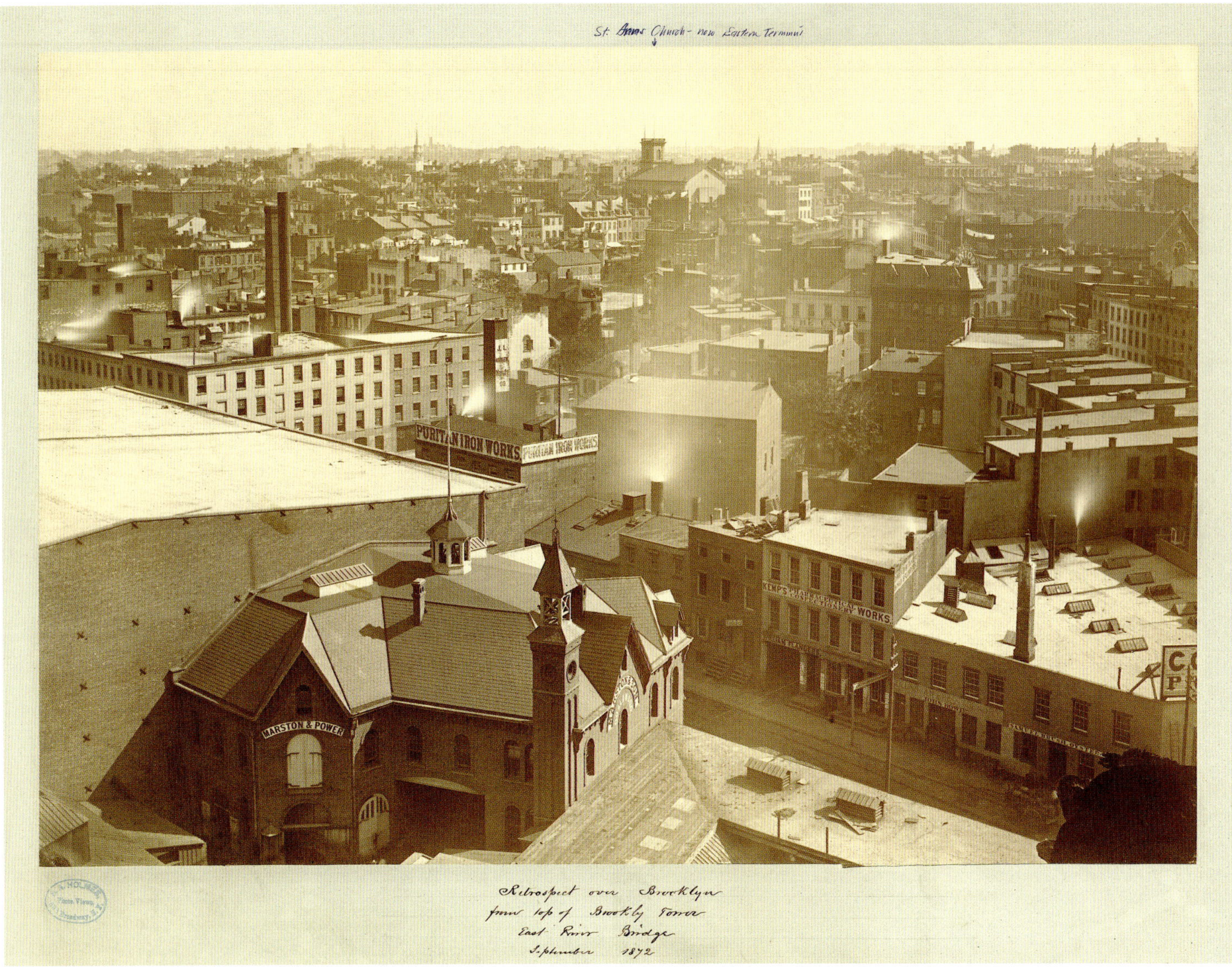

"Retrospect over Brooklyn/from top of the Brooklyn Tower/East River Bridge/September 1872." The stamp of photographer Silas A. Holmes appears at lower left. Note the ink notation at the top middle of the mat, with an arrow pointing toward the highest building on the horizon: "St. Ann's Church—now Eastern Terminal [of the bridge]." St. Ann's, rising above its neighbors, was a useful landmark and represented the site where the approach on the Brooklyn side would end—the eastern terminus of the bridge. In the middle distance, to the right of the centerline, the building with a white facade and "The Union" painted across its front was the headquarters of Brooklyn's *Union* newspaper—where the Bridge Company offices were conveniently located. RPI.

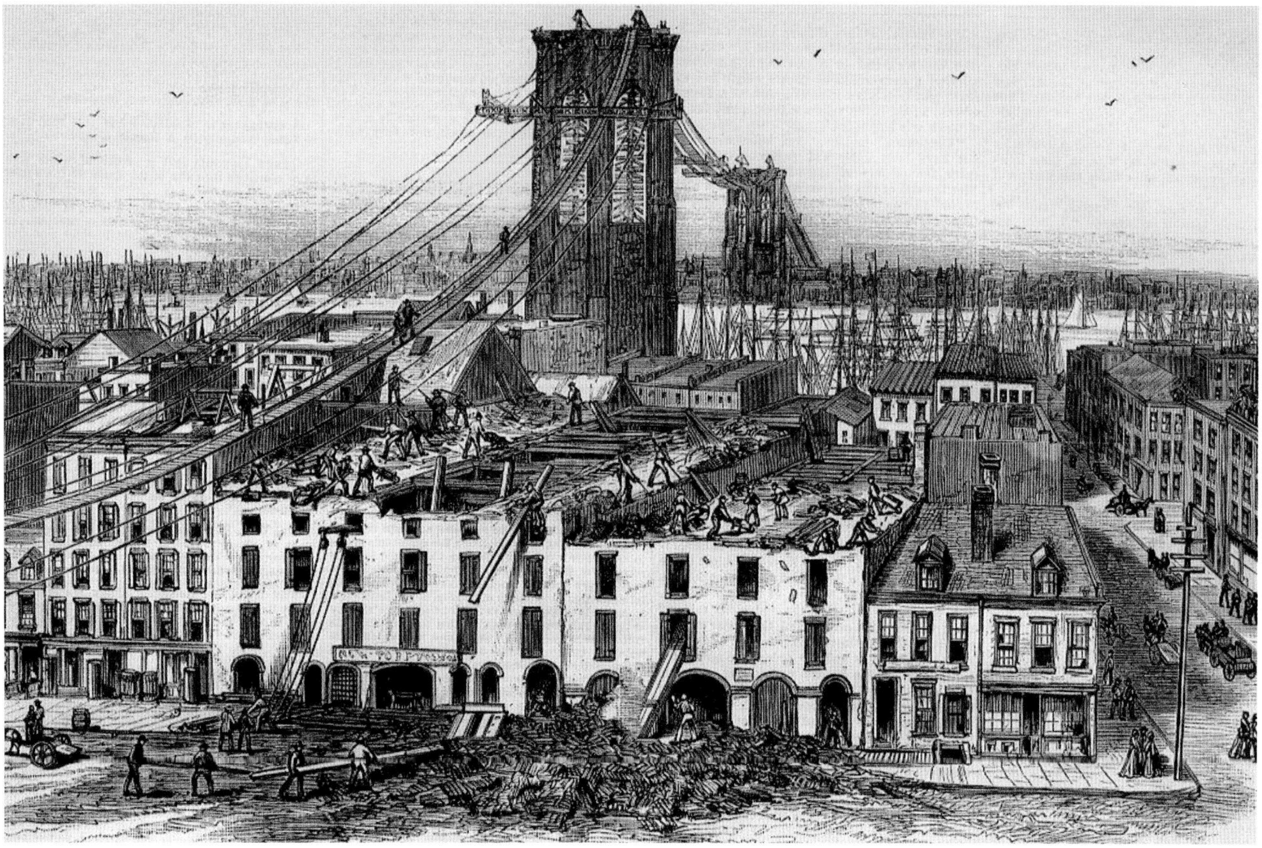

"New York City—Tearing Down Buildings on Water Street to Make Way for the New York Approach to the East River Bridge," *Frank Leslie's Illustrated Newspaper*, July 1877. In this woodcut, the towers are complete, the footbridge is in place, the cable wires are being spun, and land needs to be cleared for the impending work on the approaches. When presenting his plans for the bridge, John Roebling wrote rather caustically that the granite and limestone of the approaches would "greatly beautify and improve this part of the city, which appears to need it more than any other" Author's collection

Earlier, workers had reduced the heights of hundreds of derelict buildings along the bridge's centerline that the Bridge Company had purchased, reinforcing their lowered roofs to stabilize them. This reduction in height addressed an ongoing concern: a fire breaking out in an abandoned building could endanger the bridge construction above. Master Mechanic E. F. Farrington explained early precautions taken: "Those parts of the work between the anchorages and the towers are well protected from fires which may occur in buildings beneath the bridge floor. In New York the height of the buildings has been reduced, and the roofs have been made fire proof by rebuilding them of iron beams and brick arches. There are fifty feet in the clear between these roofs and the floor of the bridge. This distance is also maintained in Brooklyn."

This lowering of buildings diminished the threat of any fire reaching the bridge above them, allowing work to proceed overhead, without having to completely demolish the structures. At the time, this let the workers focus on the bridge construction above, having only done the minimal amount of clearance. But now, as construction of the approaches was set to begin, full removal of the unoccupied buildings could no longer be deferred.

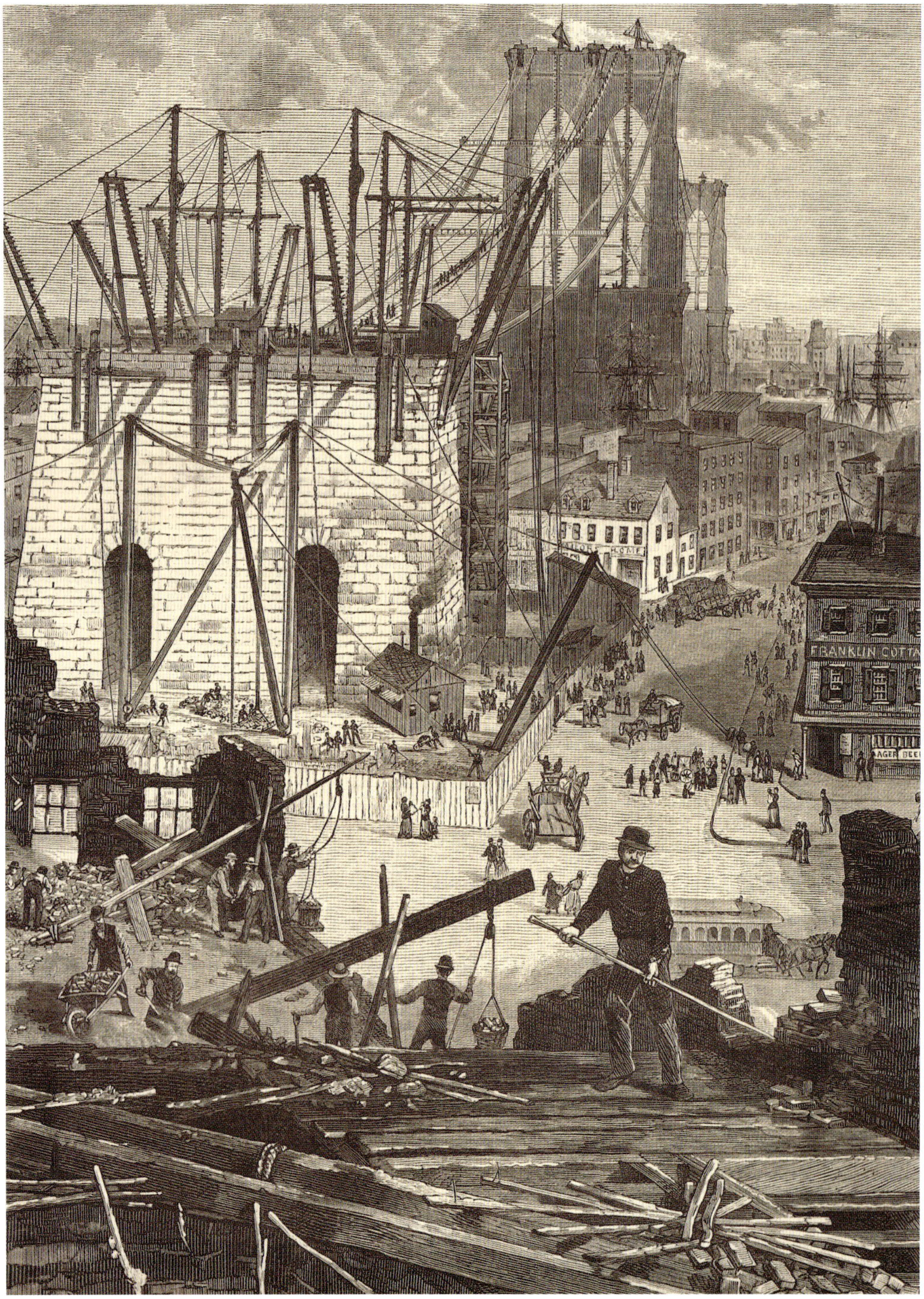

"The East River—Demolition of Buildings for the New York Approaches—Drawn by W. P. Snyder," *Harper's Weekly*, November 24, 1877. The work on the New York anchorage is almost complete as the towers rise in the distance and land is being cleared for the New York approach in the area of Franklin Square, at Franklin and Cliff Streets. The anchorage, shown at left with derricks on top, was built on the site of the United States' first Executive Mansion. Located at 1 Cherry Street, it was occupied by President George Washington from April 1789 to February 1790, while New York City was the nation's capital. Author's collection.

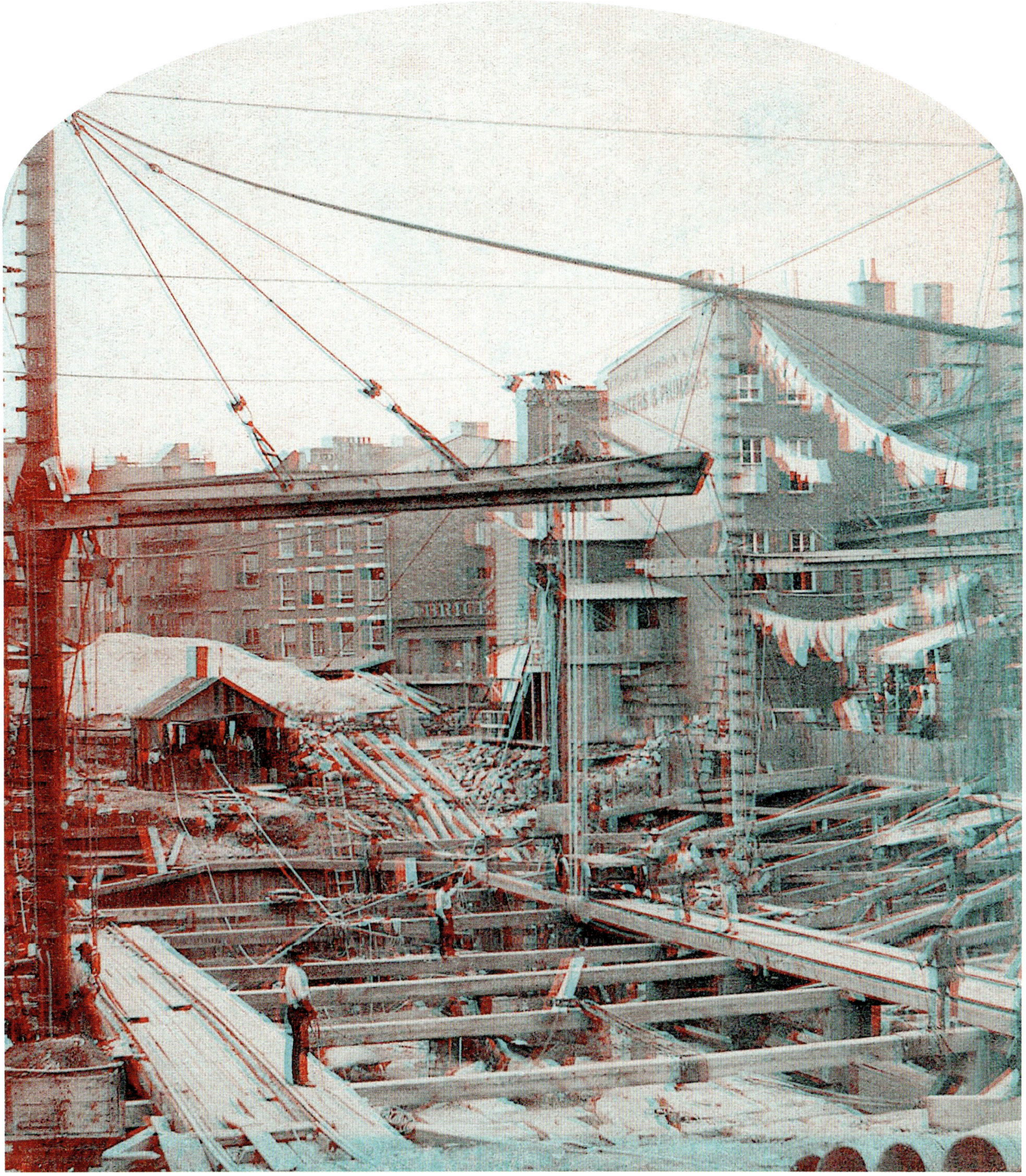

3D This stereoview by Hall Brothers, here converted into an anaglyph, captures the narrow confines in which a lot of the construction work on the bridge—including the approaches—was done. Courtesy of the Jeffrey Kraus Collection.

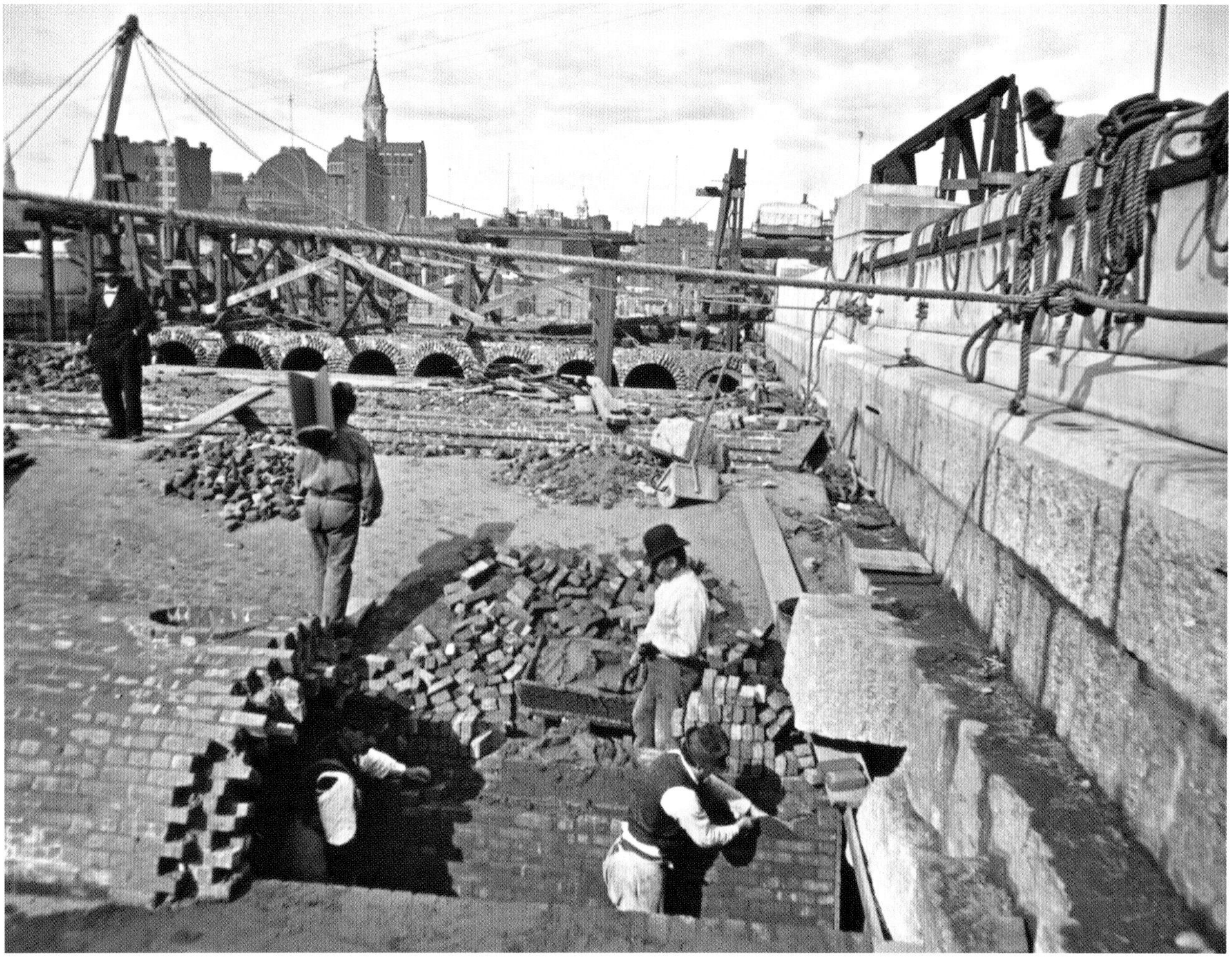

Half-stereoview (one of the two slightly different images of a stereoview) showing work being done along the New York approach. A hod carrier walks away from the camera, likely having just delivered bricks to the bricklayers. At left is a man in a suit, perhaps a supervisor. In the left foreground is a partially built brick arch, and along the right side are granite blocks, already a part of the New York approach. Brick and granite were the two primary materials of the approaches. Courtesy of the Michael E. Ach Collection.

Once the demolition was complete, erecting brick vaults was a first step in the building of the approaches. These would provide strength, allowing the approaches to bear the weight of the various users of the bridge. After these vaults were completed, they were faced and finished with granite blocks, enhancing the rhythm and elegance of the arches, characteristics that still distinguish the approaches today.

A craneway, or wooden railway, was built to transport the heavy granite blocks to where they were needed along the approaches. Washington Roebling described these temporary wooden rails high atop their lumber bases as " very curious." He explained the setup: "An elevated railroad was built just over the place where the walls were to stand. On this lofty railroad ran a very accommodating engine, that not only picked up the big stones from the trains in the streets, but lifted each block in the air and carried it to just the place where the masons wished it laid."

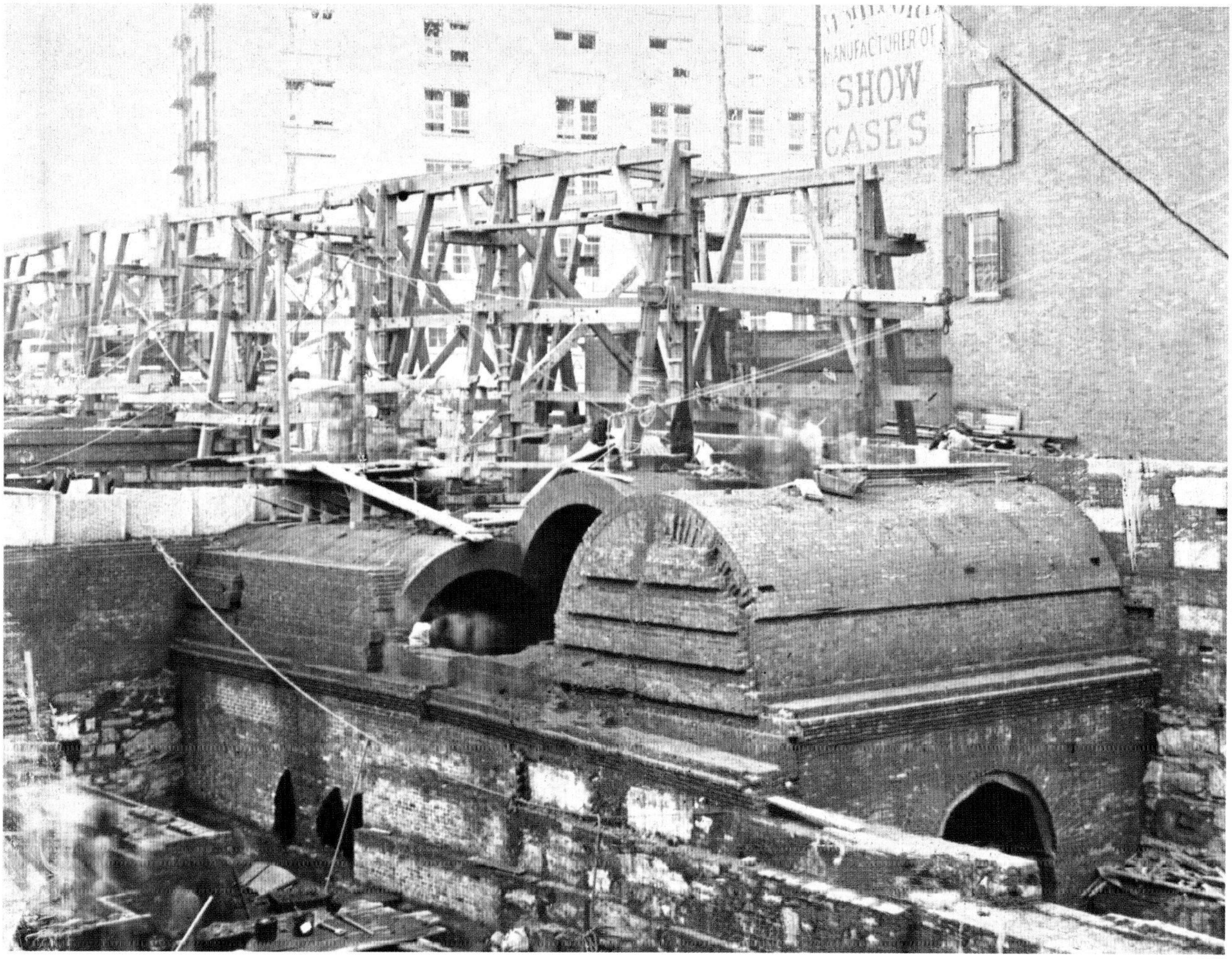

This print, from a lantern slide photograph, offers a good view of the vaulted brickwork and the wooden craneway, or railway, mounted above it. Author's collection.

Temporary wooden frames, or falsework, were built to hold the granite blocks in their assigned places. The joints between the stones were then filled with mortar in order to fix the blocks in place. Once the mortar set, the wooden falsework was removed. Construction photographs capture the wonderful patterns created by these intricate wooden frames.

Work on both the Brooklyn and New York approaches was completed in July 1882. Aside from their aesthetic appeal, the brick and granite vaults underneath the bridge floor created large spaces for storage and commercial enterprises.

Just after the bridge opened in May 1883, critic Montgomery Schuyler wrote in *Harper's Weekly* that he was impressed by the "'bits' of buildings" that the approaches framed in the distance: "We get what is to be got nowhere else in our rectangular city."

Meanwhile, work on the suspended superstructure—the deck of the bridge—continued. It had begun early in 1881, simultaneous with the work on the approaches.

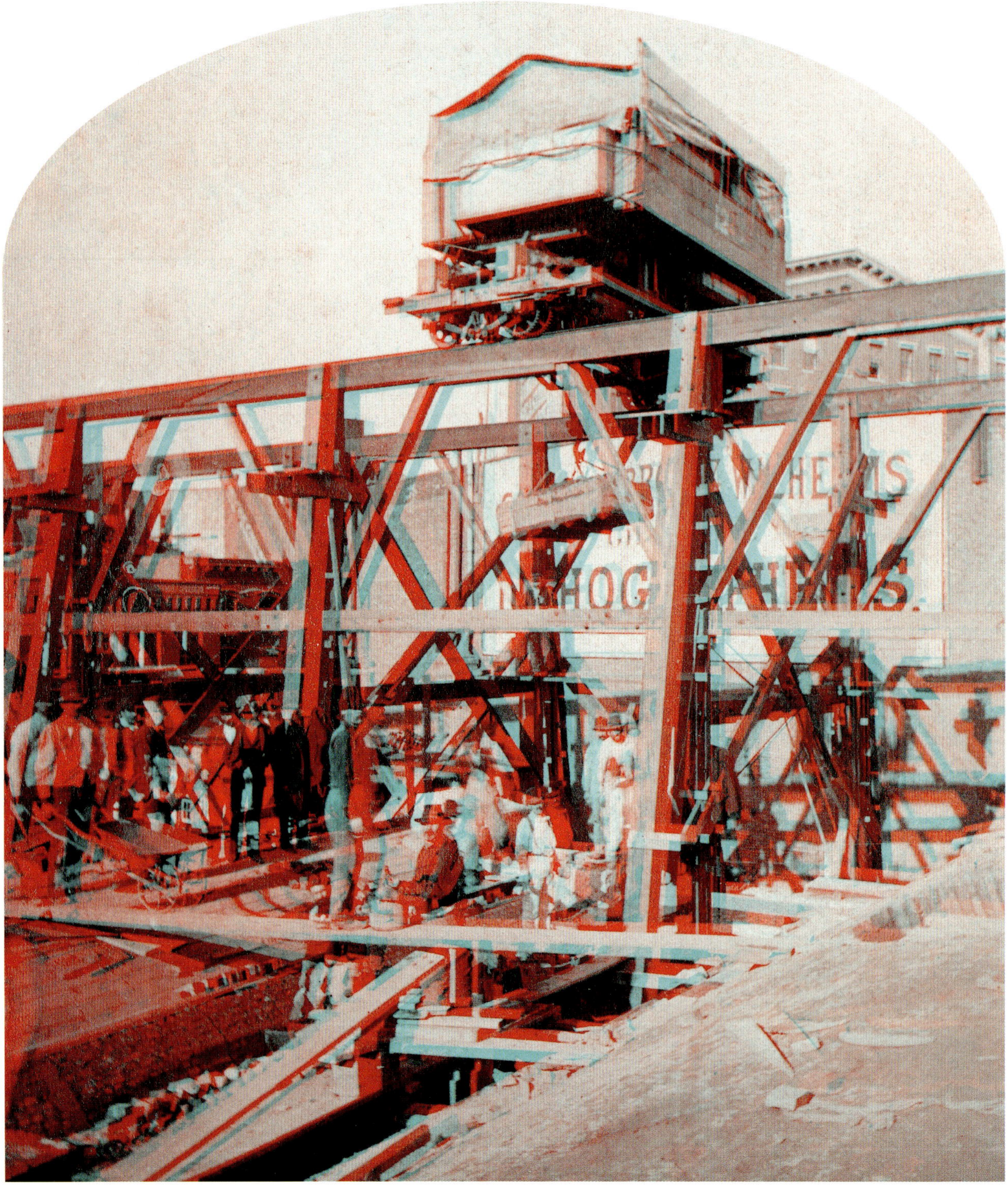

3D A covered steam-powered car, marked "12" on its side, was part of the wooden rail system used to move supplies around the massive work site. Note the construction materials hanging by a hook beneath the car. Painted on the wall behind it is an advertisement for a lithographic company, the name of which is only partially visible. Courtesy of the Michael E. Ach Collection.

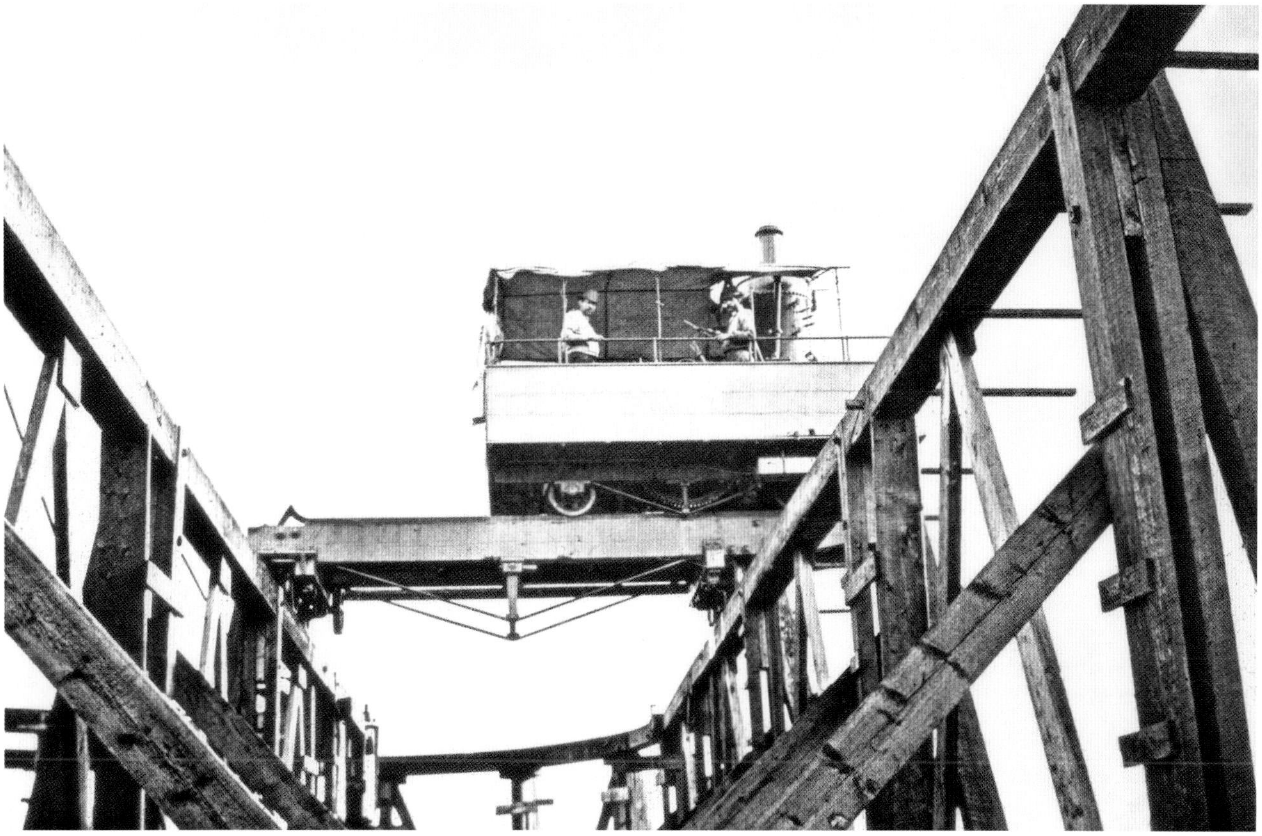

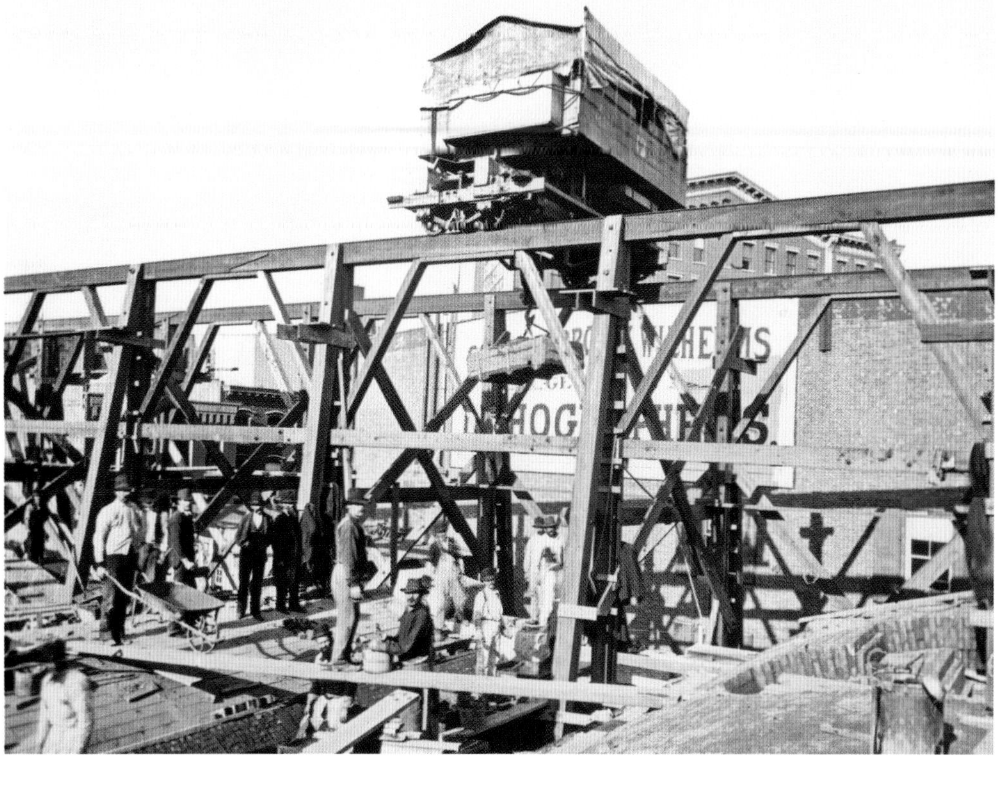

ABOVE: Workers inside a car atop the craneway, in this print from a lantern slide photograph, ca. 1880. These cars moved along tracks, both toward and away from the photographer's position, as well as side to side, allowing them to deliver the granite stones to their final placements. Author's collection.

LEFT: Print from a lantern slide of the image shown in the anaglyph on the opposite page. Author's Collection.

This photograph of the framework supporting a craneway was taken along the New York approach, looking east toward the bridge; at the far right edge of the image is the Franklin Square elevated railroad station, just below the bridge tower. Note the bottom of a car atop the tracks, just barely visible at the top edge of the image. Two parallel craneways are shown here, on either side of the approach: one to transport and position materials on the north side (at left) and the other to work the south side (at right). Granite blocks rise in the foreground, and rounded wooden falsework has been installed to support the granite arches that are about to be built. RPI.

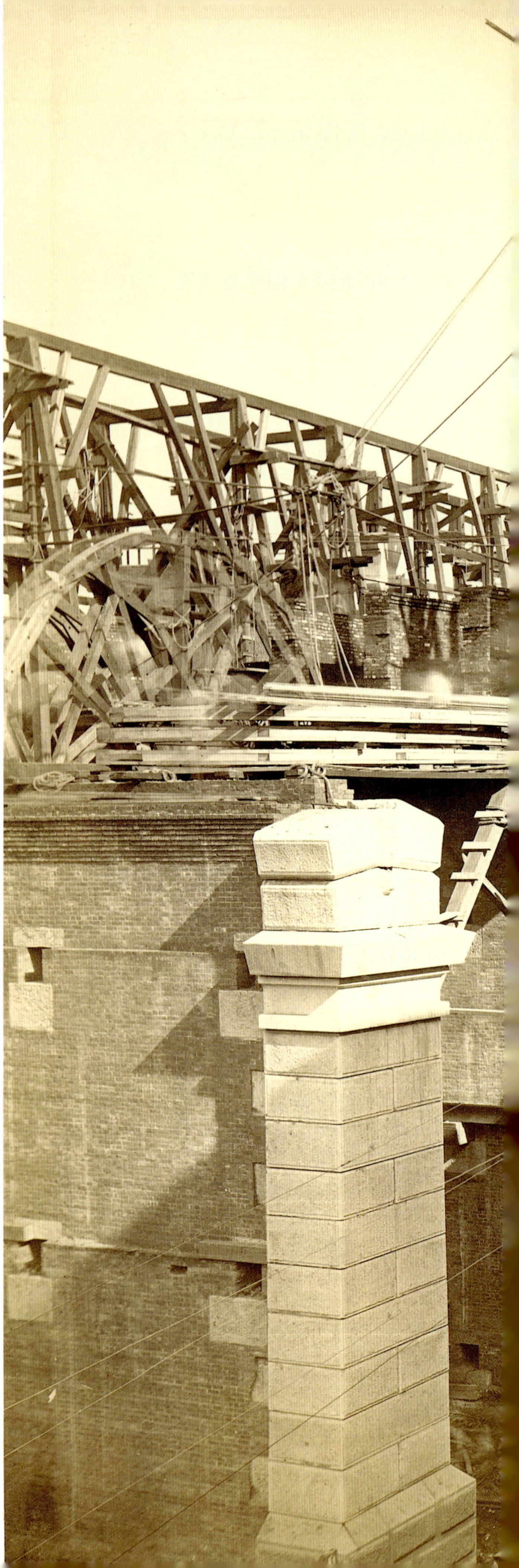

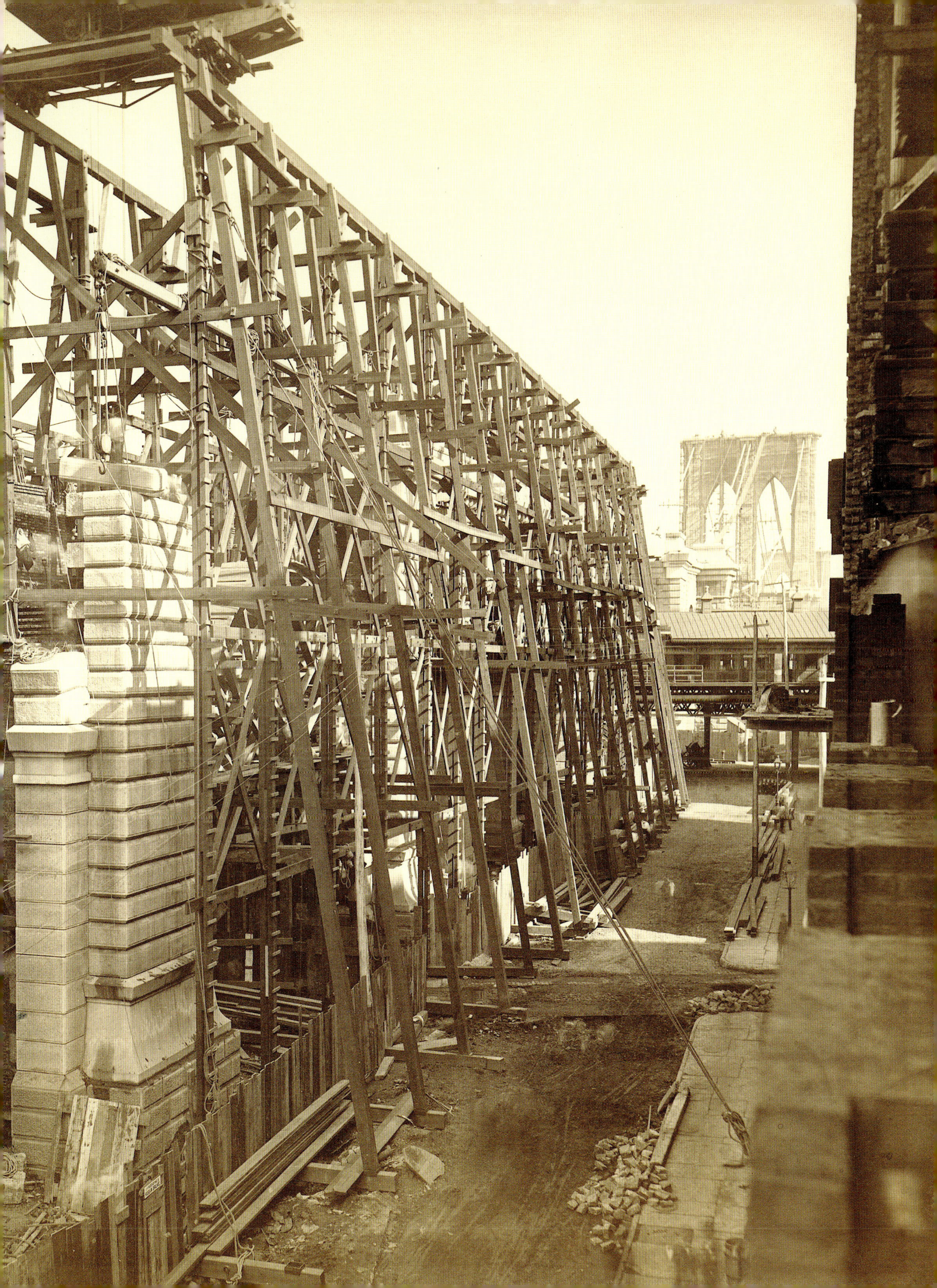

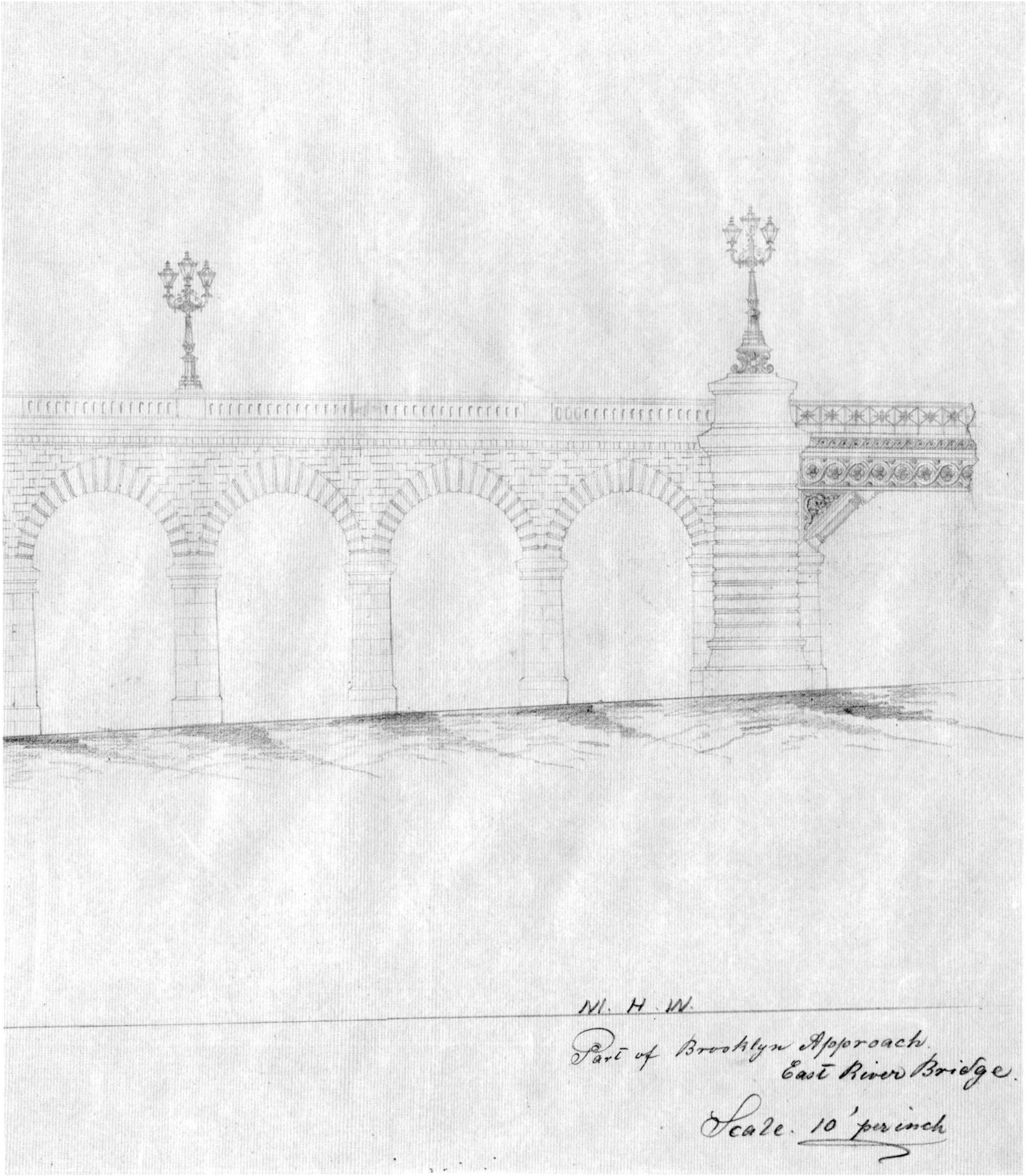

"Part of Brooklyn Approach/East River Bridge," ca. 1876. A detailed drawing of the Brooklyn approach, the section of the bridge that allows users on and off the suspended spans. It shows the bridge's distinctive granite arches and, and at far right, an iron span supporting the bridge roadbed that would extend over a Brooklyn street. Note the elaborate lighting fixtures that were contemplated to enable night use of the bridge. Ultimately, less elaborate and less expensive fixtures were installed. RPI.

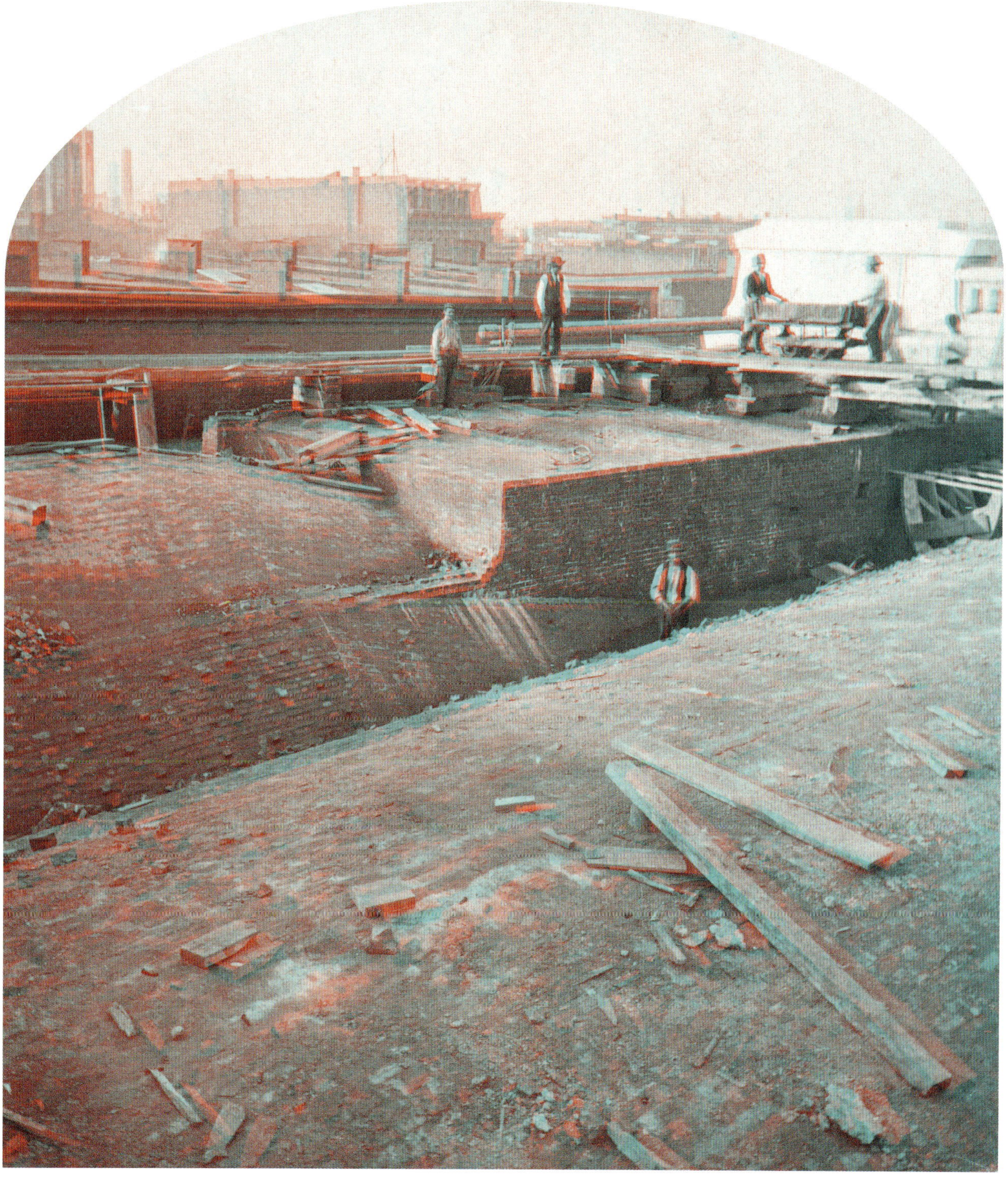

3D Anaglyph showing workers atop the completed brick vaults of one of the approaches. This image captures the scale of the vaults. Courtesy of the Michael E. Ach Collection.

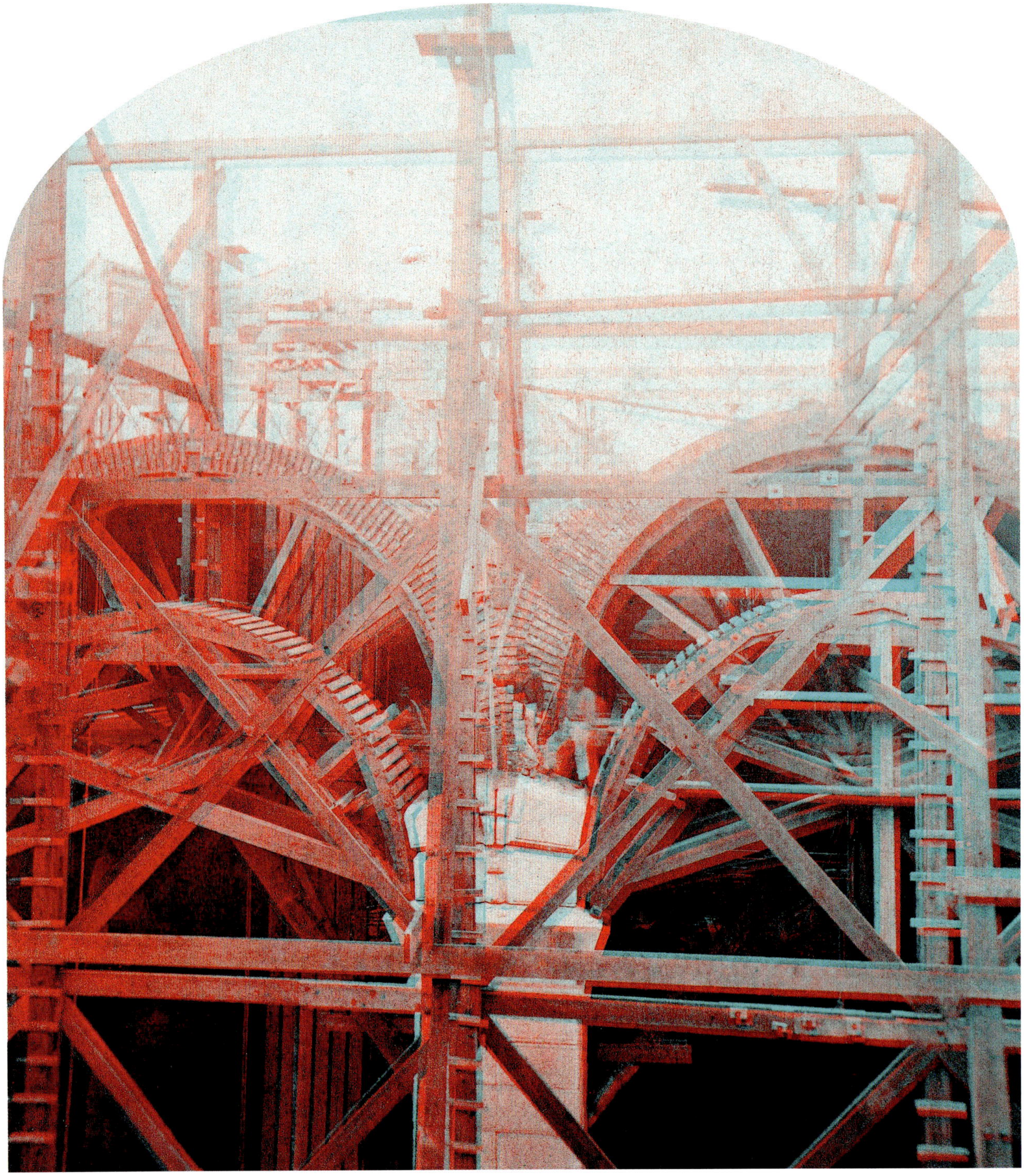

3D An anaglyph of the wooden frames, or falsework, that temporarily supported the granite arches of the approaches. After the granite blocks were positioned on the frames, masons mortared the joints, the mortar hardened, and the falsework was dismantled. At far left, a worker wearing suspenders climbs a ladder. Courtesy of the Michael E. Ach Collection.

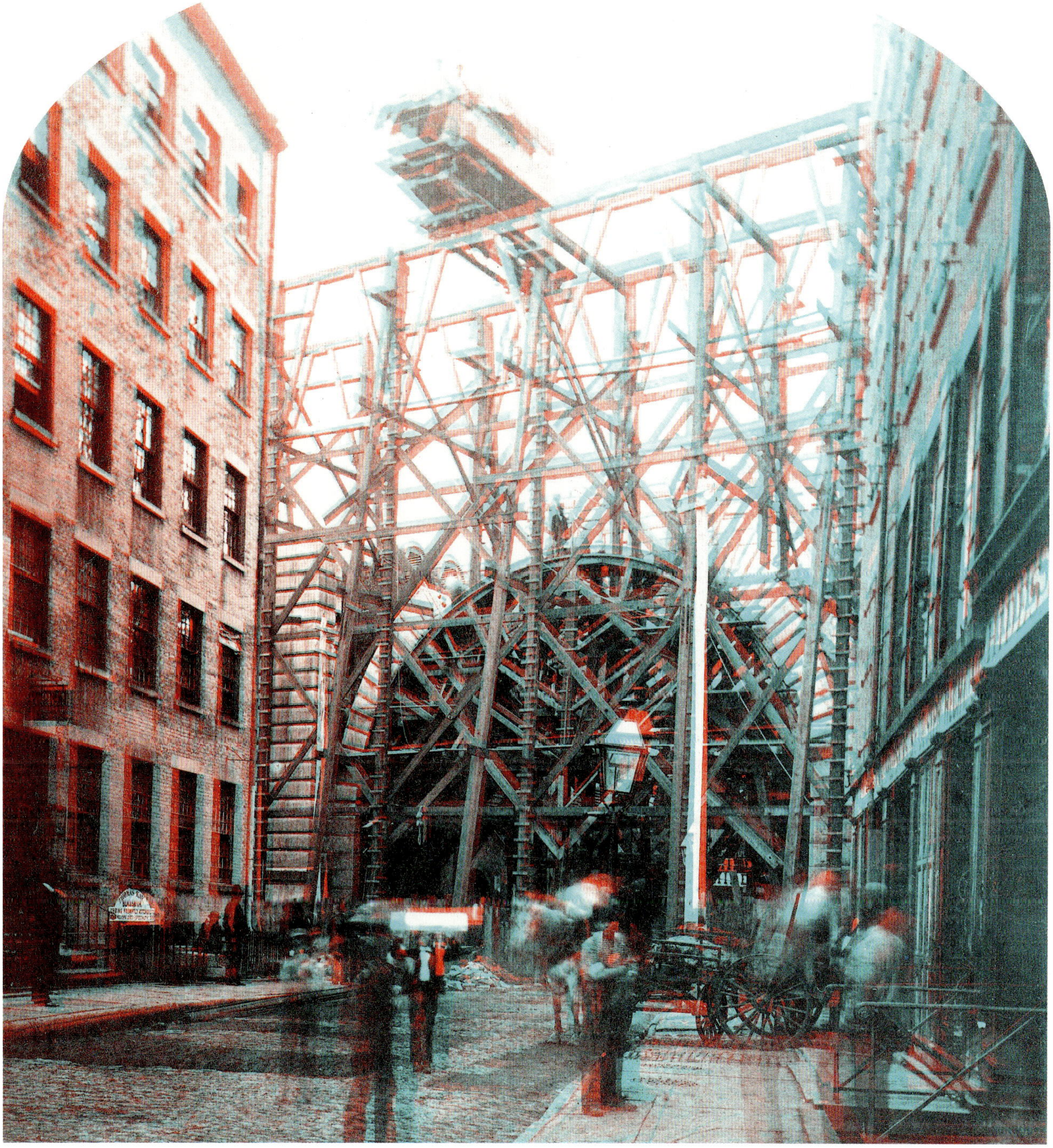

3D An anaglyph showing a craneway towering over the granite facing for one of the approach arches. This image appears to have been taken in Manhattan: at left, a sign for Thomas Kay, Blacksmith, "Jobbing Promptly Attended To, Jewelers Dies Specialty," and with the number "99," is visible with magnification. A Thomas Kay is listed as a blacksmith in the New York City directories for 1876, 1877, and 1880 and is recorded as living at 26 James Street in Manhattan—just three short blocks from where the Brooklyn Bridge was being built. Because his business address is not listed, the name of the street in the photograph remains a mystery. Courtesy of the Michael E. Ach Collection.

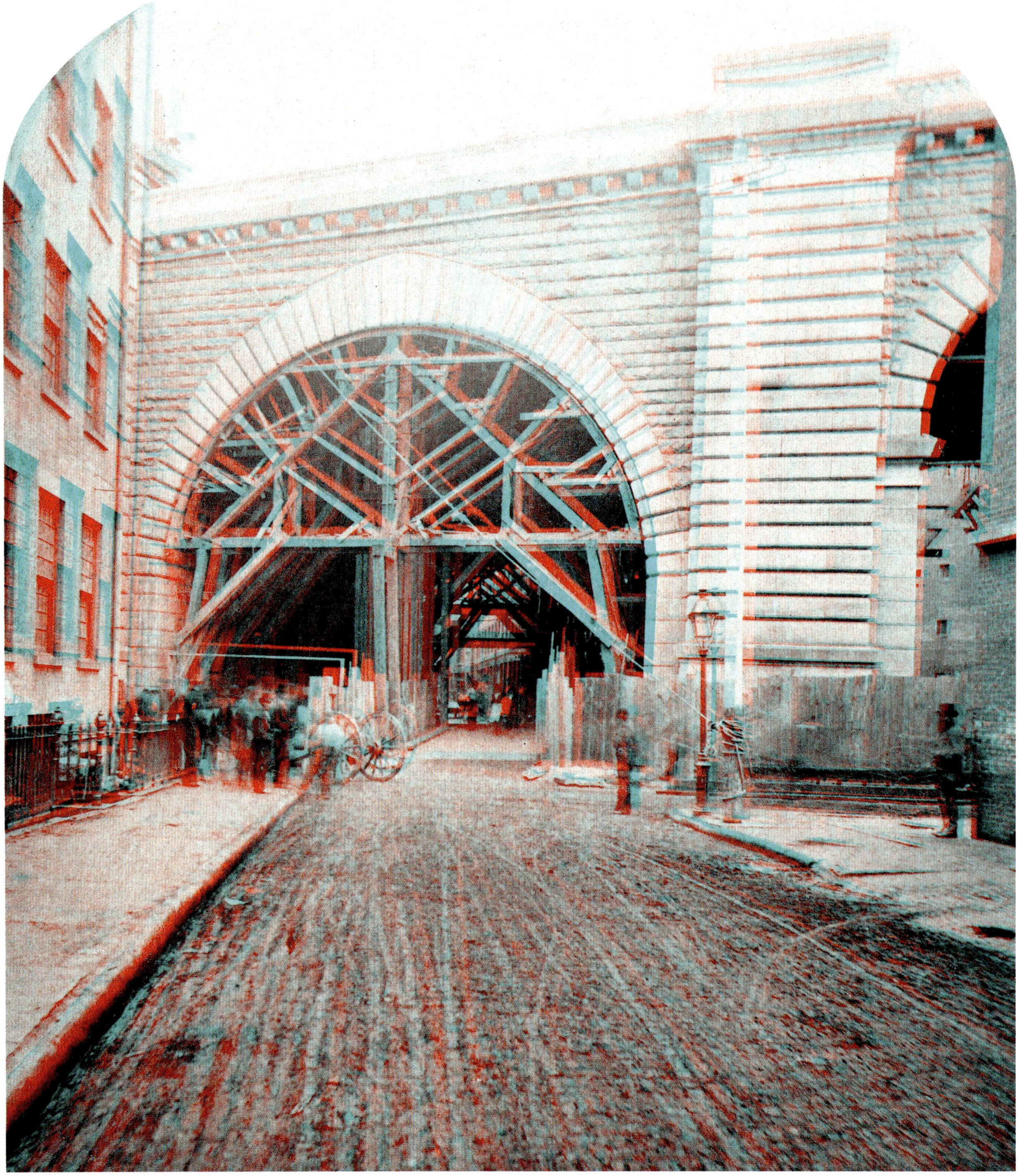

3D This anaglyph shows the wooden framework temporarily supporting a granite arch along one of the bridge's approaches. This image captures the geometric beauty of these wooden forms. Note that this view captures the same section of the approach as that on the previous page, but here the craneway and scaffolding have been removed. Courtesy of the Michael E. Ach Collection.

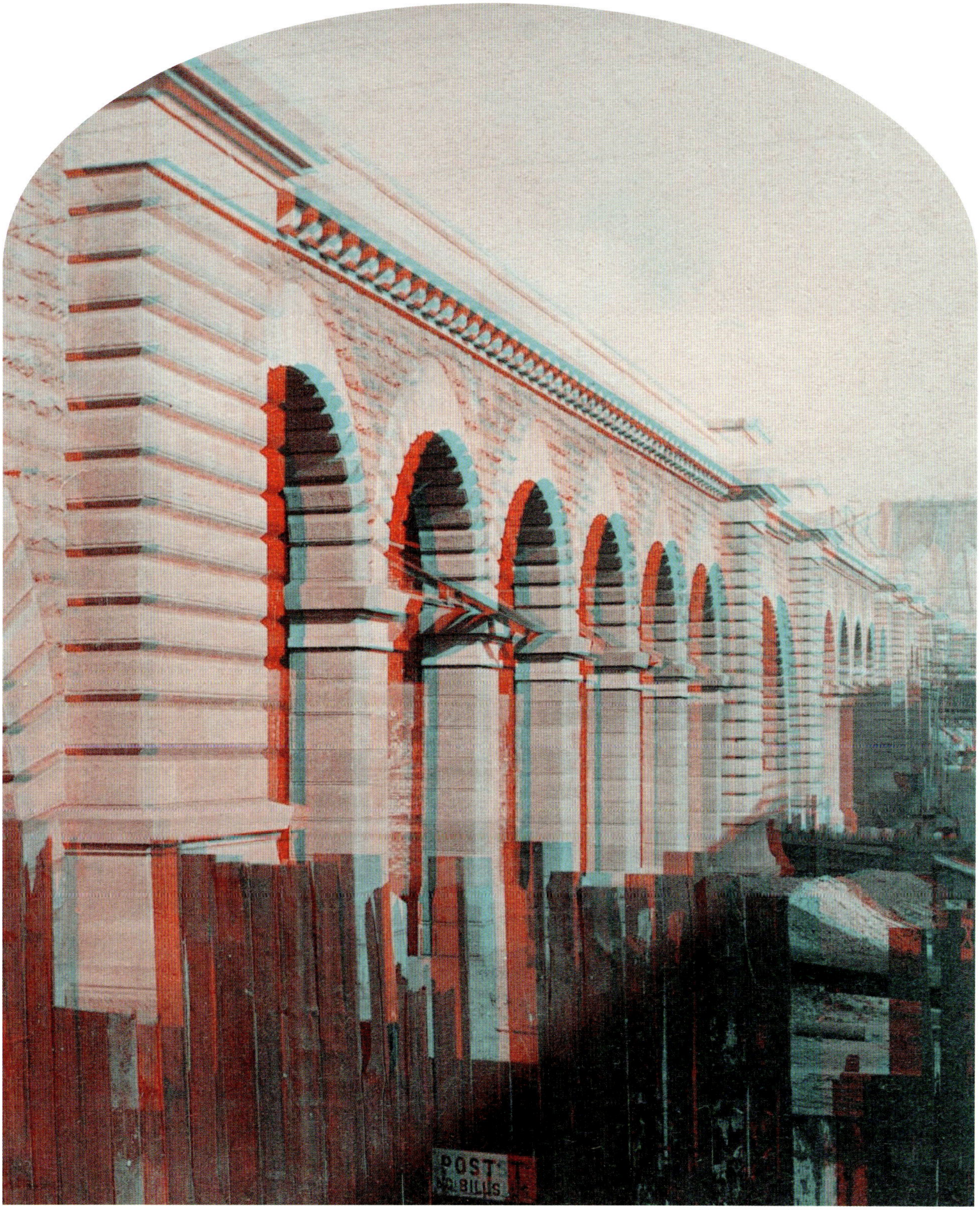

3D The New York approach to the bridge, with its characteristic straight line, ca. 1882. This approach was about one-third longer than the one in Brooklyn. Note at far right the New York tower in the distance. Courtesy of The Green-Wood Historic Fund, Jeffrey I. Richman Stereoview Collection.

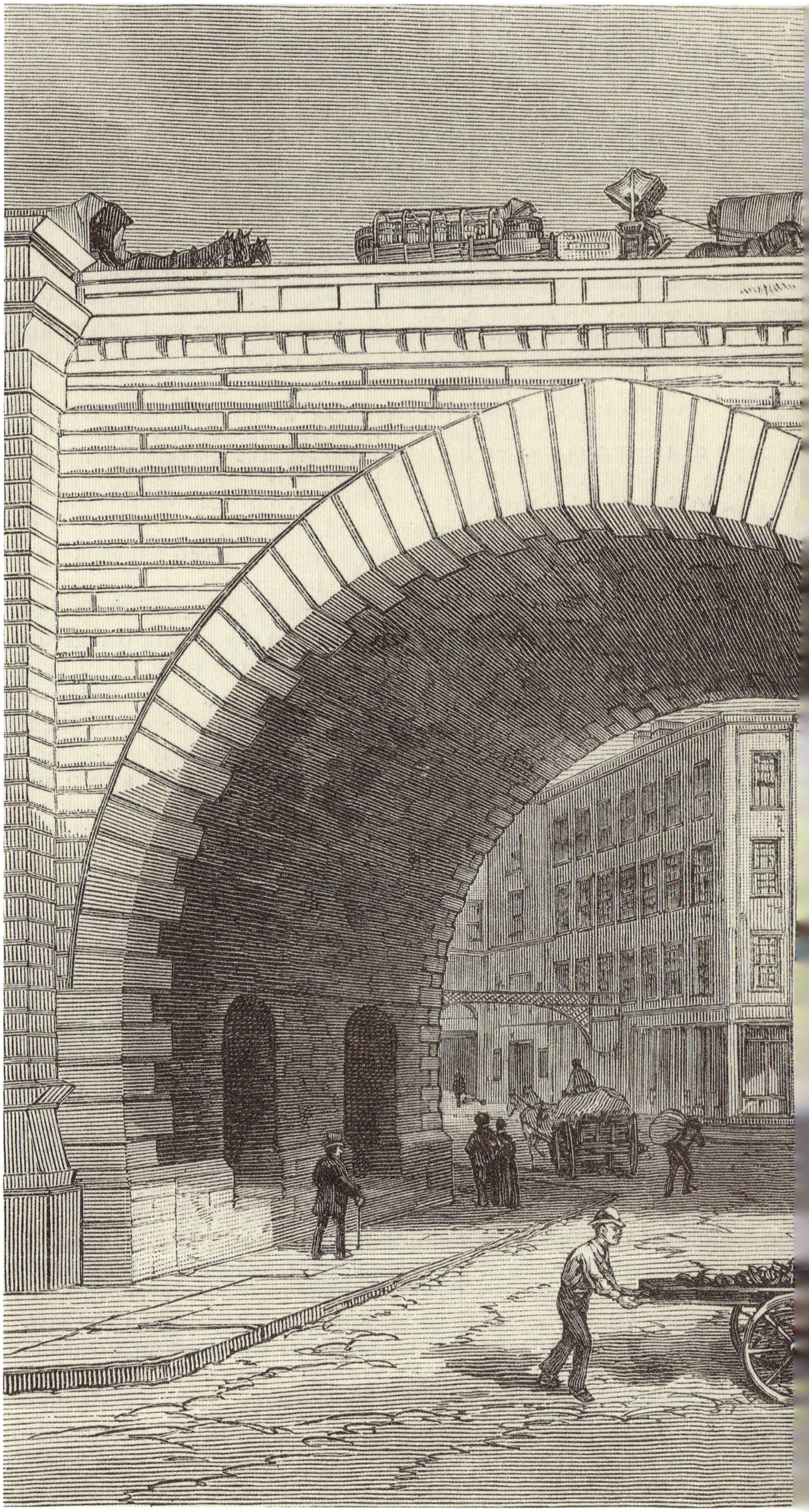

"Archway Under Approach, New York Side—Drawn by Charles Graham," *Harper's Weekly*, May 26, 1883. A street view framed by an arch was unusual for New York City at the time. Author's collection.

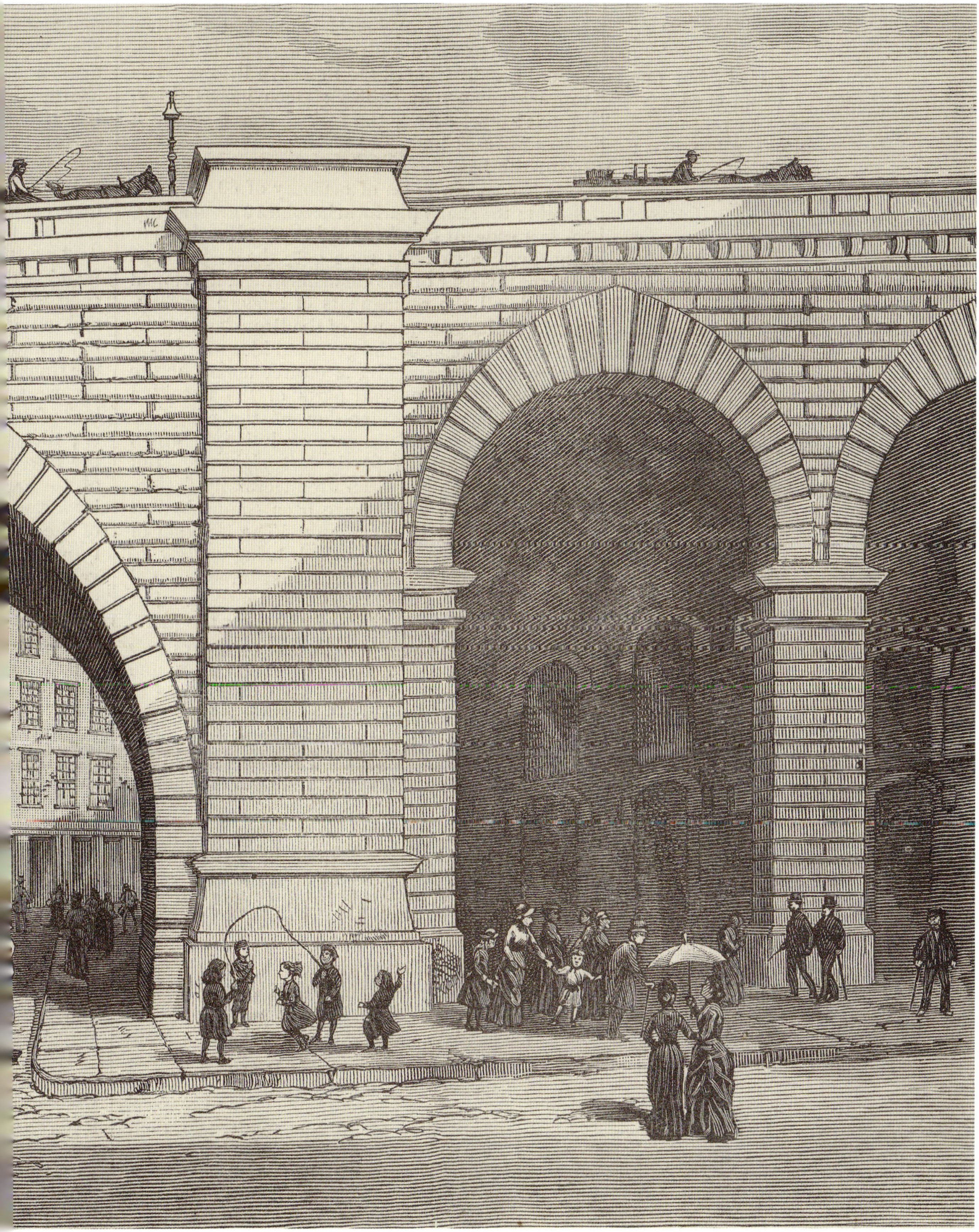

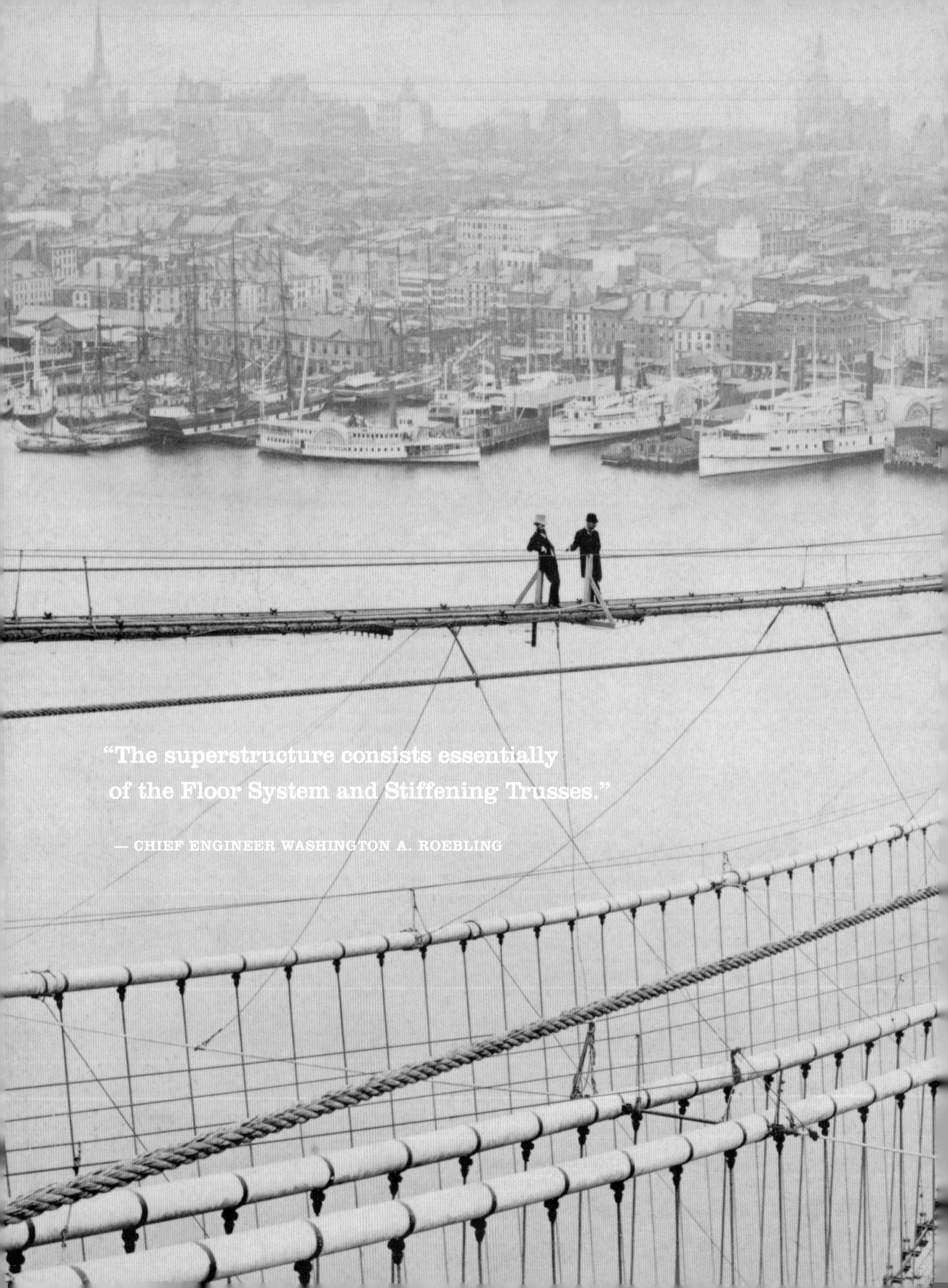

"The superstructure consists essentially of the Floor System and Stiffening Trusses."

— CHIEF ENGINEER WASHINGTON A. ROEBLING

CHAPTER 9

Suspended Superstructure

1881–1883

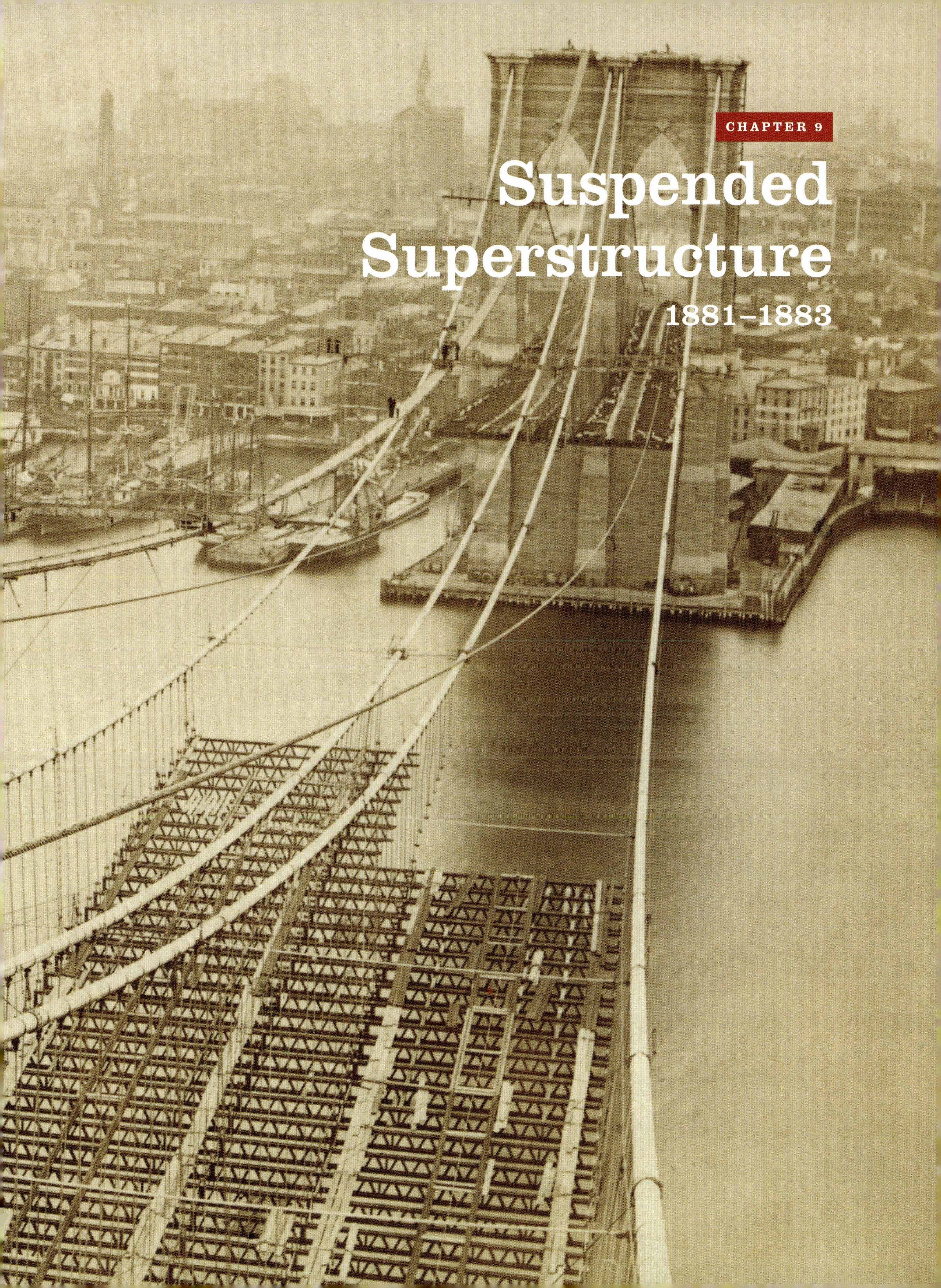

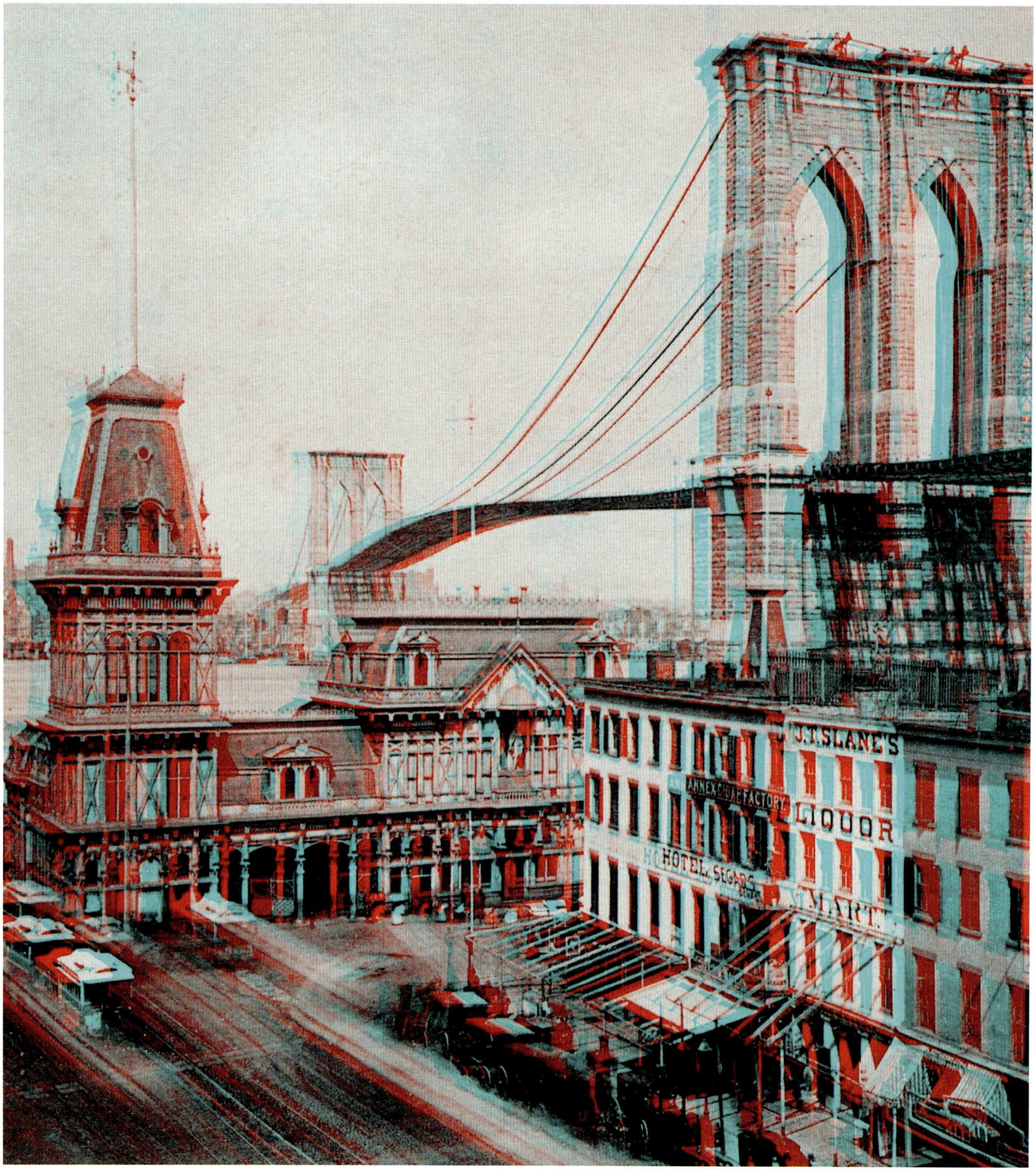

3D An anaglyph view from the Brooklyn side of the bridge, with the Fulton Ferry Terminal at left, ca. 1881. Suspender wires are just barely visible hanging down from the cables. Courtesy of the Jeffrey Kraus Collection.

PREVIOUS SPREAD: A view looking toward New York City, showing two well-dressed men standing rather nonchalantly on the footbridge, at center left. This photograph was taken in 1881, when the work of constructing the steel deck of the central bridge span was in progress.

The success of a suspension bridge depends on the "suspended superstructure," as Washington Roebling called it. This is the metal framing for the bridge's roadbed, the part of the bridge upon which all users would cross. The superstructure includes the framing along both edges of the roadbed, north and south sides. This framing helps give the bridge its necessary stiffness and strength.

The roadbed is suspended over the central river span (between the two towers) and the two suspended land spans (between each tower and its respective anchorage). The part of the roadbed's metal framing that runs through the Gothic Revival arches of both towers is not suspended from the cables. Rather, it rests directly on the towers' stones. Similarly, at the anchorages, because the bridge's four cables dip below the bridge floor, that section of the roadbed is not hung from the cables; it rests instead on top of cast iron steps that are clamped to the cables.

According to Washington Roebling's specifications, he intended the superstructure to "consist of an iron framing eighty-five feet in width from out to out," or edge to edge, dictating how wide the roadways could be. Washington also specified that the superstructure be "suspended from four main cables by wire rope suspenders attached to iron floor-beams, which are placed seven feet six inches apart." Ultimately, Washington decided to use floor beams of steel rather than iron. He had learned from his father that the stiffer and heavier he could make a bridge's roadway, the more firmly it would stand against wind gusts and the better it would support heavy traffic. Though John Roebling was not the first engineer to understand this, he improved on the concept, developing a system of diagonal "stays" that substantially enhanced the stiffness—and, therefore, the strength—of his bridges. Washington further strengthened the diagonal stays by adding strong trusses arrayed in the shape of triangles. These trusses are twelve feet high and run above the roadbed, alongside its outer edge.

When the bridge trustees learned of Washington's plans to add significant weight to the superstructure, they demanded that he explain this change. In response, in a letter dated January 9, 1882, Washington explained that the additional weight would give greater strength to the superstructure and the bridge. He further wrote: "When I considered to make this change, it was not so much owing to personal solicitation as to the reflection, 'Of what benefit had it been to erect this bridge at a vast expense unless we use it for every possible purpose to which the structure will lend itself.'" Washington assured the trustees that there was no cause for concern; given the strength of the cables, they were very much capable of supporting the superstructure with his modification. As he explained, even with the additional weight, it was more likely that the cables would pull the massive anchorages over before they would snap—and neither was going to happen. The bridge was designed to be strong enough that if, in the future, a railroad were to be built across it, it could bear that weight easily.

In addition, diagonal stays were installed between the suspender wires to further increase the strength and stability of the bridge. "Wind braces"—wires designed to prevent the bridge from vibrating—were also installed. These were attached to one side of each tower, ran underneath the deck, then attached to the tower's opposite side to further stiffen the central river and land spans. In the end, Washington's modifications increased the strength of the bridge and ensured that the superstructure would last for fifty years before any major upgrades were needed.

Because the bridge was being built over a navigable waterway, the height of the superstructure above the East River was determined by Army engineers, whose approval was needed before construction could begin. After some study, they specified that the river span, at its center, had to be no lower than 135 feet above the "mean high water of spring tides." This would place the bridge high enough that only the tallest of sailing ships would have to trim their topmasts to pass underneath. This was five feet higher than Washington's design and required that he adjust his plans.

It was, of course, necessary that the bridge floor rise at a gentle grade so that horses would be able to pull their loads up the incline. In order to raise the center span five feet higher, Washington added sixteen feet to its length—from 1,600 to 1,616 feet, tower to tower. This slightly increased the arc of the bridge so that it was at the required height above the river. Because the Army engineers had also mandated that nothing hang below the river span, Washington shifted the trusswork for this section of the bridge entirely above the roadway. He also added five feet to the width of the roadway, a change that would have pleased his father, who had often bragged that his bridge would be broader than Broadway. The grand structure John Roebling had envisioned was going to be even wider than he had hoped.

Once the four cables were lowered onto their saddles atop the towers and fully wrapped along their lengths, workers began to affix the suspender wires from which the spans would hang. Given the weight of the superstructure, the integrity of the suspenders and the iron bands that securely attached them to the cables was crucial. An article published in *Scientific American* on May 21, 1881, described the suspenders and the bands used to attach them to the cables:

> From below the suspenders look like spider lines; they are, however, stout ropes of steel wire, from 1⅝ to 1¾ inches in diameter, and able to sustain a weight of 50 tons or more each, or five times the heaviest load likely ever to fall upon them. The bands by which the suspenders are attached to the cables above are of wrought iron, five-eighths inch thick by 5 inches wide.

The first step in attaching the suspenders to the cables was blacksmith work: heating metal bands on a forge so they could be stretched around the cables until they fit tightly. In order to do this, men were stationed on "wrapping buggies," platforms that hung from the cables. These buggies were large enough to accommodate several workers and a small forge. Once one of the metal bands had been attached to a cable, one end of a twisted suspender wire was then connected to it. The suspender wire was then dropped down, and workers connected it to one of the steel crossbeams that would form the bridge deck. These deck beams—each assembled from two lengths of steel riveted together for a total length of eighty-five feet—were then swung into position, perpendicular to the direction of the bridge. Master Mechanic E. F. Farrington noted, "It is customary to hang all the suspenders to the cables before beginning to suspend the beams, but here only a few were attached at first; as it was found it would be easier to put them up as the floor progressed." The workers then repeated these steps until the bridge deck was complete. After all the suspender wires were attached and the beams positioned in place, the diagonal stays were installed, creating the Brooklyn Bridge's characteristic weblike appearance.

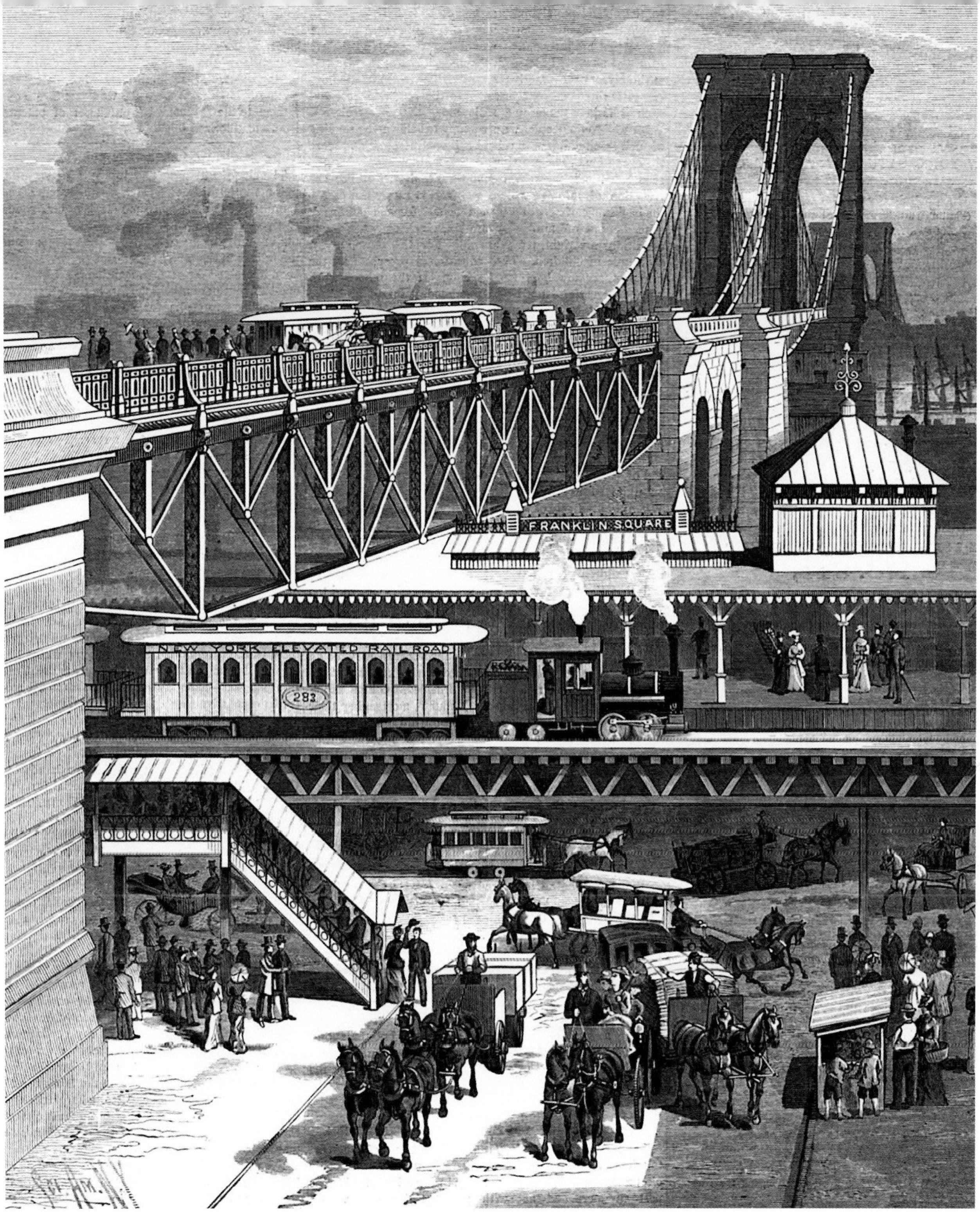

"The East River Bridge—Street Bridge at Franklin Square," *Scientific American*, August 5, 1882. This elevated railroad station is topped by a sign identifying it as Franklin Square, a stop in Manhattan. At left, underneath the bridge deck, trusses support a span that stretches from the New York anchorage toward the New York tower. The Roeblings had to modify their plans and move the trusses supporting the central span of the bridge (running between the towers) up above the span, rather than below it, because of the dictates of the Army engineers. For the Franklin Square bridge, though, the trusses were left underneath the roadbed because it spanned a road and would not impede river traffic. The artist who recorded this scene was a poor observer. Visible at the upper left of this image, across the East River from this Manhattan scene, are the easily identifiable New York Tribune Building and the shot tower—both of which were in Manhattan, behind the viewer, and, therefore, could not have been in the distance. Oops! Courtesy of Richard Haw.

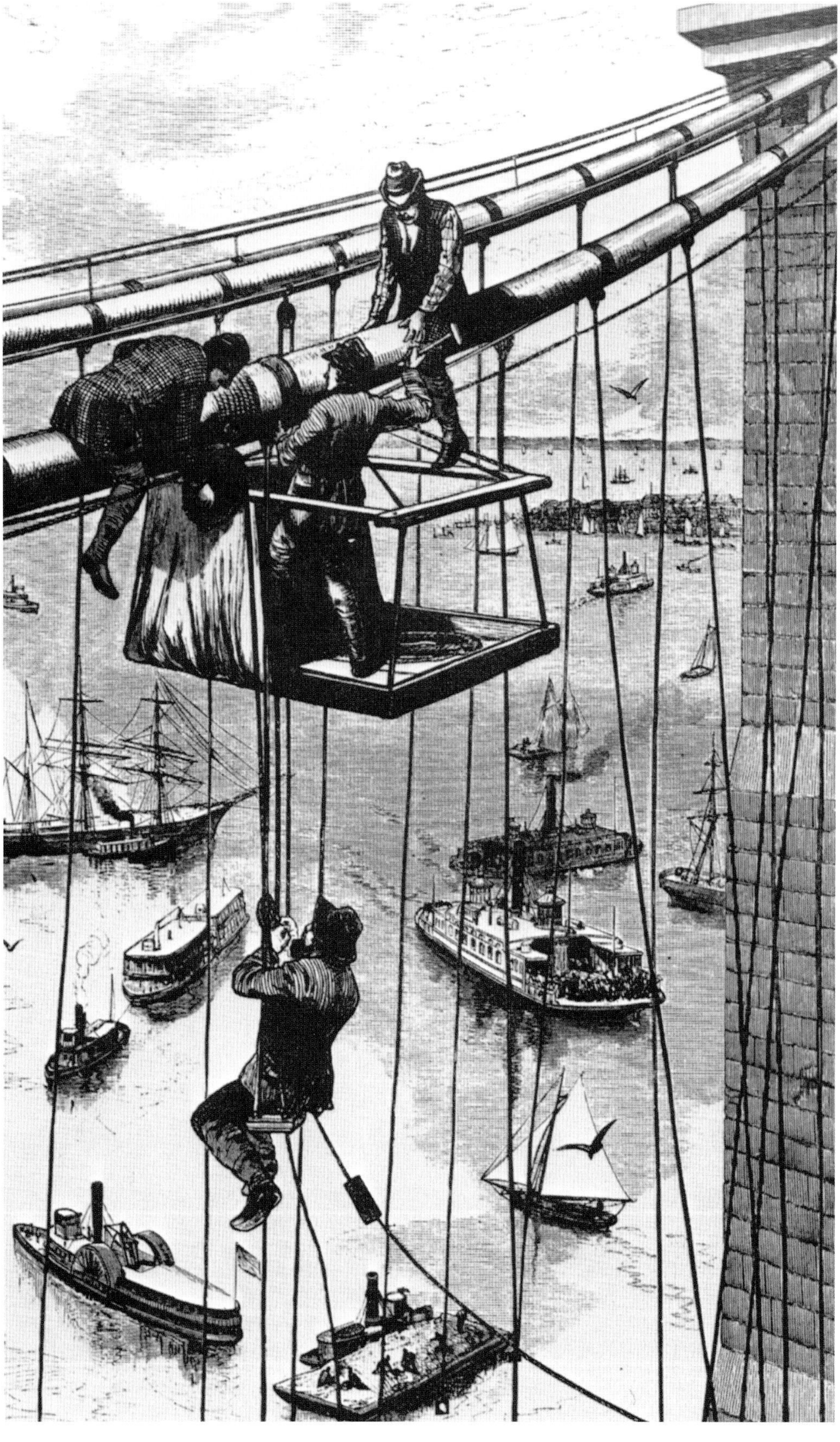

A woodcut showing a team of men hard at work attaching the top end of a suspender to the cable. The floor beams were then attached to the bottom end of the suspenders. Author's collection.

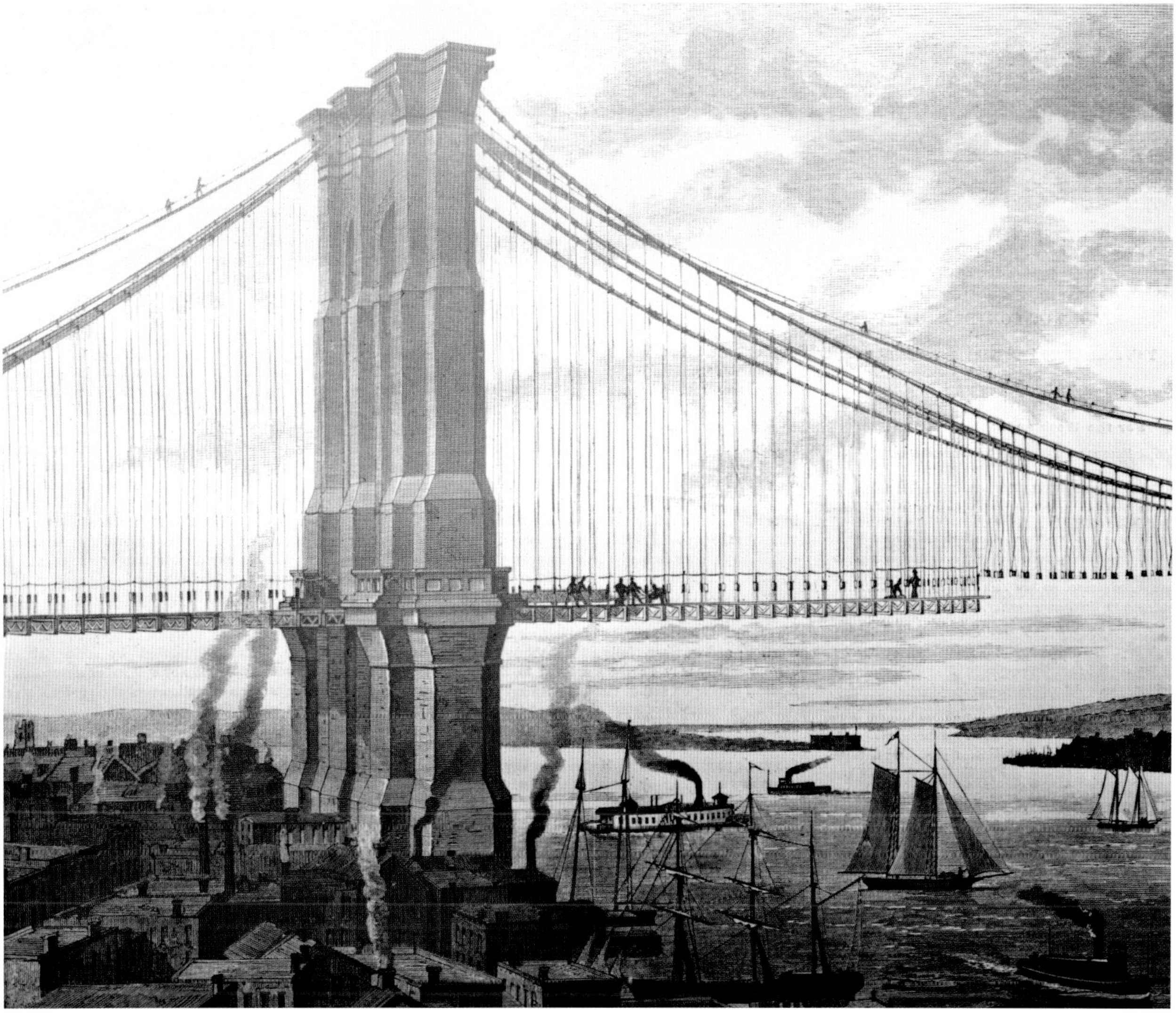

ABOVE: "Suspending the Floor Beams of the East River Bridge," *Scientific American*, May 21, 1881. This woodcut shows the workers assembling the floor beams of the superstructure. The suspender wires, at left, have already been attached to the cables above and the floor beams below; at right, the suspender wires dangle in the air, yet to be attached to steel beams. The view looks south from Brooklyn, including the Brooklyn tower and, in the distance, Governor's Island (and Castle William). Across the open water, at right, is the southern tip of Manhattan, with Staten Island behind. In order to balance the strains on the towers and cables, the same work was proceeding simultaneously on the New York side. Note the people walking on the footbridge at top. Courtesy of Richard Haw.

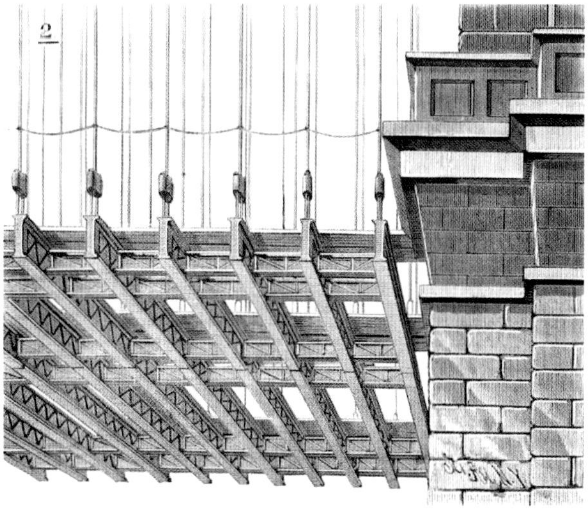

RIGHT: "The Floor Beams from Below," *Scientific American*, May 21, 1881. This woodcut shows the steel beams that form the bridge deck upon which the roadway runs. Also shown are the suspender wires that drop down from the cables above, connecting the steel beams to the cables. Part of one of the bridge towers is visible at right. Courtesy of Richard Haw.

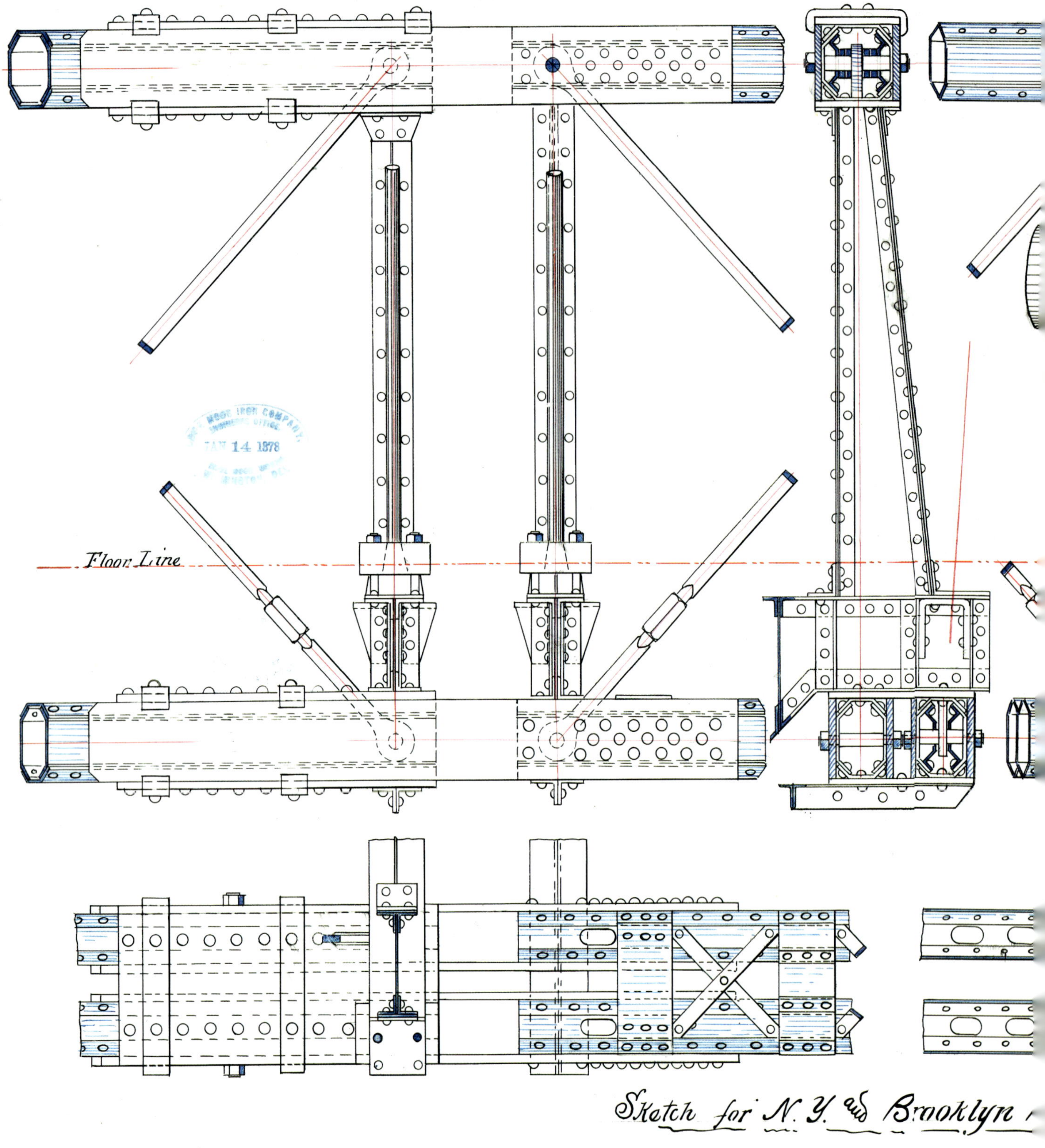
JAN 14 1878
Floor Line
Sketch for N. Y. & Brooklyn

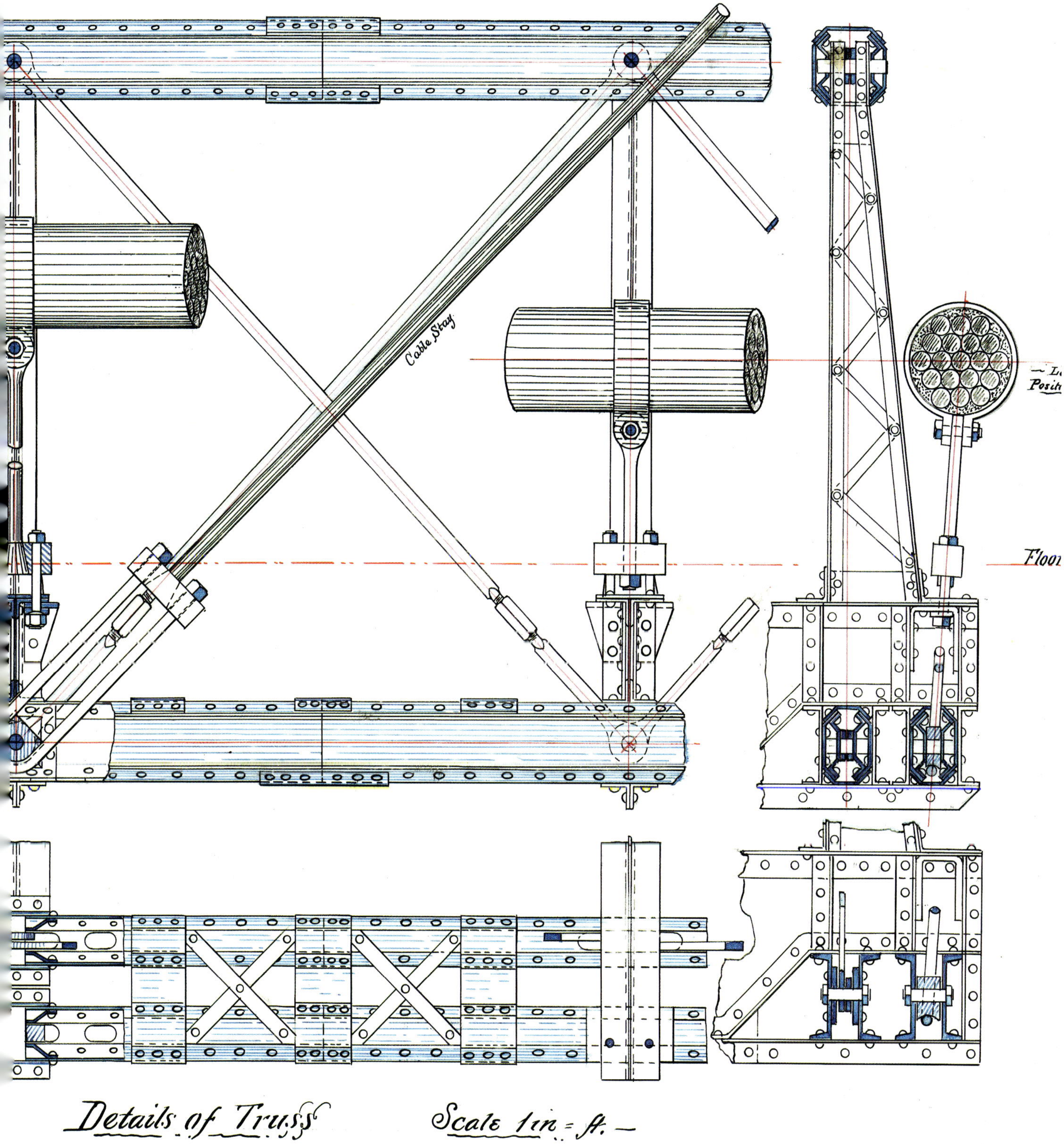

"Sketch for N.Y. and Brooklyn Bridge, Details of Truss," stamped January 14, 1878, by the Edge Moor Iron Company of Wilmington, Delaware. The metal trusses added stiffness to the bridge, making it stronger. RPI.

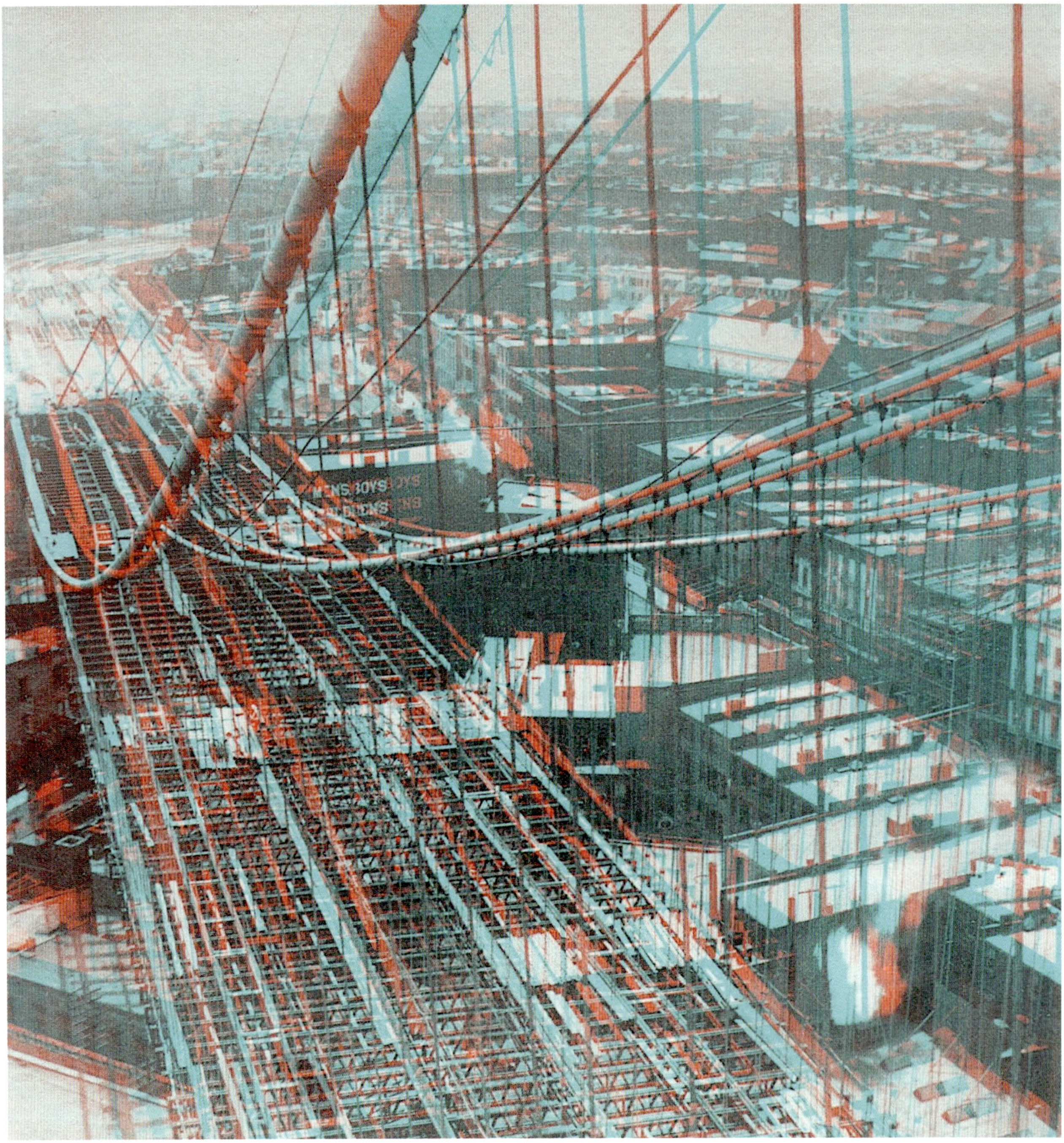

3D "Brooklyn, from the Brooklyn pier, Suspension Bridge," ca. 1881. Parallel lines of boards, running the length of the bridge spans, have been placed across the steel beams to allow the workers to move safely around the construction site. The beams would soon be covered entirely with wooden boards to form the roadbed. Note how the approach turns to the right in the distance; this identifies it as the Brooklyn approach, since the New York approach ran straight. Brooklyn's Fulton Street is at right, running parallel to the bridge. Author's collection.

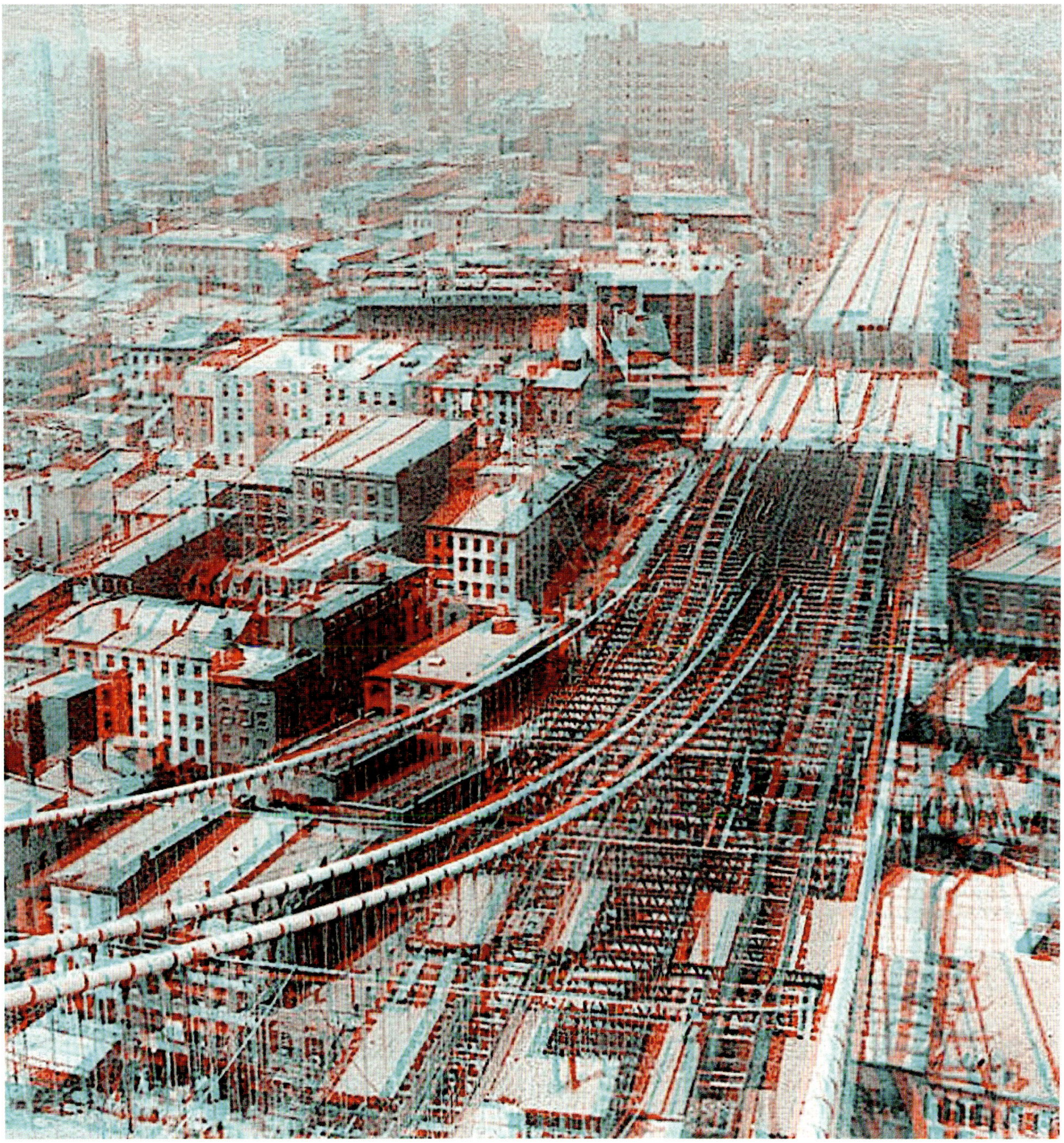

3D Looking down from the New York tower at the deck and the anchorage, with the straight approach stretching to the west, ca. 1882. This anaglyph shows the steel beams for the floor decking fully in place, with some temporary wood boards laid down for the workers to walk on. Note how the cables, as they near the anchorage, disappear below the deck; there, cast-iron steps were installed atop the cables to support the steel beams of the deck. Author's collection.

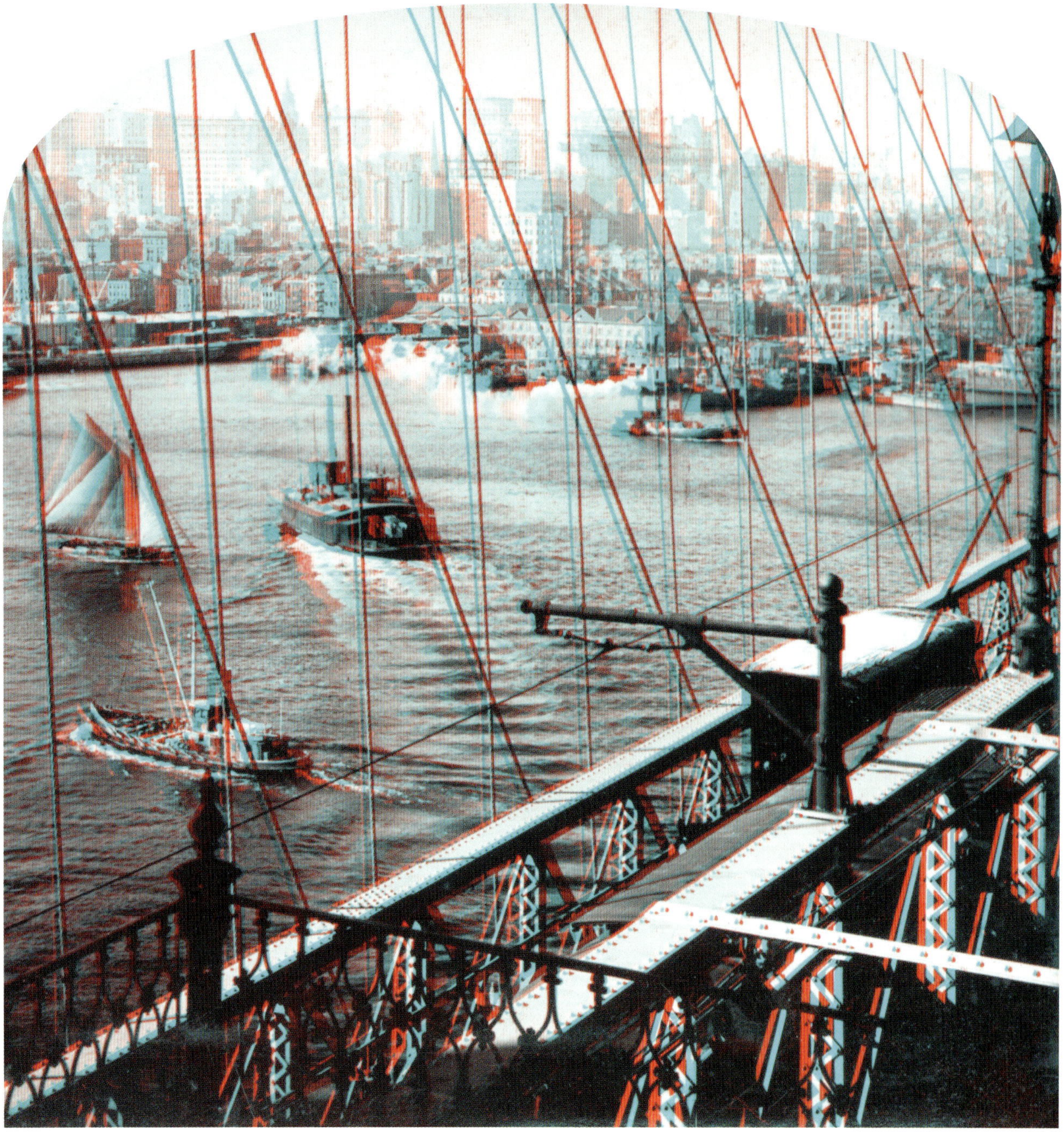

3D "New York Skyscrapers Seen From Brooklyn Bridge," anaglyph of a stereoview by Keystone View Company, ca. 1895. Looking southeast toward Manhattan, with the suspenders and superstructure of the bridge—including the trusses—in the middle ground. Note the railing for the promenade in the foreground and, at right, a wagon on its way across the bridge. Courtesy of The Green-Wood Historic Fund, Jeffrey I. Richman Stereoview Collection.

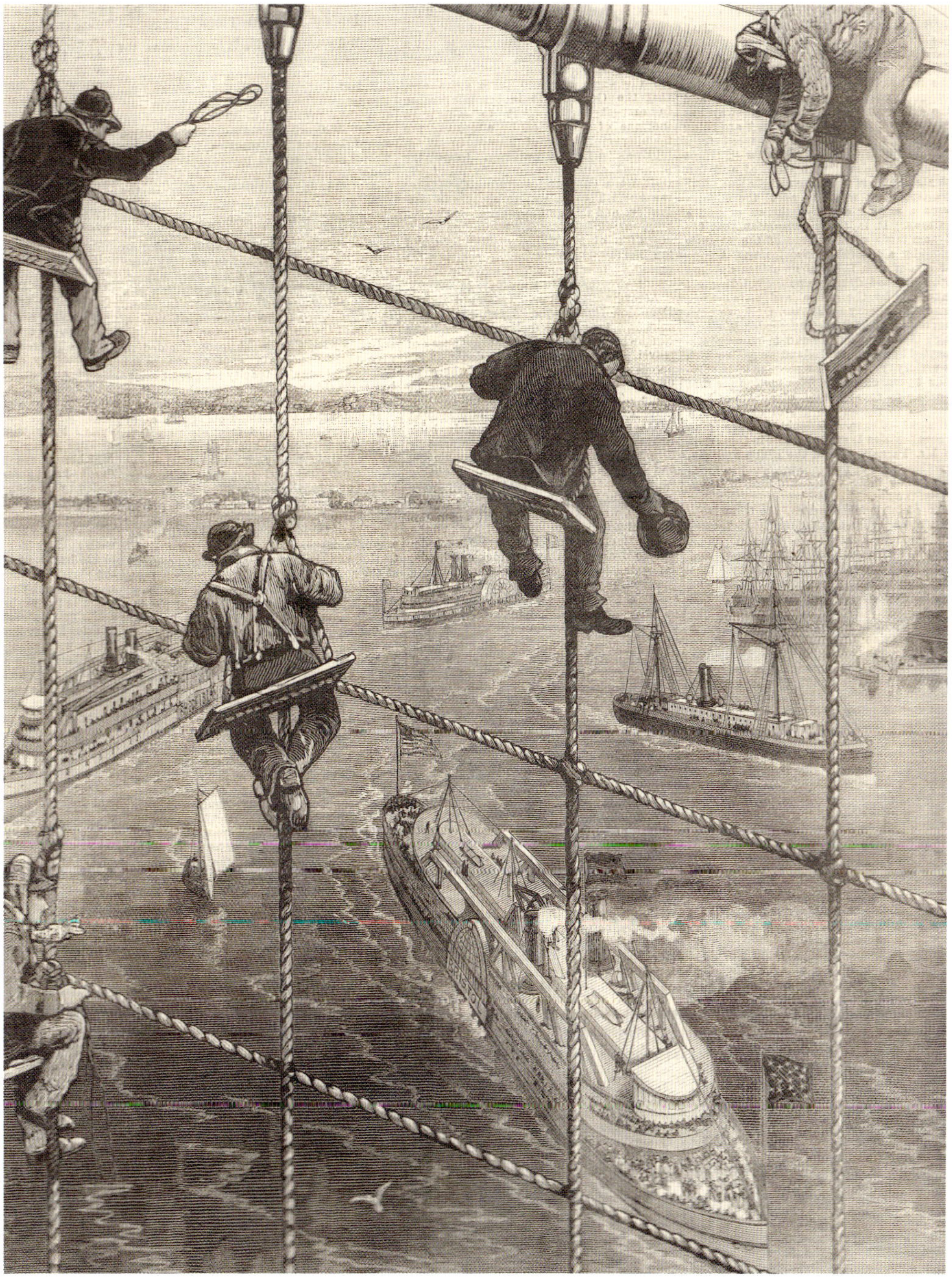

"New York—Completing a Great Work—Lashing the Stays of the Brooklyn Bridge," *Frank Leslie's Illustrated Newspaper*, April 28, 1883. This woodcut was published late in the bridge's construction—less than a month before it would open. It shows workers hanging high above the East River in boatswain's chairs, tying the bridge's suspenders and stays together. The accompanying text describes the process: "Each rigger is provided with a large pail of marlin, or tarred line, with which he lashes the stays and suspenders together until the little diamond shaped spaces are perfect in form. This is done so that when the superstructure is lowered by means of the screws at the base of the suspenders the strain will be equally distributed. The marlin lashings are temporary and will shortly be replaced by permanent iron clamps." Author's collection.

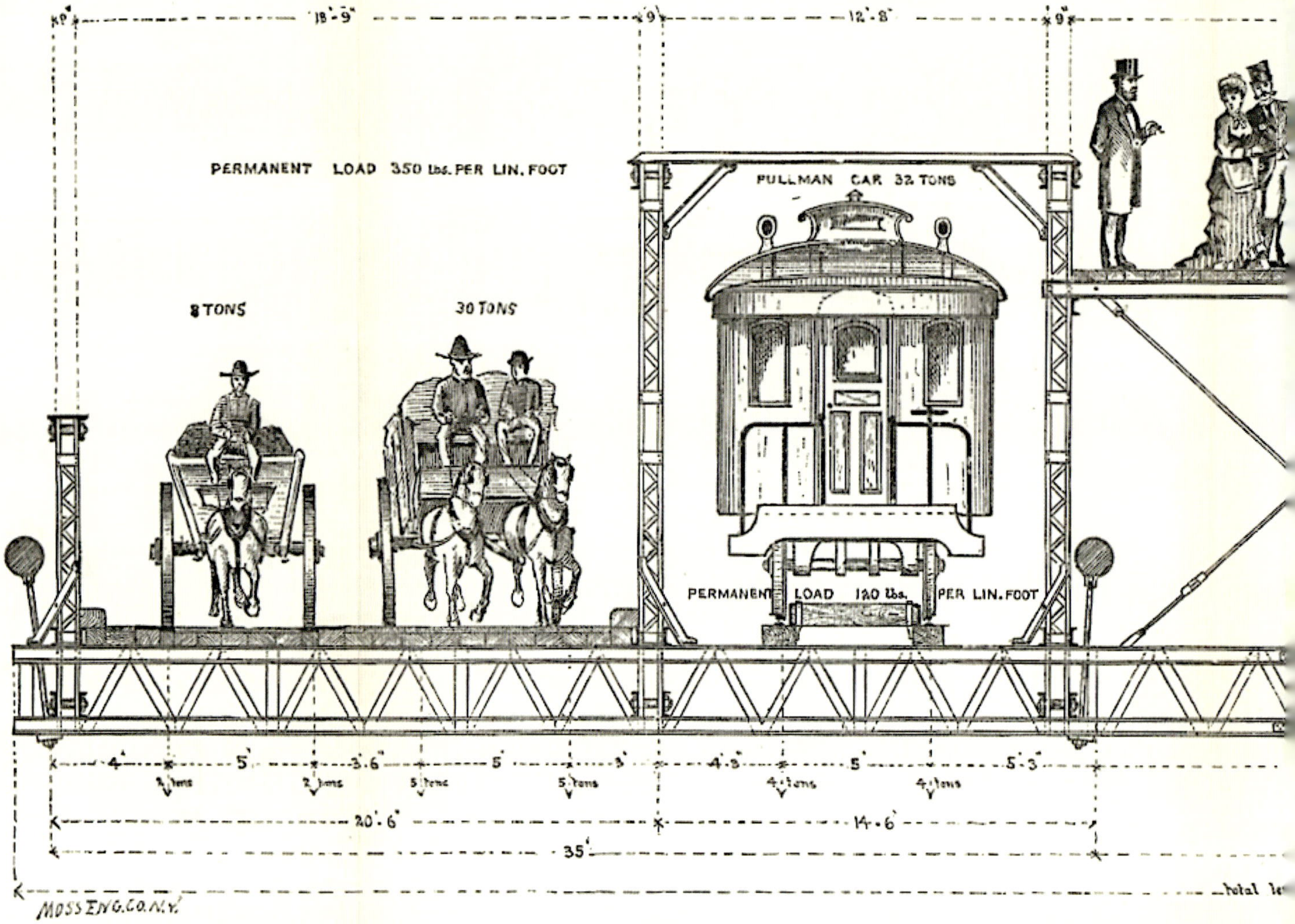
PERMANENT LOAD 350 lbs. PER LIN. FOOT
8 TONS
30 TONS
PULLMAN CAR 32 TONS
PERMANENT LOAD 120 lbs. PER LIN. FOOT
18'-9"
12'-8"
20'-6"
14-6
35'
MOSS ENG. CO. N.Y.

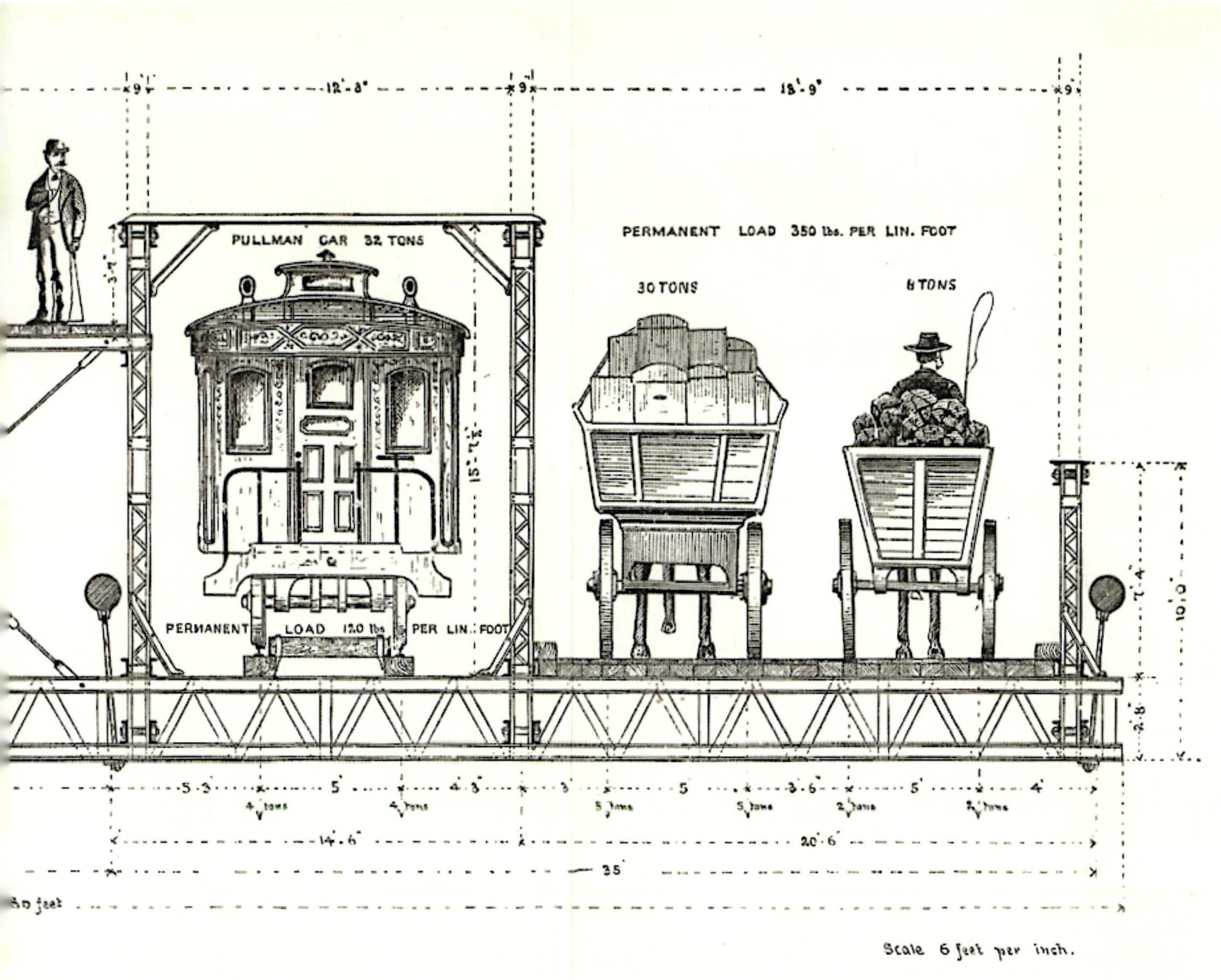

As depicted in this drawing, published in *New-York and Brooklyn Bridge Proceedings, 1867–1884*, the deck of the bridge would be divided into five sections: two outer lanes in each direction for carriages, wagons, and horse and rider; two railroad tracks (one in each direction), and an elevated pedestrian promenade at center. They all sit on the steel-beam deck shown below them. Author's collection.

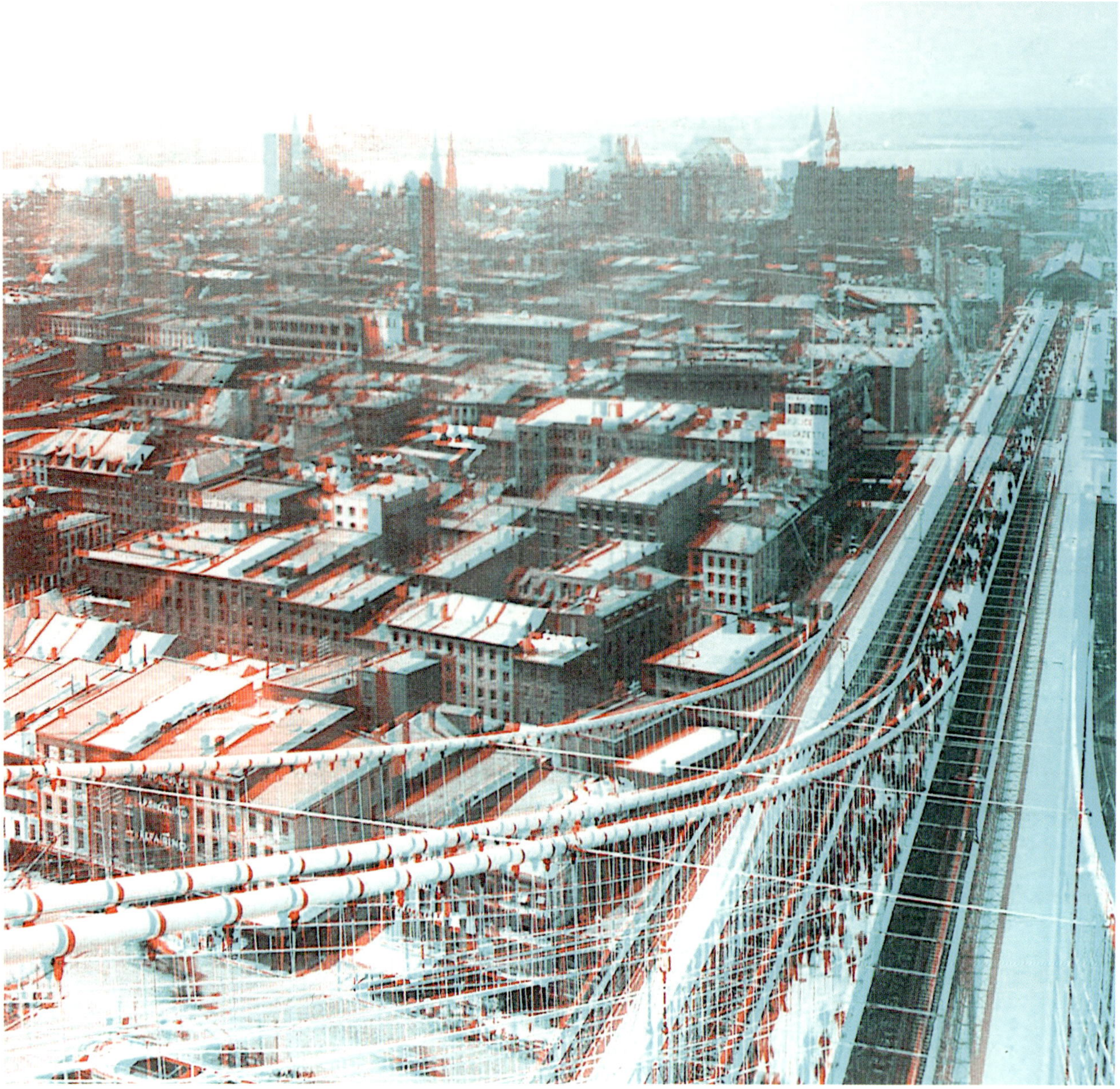

3D An anaglyph showing the view from atop the bridge's New York tower, looking west across Manhattan, ca. 1885. The five parallel sections of the roadbed are clearly seen atop the New York approach, with the railway terminal—completed after the bridge opened—in the distance. Courtesy of the Jeffrey Kraus Collection.

OPPOSITE: "Wonder of the Age, Brooklyn Bridge from the World Building, New York, U.S.A," looking east toward Brooklyn from a Manhattan skyscraper opposite City Hall—half of a stereoview published by Strohmeyer & Wyman in 1898. At lower left is the bridge's New York rail terminal, completed and in operation for fifteen years by the time this stereoview was issued. Courtesy of The Green-Wood Historic Fund, Jeffrey I. Richman Stereoview Collection.

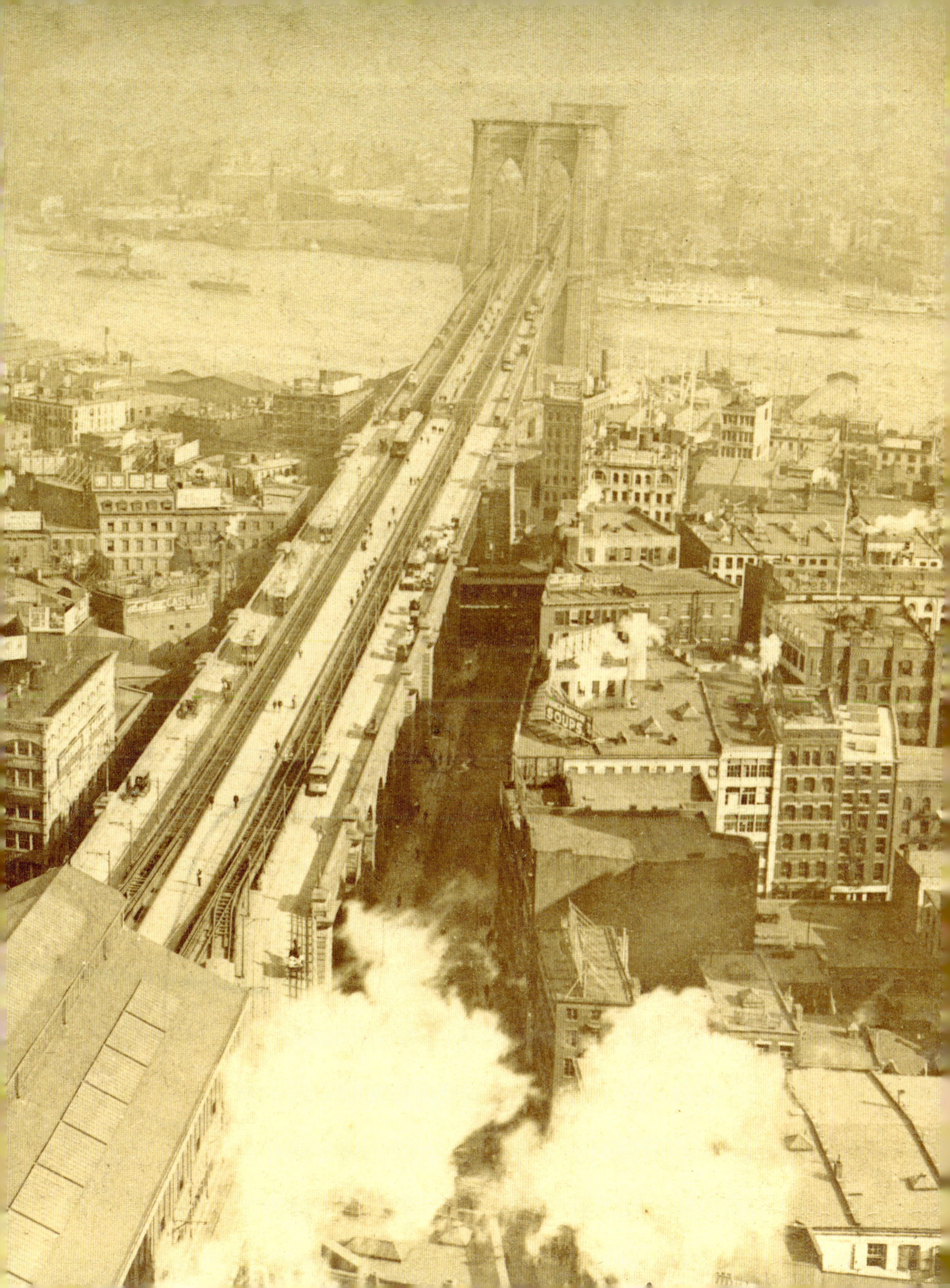

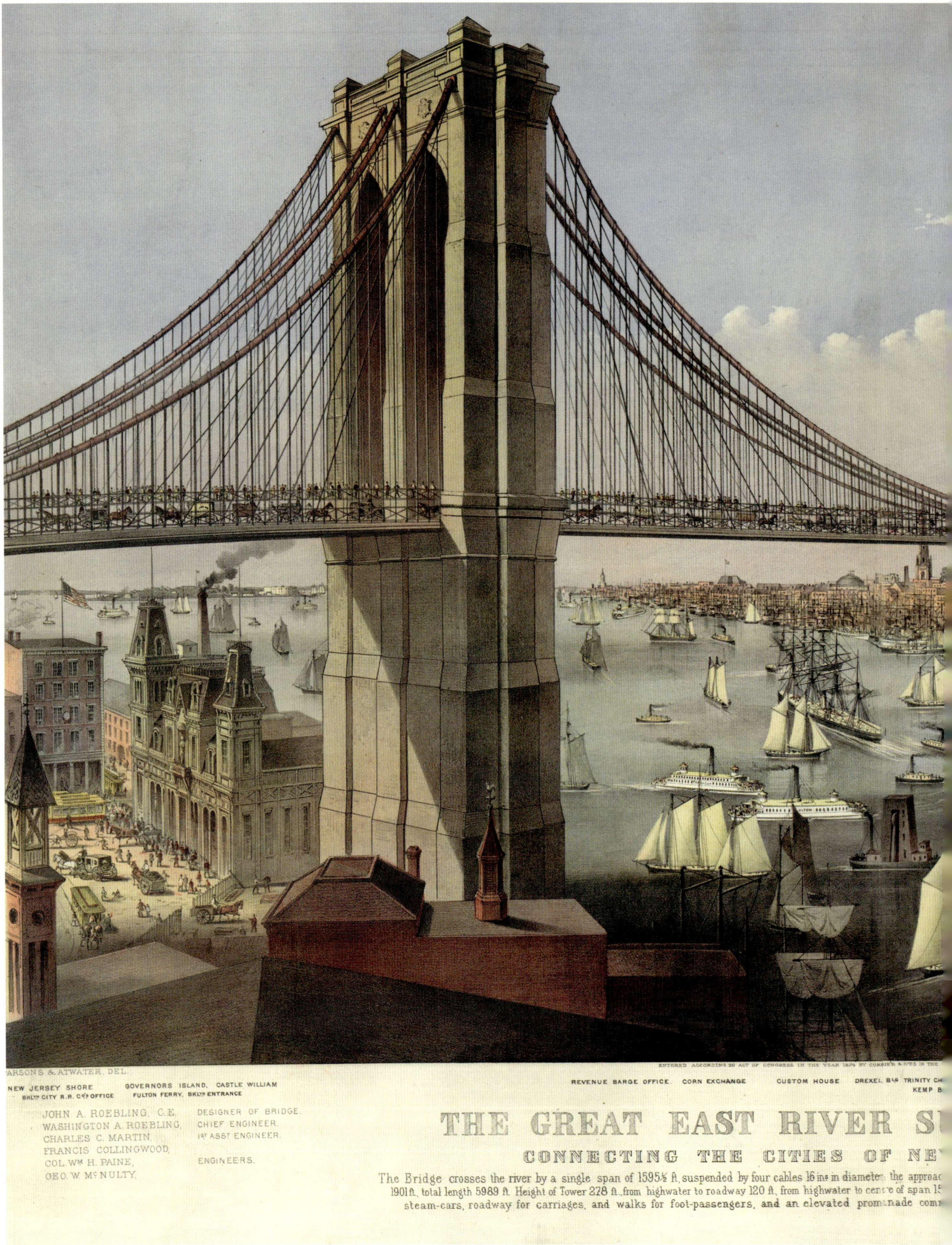

PARSONS & ATWATER, DEL.
ENTERED ACCORDING TO ACT OF CONGRESS IN THE YEAR 1874 BY CURRIER & IVES IN THE
NEW JERSEY SHORE
BKLYN CITY R.R. CO'S OFFICE
GOVERNORS ISLAND, CASTLE WILLIAM
FULTON FERRY, BKLYN ENTRANCE
REVENUE BARGE OFFICE.
CORN EXCHANGE
CUSTOM HOUSE
DREXEL BLG
TRINITY CH
KEMP B
JOHN A. ROEBLING, C.E. DESIGNER OF BRIDGE.
WASHINGTON A. ROEBLING, CHIEF ENGINEER.
CHARLES C. MARTIN, 1ST ASST. ENGINEER.
FRANCIS COLLINGWOOD,
COL. WM. H. PAINE,
GEO. W. MCNULTY, ENGINEERS.
THE GREAT EAST RIVER S
CONNECTING THE CITIES OF NE
The Bridge crosses the river by a single span of 1595½ ft, suspended by four cables 16 ins in diamete the approac
1901 ft, total length 5989 ft. Height of Tower 278 ft., from highwater to roadway 120 ft, from highwater to centre of span 15
steam-cars, roadway for carriages, and walks for foot-passengers, and an elevated promenade com

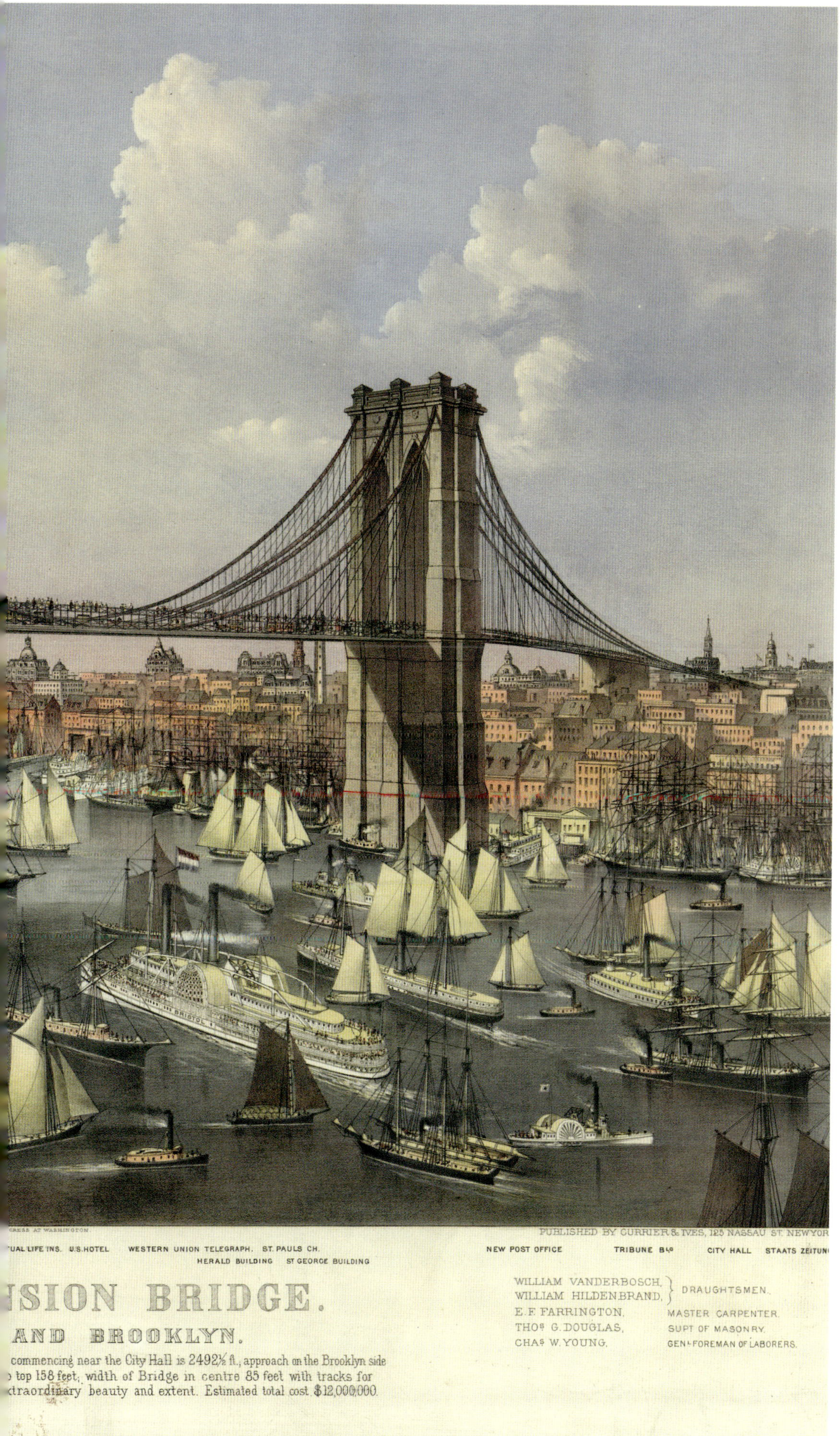

"The Great East River Suspension Bridge./Connecting the Cities of New York and Brooklyn." This lithograph, published by Currier & Ives in 1874, is a view from the Brooklyn side, looking southwest. It shows the bridge as it was expected to look when it was completed. The text provides detailed weights and measurements as well as the names of the men who led the construction.

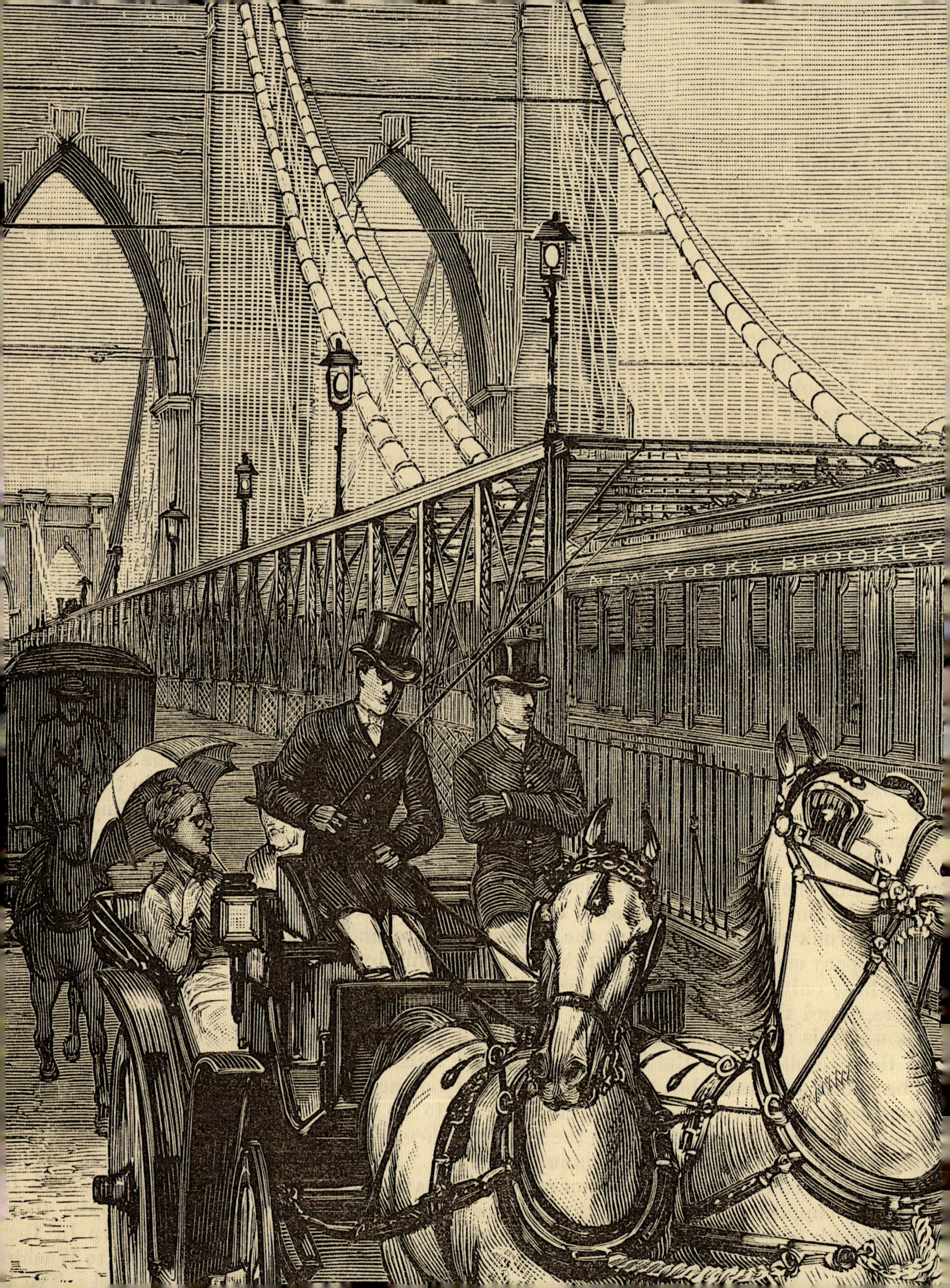
NEW YORK & BROOKLY

CHAPTER 10

Railways and Terminals

1882–1883

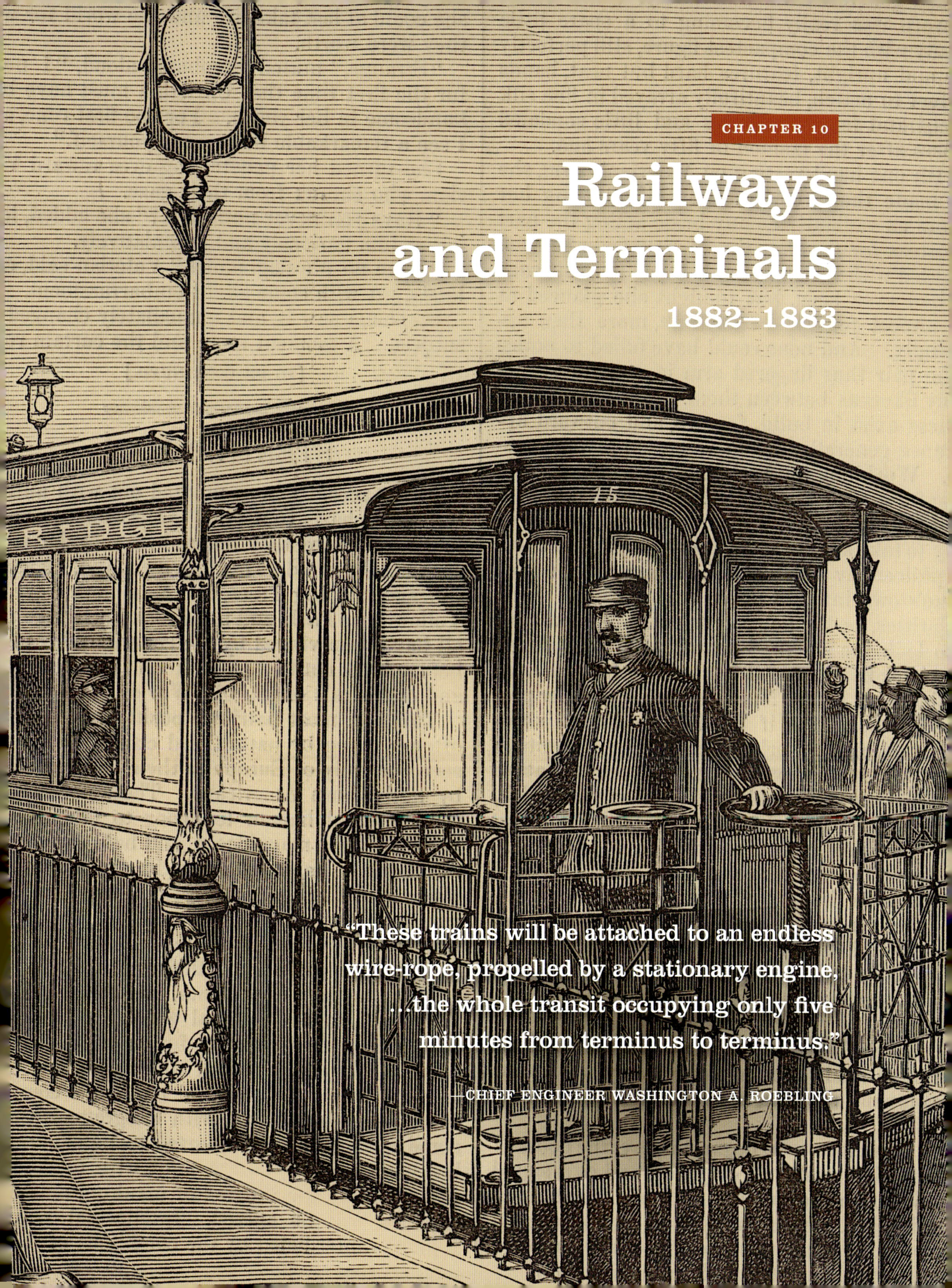

"These trains will be attached to an endless wire-rope, propelled by a stationary engine, ...the whole transit occupying only five minutes from terminus to terminus."

—CHIEF ENGINEER WASHINGTON A. ROEBLING

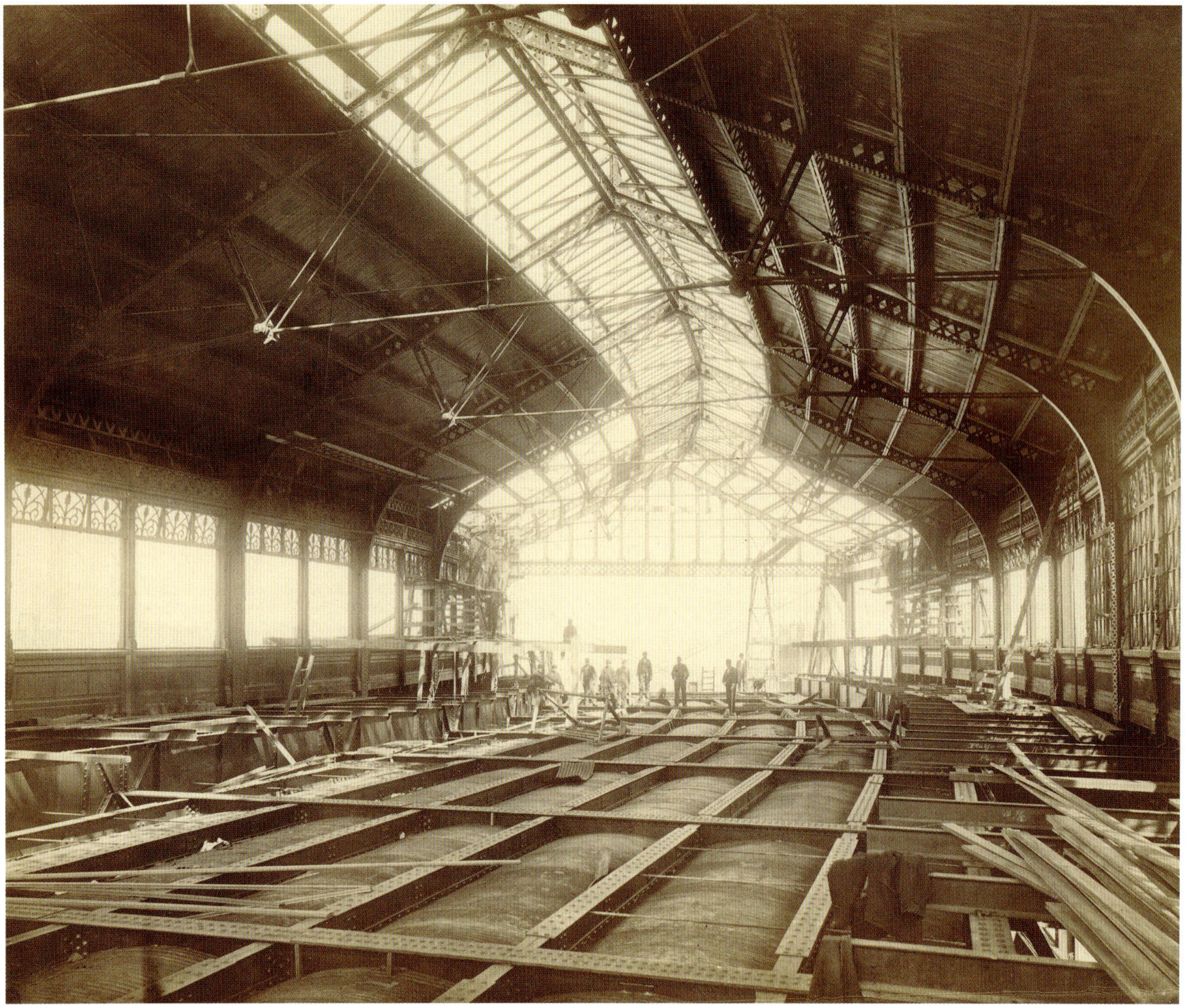

A notation on this photograph's mat identifies it as the bridge's Brooklyn terminal, ca. 1883, and the bend in the rails confirms that; those at the New York terminal were straight. Note the parallel beams laid out in three pairs, for the three sets of rails, stretching toward the workers, two on the outside for passengers to enter and exit the train cars, and a third in the middle for the storage of additional cars. RPI.

PREVIOUS SPREAD: "The Cars as They Will Be on the Brooklyn Bridge—Drawn by Harry Ogden," *Harper's Weekly*, June 16, 1883. An illustration of things to come. The railway across the bridge would not begin operating until September. Author's collection.

The Brooklyn Bridge's first chief engineer, John A. Roebling, had early on planned for a rail shuttle to cross the bridge on two tracks—one going east, the other going west. The tracks would be positioned between the outer roadways and the pedestrian promenade. John conceived of this feature before his death in 1869, but it was not until May 6, 1878, that the bridge's board of trustees approved a resolution to appoint a committee to look into "the construction or running of a rapid transit or other steam roads to either terminus of the Bridge," with the goal "that the Bridge shall fully realize the main object of its construction, the facilitation of intercourse between the two cities."

In a report to the bridge trustees, Washington Roebling described how the railway would operate across the Brooklyn Bridge. As per his father's plans, there would be a stationary engine, "located on the Brooklyn side, underneath the floor," that propelled an endless wire-rope, running in a loop from one end of the bridge to the other and pulling train cars at "speeds of twenty miles per hour." John Roebling had determined that the trains could travel safely at speeds of thirty or even forty miles per hour and that the trains could be as long as ten cars, with each car carrying one hundred passengers. The two tracks allowed one train to proceed east to Brooklyn, while a second one traveled simultaneously in the opposite direction. The tracks, according to John, would operate "like an inclined plane." It was an "operation perfectly simple and perfectly well understood." John added, "There is no novel feature and no experiment involved in its arrangement."

Much like skiers grasping a towrope, the train cars started or stopped by grabbing or letting go of the continuously moving wire rope below. Washington wrote that the grabbing mechanism was "an ingenious arrangement for attaching cars to a moving rope, devised by [Assistant Engineer] Col. W. H. Paine," one that had been "successfully at work for more than a year on the Sutter Street Railways in San Francisco." Paine's wire-catching invention for these cable cars created a smooth transition from standing still to motion, and vice versa, avoiding violent jerking that would jar the passengers.

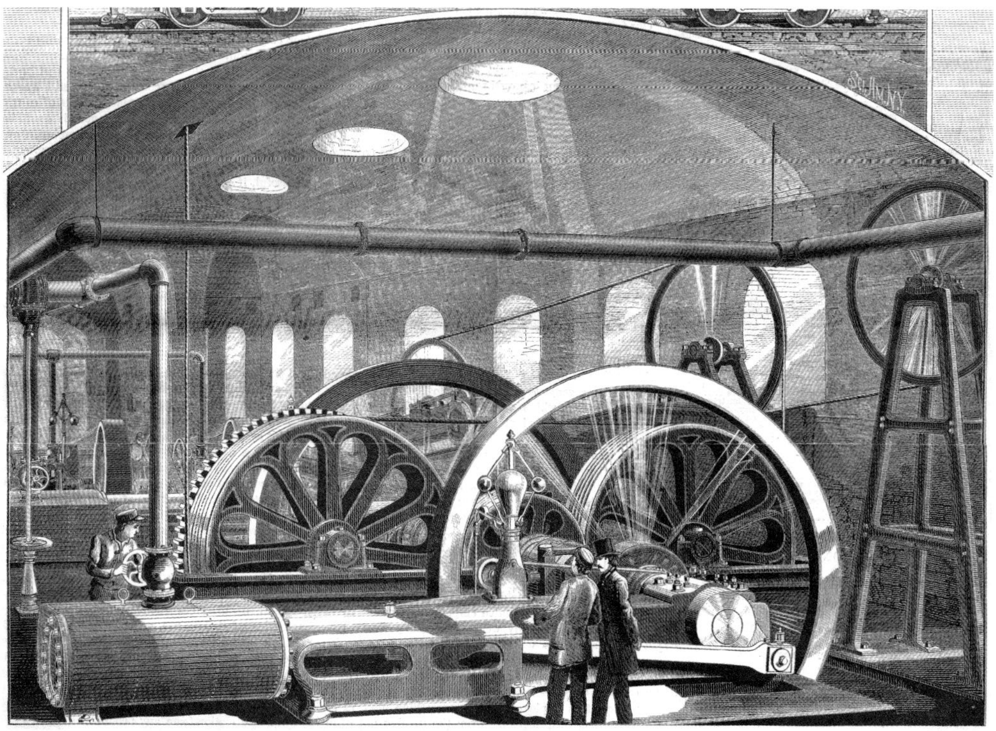

"The New York and Brooklyn Bridge—Brooklyn Station and Engine Room," *Scientific American*, June 2, 1883. These engines powered the cable that moved the rail cars.

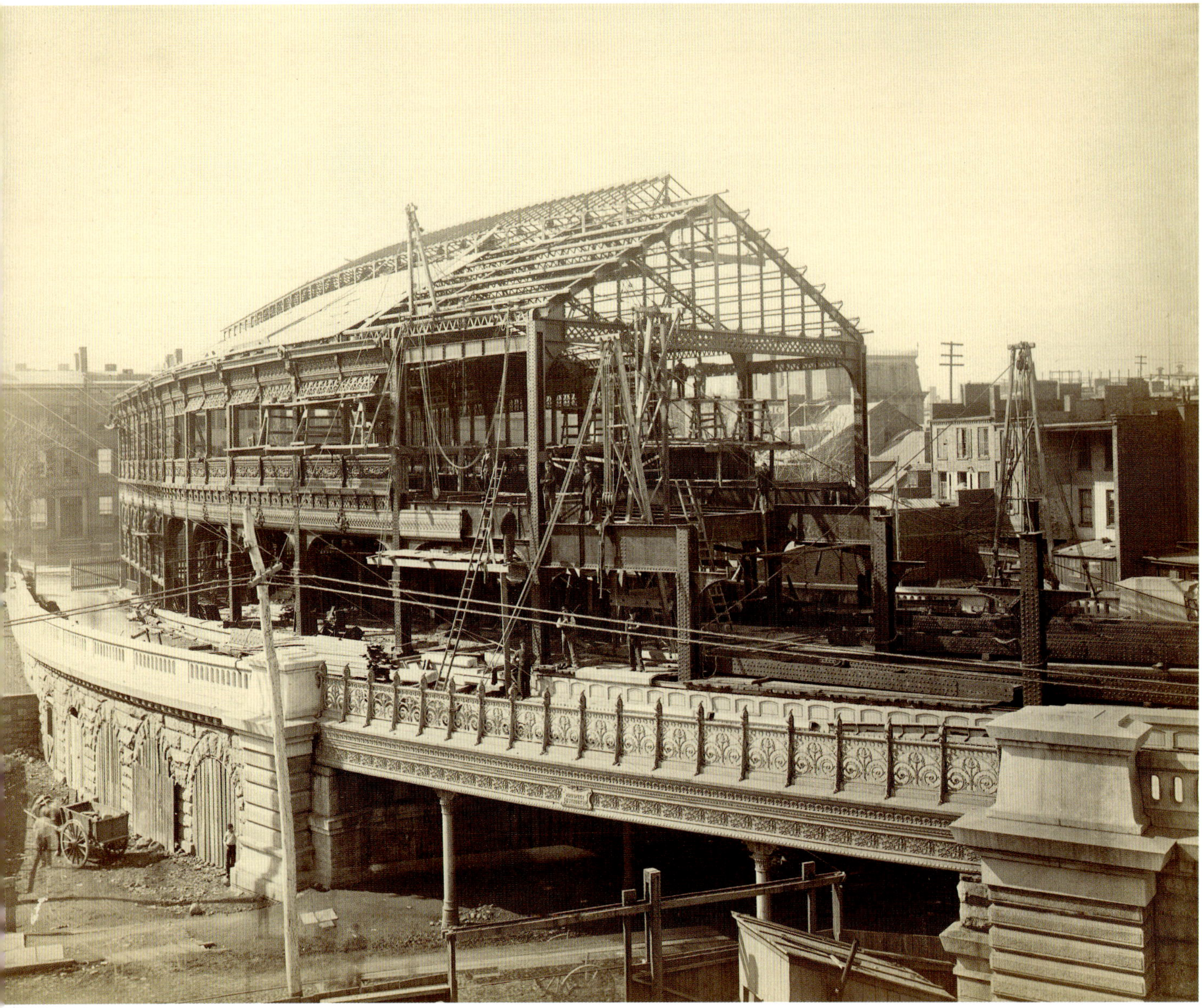

"Brooklyn Station Building." This photograph, looking east, likely dates from early 1883. It shows the characteristic curve of the bridge's Brooklyn terminal, with the rails continuing off to the right on top of the approach, toward the river. Note the iron span in the foreground, with its elaborate cast-iron railings; it crosses Prospect Street, identified by the decorative sign mounted midway in the span. RPI.

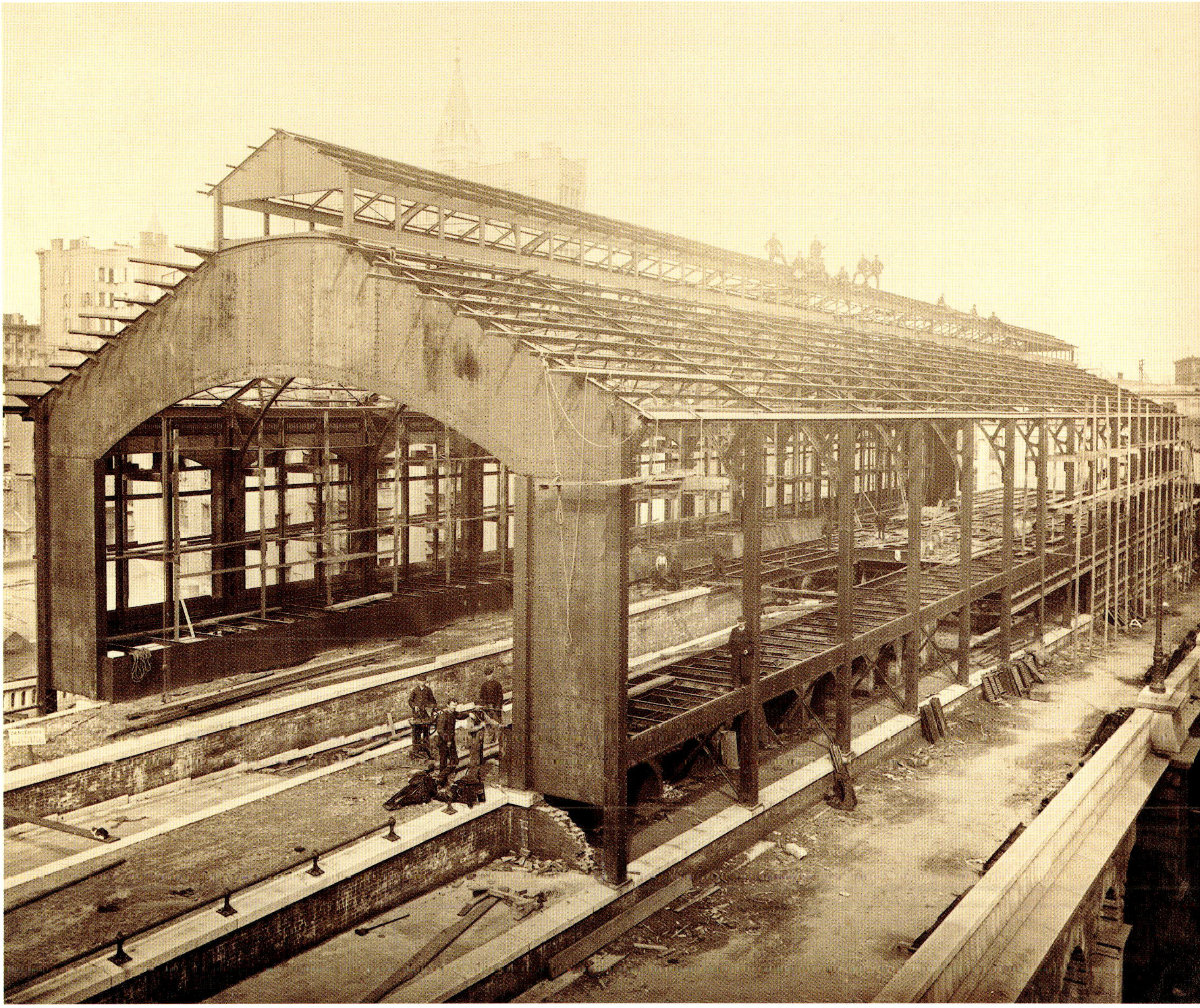

"N.Y. Station, East River Bridge, F. Collingwood, Eng[inee]r." On back, in pencil: "To F. Collingwood, Esq./Eng[inee]r. N.Y. Approach/With Compliments of Jones and Benner." This photograph, ca. early 1883, shows the terminal end facing the bridge. Not impressed by the terminals, prominent critic Montgomery Schuyler described them in *Harper's Weekly* as "grossly illiterate and discreditable to the great work." MCNY.

"The Great Suspension Bridge Between New York and Brooklyn. The Bridge Entrance, New York Side," *Scientific American*, April 6, 1878. Note that this image was published five years before the bridge opened and involved some guesswork. The railroad cars are shown, but the railway terminal that was to be built at this location is nowhere to be seen.

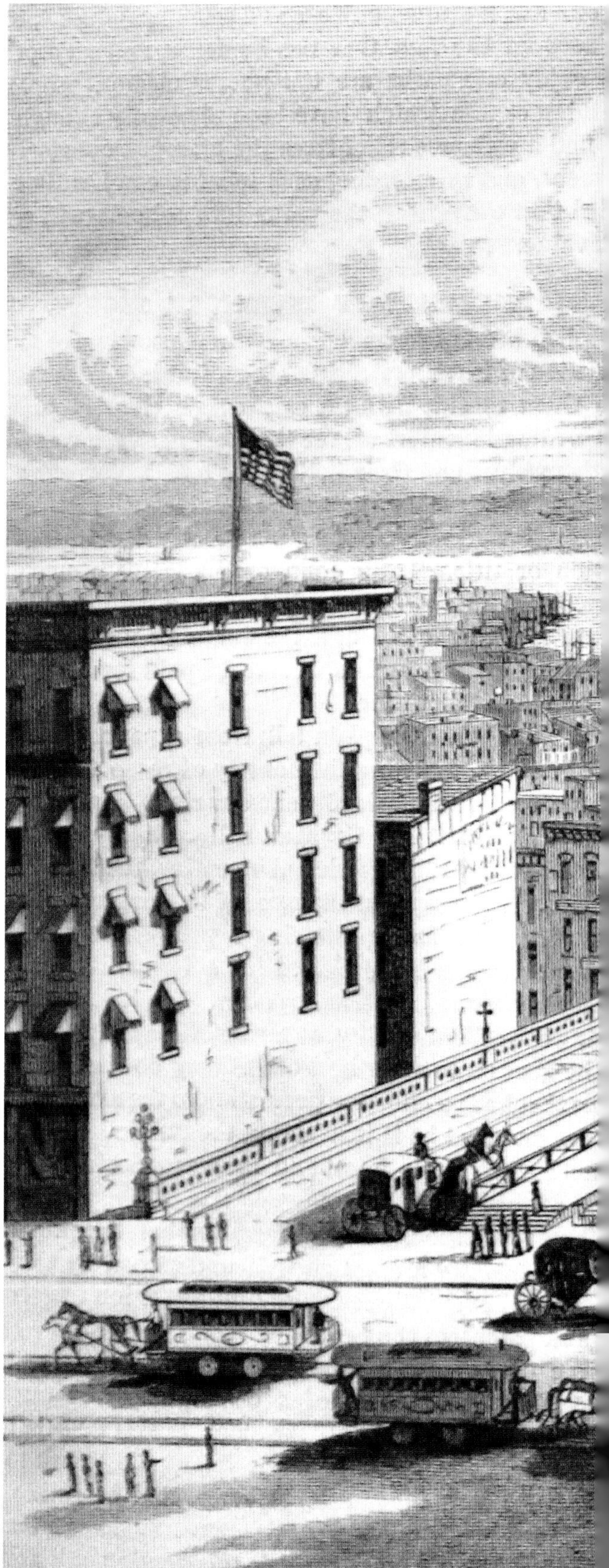

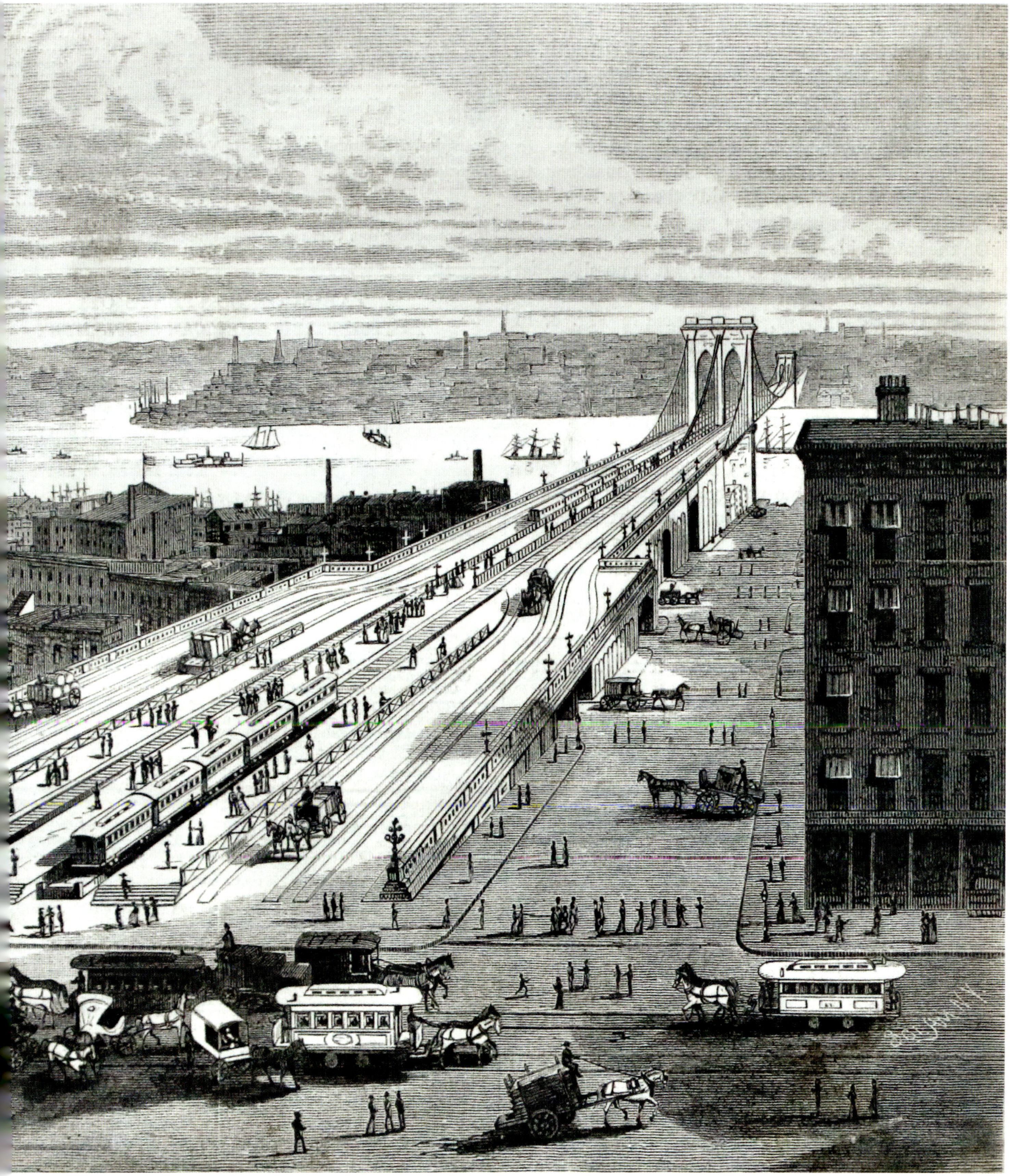

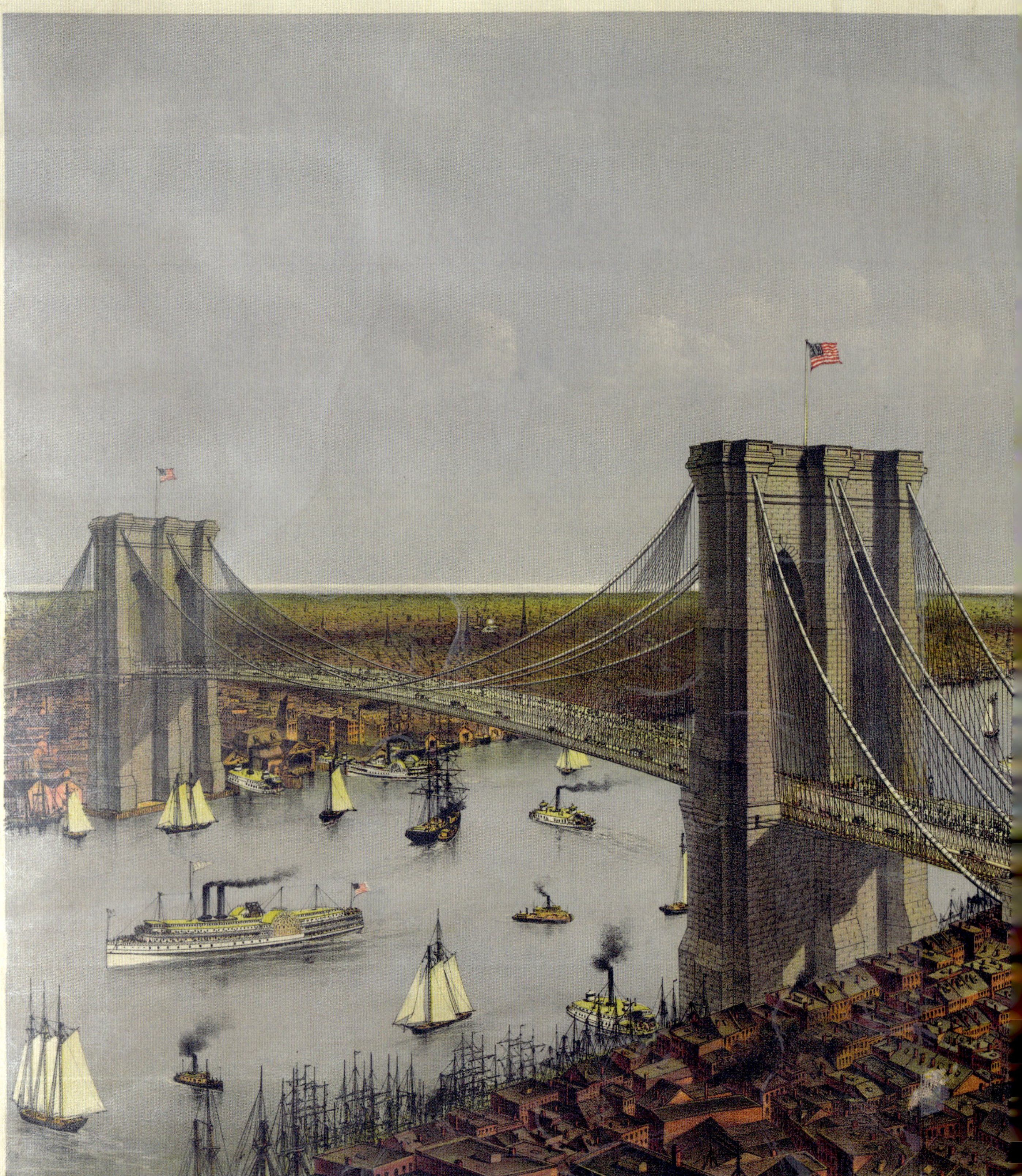

PUBLISHED BY CURRIER & IVES, 115 NASSAU ST. NEW YORK. COPYRIGHT, 1885, BY CURRIER & IVES, N.Y.

855B

Brooklyn Tower. Fulton Ferry, Brooklyn. Atlantic Ocean (in the distance) New York Tower.

GRAND BIRDS EYE

THE GREAT EAST RIVER S

CONNECTING THE CITIES OF NEW

Showing also the splendid pan

THE BAY AND THE PORT OF

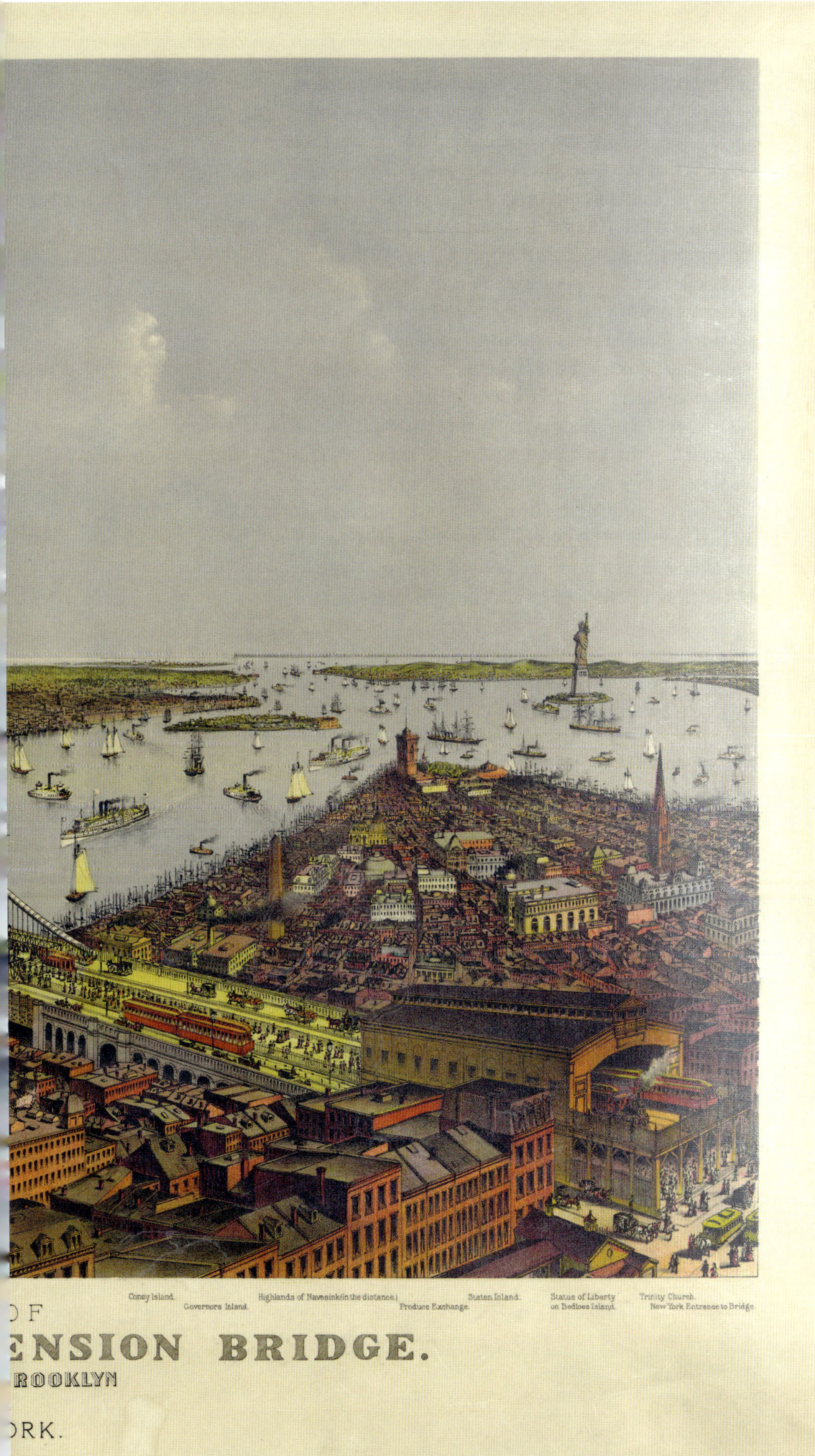

"Grand Birds Eye View of/The Great East River Suspension Bridge./ Connecting the Cities of New York & Brooklyn/ Showing also the splendid panorama of/ the Bay and the Port of New York." Looking southeast from Manhattan toward Brooklyn. Currier & Ives, 1885.

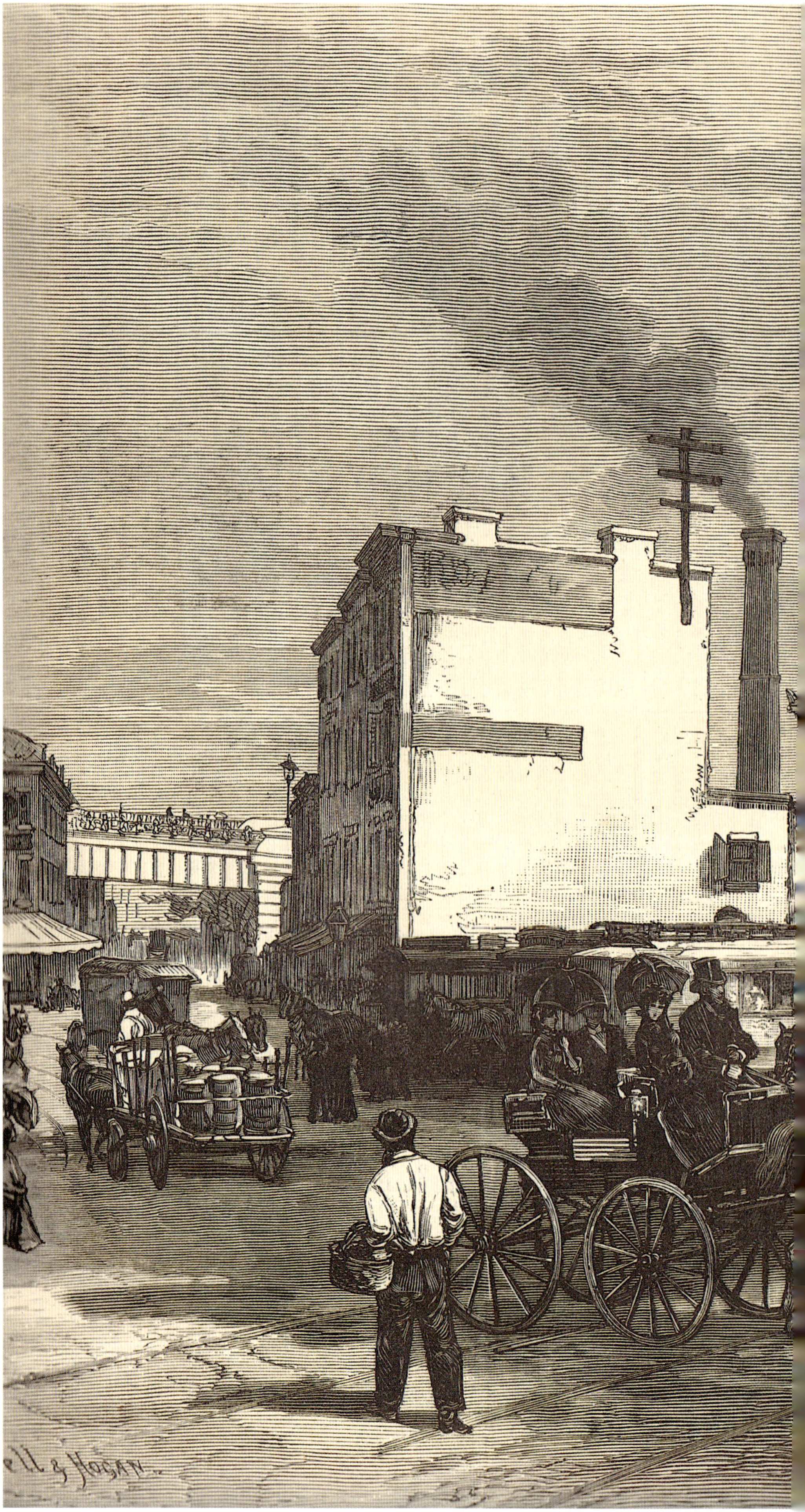

"The New York and Brooklyn Suspension Bridge—Brooklyn Entrance—Drawn by Schell and Hogan," *Harper's Weekly*, May 26, 1883. This woodcut was published just days after the opening of the bridge—but months before the train terminals would be completed and in operation. Still, at center left, the smokestack for the engines that would power the cable wire moving the railroad cars is puffing away, and in the distance between the two buildings at left, one of the approach's iron spans is jammed with pedestrians. This terminal would house the bridge's opening ceremonies, attended by six thousand invited guests. Author's collection.

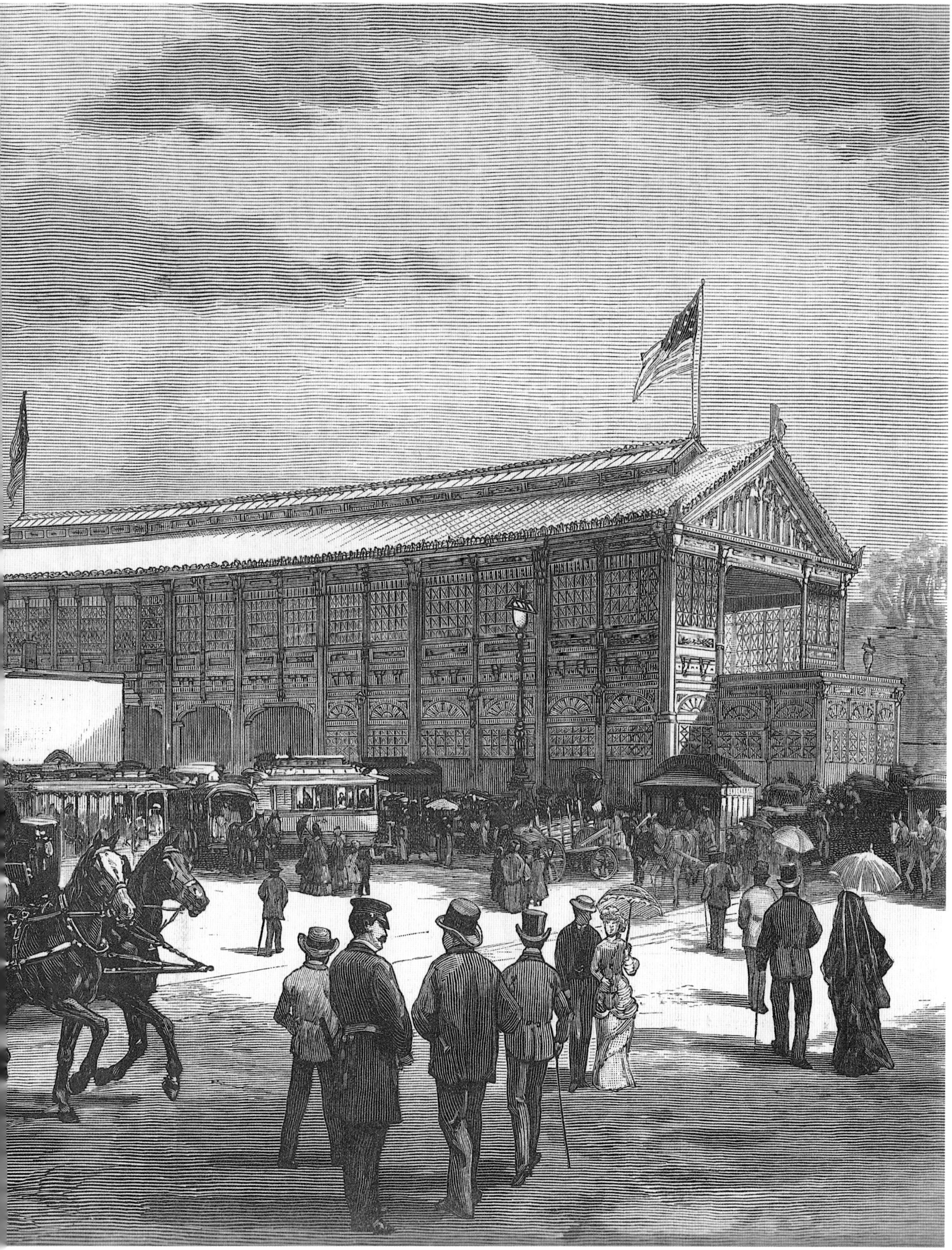

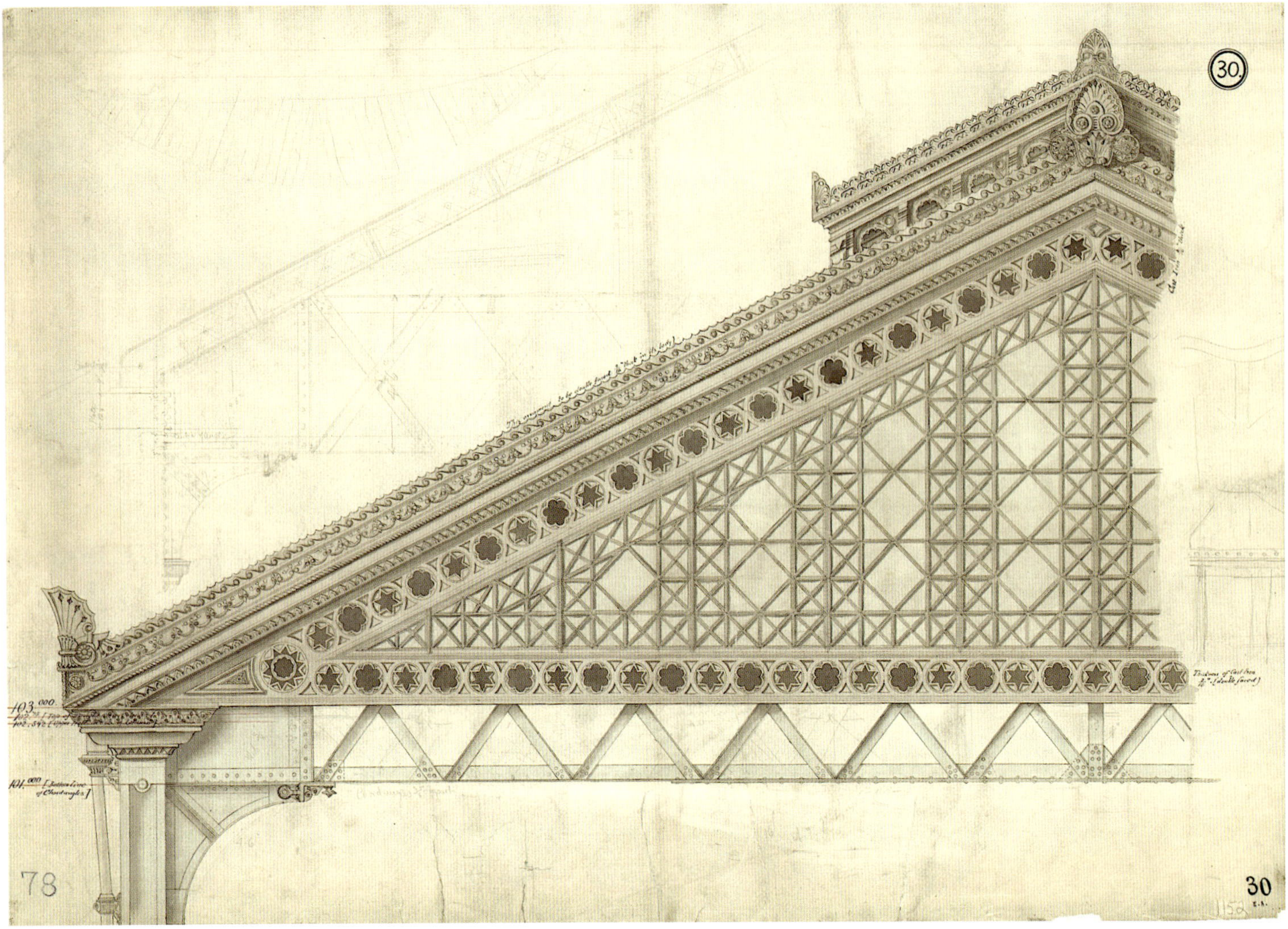

Detail of a drawing showing the Brooklyn terminal's gabled iron roof. NYC Municipal Archives.

The bridge's rail system functioned smoothly. Trains ran around the clock, and, in the first year of their operation, 9.2 million passengers rode across the bridge in one of the train cars. By 1885, just two years after the railway's opening, it had carried a total of 20 million passengers. Ultimately, the rail shuttle would carry hundreds of millions of passengers across the Brooklyn Bridge before it ceased operations in 1944. Shortly thereafter, the iron railway terminals were torn down.

While the railway and terminals were not yet operational when the opening day ceremonies took place on May 24, 1883, the bridge itself was ready for foot and vehicular traffic on that date. After years of planning and fourteen years of construction, it was time to open the Brooklyn Bridge.

OPPOSITE: "The Jam at the Entrance to the Brooklyn Bridge—Drawn by Thure de Thulstrup," *Harper's Weekly*, August 29, 1885. Just two years after the bridge opened, it was already hugely popular. Author's collection.

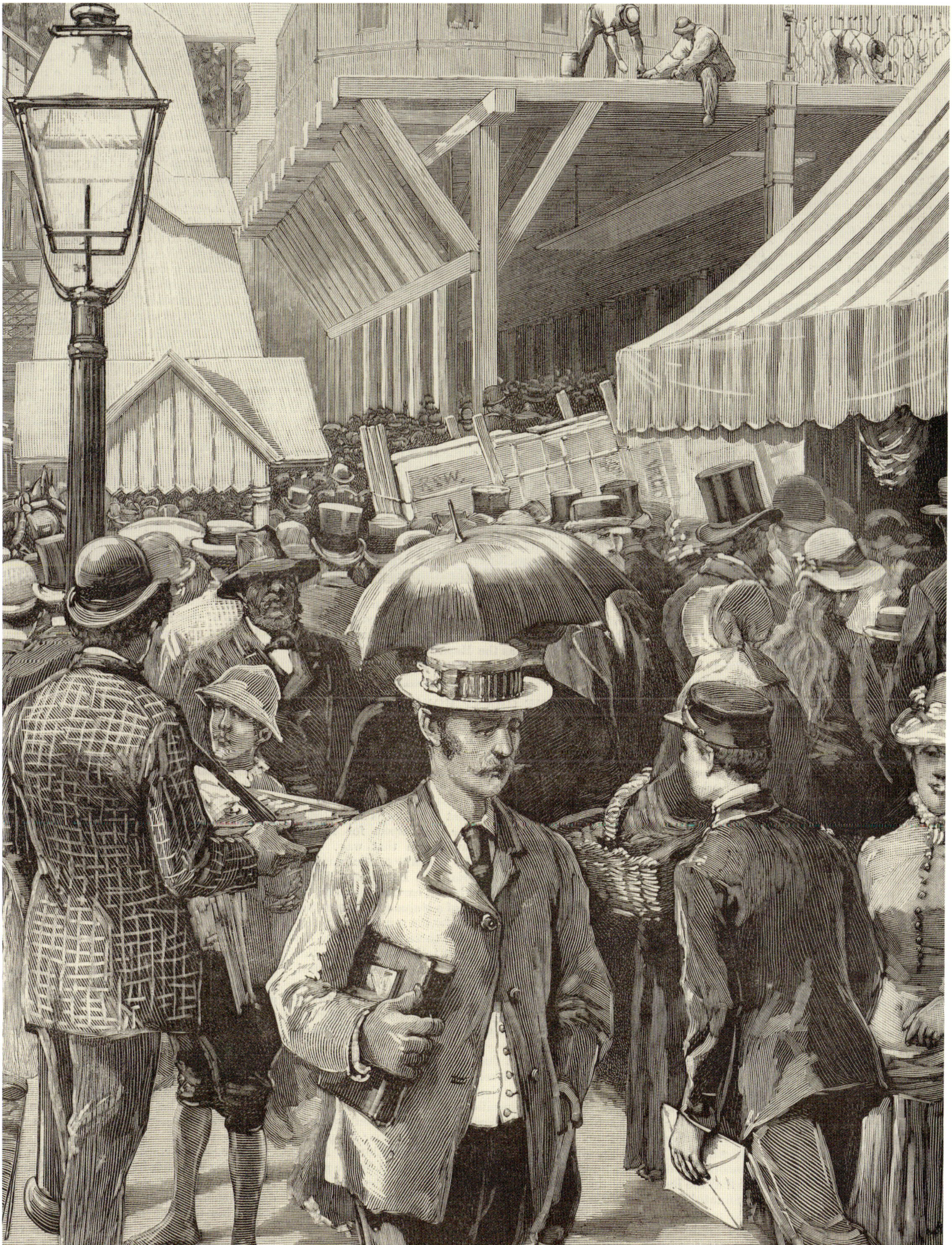

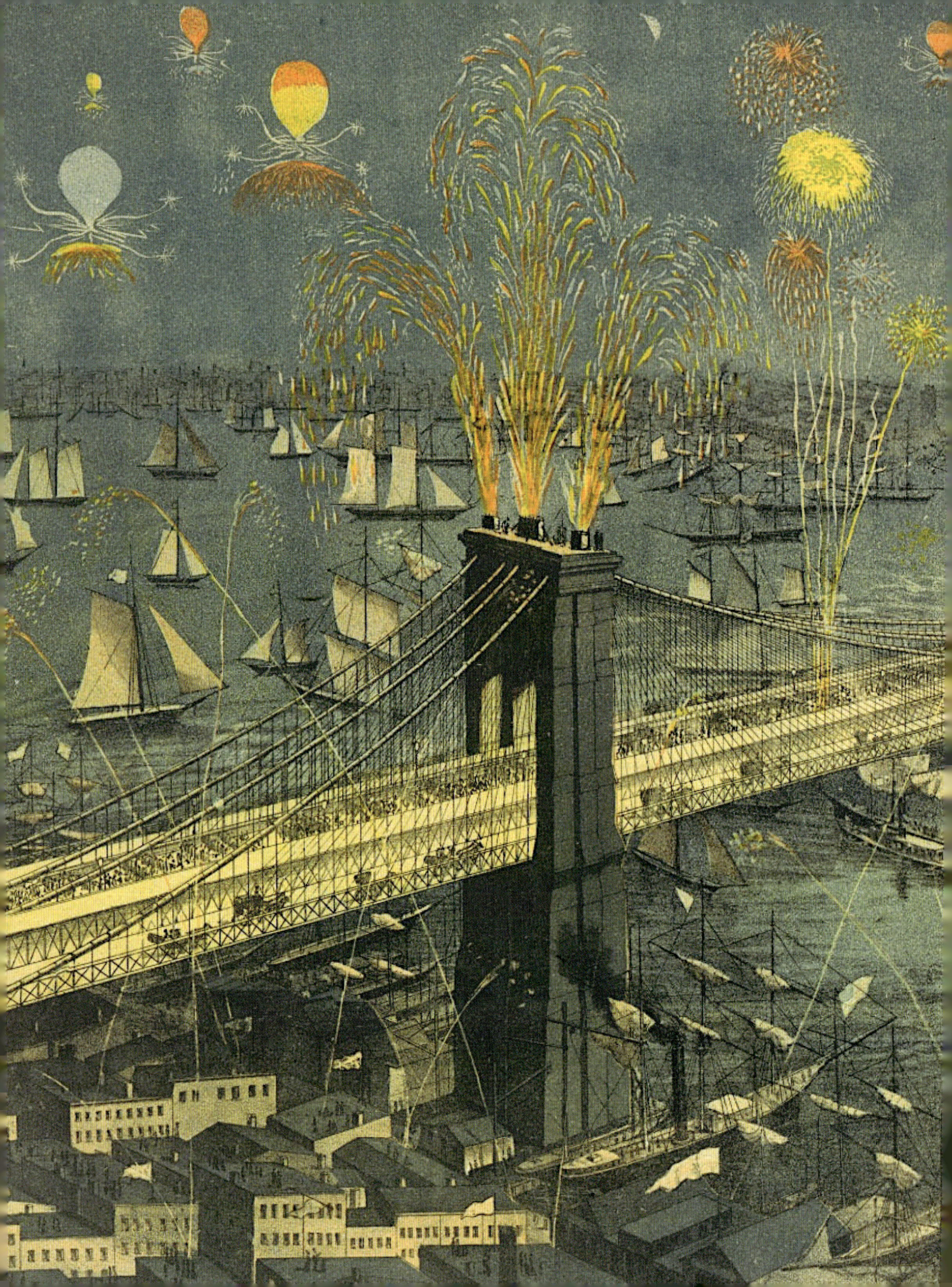

CHAPTER 11

Opening Day

May 24, 1883

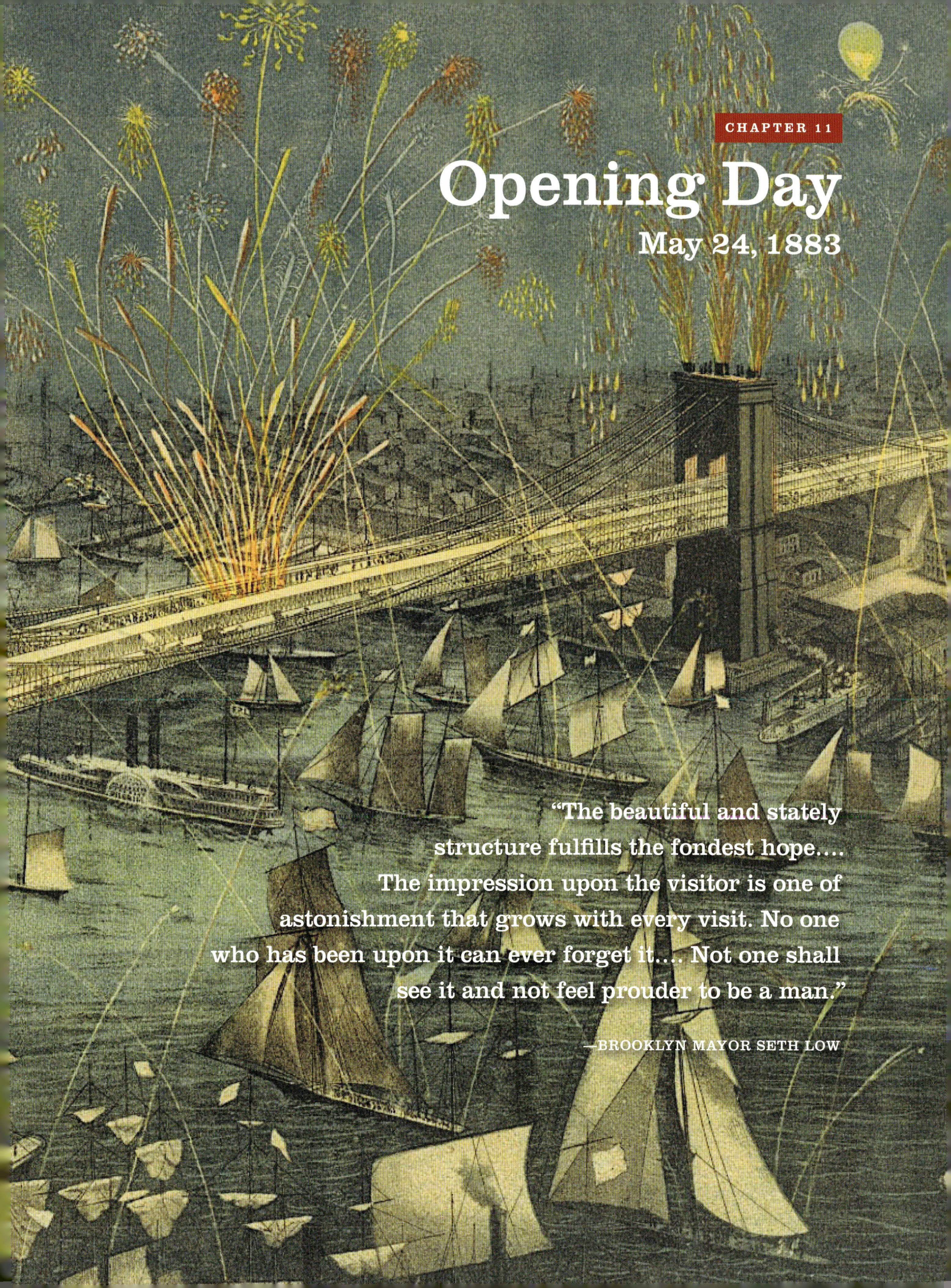

"The beautiful and stately structure fulfills the fondest hope.... The impression upon the visitor is one of astonishment that grows with every visit. No one who has been upon it can ever forget it.... Not one shall see it and not feel prouder to be a man."

—BROOKLYN MAYOR SETH LOW

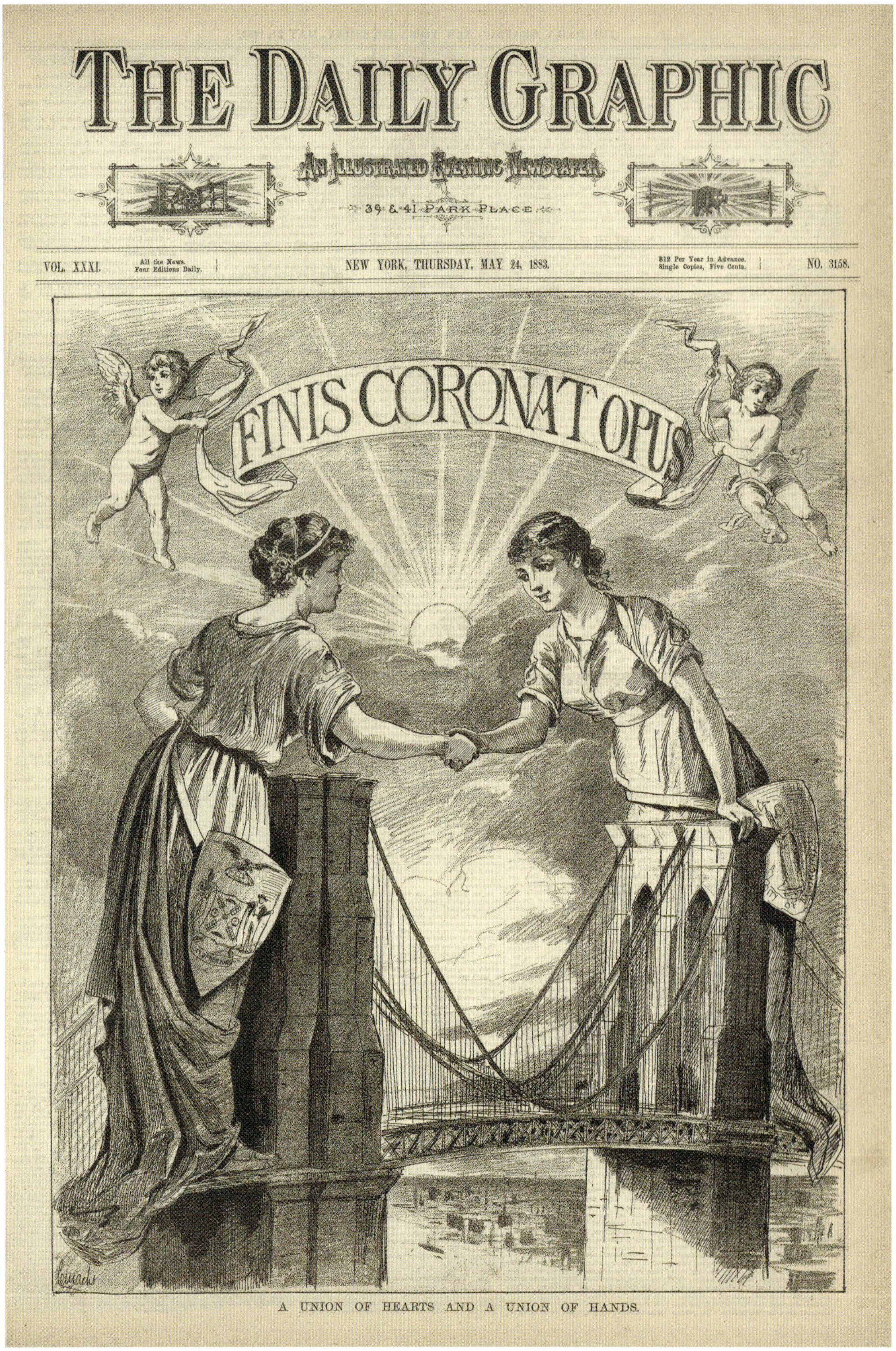
THE DAILY GRAPHIC

AN ILLUSTRATED EVENING NEWSPAPER

39 & 41 PARK PLACE

VOL. XXXI. | All the News. Four Editions Daily. | NEW YORK, THURSDAY, MAY 24, 1883. | $12 Per Year in Advance. Single Copies, Five Cents. | NO. 3158.

A UNION OF HEARTS AND A UNION OF HANDS.

"A Union of Hearts and a Union of Hands," an illustration that appeared on the cover of *The Daily Graphic*, May 24, 1883, depicting the cities of Brooklyn and New York as women shaking hands. Courtesy of Richard Haw.

PREVIOUS SPREAD: "Bird's-Eye View of the Great New York and Brooklyn Bridge and Grand Display of Fire Works on Opening Night." A lithograph published by A. Majors, 1883. Courtesy of The Metropolitan Museum of Art.

As the bridge's opening day—May 24, 1883—approached, there was still some work to be done. Installation of the electric lights was not yet complete, and the railway terminal buildings were months away from being finished. But the bridge itself was ready for its opening.

The Brooklyn Bridge had taken fourteen years to build—almost three times as long as John Roebling, its first chief engineer, had predicted. It had cost $15 million—more than twice the amount he had anticipated. During the building of the bridge, between twenty and forty workers—including John Roebling himself—had died. But, in the end, the bridge surpassed all expectations.

Opening day was planned as the largest celebration New York City had seen since 1825, when the Erie Canal opened, completing the water route between the Great Lakes and New York Harbor. The day would be one of triumph and jubilation. The excitement had been building since 1869, when construction of the bridge had begun. Manhattanites and Brooklynites had been watching its progress for over a decade, and they were ready to walk across it. Thirteen thousand tickets, printed by Tiffany & Co., invited a lucky few to the opening ceremonies. Another seven thousand tickets allowed access to the bridge on the opening day. *The New York Times* predicted that, for Brooklyn, the bridge's opening would be "the greatest gala day in the history of that moral suburb." Brooklyn even designated the opening day as an official holiday—"The People's Day"—with schools and most businesses closed.

Early that Thursday morning, the docks and the buildings along the river were packed. At 9:00 a.m., workers took down the fence at Chatham Street on the Manhattan side of the bridge, and a line of policemen moved into position to restrain the crowds. The nearby streets were jammed, and rooftops were crowded. The East River was filled with ships of all descriptions, each weighed down with as many onlookers as they could carry. The North Atlantic Squadron of the United States Navy had sailed into New York Harbor to join the festivities. Its guns, along with those at the Brooklyn Navy Yard and on Governors Island, fired salutes. The few businesses on either side of the river that remained open were only sparsely staffed, aside from the liquor stores and bars, which were overflowing with customers.

At Brooklyn's City Hall, just after noon, infantrymen, some in coats of red, others in coats of blue, United States artillerymen and Marines, as well as city officials and bridge trustees, began to march toward the bridge. Emily Roebling and her invited guests followed in carriages, cheered by the crowds that lined the procession. Most of these people, including Emily and her guests, stopped at the unfinished Brooklyn railway terminal, where the ceremonies were to be held, while the leading dignitaries, accompanied by New York's 23rd Regiment, continued onto the bridge, marched across it, then headed to City Hall Park in Manhattan, where they were greeted by New York officials and other distinguished guests.

Having arrived there at about 1:30 p.m., they were soon joined by President Chester A. Arthur, Governor Grover Cleveland of New York (a future president), and Mayor Franklin Edson of New York City. The press of the crowd against the police lines was tremendous. The dignitaries reassembled, and led by the famed 7th Regiment of the New York State National Guard, with its band playing and flags flying, they headed up onto the bridge and back toward Brooklyn. As they walked, the throngs of ticket holders, already on the bridge, cheered. When the procession reached the Brooklyn tower, they paused as Mayor Seth Low of Brooklyn offered an official greeting, the Marines presented arms, and first one Navy gun, then a host, fired in salute. Steam whistles shrieked, bells rang, and crowds cheered.

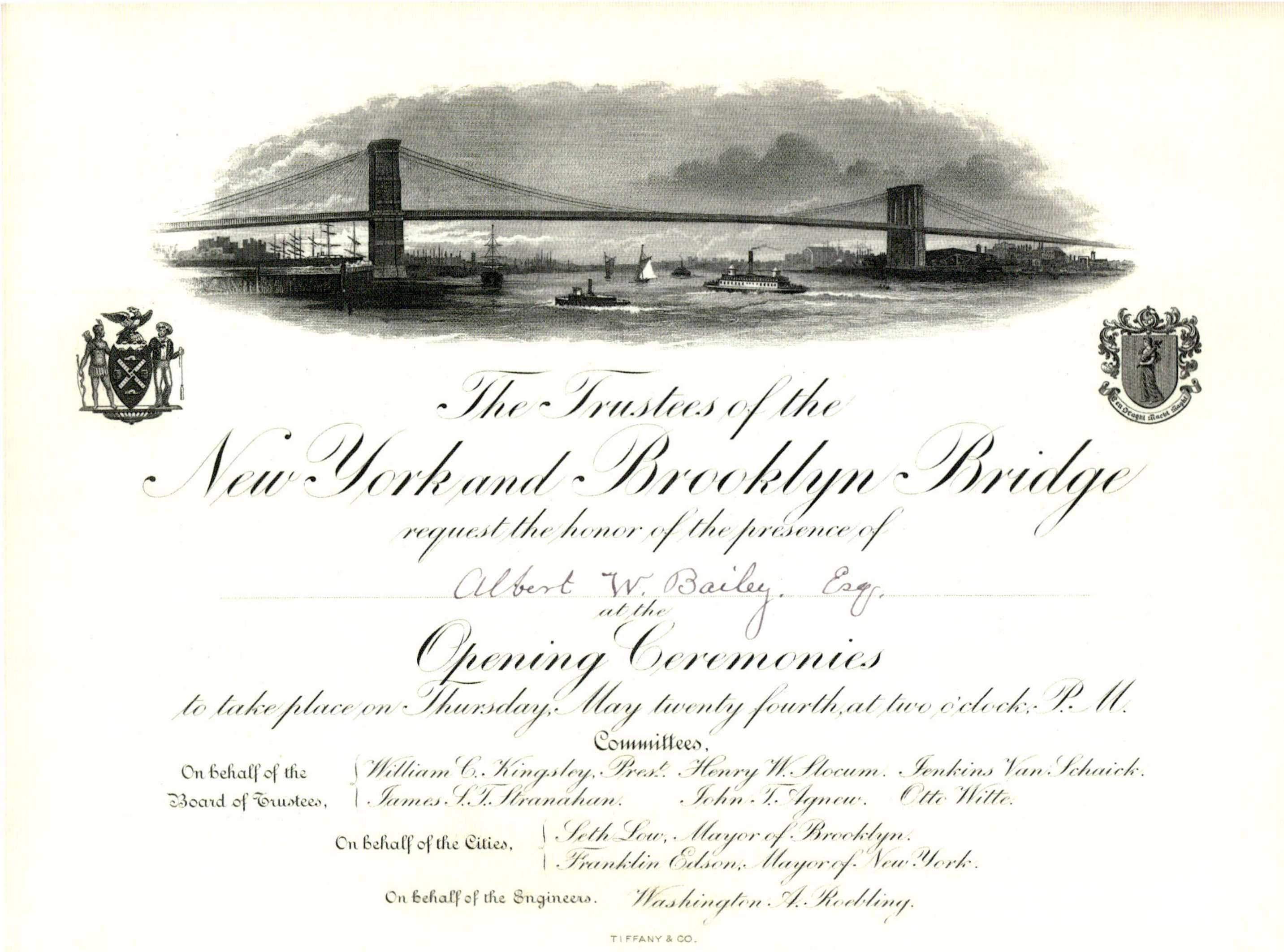

The Trustees of the
New York and Brooklyn Bridge
request the honor of the presence of
Albert W. Bailey, Esq.
at the
Opening Ceremonies
to take place on Thursday, May twenty fourth, at two o'clock, P. M.

Committees,

On behalf of the Board of Trustees, William C. Kingsley, Prest. Henry W. Slocum. Jenkins Van Schaick.
James S. T. Stranahan. John T. Agnew. Otto Witte.

On behalf of the Cities, Seth Low, Mayor of Brooklyn.
Franklin Edson, Mayor of New York.

On behalf of the Engineers. Washington A. Roebling.

TIFFANY & CO.

Invitation to the opening of the Brooklyn Bridge, made out to one Albert W. Bailey, who at the time was a lawyer living on Monroe Place in Brooklyn Heights. Courtesy of The Green-Wood Historic Fund.

TRUSTEES AND OFFICERS

OF THE

NEW YORK AND BROOKLYN BRIDGE.

TRUSTEES.

NEW YORK.	BROOKLYN.
JOHN T. AGNEW,	WILLIAM C. KINGSLEY,
JOHN G. DAVIS,	WILLIAM MARSHALL,
J. ADRIANCE BUSH,	HENRY W. SLOCUM,
HENRY CLAUSEN,	JAMES S. T. STRANAHAN,
THOMAS C. CLARKE,	ALFRED C. BARNES,
CHARLES MACDONALD.	ALDEN S. SWAN,
H. K. THURBER,	OTTO WITTE,
JENKINS VAN SCHAICK,	JAMES HOWELL,
FRANKLIN EDSON, Mayor, *Ex-officio.*	SETH LOW, Mayor, *Ex-officio.*
ALLAN CAMPBELL, Comp., *Ex-officio.*	AARON BRINKERHOFF, Comp., *Ex-officio.*

JOHN T. AGNEW, Chairman Executive Committee.

OFFICERS.

WILLIAM C. KINGSLEY, President.

J. ADRIANCE BUSH, Vice-Pres. OTTO WITTE, Treasurer.

ORESTES P. QUINTARD, Secretary.

CHIEF ENGINEER.

WASHINGTON A. ROEBLING.

ASSISTANT ENGINEERS.

CHARLES C. MARTIN,
FRANCIS COLLINGWOOD,
SAMUEL R. PROBASCO,
WILLIAM H. PAINE,
GEORGE W. McNULTY,
WILHELM HILDENBRAND.

A page from the program for the Brooklyn Bridge opening day ceremonies, May 24, 1883, listing the names of the trustees, officers, chief and assistant engineers. Courtesy of The Green-Wood Historic Fund.

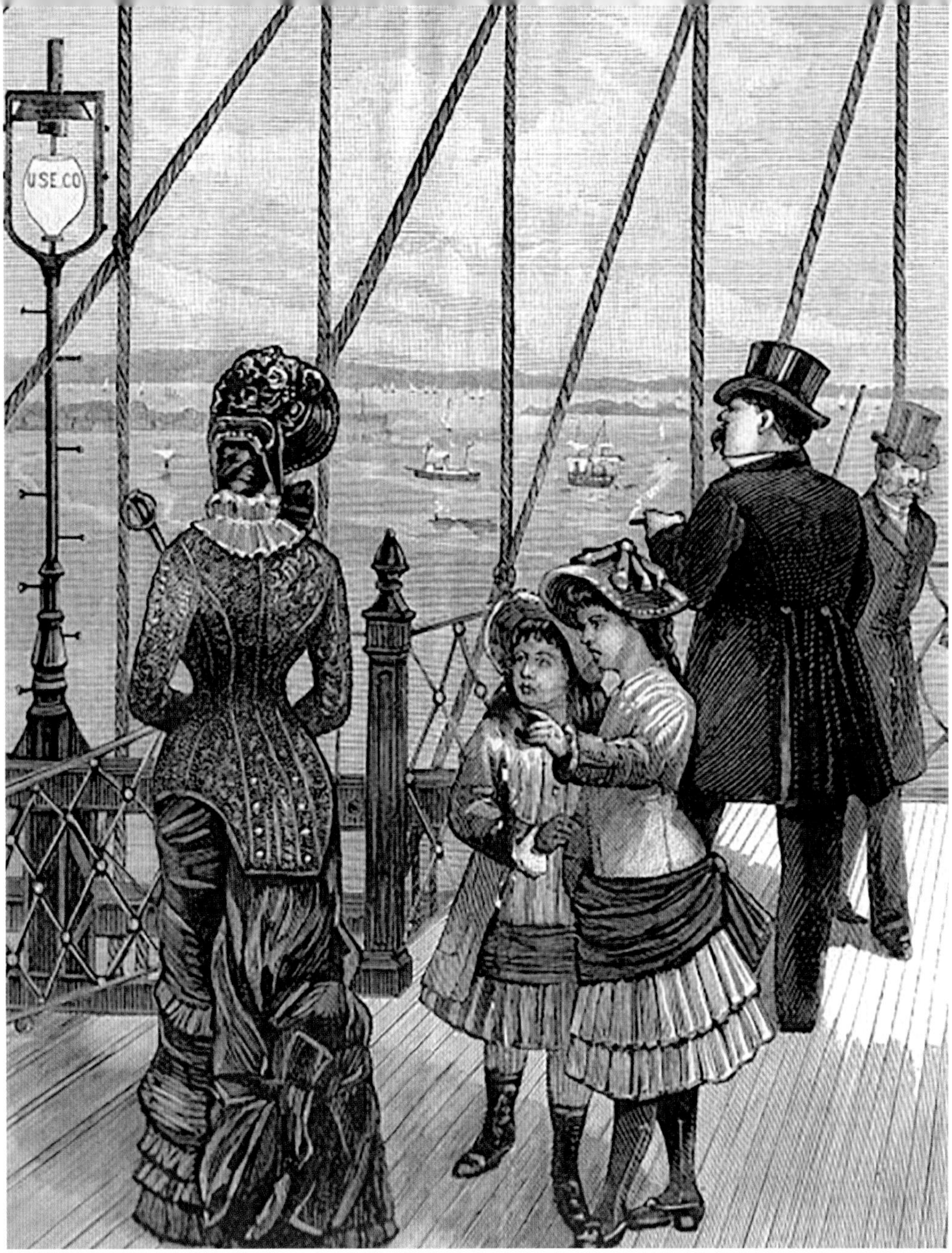

"The Tower Promenade." This woodcut, published in *Frank Leslie's Popular Monthly*, July 1883, shows a family lucky enough to have received tickets to walk across the Brooklyn Bridge on its opening day: May 24, 1883. Courtesy of Richard Haw.

The *New York Sun* wrote, "The climax of fourteen years' suspense seemed to have been reached, since the president of the United States of America had walked dry shod to Brooklyn from New York."

The procession then continued to the Brooklyn railway terminal, where six thousand people, packed inside for the formal opening ceremonies, waited expectantly. Huge American flags framed the scene. President Arthur was greeted by wild cheers. Raised platforms accommodated the president, governor, and other civic leaders. The men spoke of peace and science and other lofty ideas. Abram Hewitt, who had served as a bridge trustee—and had given Chief Engineer Washington Roebling (notably absent at this triumphant moment) a hard time during the construction—spoke on behalf of New York City, and the Reverend Dr. Richard Storrs spoke for Brooklyn. With no means of amplification available and the terminal open at its west end, only those closest to the speakers heard what was said. That did not deter Reverend Dr. Storrs, who rambled on for an hour.

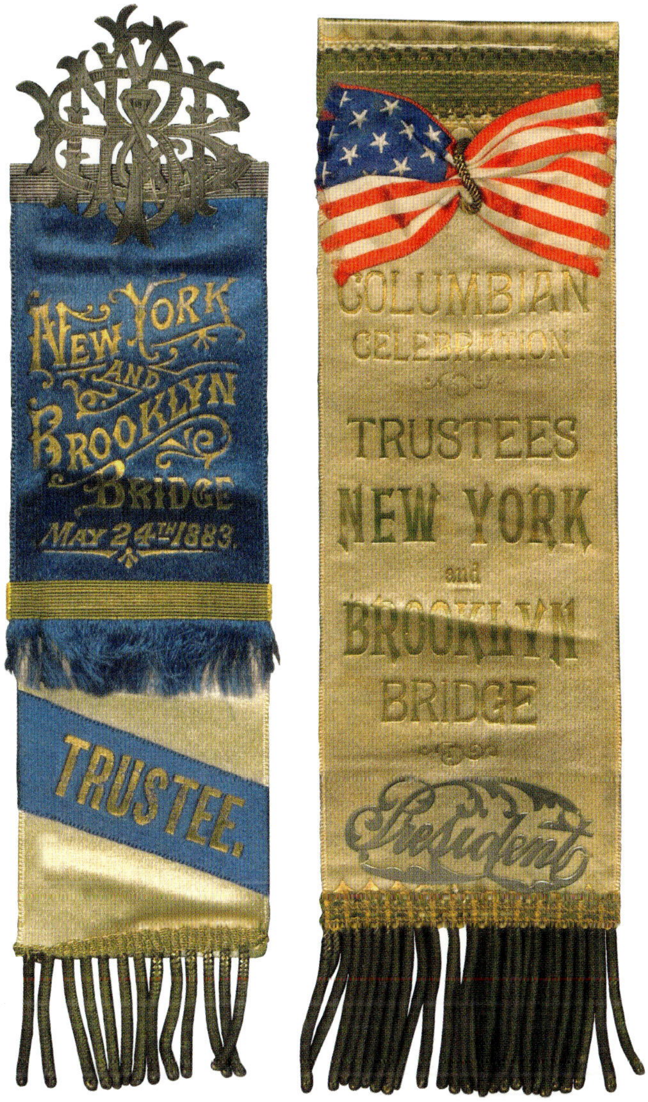

RIGHT: James Howell, a trustee of the Brooklyn Bridge (who also served as mayor of the City of Brooklyn), wore this blue ribbon at its opening in 1883. He wore the ribbon at right—identifying him as president of the Brooklyn Bridge—in 1892 for New York City's Columbian Celebration. Courtesy of The Green-Wood Historic Fund.

BELOW: A ticket to the opening ceremonies held on May 24, 1883. Issued by what was by then known as the New York and Brooklyn Bridge Company. The tickets were printed by Tiffany & Co. Courtesy of The Green-Wood Historic Fund.

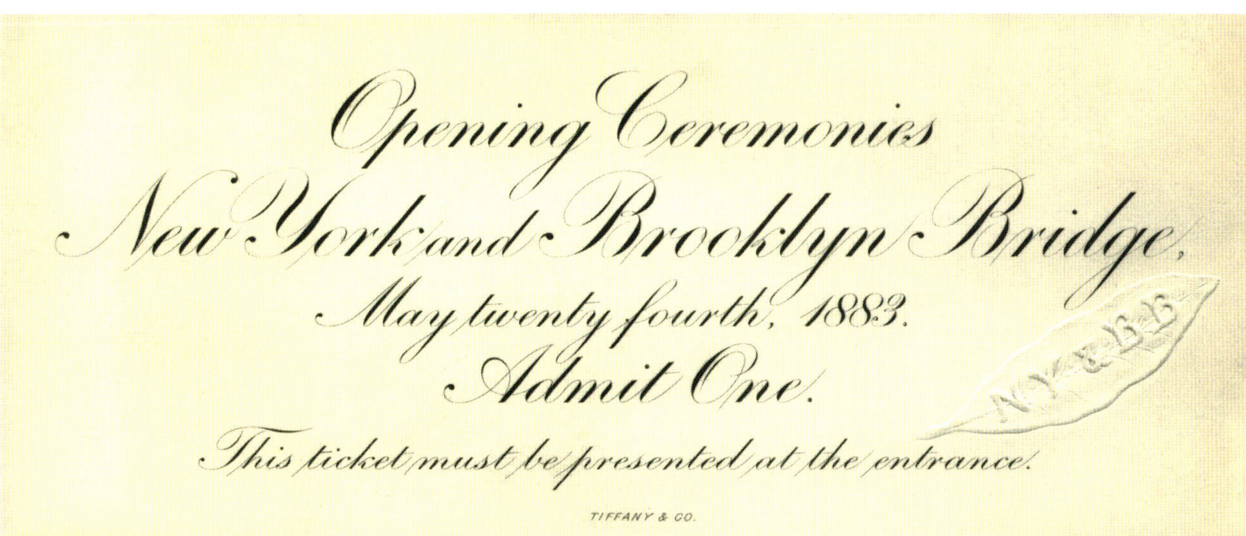

Opening Ceremonies

New York and Brooklyn Bridge,

May twenty fourth, 1883.

Admit One.

This ticket must be presented at the entrance.

N.Y. & B.B.

TIFFANY & CO.

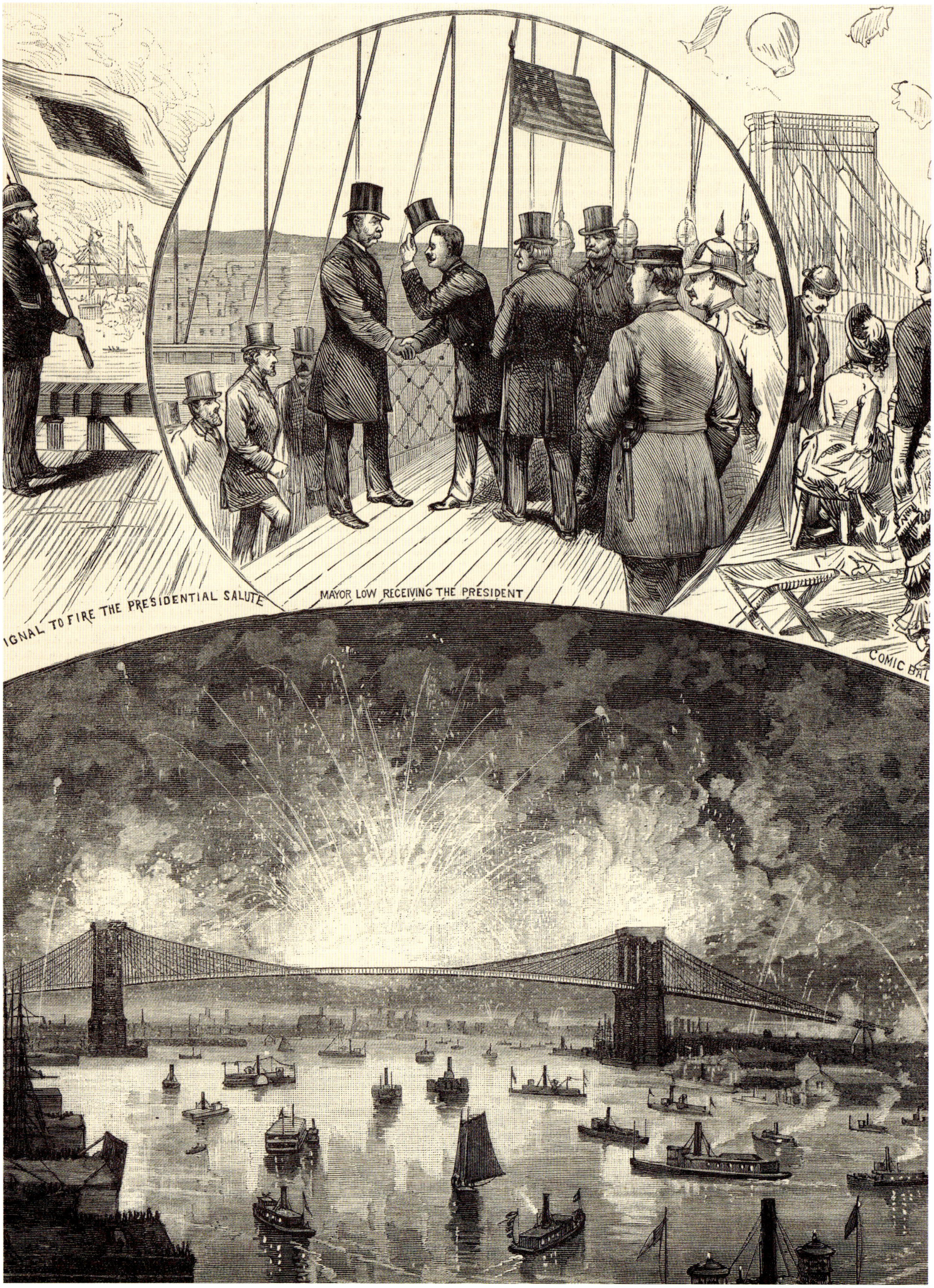

IGNAL TO FIRE THE PRESIDENTIAL SALUTE
MAYOR LOW RECEIVING THE PRESIDENT
COMIC BAL

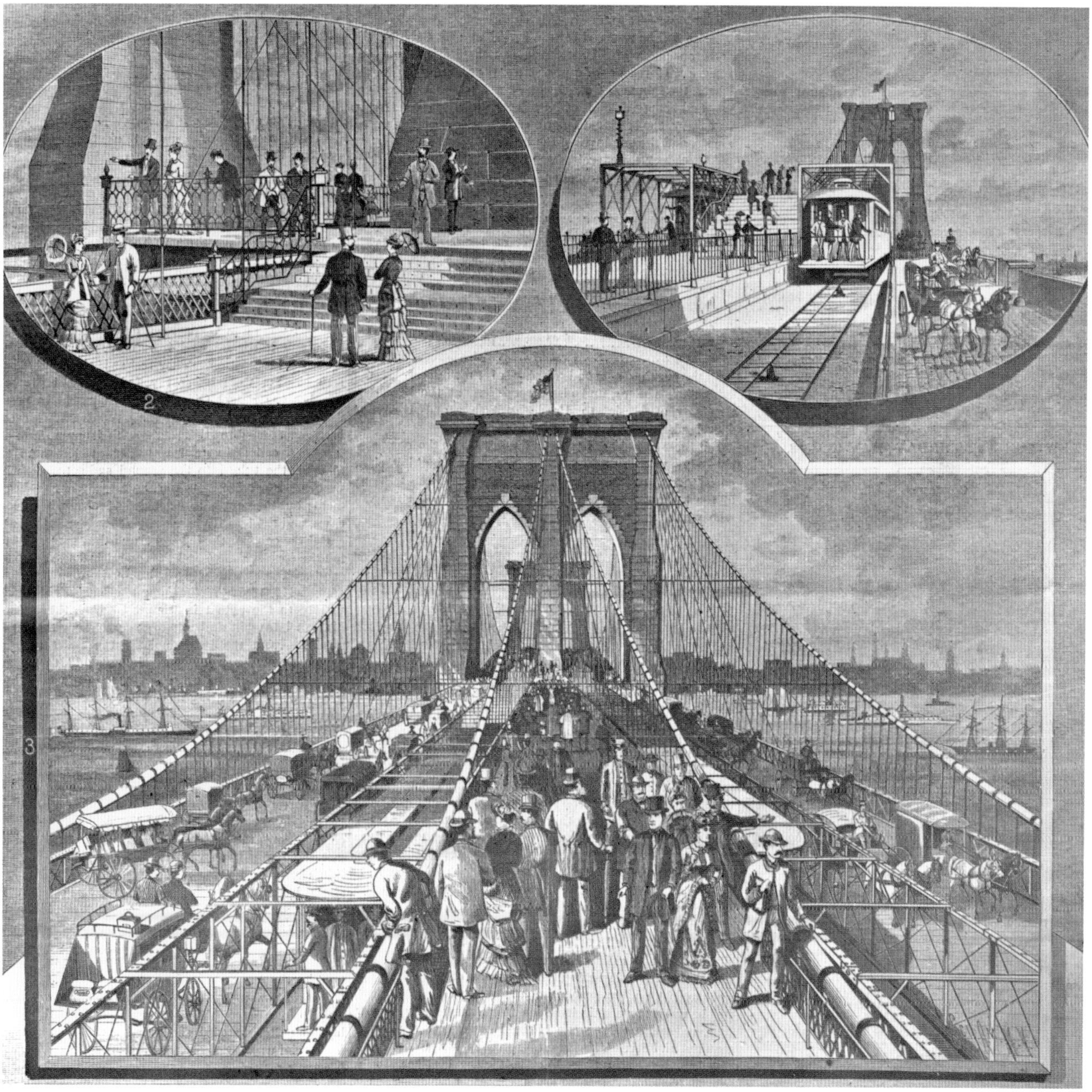

ABOVE: "Opening of the Great Suspension Bridge Between New York and Brooklyn, carriageway, railway, and promenade," *Scientific American*, May 26, 1883. The bottom view is from the Brooklyn side, looking toward Manhattan. The railway to either side of the promenade was not in operation when this was published. Note that the vehicular traffic is drawn incorrectly—the directions are opposite to the reality.

OPPOSITE: Top, left to right: "The Signal to Fire the Presidential Salute," "Mayor Low Receiving the President," and "Comic Balloons," *Frank Leslie's Illustrated Newspaper*, June 2, 1883.

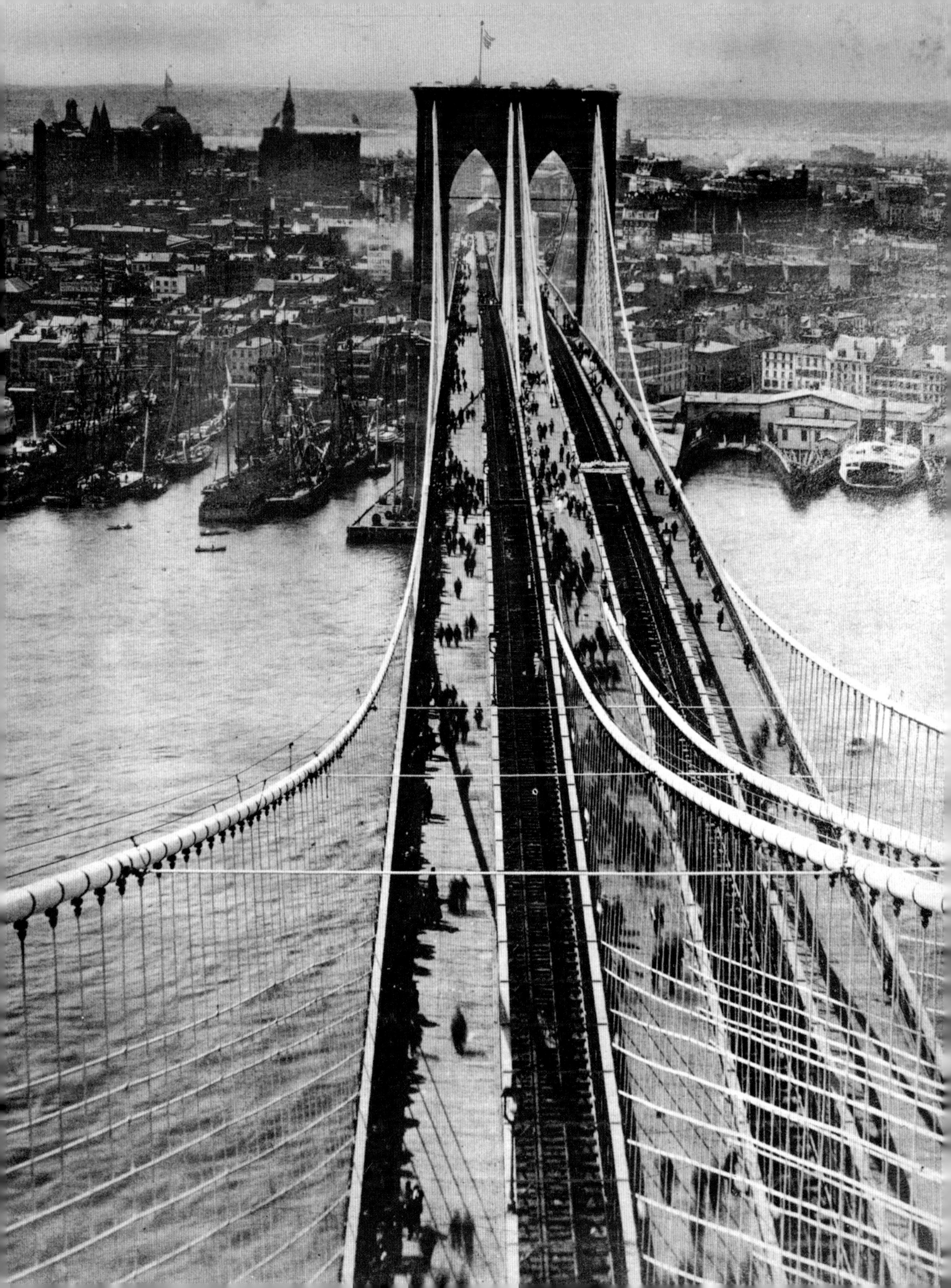

A token issued in 1889—six years after the bridge opened—to commemorate the hundredth anniversary of the inauguration of President George Washington. It shows an image of the Brooklyn Bridge with the label "The Eighth Wonder," implying "of the world." This commemorative token demonstrates how quickly the Brooklyn Bridge became an American icon. Author's collection.

On June 2, 1883, a week after the Brooklyn Bridge's opening, *Harper's Weekly* expressed its view of the appropriateness of the grand opening:

> But the bridge, as a work of engineering, is so distinct an advance upon anything that has been done before in the same direction, and its influence will be so important upon the relations of the communities which it unites, that it is worthy of the great pageant with which it was opened to the public, and the exultant oratory in which its completion is celebrated.

An editor at *Scientific American* was greatly impressed by the new structure, describing the bridge as a "marvel of beauty viewed from the level of the river. In looking at its vast stretch, not only over the river between the towers, but over the inhabited, busy city shore, it appears to have a character of its own far above the drudgeries and exactions of the lower business levels."

Montgomery Schuyler, one of the leading critics at the time of the opening, wrote in *Harper's Weekly* that the bridge was not impressive as a work of art, but nevertheless that it was "one of the greatest and most characteristic" projects of the nineteenth century. He continued: "It so happens that the work which is likely to be our most durable monument, and to convey some knowledge of us to the most remote posterity, is a work of bare utility; not a shrine, not a fortress, not a palace, but a bridge."

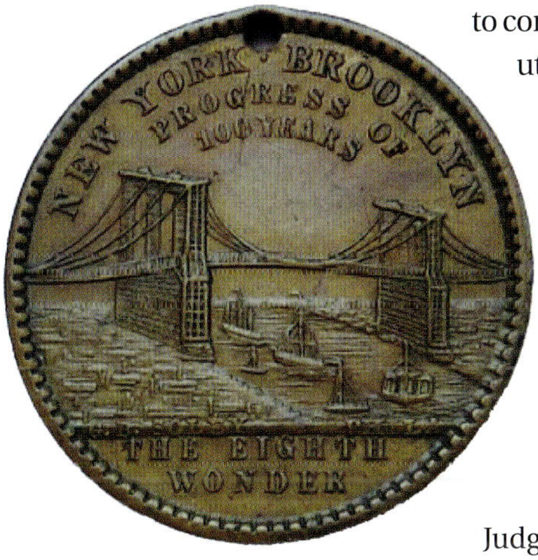

In *The Great Bridge,* David McCullough lauded the importance of the Brooklyn Bridge: "It can be seen as the beginning of modern New York—of monumental scale, of structural steel—or the end of old Brooklyn. It is all solidity and permanence, the very reverse of rootlessness. It says, perhaps, as does nothing else built by Americans before or since, that we had come to stay."

And Master Mechanic E. F. Farrington gazed into the future and accurately predicted that Brooklyn, because of the bridge, would now become a bedroom community for New York City:

> Judging from the past, probabilities point to an early settlement of the vast area lying east of the present city limits of Brooklyn; and there is no reason to suppose that the relative present condition of New York and Brooklyn will be essentially changed. New York will remain as now, the great mart of trade, commerce and manufactures; while Brooklyn will expand into a charming collection of comfortable homes. Then the use and purpose of the great bridge will be developed.

OPPOSITE: This image was published in 1983 in Mary J. Shapiro's *A Picture History of the Brooklyn Bridge,* with the caption: "Opening Day, May 24, 1883." Given that the carriageways (the outside lanes on the bridge) were open to pedestrians only on May 24, it is very likely that this image does indeed capture the first day of public access to the bridge. Though David McCullough, in *The Great Bridge,* reports that, on that day, "[u]p on the New York tower a lone photographer was busily at work under a black hood," this photograph was taken from the Brooklyn tower, looking toward Manhattan, as the easily identifiable buildings in the distance demonstrate. In any event, the enterprising photographer who lugged his camera to the top of the bridge tower captured history that day.

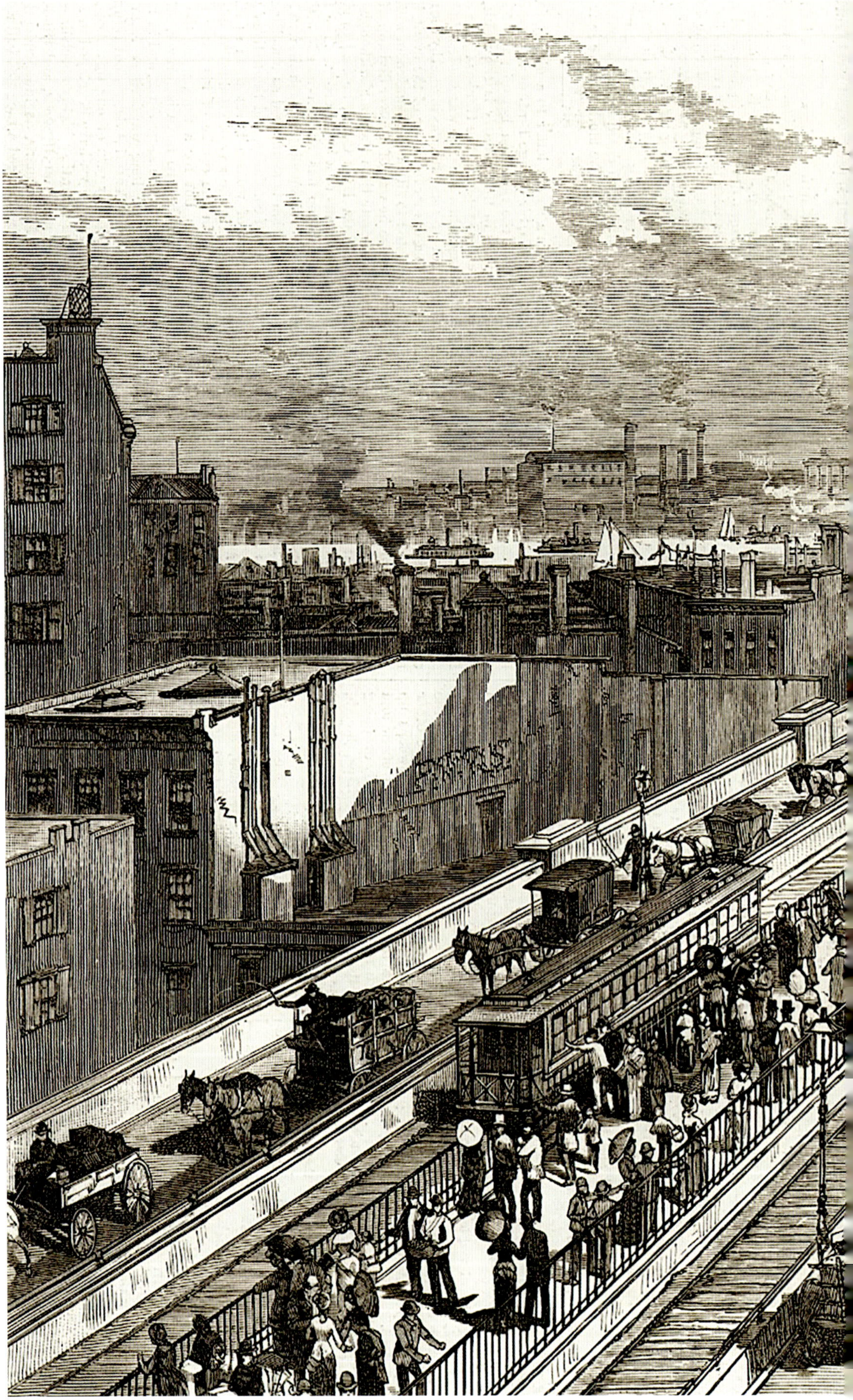

"The New York and Brooklyn Suspension Bridge—The Approach on the New York Side," *Harper's Weekly*, May 26, 1883. This woodcut envisions the bridge as fully operational, with its two outer sections for wagons and carriages, two tracks devoted to a railway carrying people from one end of the bridge to the other (which in fact did not open until September 1883), and the elevated pedestrian promenade in the middle. Author's Collection.

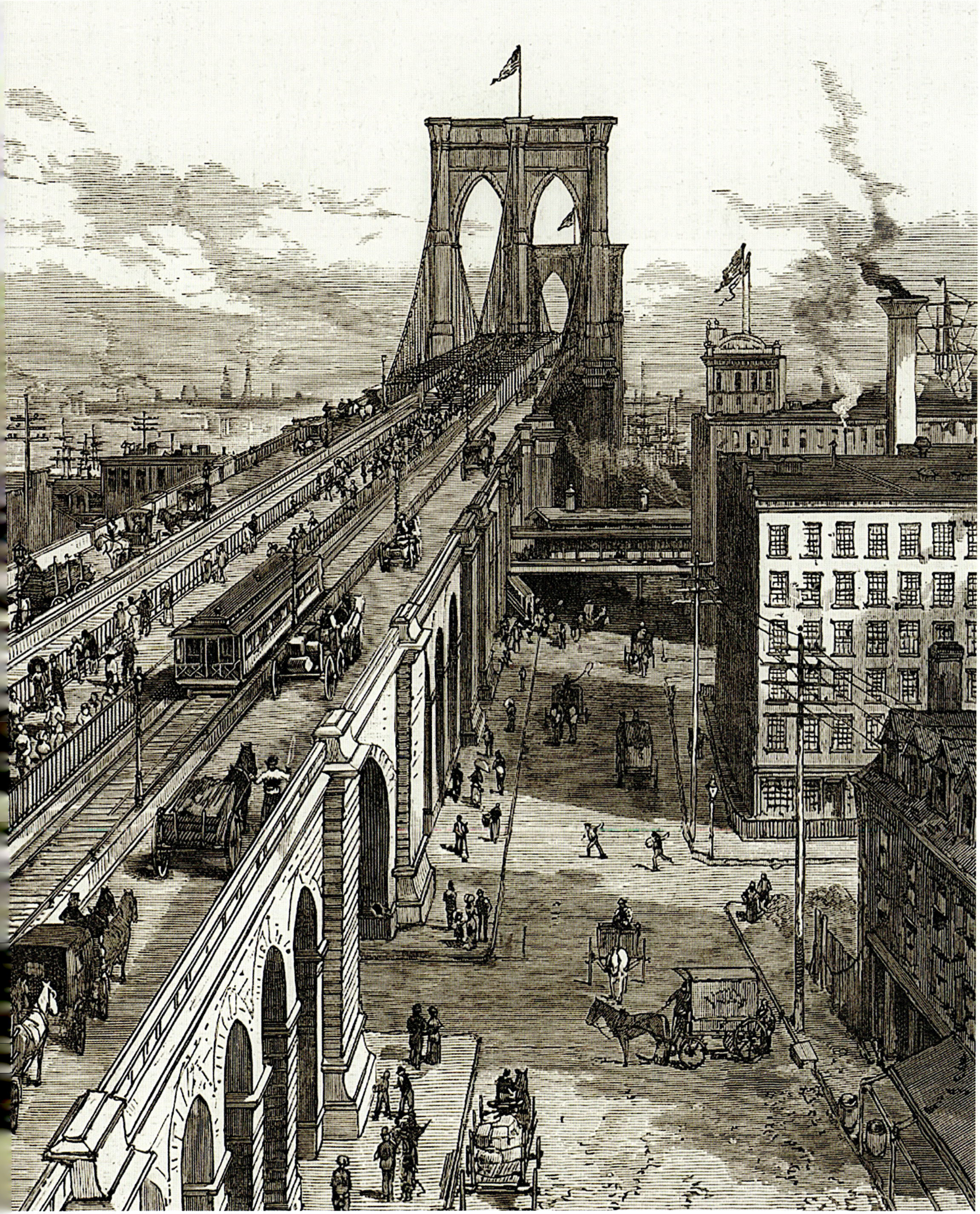

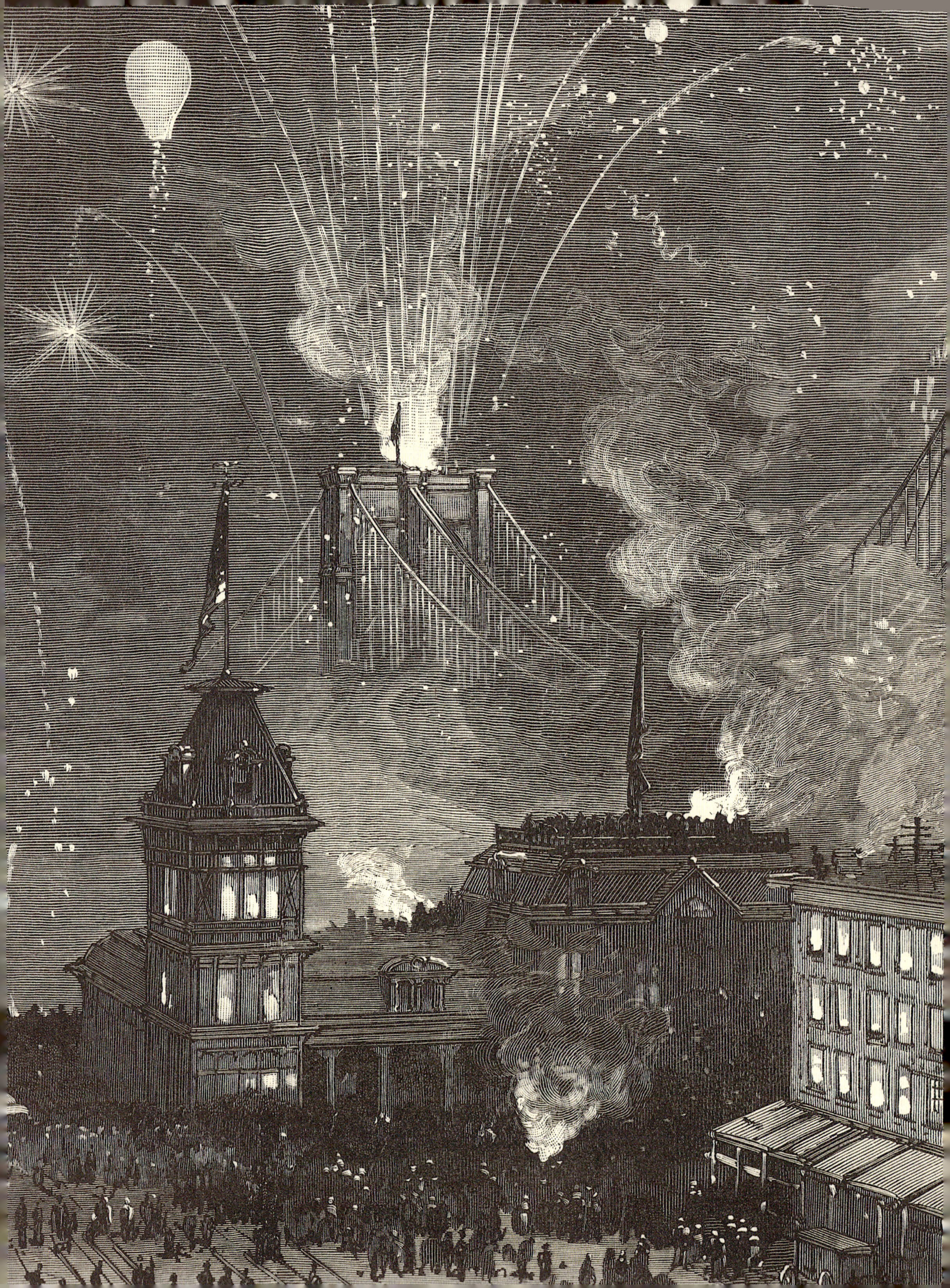

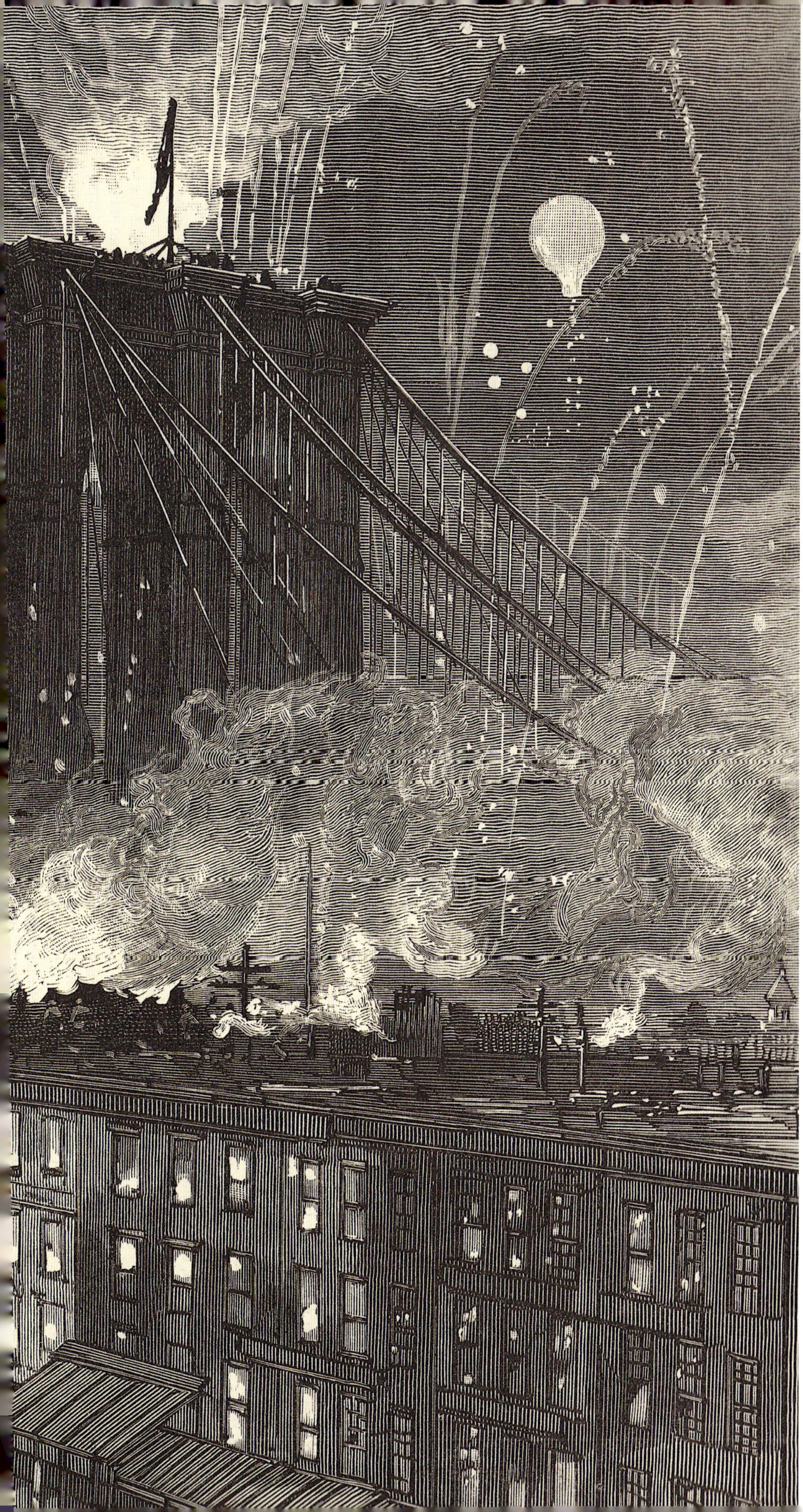

"The Great Bridge—Fireworks and Illumination, From the Brooklyn Side—Drawn by Charles Graham," *Harper's Weekly*, June 2, 1883. The bridge on opening night, May 24, 1883. Note the balloons floating at left and right. They were used to launch some of the fireworks. Large crowds gathered on the streets and rooftops to join in the celebration. Author's collection.

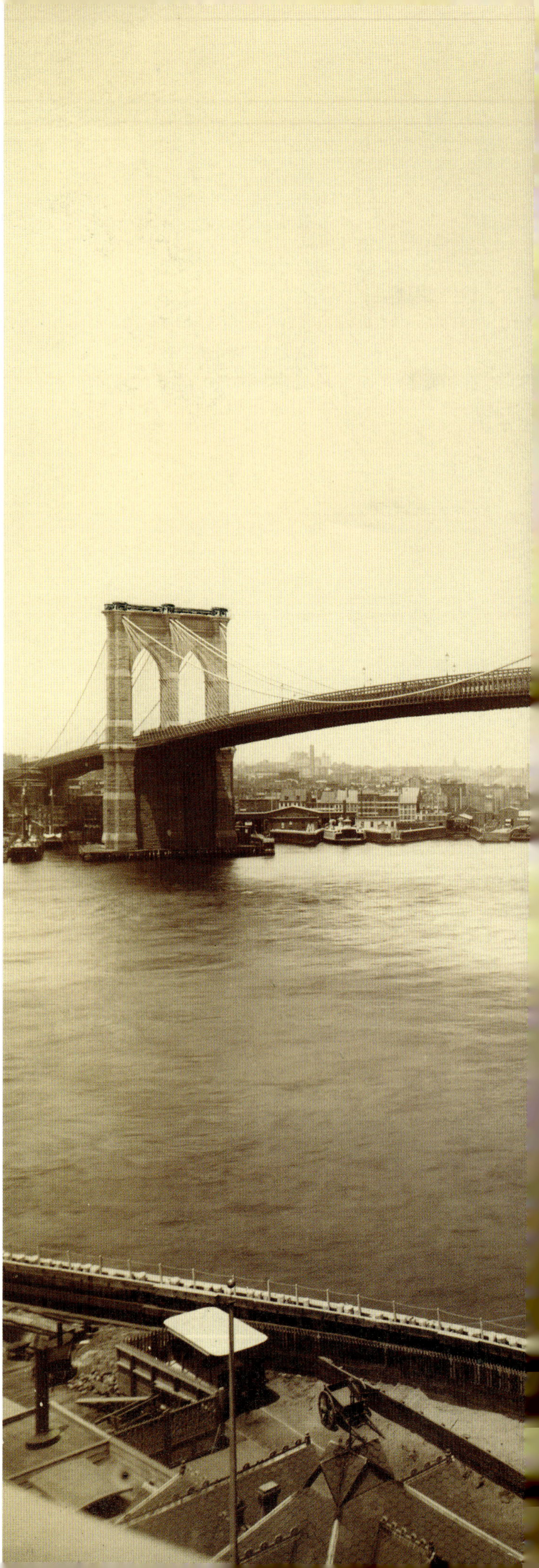

Photograph of the Brooklyn Bridge, looking northwest from Brooklyn, ca. 1900. Note the ferry boats in the foreground. RPI.

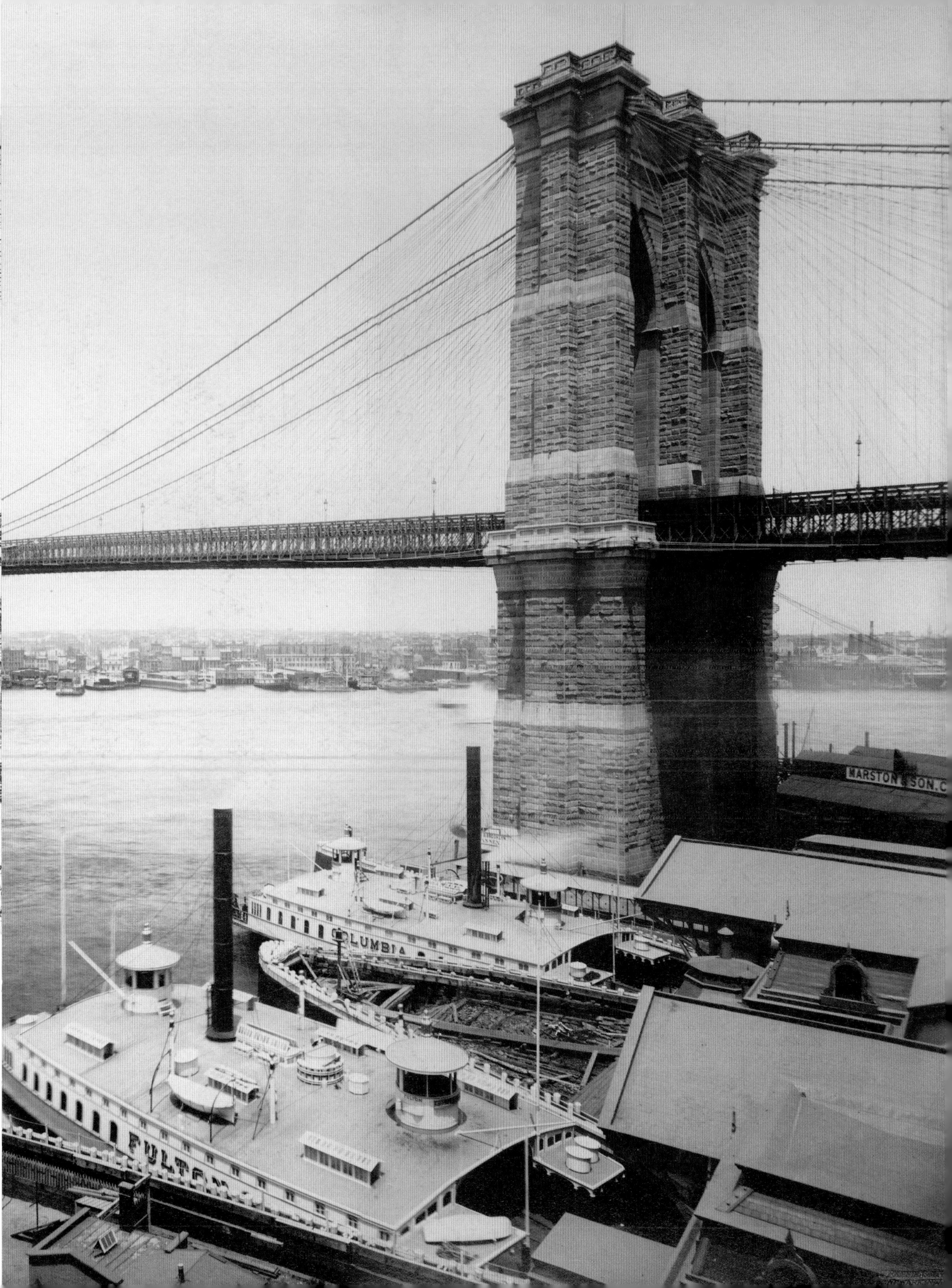
MARSTON & SON. C
COLUMBIA

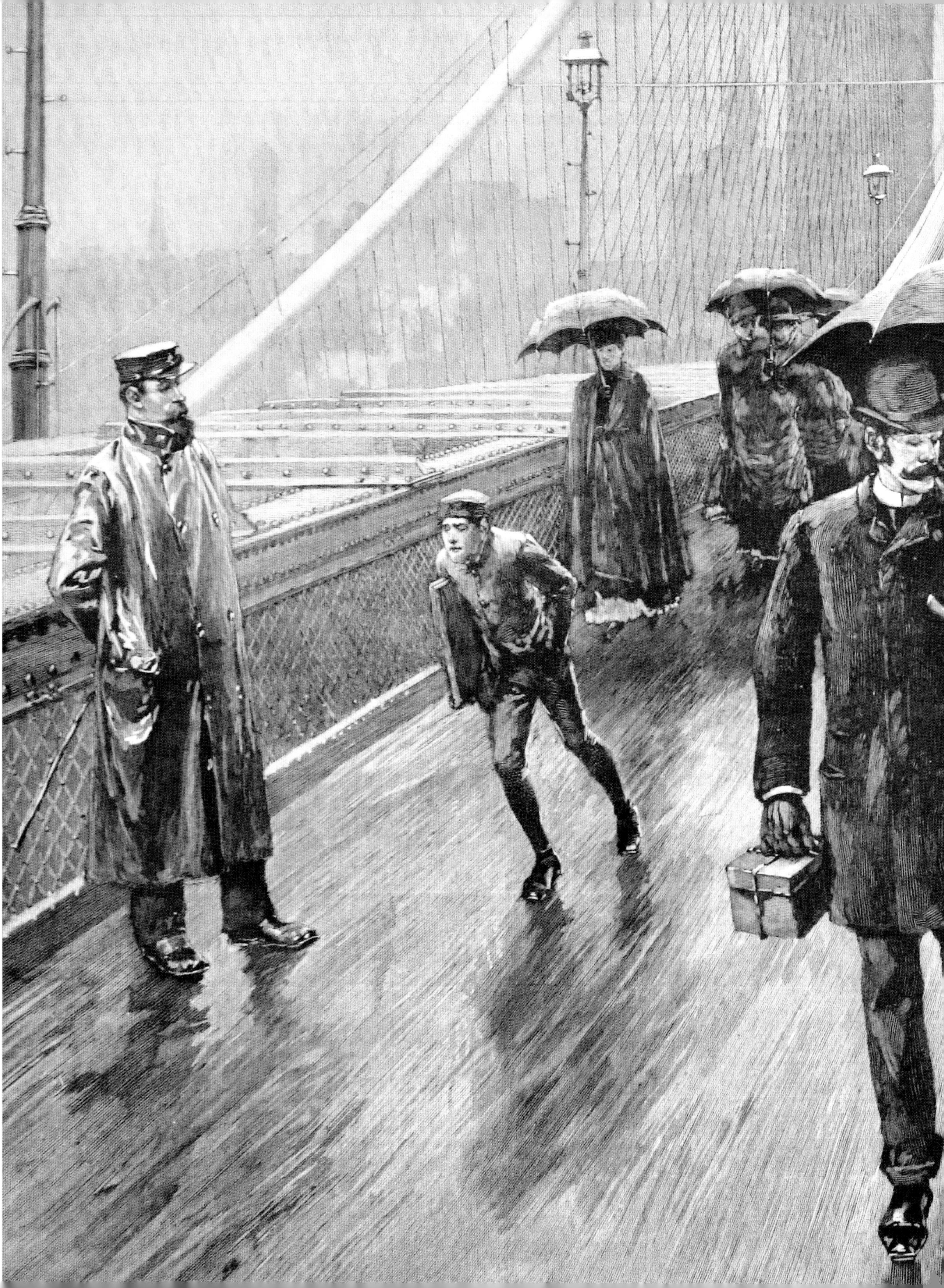

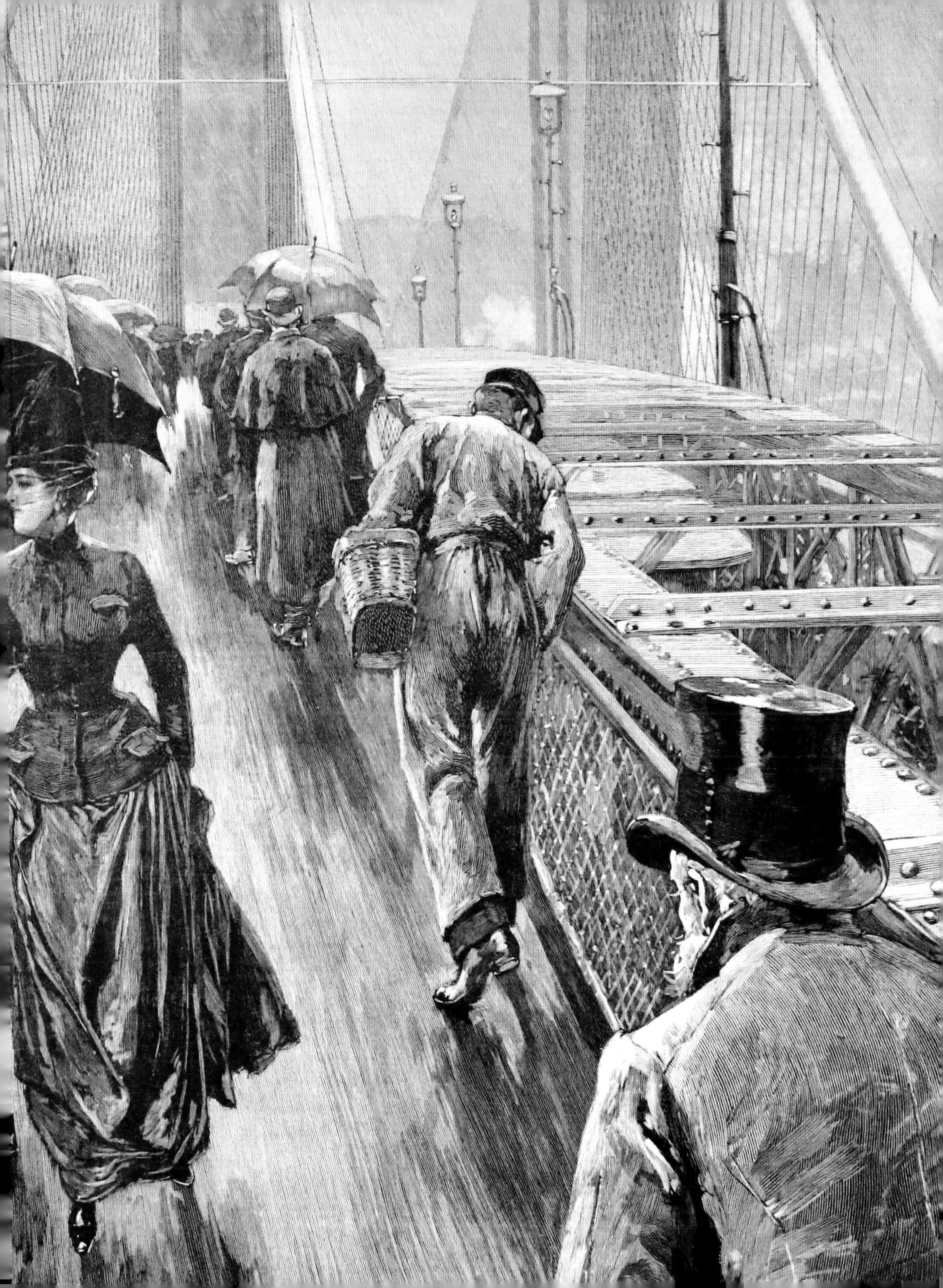

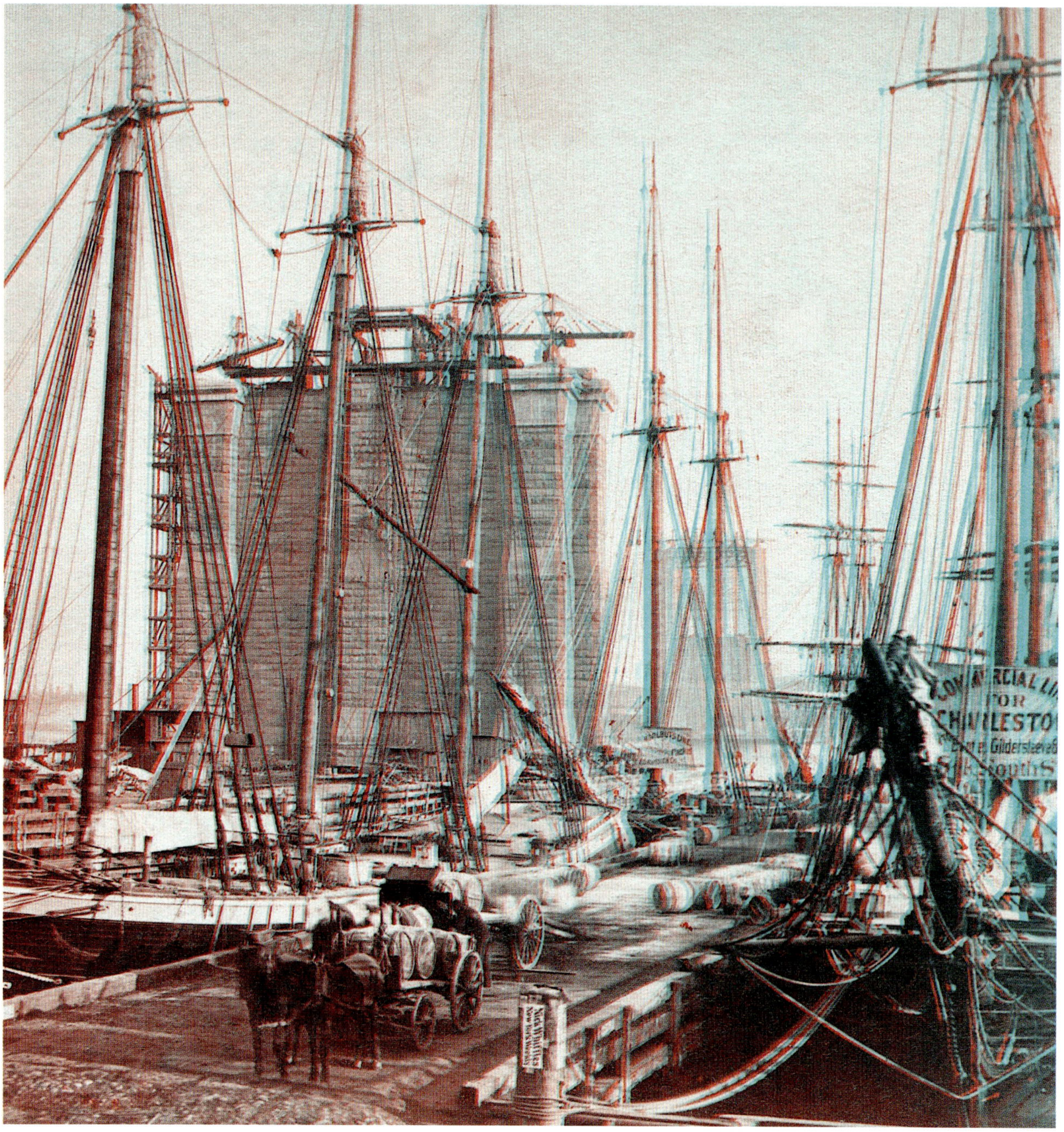

3D Anaglyph of the New York docks and the two bridge towers, derived from a stereoview by Gustavus W. Pach, ca. 1875, the year the Brooklyn tower in the distance was completed. Courtesy of the Michael E. Ach Collection.

Appendix: Photographers

The photographs in this book, supplemented by original drawings and old woodcuts, provide an extraordinary record of the construction of the Brooklyn Bridge. They also give us insight into the practice of photography from 1869, when work on the bridge began, to 1883, when it opened. Three men who greatly contributed to this photographic record are known to us: Silas A. Holmes, Robert B. Talfor, and Gustavus W. Pach. However, a number of these images have no identifying mark, and the photographers who created them remain unknown.

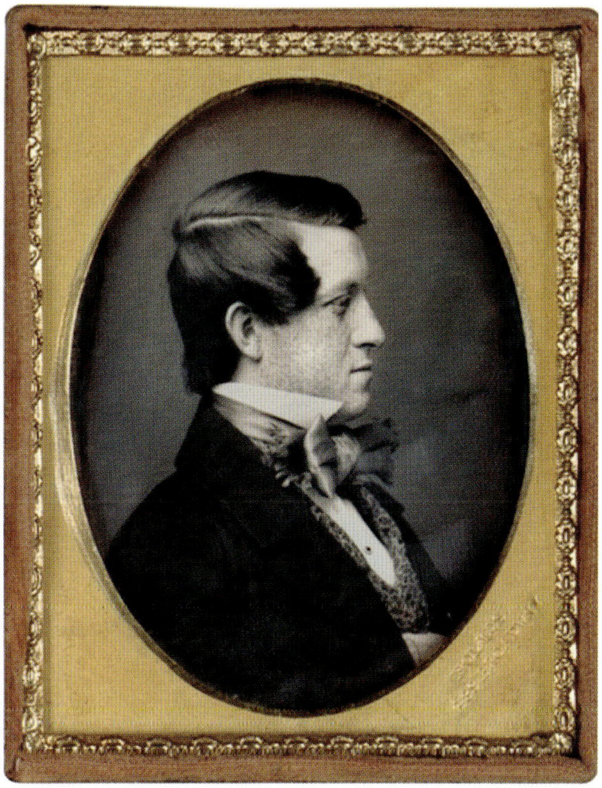

Self-portrait (a ninth-plate daguerreotype) of photographer Silas A. Holmes, ca. 1854.

SILAS A. HOLMES

Most of the construction photographs that can be attributed to a specific photographer were taken by Silas A. Holmes. Some of the images are marked with his ink stamp. Others, though unmarked, are mounted like the photographs that bear his stamp. Holmes appears to have worked specifically for the Bridge Company on some, and perhaps all, of these photographs. Three images in the Roebling Collection at Rensselaer Polytechnic Institute, taken by Holmes and bearing his stamp, are of original drawings of the bridge. Ostensibly these photographic reproductions were paid for by the Bridge Company so that assistant engineers and other staff would have copies of the drawings necessary for their work.

Holmes was born in 1820. As early as 1848, he is listed in Manhattan directories as a maker of daguerreotypes. His 1851 advertisement in the *New-York Daily Tribune* touted the two skylights illuminating his studio, his long experience, and his moderate prices. Later that year, in the same newspaper, he informed the public that he had sold forty thousand daguerreotypes—an extraordinary number, considering that they could not be mass-produced. He sold his work at his fourth-floor "Depot of Art" at 289 Broadway, at the corner of Reade Street, just one block north of City Hall Park in Manhattan. *Craig's Daguerreian Registry*, a respected directory of American photographers working in the 1860s, last revised by John S. Craig in 2003, reports that Holmes's business was located there from 1848 to 1859, before it moved to 691 Broadway, between Third and Fourth Streets, where it remained through 1860.

Holmes was also a pioneer of new photographic techniques. On May 30, 1854, he was granted patent number 10,987 for a device to take stereoscopic (as well as non-stereoscopic) daguerreotypes. It was referred to in Holmes's advertisements as "the Double Camera for taking two portraits at once." The device was an important photographic development. It eliminated the need to slightly move a camera to get the two different angles of view necessary to create a three-dimensional (or stereoscopic) effect, and it also made the photographic process more efficient by allowing for the capture of two images with just one exposure

A paper sleeve (below) inside which Silas A. Holmes's ferrotype photograph (right) was sold.

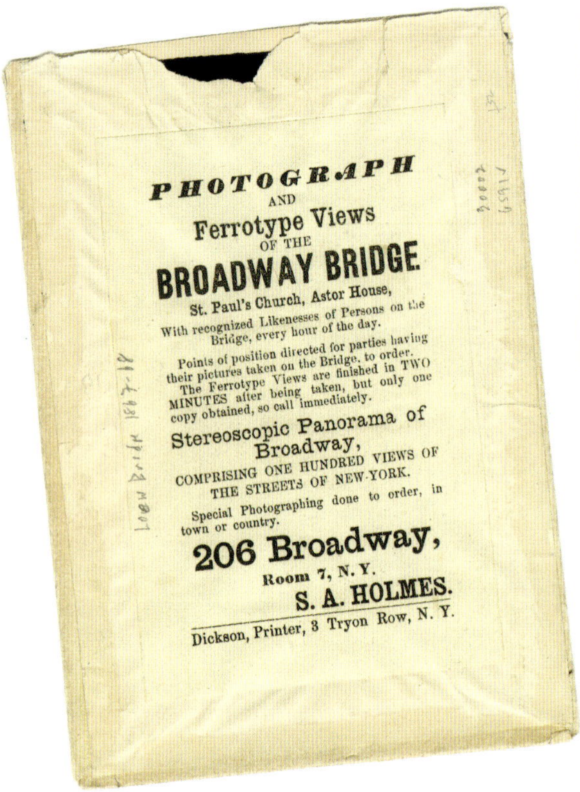

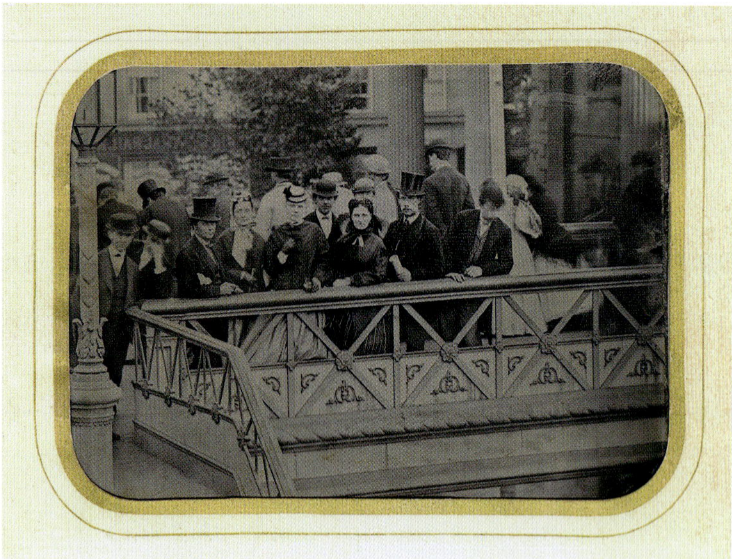

Above is one of the many photographs that Holmes took of people on the Loew Bridge, the short-lived (1867–1868) pedestrian walkway crossing over lower Broadway. Note that the ferrotype process produces a reversed image: the columns of St. Paul's Chapel, on the west side of Broadway, as seen here, would seem to indicate that this image was shot facing south, when, in fact, it was taken from the fourth-floor window of Holmes's studio, looking northwest. Ferrotypes, also called tintypes or melainotypes, were created inexpensively by coating a thin sheet of iron with chemicals that, when exposed to light, quickly create a lasting image. Both courtesy of the Jeffrey Kraus Collection.

(even when a stereoscopic effect was not desired). As described in *Wilson's Cyclopaedic Photography*, Holmes also pioneered the lamprotype, a photograph created by buffing an ambrotype (an early type of photograph that captured an image on a glass plate) in order to "take off the brown, somber look of the plate and give an intense, soft and clear texture to the picture."

Throughout the years of construction of the Brooklyn Bridge, Holmes made his living as a photographer in Brooklyn and Manhattan. According to the federal censuses of 1860 and 1870, he lived in Brooklyn's Eleventh Ward with his wife, Ann, and their two young daughters, Isabella and Florence. He continued to work as a daguerreotypist, still using the earliest of photographic techniques even after they became archaic. By 1878, the New York City Directory lists his studio at 526 Broadway, near Bleecker Street in Manhattan, and reports that he was living on Long Island. As of 1882, he had moved to 2098 Atlantic Avenue, between Howard and Saratoga Avenues, in the Brownsville section of Brooklyn. Per the 1884 New York City Directory, published the year after the Brooklyn Bridge was completed, Holmes was still working as a photographer.

Just before construction on the bridge began, Holmes had created a niche for himself in the photography business by offering to take portraits of pedestrians crossing the short-lived (1867–1868) Loew Bridge in Manhattan. The bridge—one of the wonders of New York City at the time—was a pedestrian walkway over Broadway at Fulton Street, just south of Holmes's photography studio at 289 Broadway.

3D

OPPOSITE: A photograph from *Holmes' Stereoscopic Panorama*, ca. 1860, looking south on Broadway, at Eleventh Street in Manhattan. At right is the St. Denis Hotel, where in 1877 Alexander Graham Bell demonstrated his great invention: the telephone. Grace Church was just to the left of this scene. Author's collection.

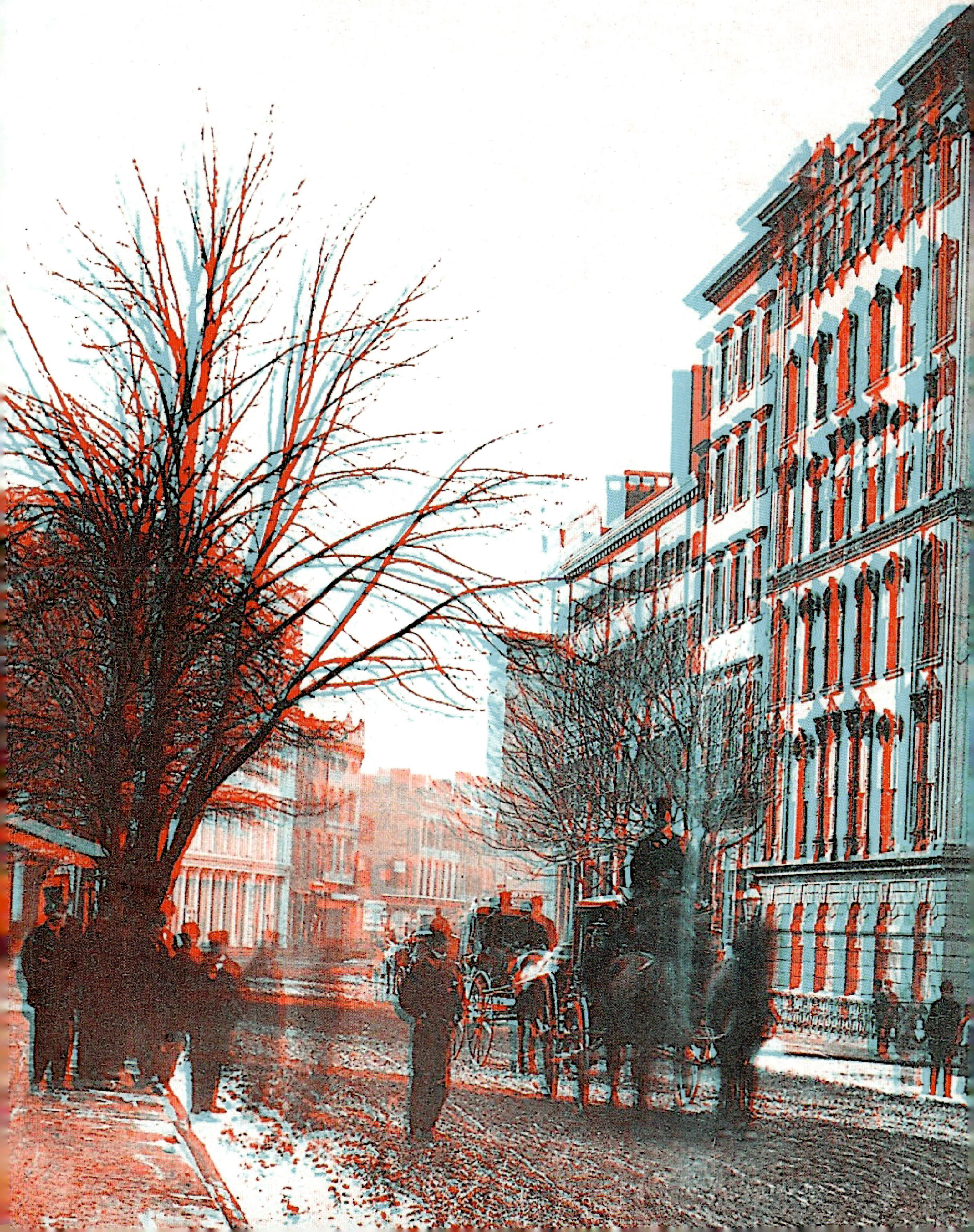

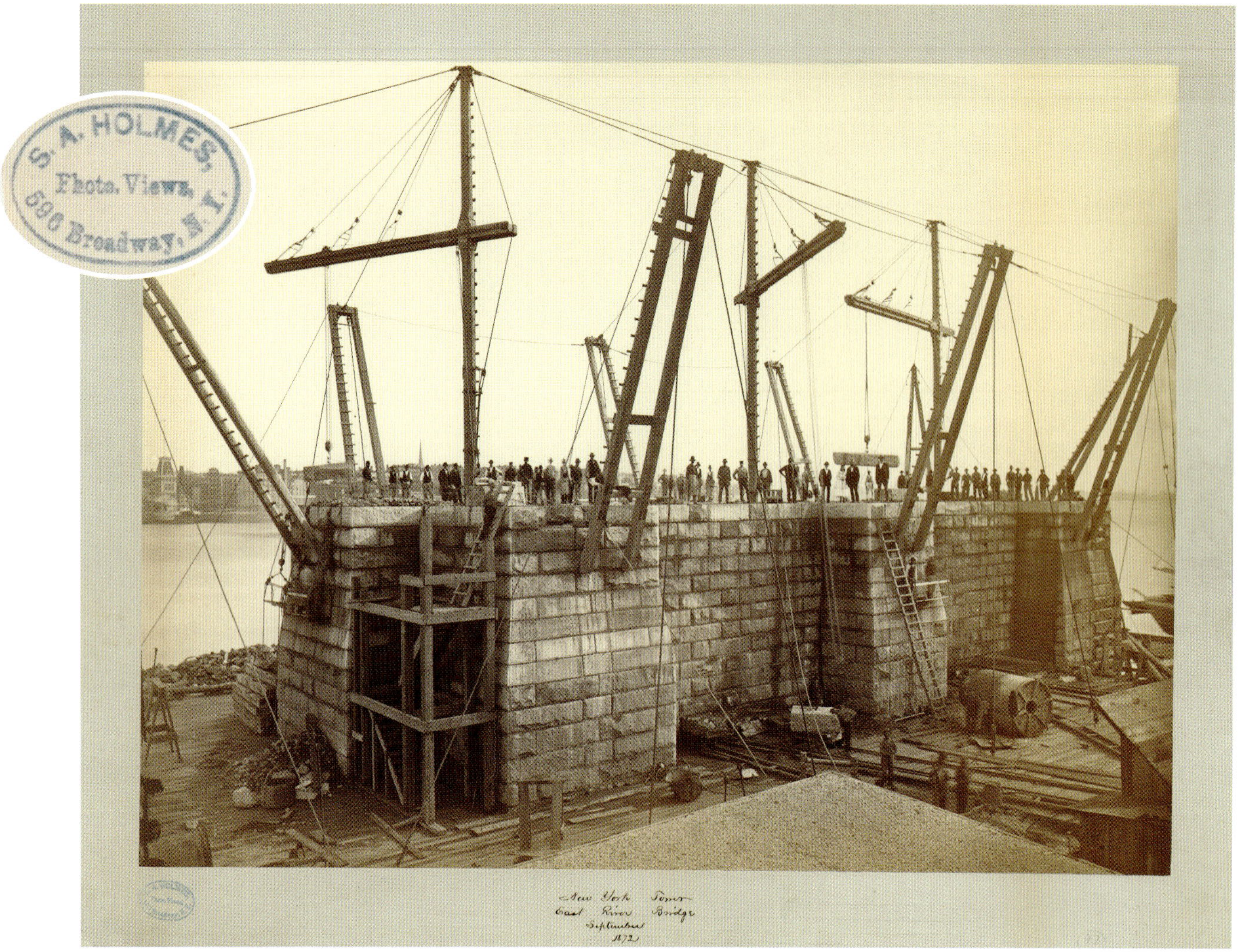

"View of Inner Upper Angle of New York Pier/39th Course of Masonry Completed/Sept. 21st. 1872." A further notation: "Height above tide: 32 [feet] 6 [inches]." The mount bears the stamp of photographer Silas A. Holmes, enlarged at upper left, and identifies the photograph as his. That same stamp appears on the mounts of several of the Brooklyn Bridge photos in the collections of RPI and MCNY. RPI.

From the late 1830s into the 1860s, Broadway near City Hall was New York City's fashionable shopping area. Business was so good, and Broadway so jammed, that one observer reported that it took "more skill to cross Broadway than to cross the Atlantic in a clamboat." John Genin's hat shop was on the east side of Broadway, between Fulton and Ann Streets. His competitor, a man named Knox, had his store on the more expensive west side of the street, where the shop windows were lit by the morning sun. Genin lobbied for a pedestrian bridge across Broadway because he wanted people to easily cross over the busy thoroughfare to patronize his store. His campaign was successful, and the walkway was built. Not surprisingly, Knox was unhappy. He, in turn, convinced local authorities to tear down the bridge, just a year after it was built.

The pedestrian bridge is now a little-remembered footnote to New York City's history, but during the time that it arched over Broadway, it provided a great business opportunity for Silas Holmes. The Loew Bridge was less than a block north of his studio, and in its fourth-floor window Holmes positioned a camera to capture "recognized Likenesses of Persons on the Bridge, every hour of the day."

Holmes was also actively photographing the streets and buildings of lower Manhattan. He created an ambitious photographic record of what he described as "the Beauties and Grandeur of Broadway . . . from Union Square to the Battery." Comprising over one hundred views "connecting every block and building," *Holmes' Stereoscopic Panorama* was sold from his retail space, the American Museum Picture Gallery.

Around the same time, Holmes advertised that he was available for "special photographing done to order, in town or country." It appears that someone involved in the building of the Brooklyn Bridge hired him to record its construction. Indeed, the details of Holmes's photos demonstrate that he was more than someone who had appeared at the construction site with his camera and tripod and just started shooting. Rather, he appears to have held the status of the official—or at least favored—photographer of the bridge's construction. A number of the photos in this book show "ghosting," or blurring, of the workers, indicating that they were moving about doing their jobs, oblivious to a photographer's efforts to capture the scene. However, in several of Holmes's images, both the workers and supervisors posed for him, standing still for the required long exposure. Thus it appears that Holmes was given the luxury of carefully composing his photographs, orchestrating the men within the frame so that they were fully visible and neatly arranged before he exposed the image. At Holmes's direction, it seems, the workers might have hoisted a granite block here and there—even left one hanging in midair—making a photograph that much more dramatic.

ROBERT B. TALFOR

Born about 1842 in England, Robert B. Talfor immigrated to the United States with his family when he was four years old. On May 14, 1861, with the Civil War having begun just a month earlier, Talfor, single and a clerk, enlisted at Brooklyn's Seventeenth Ward in the famed 5th New York Infantry, also known as Duryée's Zouaves, under the command of Colonel Abram Duryée. On June 10, 1861, while on picket duty at Fortress Monroe, Virginia, Talfor wrote a letter to a friend detailing the death of a comrade killed in action at the Battle of Big Bethel. Talfor, on duty for three consecutive days without sleep, described himself as "one who will fight for the Union."

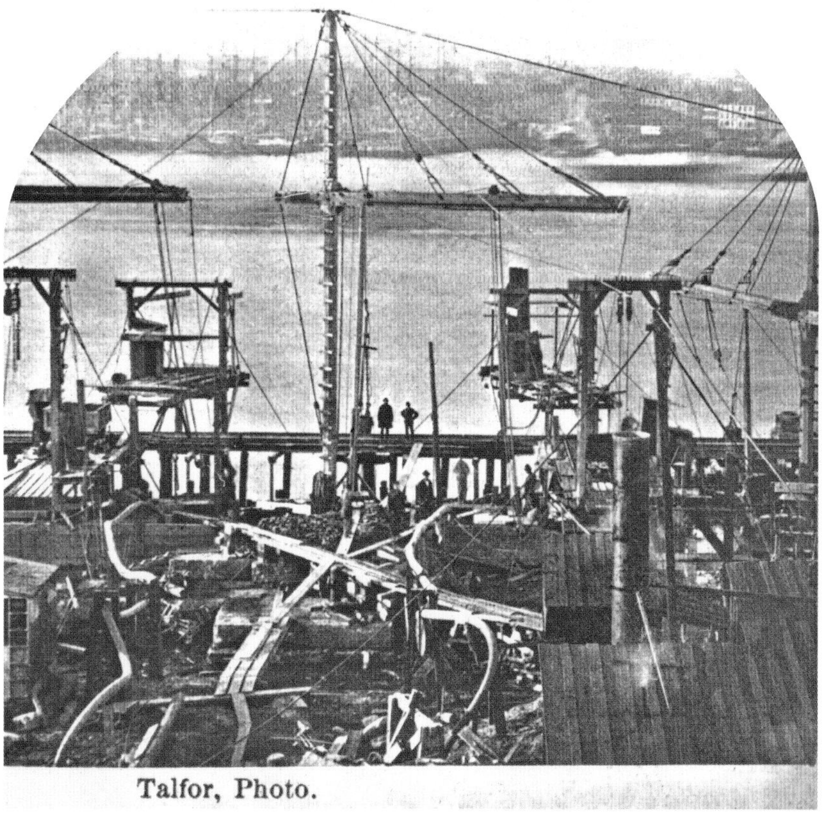

Work in progress on the Brooklyn caisson, likely May or June 1870, photographed by Robert Talfor. It is so dated because no stones for the tower have yet been laid atop the caisson. This print is one half of a stereoview that was captured on a lantern slide. The wooden frames for the clamshell buckets, with the pulleys hanging down from the crosspieces of each frame, are at right and far left. Author's collection.

After serving his full two-year enlistment, part of which he spent as a surveyor, he mustered out on May 14, 1863. In 1865, as the Civil War was ending, Talfor was working for the United States Army as a clerk and living back in Brooklyn's Seventeenth Ward with his parents. By then, he was a naturalized citizen. Just one year later, in 1866, Talfor launched his career as a photographer. According to the Internal Revenue Service assessment lists, he was working in the Greenpoint section of Brooklyn, his studio at the intersection of Greenpoint and Washington Avenues. That year, he was assessed a tax of $15.40 on a business valuation of $308.

In 1870, Talfor photographed the work in progress atop the Brooklyn caisson, as the Brooklyn tower was about to rise above it. (See pages 80–81.) Then, in 1871, he photographed the New York caisson just before its May launch. (See pages 84–85.) For each of these images, Talfor proudly had his name printed on the mat.

In the 1870 federal census, Talfor is recorded as a photographer still living in Brooklyn's Seventeenth Ward with his parents. In 1871, the Brooklyn directory lists his place of work as West and Kent Streets in Greenpoint. Then, in 1873, Talfor visited Louisiana, where he produced the item for which he is best known: a hand-colored album, *Photographic Views of the Red River Raft*. Its fifteen images document the United States Army Corps of Engineers removing the "Great Raft," a blockage of logs along that waterway. In 2018, a copy of that album sold at auction for $93,750.

The 1875 Brooklyn Directory again records Talfor working as a photographer in the same ward, though by now he was married to a woman named Maria. But then, in the directory for 1877, his profession is listed as "engineer" for the first time, before reverting to "photographer" in the directories for 1879 and 1880, when Talfor was working in Texas.

In 1882, the directory for Houston, Texas, records Talfor as the "U.S. engineer in charge of improvements on Buffalo Bayou"—a usually slow-moving stream that was prone to flooding. In July 1885, Talfor was still serving as an assistant engineer in the Engineer Department of the United States War Department, assigned to improve Texas's rivers and harbors. That work continued into 1898. By the time of the 1900 census, he was living in Galveston, Texas. It is not known when he died or if he had continued to take photographs later in life.

GUSTAVUS W. PACH

Born in 1848, Gustavus W. Pach was a German immigrant who, with his brothers, was brought to New York City as a child by their parents. Pach was first listed as a photographer in the New York City Directory of 1866–1867. Due to health concerns, he moved his business to New Jersey for the cleaner air, but by the time the New York City Directory was published for 1871–1872, Pach was back in Manhattan.

During most of the years of the Brooklyn Bridge construction, Pach had his own photography studio (1871–1877 at 858 Broadway, then 1877–1881 at 841 Broadway). From 1881 through the bridge's completion in 1883, he continued to work at 841, but his brothers Godfrey (sometimes referred to as Gotthelf) and Oscar joined him as "G. W. Pach & Bros."

In the 1870s, apart from his own studio, Gustavus also ran photographic studios with his brothers at prominent colleges, including Yale, Harvard, West Point, Columbia, Vassar, and Princeton, where they

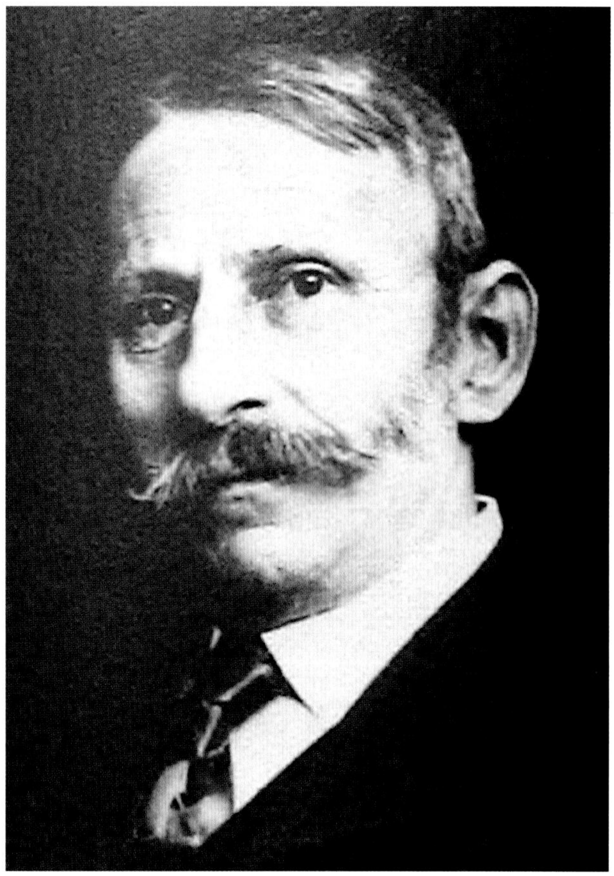

Photographer Gustavus W. Pach, ca. 1900.

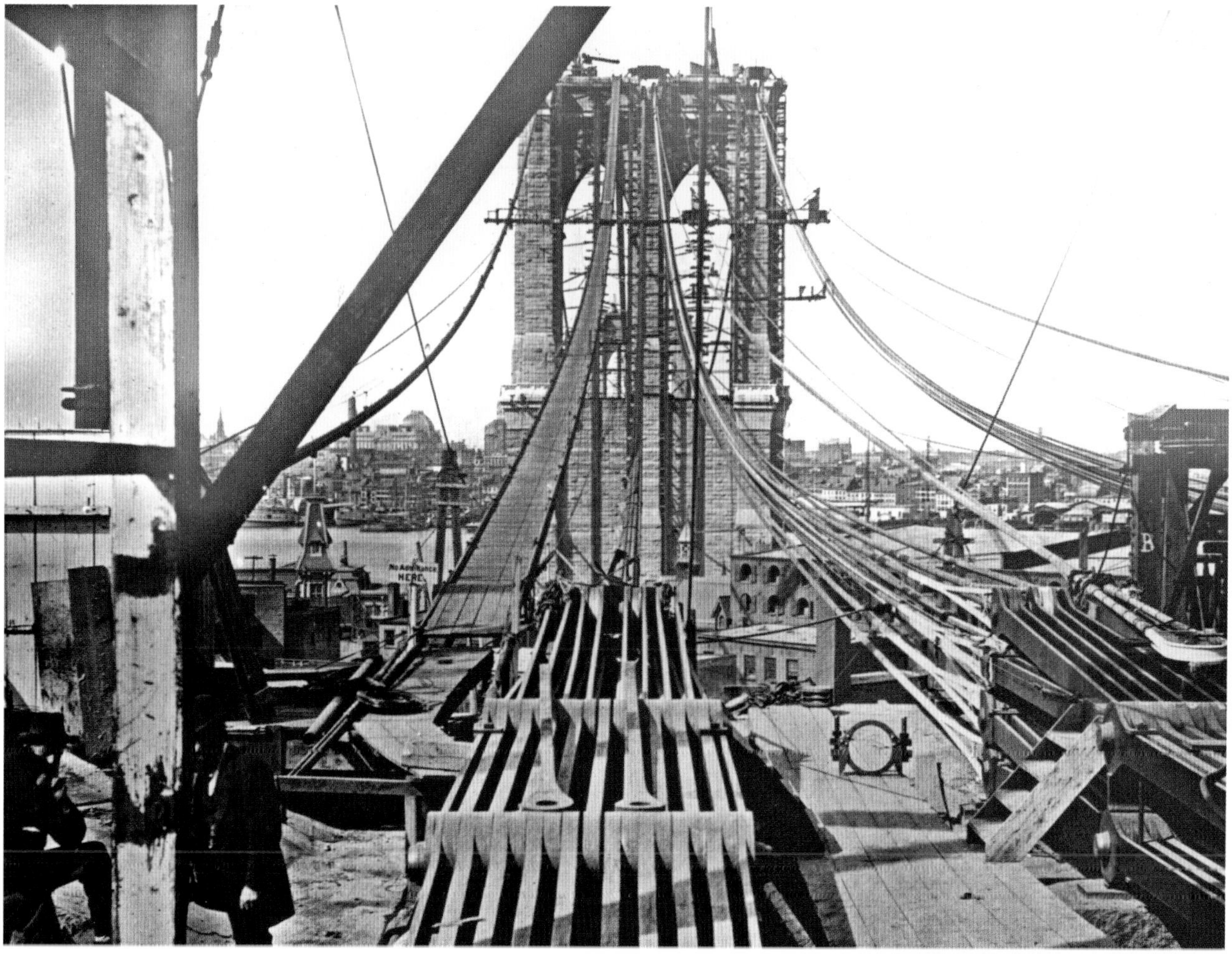

A lantern slide photograph taken by Gustavus W. Pach from atop the Brooklyn anchorage, looking west toward Manhattan, with the footbridge and the Brooklyn tower visible in the mid-distance. At left, the sign at the entrance to the footbridge reads, "No Admittance HERE." This image was taken in the same photo session as that of the trustees and engineers that appears in Chapter 1 (pages 32–33). Note that both photographs were shot from the same general location, though the camera has been moved a few feet to the left here. There are two men in the far left foreground. Assistant Engineer C. C. Martin stands second from left and is here wearing the same hat, suit, and watch chain that he wore in the other photograph. Written in ink on the mount of the photograph at the Museum of the City of New York: "Showing Footbridge-Anchor Bars (in part) and a Finished Strand of Cable ready to lower into position. Oct. 25th 1878." On the back of that same print is a note in pencil that appears to have been added well after this image was taken: "Photo by G. W. Pach, photographer's stamp was cut off edge of [mount] board." Author's collection.

took portraits of students, many of whom were athletes. They had Pach studios in Long Branch and Ocean Grove, New Jersey. The Pachs were hired by the Metropolitan Museum of Art to memorialize its collections. In 1876, the brothers exhibited their work at the Centennial Exposition in Philadelphia, a prestigious venue where the best in America displayed their wares. The Pach brothers also showed their photographs at annual professional gatherings, including the American Institute of the City of New York and the Photographers Association of America.

According to the 1880 census, the Pach family—parents and four sons—were living together in "Ocean," New Jersey. Augustus, Oscar, and Godfrey were listed as photographers, brother Barney as a baker. Their brother Morris, a cigar maker, lived nearby.

An article in *The New York Times* in 1895 gives us an idea of the size of the Pach Brothers' New York City operation. It reported that thirty employees and twenty patrons were present when a fire started at Pach Brothers' 935–937 Broadway studio. By then, the firm was widely known for its fine photographs of celebrities, politicians, actors, and events. In 1903, Gustavus Pach shifted all of his business to the Pach Brothers' New Jersey studios. He died the following year in that state. The Pach photographic empire remained in business for more than a century, thriving into the 1990s, when their studio in midtown Manhattan finally closed.

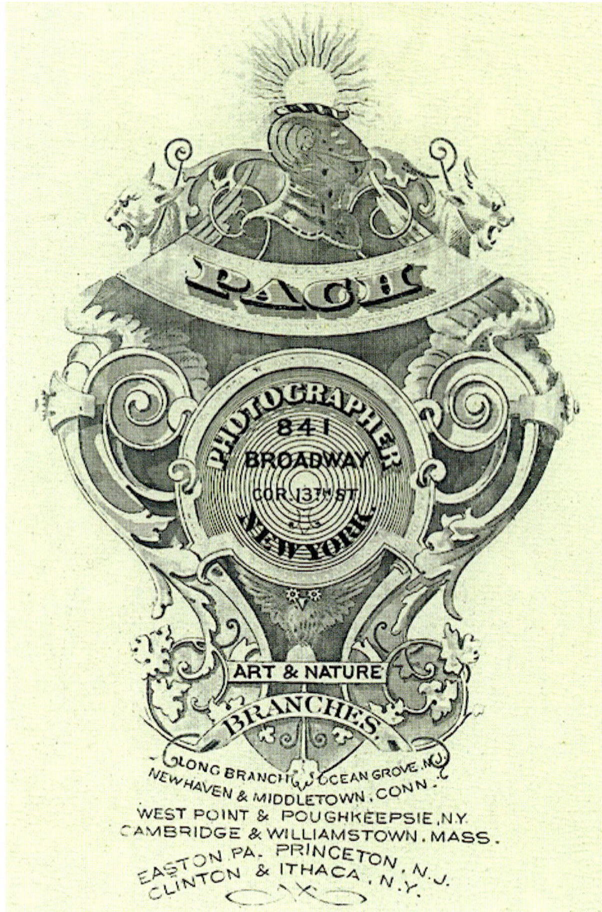

The back of a cabinet card that dates between 1877 and 1881, when Gustavus W. Pach's studio was located at 841 Broadway. Note the many branches of his very successful photography business.

The photographic record of the building of the Brooklyn Bridge is spectacular. This book was only possible because of these extraordinary images, and much of the credit for them goes to photographers Silas A. Holmes, Robert B. Talfor, and Gustavus Pach. But credit also goes to the anonymous artists responsible for many of the images that grace the pages of this book. Taken together, this rich record allows us to step back in time and witness the building of the Brooklyn Bridge, a historic feat of engineering, a true American icon, and one of the wonders of the world.

NOTES ON A FEW PHOTOGRAPHS

Around 1880, when the photograph below was taken, exposures typically had to be long, making it difficult to capture an unblurred image, to freeze movement. Both of the men in this shot were moving when the photographer exposed his glass plate. As a result, the man in the foreground at the far left appears particularly blurred, like a ghost.

The two images on the following pages—one a stereoview, the other a glass lantern slide photograph—help us understand the state of photography during the years 1869–1883, when the Brooklyn Bridge was being built. Photographic images in that era were being published in various formats: *cartes de visite*, mounted on cards the size of a "calling card;" cabinet cards, which featured larger photographs, about four by six inches; large mounted prints; glass lantern slides; and 3D stereoscopic views. As these two images demonstrate, the differences in format could affect what the viewer saw.

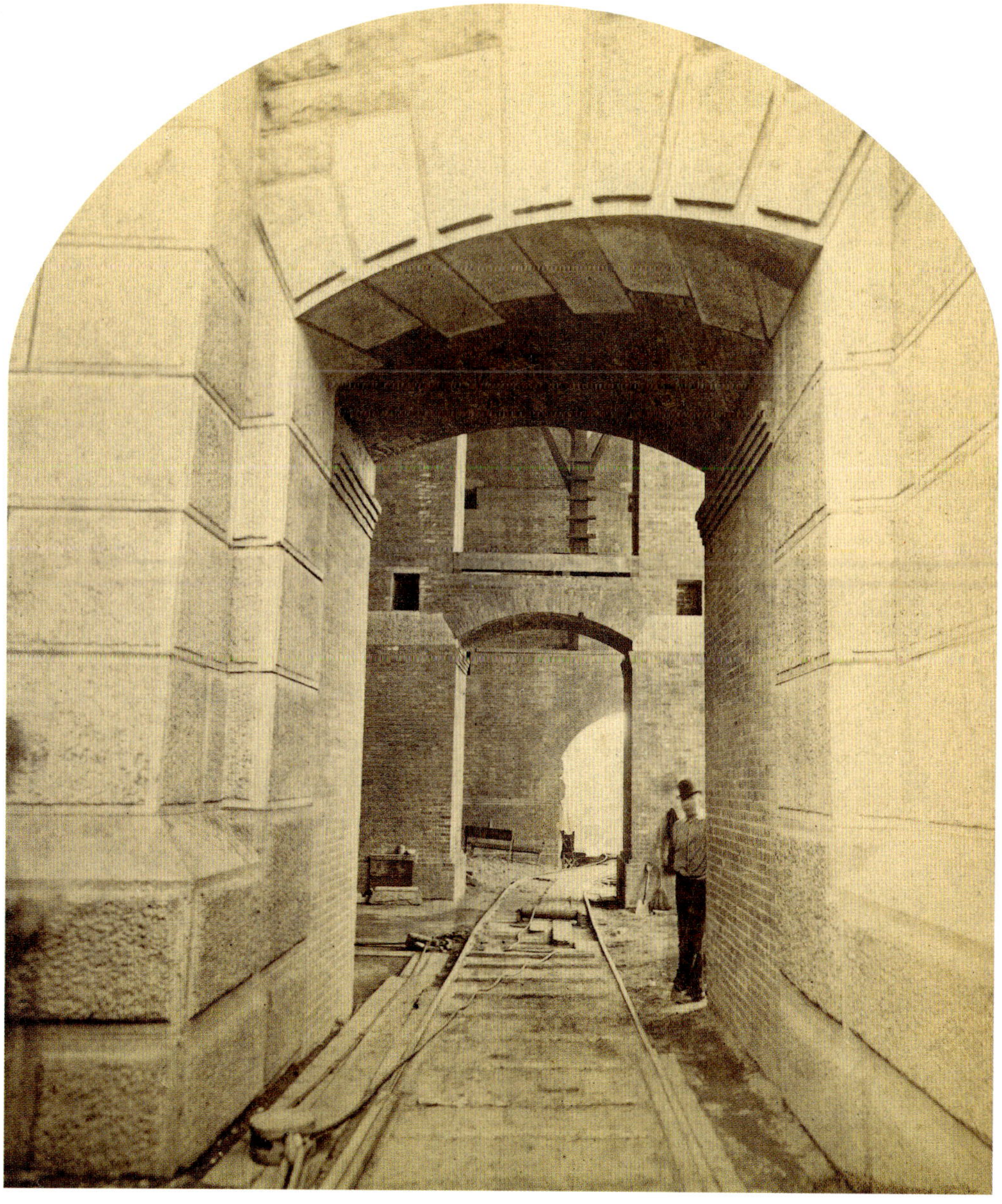

This is a half stereoview. Note the worker in the passageway at right and the ghost image of the worker at left. Both men moved during the long exposure, resulting in these effects. Courtesy of the Michael E. Ach Collection.

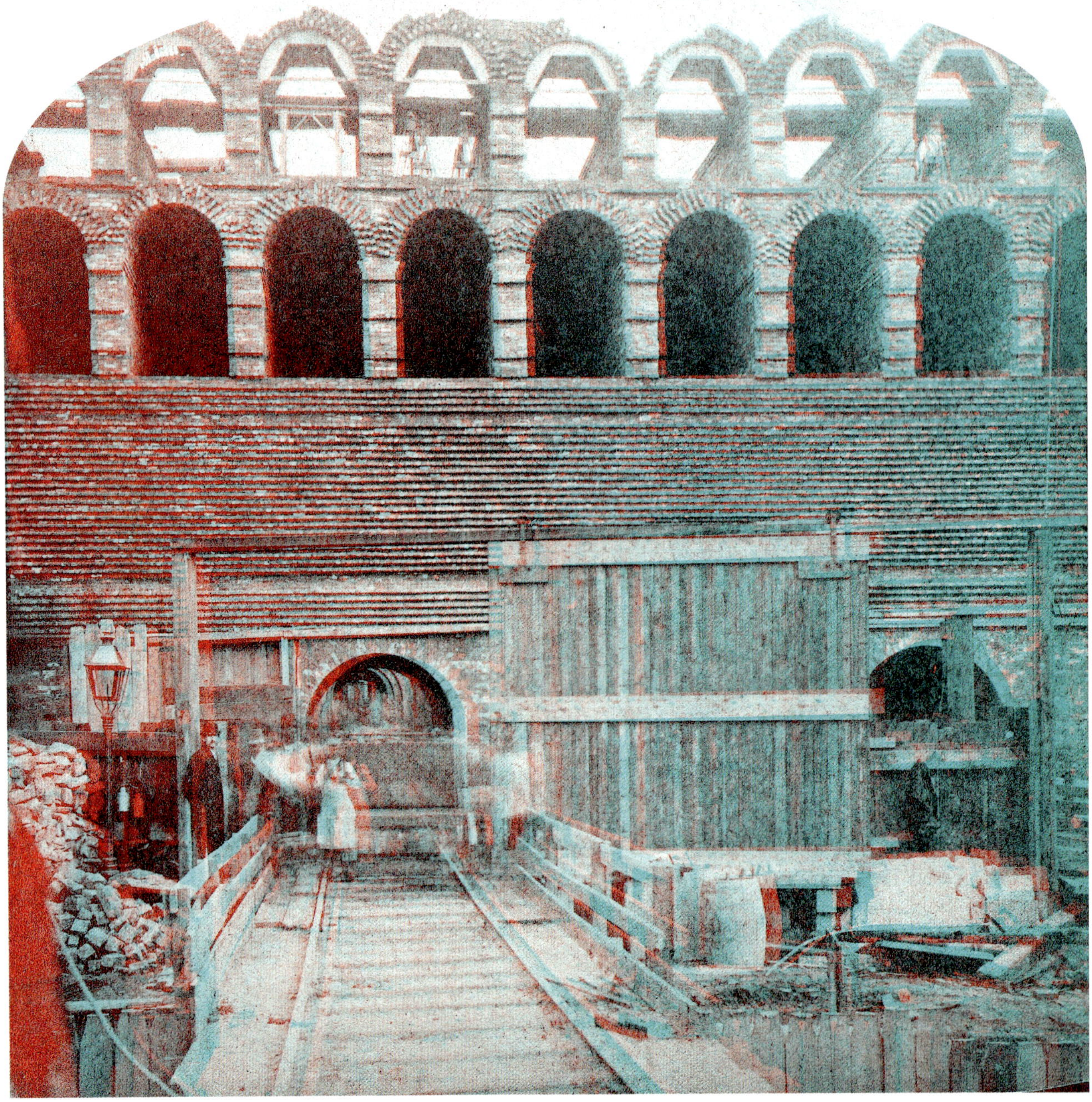

3D An anaglyph of the brick arches being constructed to support one of the approaches. The rails stretching toward the photographer allowed for easy movement of heavy construction material on carts. The men captured in this image were on the move as the photographic plate was being exposed. As a result, they appear ghostly. Courtesy of the Michael E. Ach Collection.

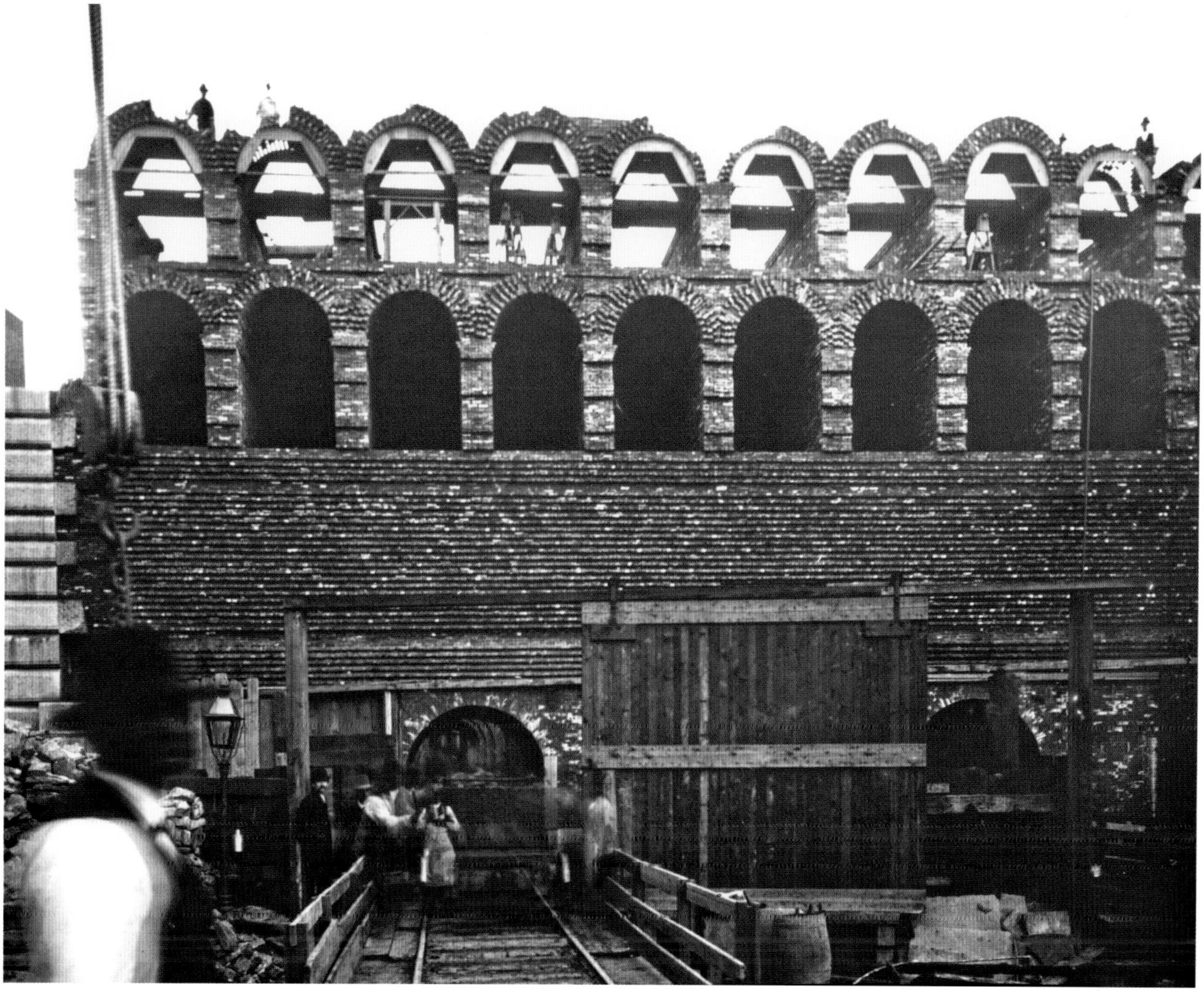

This original lantern slide photograph was used to create a stereoview that was in turn morphed into the anaglyph at left. An examination of both demonstrates how an image can lose historical value in a different format due to cropping. Compare this image with the anaglyph. Here, a man of color, wearing a hat, appears in the frame, in the lower left corner. This passerby is barely visible in the anaglyph on the opposite page. Author's collection.

FOLLOWING PAGE: Cover of publication printed on the occasion of the opening ceremonies for the Brooklyn Bridge, then called "New York and Brooklyn Bridge." Author's collection.

Opening Ceremonies

of the

New York and Brooklyn Bridge

May 24th. 1883.

Selected Bibliography

I list here works that have been of use in the creation of this book. This bibliography is by no means a complete record of all of the sources I have consulted.

Barnard, Charles. "The Brooklyn Bridge." *St. Nicholas Magazine*, July 1883.

Blanck, Maggie. Brooklyn Bridge Workers. 2016. http://www.maggieblanck.com/Occupations/BrooklynBridgeWorkers.html.

Brooklyn Museum. *The Great East River Bridge, 1883–1983.* New York: Brooklyn Museum, 1983.

Collingwood, Francis. "Notes on the Masonry of the East River Bridge." *Transactions of the American Society of Civil Engineers*, 1877, Vol. VI, Issue 1: 7–27.

Conant, William. "Brooklyn Bridge Anchorage." *Harper's Monthly*, May 1883: 925–946.

Edwards, Richard. *Industries of New Jersey: Essex County Including City of Newark.* New York: Historical Publishing Company, 1882.

"The East River Bridge." *Scientific American*, March 10, 1877: 1.

Farrington, E. F.; Cooper Union for the Advancement of Science and Art. *Concise Description of the East River Bridge: With Full Details of Construction: Being the Substance of Two Lectures Delivered at the Cooper Union, March 6 and 13, 1880, by E. F. Farrington, Master Mechanic.* New York: C.D. Wynkoop, printer, 1881

Haw, Richard. *Art of the Brooklyn Bridge: A Visual History.* New York: Routledge, 2008.

Haw, Richard. *Engineering America: The Life and Times of John A. Roebling.* New York: Oxford University Press, 2020.

Hildenbrand, Wilhelm. "Cable-Making for Suspension Bridges, With Special Reference to the Cables of the East River Bridge." *Van Nostrand's Magazine*, 1877.

McCullough, David. *The Great Bridge.* New York: Simon & Schuster, 1972.

Report of the Committee Appointed by the Board of Trustees, Consisting of the Mayors of New York and Brooklyn, Together with the Report of the Accountants to Said Committee, in the Matter of an Examination of the Financial Affairs of the New York and Brooklyn Bridge, December 31, 1883. Brooklyn: Brooklyn Eagle Book and Job Printing Dept., 1884.

Roebling, Washington Augustus, and C. C. Martin. *General Specifications for the Superstructure of Street Bridges, Brooklyn Approach, East River Bridge,* June 1880. Roebling Collection, Rensselaer Polytechnic Institute.

Roebling, Washington Augustus. *Pneumatic Tower Foundations of the East River Suspension Bridge.* New York: Averell & Peckett, Printers, 1873.

———. *Specifications for Granite Face, Arch, and Other Stone, Required for the New York and Brooklyn Bridge Approaches to the East River Bridge,* 1877. Roebling Collection, Rensselaer Polytechnic Institute.

———. *Specifications for the Iron Work of the Suspended Superstructure of the East River Bridge,* 1878. Roebling Collection, Rensselaer Polytechnic Institute.

———. *Specifications for the Rope Driving Machinery for Operating the Railway of the New York and Brooklyn Bridge,* 1882. Roebling Collection, Rensselaer Polytechnic Institute.

———. *Specifications for the Wire Ropes for the East River Bridge,* 1876. Roebling Collection, Rensselaer Polytechnic Institute.

———. *Specifications for Wire Rope Sockets Required for the East River Bridge,* 1877. Roebling Collection, Rensselaer Polytechnic Institute.

———. *Specifications for Saddles and Saddle-Plates for the Brooklyn Tower of the East River Bridge,* 1874. Roebling Collection, Rensselaer Polytechnic Institute.

———. *Specifications for Steel Cable Wire for the East River Suspension Bridge,* 1876. Roebling Collection, Rensselaer Polytechnic Institute.

———. *Specifications for Timber for New York Anchorage, East River Bridge,* 1875. Roebling Collection, Rensselaer Polytechnic Institute.

Sayenga, Don. "The Brooklyn Bridge Experiment." *Wire Rope News & Sling Technology,* December 2018.

Shapiro, Mary J. *A Picture History of the Brooklyn Bridge.* New York: Dover, 1983.

Steele, Chris. "Silas A. Holmes." Pioneering American Photographers, 1839–1860. https://pioneeramericanphotographers.com, 2018.

Stewart, Elizabeth C., ed. *Guide to the Roebling Collections at Rensselaer Polytechnic Institute and Rutgers University.* Troy, NY: Friends of the Folsom Library, Rensselaer Polytechnic Institute, 1983.

Wagner, Erica. *Chief Engineer: Washington Roebling, The Man Who Built the Brooklyn Bridge.* New York: Bloomsbury, 2017.

Wilson, Edward L. *Wilson's Cyclopaedic Photography: A Complete Handbook of the Terms, Processes, Formulae and Appliances Available in Photography.* New York: Edward L. Wilson, 1894.

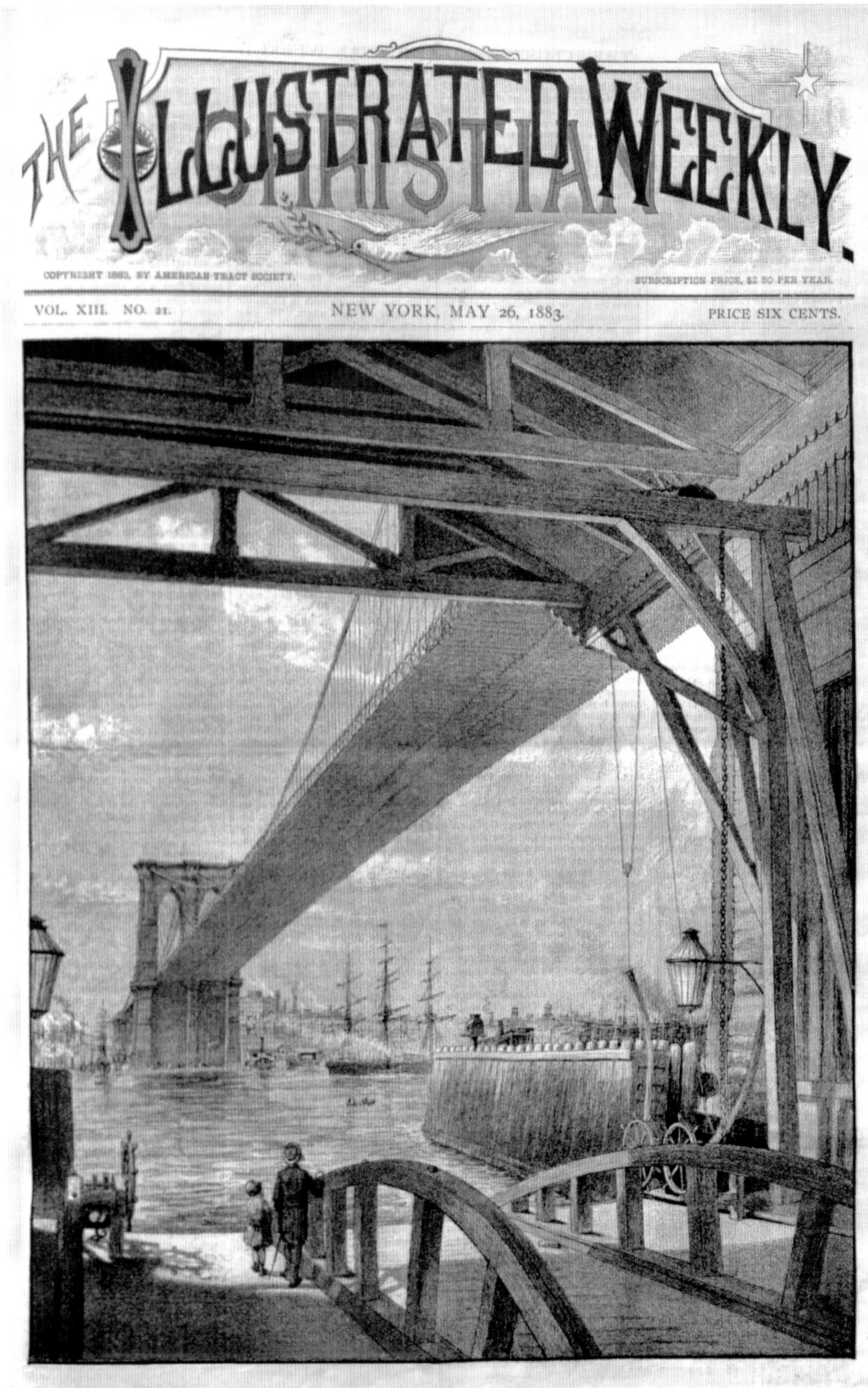

"Underneath the Great Bridge. View from Fulton Ferry, Brooklyn,"
Illustrated Christian Weekly, May 26, 1883.

Index

Numbers in italics refer to the page numbers of illustrations.

Building the Brooklyn Bridge 1869–1883:
An Illustrated History, with Images in 3D

Editor: Beth Daugherty
Cover and book design: Kathryn Hart

Typeset in Utopia, Univers, and Clarendon
Printed on 128 gsm Gold East Matt Art

First edition

Printed and bound in China by Pimlico Book International

ISBN: 978-1-7356001-2-3 (cloth)

Library of Congress Control Number: 2021908802

Published by Bauer and Dean Publishers, Inc.
P.O. Box 98, Times Square Station
New York, NY 10108
www.baueranddean.com

In association with The Green-Wood Historic Fund, Brooklyn, New York

Distribution by ACC Art Books (orders processed by National Book Network)
6 West 18th Street, suite 4B
New York, NY 10011
Tel (212) 645-1111
ussales@accartbooks.com

ENDPAPER: "Bird's Eye View of the City of New York," published by L. W. Schmidt, 24 Barclay Street, New York, ca. 1875. The Brooklyn Bridge spans the East River, just left of center.

FRONTISPIECE: "Bird's-Eye View of the Great New York and Brooklyn Bridge and Grand Display of Fire Works on Opening Night." A lithograph, published by A. Majors, 1883.

IMAGE OPPOSITE CONTENTS: "Under the Tower—From the Water-color Painting by F. Hopkinson Smith," *Harper's Weekly*, February 18, 1882. The view shows the deck of the bridge being hung from the completed cables. The edge of the New York anchorage is at far left. A photograph taken from the opposite direction is on page 139. Author's collection.

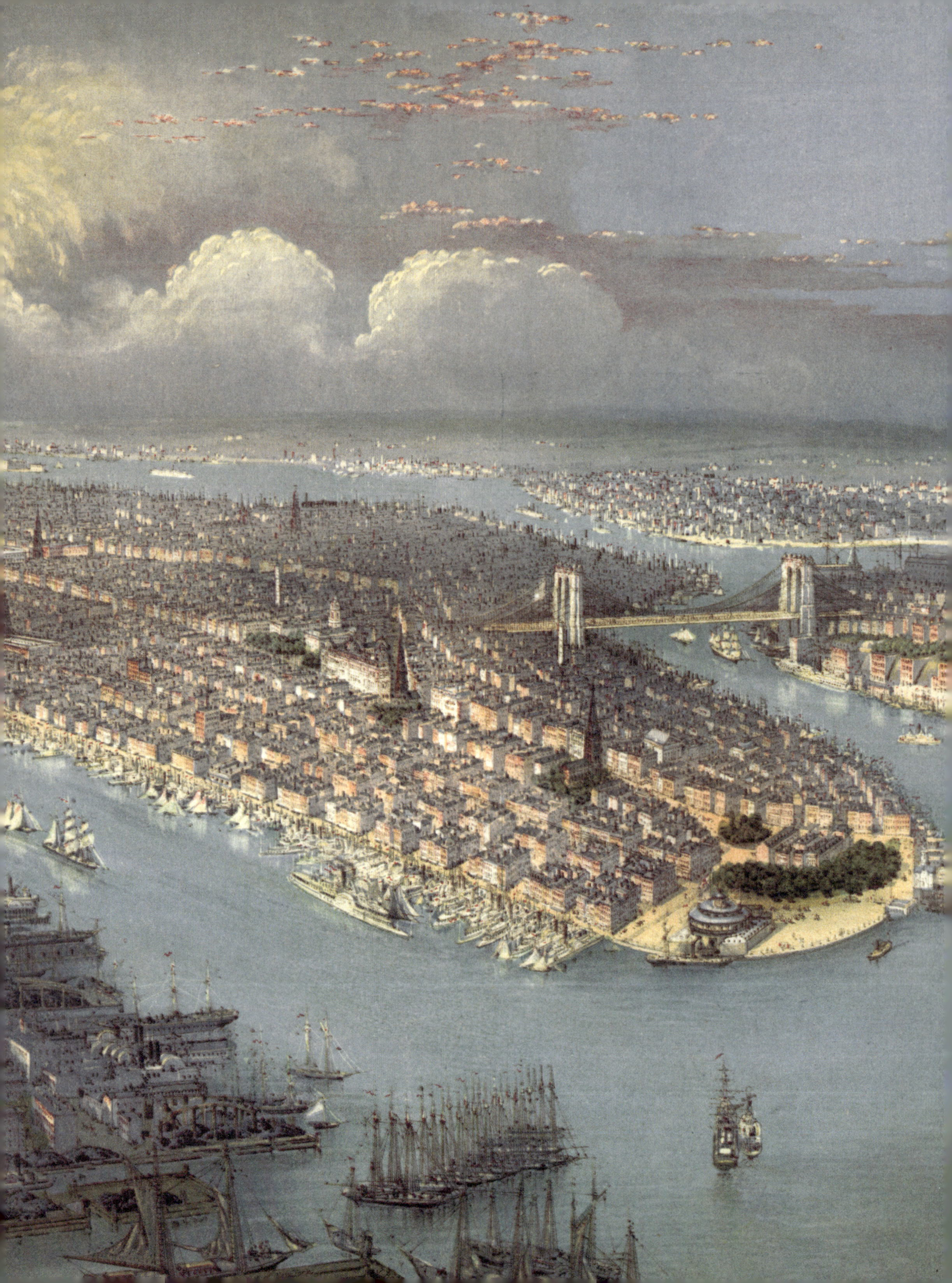